SEMICONDUCTOR MATERIAL AND DEVICE CHARACTERIZATION

SEMICONDUCTOR MATERIAL AND DEVICE CHARACTERIZATION

DIETER K. SCHRODER
Arizona State University
Tempe, Arizona

A WILEY-INTERSCIENCE PUBLICATION

John Wiley & Sons, Inc.

NEW YORK / CHICHESTER / BRISBANE / TORONTO / SINGAPORE

7240405

PHYS

Library of Congress Cataloging in Publication Data:

Schroder, Dieter K.
    Semiconductor material and device characterization/Dieter K.
    Schroder.
        p. cm.
    "A Wiley-Interscience publication."
    Includes bibliographical references.
    ISBN 0-471-51104-8
    1. Semiconductors. 2. Semiconductors--Testing. I. Title.
QC611.S335   1990
621.381'52--dc20
                                                89-24881
                                                    CIP

Printed in the United States of America

10 9 8 7 6

PREFACE

When several years ago, I prepared to teach a course on Semiconductor Material and Device Characterization, I found no suitable book for such a course. Earlier books on characterization did not contain the breadth of modern characterization techniques and were out of print. Not having a textbook, I used papers, review papers, chapters in books, out-of-print books from the library, and developed my own course notes. Today I also hold a three-day short course on this topic for which I developed related notes. I teach industrial courses from time to time and have run into the same problem of lack of book and thus have had to rely on reprints and notes. From this experience and from discussion of this problem with my industrial and academic colleagues, I have become very much aware of the need for a textbook covering modern semiconductor characterization techniques. This book grew out of my course notes, which have benefited from discussions with students and colleagues who have helped to clarify points of confusion.

In the current semiconductor literature there are many books on the physics of semiconductor devices, and even a number of books on processing of semiconductors. There are several books on semiconductor device and circuit design and a few books on modeling of semiconductor devices and processes. But there are no comprehensive books on semiconductor characterization. The earlier books *Semiconductor Measurements and Instrumentation* by W.R. Runyan and *Characterization of Semiconductor Materials* by P. F. Kane and G. B. Larrabee are out of print. The recent book *Electrical Characterization of GaAs Materials and Devices* by D. C. Look addresses electrical measurements of GaAs very well, but does not cover the broader field of characterization addressed in this book.

All semiconductor devices and materials are characterized to a greater or lesser degree. Processes are characterized through the use of test structures. Many papers, review papers, book chapters, and specialized books exist in the field of characterization, but no one has integrated these various topics into one volume. I have attempted to do that by including the main characterization techniques of the semiconductor industry—electrical, optical, chemical, and physical—in this book.

I wrote this book with two distinct audiences in mind. One is the first- or second-year graduate student who is familiar with semiconductor device physics, knows and understands the basic semiconductor devices, and wishes to learn about semiconductor measurements. The second audience is the industrial researcher who also understands devices and who may be familiar with some characterization methods and wants to learn about others or who wants to become familiar with the wide spectrum of measurement methods found in the modern semiconductor industry. The book may even be considered a handbook to look for a specific characterization technique. Those readers interested in more detail may wish to consult some of the references. I have consulted and included more than 1300 references. These are the references I found most useful during the preparation of the manuscript. They are fairly comprehensive but obviously not all-inclusive. I did not exclude references deliberately; rather, I chose to include those that I found to be most helpful.

I have written the book from a semiconductor device point of view for the person who is reasonably familiar with the physics and operation of the major semiconductor devices—*pn* junctions, bipolar junction transistors, metal-oxide-semiconductor capacitors and transistors, solar cells, and Schottky barrier diodes. I have stressed the concepts wherever possible; in some instances I have explained the necessary material or device background for understanding certain characterization methods. But obviously there is no space to derive all device concepts, and the reader who is not familiar with the underlying concepts should consult appropriate semiconductor device physics books. I have used the contents of this book during the past seven years as a one-semester graduate course and most of the material in an abbreviated three-day short course. During the one-semester course I do not cover all the details of the book. The book contains material that is sufficiently broad to be also suitable for a two-semester course.

I chose the topics by carefully considering the range of semiconductor characterization techniques in use and by discussions with people active in the field. I have used and am familiar with many of the methods. Electrical characterization methods are by far the most prevalent. Consequently I have devoted the major part of the book to them. Optical methods are for the most part more specialized, not used as frequently, but are becoming more popular. Their non-contacting nature and high sensitivity is a decided advantage. Chemical and physical characterization methods are yet more specialized. The high spatial resolution and ability to identify elements and compounds makes them very valuable for some applications. They are usually performed by specialists or offered as services, so it is very useful to be familiar with these methods to understand their applicability and their limitations.

Many people have in one way or another contributed to this book by discussions, questions, comments, and reading of chapters. Students at ASU and attendees at short courses have helped clarify many concepts. I would especially like to thank those who have contributed directly during the writing of this book. K. Joardar, D. A. Johnson, S. H. Park, K. T. Shiralagi, and S. Visitserngtrakul from Arizona State University and Tom Shaffner from Texas Instruments read various chapters and made valuable corrections and suggestions. B. Hussain, Z. Mahdavi, I. G. Hwang, and P. S. Ku from Arizona State University and my wife Beverley checked the many references. Many discussions with Tom Shaffner and Graydon Larrabee from Texas Instruments and Ron Roedel from ASU have helped clarify numerous concepts especially in optical, chemical, and physical characterization. Several students helped with experimental data. They are acknowledged in the figure captions. My son Mark spent many hours drawing the figures, and my son Derek did some of the typing. Lastly I thank the Department of Electrical and Computer Engineering at Arizona State University for providing the atmosphere and one-semester sabbatical leave to write this book.

DIETER K. SCHRODER

Tempe, Arizona

CONTENTS

INTRODUCTION

This book is about semiconductor material and device characterization techniques. It provides a comprehensive survey of electrical, optical, chemical and physical characterization techniques. The discussion of electrical characterization methods takes up the major part of the book because these methods are the ones most commonly used. The plan of the book is the following.

Chapter 1: Resistivity Resistivity, a basic semiconductor material parameter, is a logical starting point. The chapter describes conventional four-point probe, Hall effect, and more recent non-contacting methods. Methods to profile the resistivity and to map the resistivity or the sheet resistance are also discussed.

Chapter 2: Carrier and Doping Concentration A broad range of measurements techniques is given. The most common, the capacitance-voltage method, is discussed in detail. Also covered are methods using current-voltage measurements, secondary ion mass spectrometry, the Hall effect, and some of the newer optical and thermal wave techniques.

Chapter 3: Contact Resistance and Schottky Barrier Height All semiconductor devices have contacts, and each contact has a contact resistance. The concept of contact resistance is introduced in this chapter, and the various measurement techniques to determine the contact resistance are discussed. Contact resistance owes its origin to the metal-semiconductor contact. Metal–semiconductor contact barrier height measurements are also part of this chapter.

Chapter 4: Series Resistance, Channel Length, Threshold Voltage   Material resistivity leads to device resistance influencing devices in various ways. Resistance generally degrades device performance but is an inherent property of semiconductor devices. Resistance measurements in *pn* junctions, Schottky diodes, solar cells, bipolar junction transistors, and MOSFET's are discussed. In MOSFET's there is a link between series resistance and channel length and width measurements. Techniques to determine both are examined. Finally, the measurement of threshold voltage, which is required for the interpretation of a number of the measurements, is detailed.

Chapter 5: Mobility   Mobility, another important material and device parameter, and its characterization are the subject of this chapter. The mobility depends very much on the manner in which it is measured, and it is important to understand the mobility determined by a particular characterization method and the method's limitations. The Hall effect is discussed in detail.

Chapter 6: Oxide and Interface Trapped Charge   The first five chapters deal with characterization methods suitable for a variety of devices; this chapter looks at MOS measurements, namely, oxide charge and interface trapped charge measurements. The relevant background for understanding the low- and high-frequency C-V curves of MOS capacitors is part of this chapter as is the concept and measurement of work function difference.

Chapter 7: Deep-Level Impurities   Impurities generally degrade device performance and are undesirable, but occasionally they are deliberately introduced. The chapter covers the detection and characterization of impurities and the concepts of generation and recombination. Impurities with energy levels deep within the band gap are best measured electrically. The most common of these electrical methods are based on transient capacitance and current, known as deep-level transient spectroscopy. The choice of capacitance or current depends on the sample configuration and resistivity, with capacitance being the more commonly used.

Chapter 8: Carrier Lifetime   One device property directly influenced by impurities is the recombination and generation lifetime. Since these two concepts are frequently confused we give an introduction to the concept of bulk and surface lifetime and then provide a variety of lifetime measuring techniques. The lifetime of a material or device depends very much on the method of measurement. Lifetime has one of the largest selection of characterization methods. The chapter discusses popular ones and points out their limitations. Lifetime is frequently not considered to be important for most integrated circuits and is therefore not routinely characterized, but it can be an excellent and very sensitive process characterization tool.

Chapter 9: Optical Characterization   Optical charcterization techniques fall into two broad categories: those methods routinely used (e.g., ellipsometry, optical microscopy) and those rarely used or used by specialists (e.g., Raman spectroscopy). The chapter discusses the more common optical techniques and provides the necessary background for understanding these techniques. Optical techniques are used for determining a wide variety of semiconductor parameters. For example, optical microscopy gives images of the surface, ellipsometry yields insulator thicknesses, optical transmittance is used to determine impurities such as oxygen and carbon in silicon, and photoluminescence is a very sensitive method for measuring shallow-level impurities with sensitivities approaching 10^{10} to 10^{11} cm^{-3}. Raman spectroscopy lends itself well to small-area organic contaminant measurements, infrared reflectance is routinely used for epitaxial layer thickness measurements, and optical methods are used for line-width determination.

Chapter 10: Chemical and Physical Characterization   The final chapter summarizes electron beam, ion beam, X-ray, and γ-ray techniques. These characterization techniques are typically the domain of specialists, who operate the rather complex equipment and interpret the results. Interpretation is frequently difficult. The techniques are very important in finding the spatial distribution of impurities, their identity and concentration, the composition of compounds, and other properties not amenable to electrical or optical characterization. The sensitivity of chemical and physical methods is generally poorer than electrical or optical techniques, but the spatial resolution can be extremely high, with atomic resolution possible in high-resolution transmission electron microscopy. The methods in this chapter are becoming more important with continued improvements in resolution and sensitivity.

Appendixes following some chapters give more detailed mathematical derivations and semiconductor properties like resistivity, mobility, and absorption coefficient. Symbols are defined in Appendix A. Frequently occurring expressions are abbreviated. Abbreviations and acronyms are defined in Appendix B.

A word about nomenclature. Most people in the semiconductor industry use a mixture of units rather than just mks units. We follow this practice in the book. For example, doping and carrier concentrations are given in cm^{-3}, wavelengths in μm, and insulator thicknesses in Å, mobilities in cm^2/V·s. Some people prefer nm for Å or μm. Some publications use only mks units. While it is easy enough to convert from one system to another, we have chosen to use the more common nomenclature. For example, a doping concentration of 10^{15} cm^{-3} is generally more readily understood than 10^{21} m^{-3}, although they are identical. Similarly a mobility of 1000 cm^2/V·s is more familiar to most people than 0.1 m^2/V·s.

SEMICONDUCTOR MATERIAL AND DEVICE CHARACTERIZATION

CHAPTER 1

RESISTIVITY

1.1 INTRODUCTION

The resistivity of a semiconductor ρ is defined by

$$\rho = \frac{1}{q(n\mu_n + p\mu_p)} \tag{1.1}$$

where n and p are the free electron and hole concentrations, and μ_n and μ_p are the electron and hole mobilities, respectively. The resistivity can be calculated from the measured carrier concentrations and mobilities. In order to determine the resistivity in this manner, both carrier concentrations and both mobilities must be known. For extrinsic materials in which the majority carrier concentration is much higher than the minority carrier concentration, it is generally sufficient to know the majority carrier concentration and the majority carrier mobility. The carrier concentration and mobility are generally not known, however. Hence we must look for an alternative resistivity measurement technique.

The resistivity can be measured in a number of ways. The methods range from *contactless*, through *temporary contact* to *permanent contact* techniques. We will discuss the more common methods in detail. Those methods not commonly used are briefly described, and appropriate references are given.

1.2   THE FOUR-POINT PROBE

The *four-point probe* technique is one of the most common methods for measuring the semiconductor resistivity because two-point probe methods are more difficult to interpret. Consider the two-point probe arrangement of Fig. 1.1. Each probe serves as a current and as a voltage probe. The total resistance between the two probes is given by

$$R_T = \frac{V}{I} = 2R_c + 2R_{sp} + R_s \qquad (1.2)$$

where R_c is the contact resistance at each metal probe/semiconductor contact, R_{sp} is the spreading resistance under each probe and R_s is the semiconductor resistance. The contact resistance arises from the mechanical metal probe contacting the semiconductor (contact resistance is discussed in Chapter 3). The spreading resistance accounts for the resistance encountered by the current when it flows from the small metal probe into the

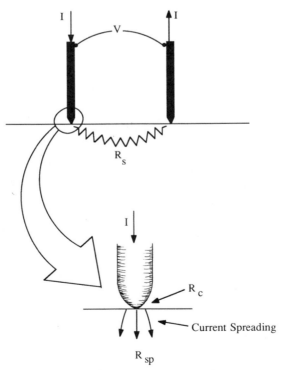

Fig. 1.1 A two-point probe showing the contact resistance R_c the spreading resistance R_{sp} and the semiconductor resistance R_s.

semiconductor. Neither R_c nor R_{sp} can be accurately calculated so that R_s cannot be accurately extracted from the measured resistance.

A solution to this dilemma is the use of four probes. Two probes carry the current and the other two probes are used for voltage sensing. The four-point probe was originally proposed by Wenner[1] in 1916 to measure the earth's resistivity, and the four-point probe measurement technique is referred to in Geophysics as *Wenner's method*. It was adopted for semiconductor wafer resistivity measurements by Valdes in 1954.[2] The probes are generally arranged in-line with equal probe spacing. But other probe configurations are possible.[3]

The use of four probes has an important advantage over two probes. Although the two current-carrying probes still have contact and spreading resistance associated with them, that is not true for the two voltage probes because the voltage is measured either with a potentiometer which draws no current at all or with a high impedance voltmeter which draws very little current. The two parasitic resistances R_c and R_{sp} are negligible in either case because the voltage drops across them are negligibly small due to the very small current that flows through them.

The potential V at a distance r from an electrode carrying a current I in a material of resistivity ρ is given by the relationship[2]

$$V = \frac{\rho I}{2\pi r} \tag{1.3}$$

For probes resting on a semi-infinite medium as in Fig. 1.2, with current entering probe 1 and leaving probe 4, the voltage V becomes

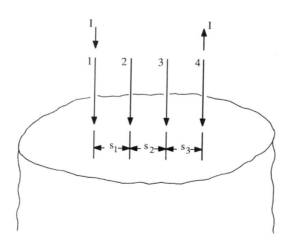

Fig. 1.2 A collinear four-point probe.

$$V = \frac{\rho I}{2\pi} \left(\frac{1}{r_1} - \frac{1}{r_4} \right) \tag{1.4}$$

where r_1 and r_4 are the distances from probes 1 and 4, respectively. For probe spacings of s_1, s_2, and s_3, as shown in Fig. 1.2, the voltage at probe 2 is

$$V_2 = \frac{\rho I}{2\pi} \left(\frac{1}{s_1} - \frac{1}{s_2 + s_3} \right) \tag{1.5}$$

and the voltage at probe 3 is

$$V_3 = \frac{\rho I}{2\pi} \left(\frac{1}{s_1 + s_2} - \frac{1}{s_3} \right) \tag{1.6}$$

The total measured voltage $V = V_2 - V_3$ becomes

$$V = \frac{\rho I}{2\pi} \left(\frac{1}{s_1} + \frac{1}{s_3} - \frac{1}{s_2 + s_3} - \frac{1}{s_1 + s_2} \right) \tag{1.7}$$

The semiconductor parameter of interest is the resistivity

$$\rho = \frac{2\pi V / I}{1/s_1 + 1/s_3 - 1/(s_1 + s_2) - 1/(s_2 + s_3)} \tag{1.8}$$

usually in units of ohm-cm, with V measured in volts and I measured in amperes. For most four-point probes the probe spacings are equal with $s = s_1 = s_2 = s_3$, and Eq. (1.8) reduces to

$$\rho = 2\pi s (V/I) \tag{1.9}$$

Optimum probe spacings are on the order of 0.5 to 1.5 mm, and they vary with sample diameter and sample thickness.[4] For $s = 1.588$ mm $= 62.5$ mils, $2\pi s$ is unity, and ρ is given by $\rho = V/I$ directly. The two most common probe spacings are 0.635 mm (25 mils) and 1.588 mm.

Semiconductor wafers are not semi-infinite in extent in either the lateral or the vertical dimension. Equation (1.9) must be corrected for finite geometries. For an arbitrarily shaped sample the resistivity is given by

$$\boxed{\rho = 2\pi s F (V/I)} \tag{1.10}$$

where F is a correction factor that depends on the sample geometry. F corrects for edge effects, for thickness effects, and for probe placement effects, and it is usually a product of several independent correction factors. For sample thicknesses greater than the probe spacing, the simple, independent correction factors contained in F of Eq. (1.10) are no longer adequate due to interactions between thickness and edge effects. Fortunately the sample thickness is generally smaller than the probe spacings, and the correction factors can be independently calculated.

1.2.1 Correction Factors

Four-point probe correction factors have been calculated by various techniques. Valdes,[2] Smits,[5] and Uhlir[6] used the *method of images*; Buehler used *complex variable theory*[7] as well as the *method of Corbino sources*[8]; Yamashita[9] solved Poisson's equation; Murashima et al.[10] utilized *Green's functions*; and Perloff[11,12] and Yamashita et al.[13] employed *conformal mapping*. We will give the most appropriate factors here and refer the reader to others where appropriate.

The following correction factors are for *collinear* or *in-line probes* with equal probe spacing, s. We write the correction factor F as a product of three separate correction factors

$$F = F_1 F_3 [\ln(2) F_2 / \pi] \tag{1.11}$$

Each of these factors can be further subdivided. F_1 corrects for sample thickness, F_2 corrects for lateral sample dimensions, and F_3 corrects for placement of the probes relative to the sample edges. A parameter that must be corrected for most practical measurement conditions is the sample thickness since semiconductor wafers are not infinitely thick. Their thicknesses are usually on the order of the probe spacing or less introducing the correction factor[14]

$$F_{11} = \frac{t/s}{2 \ln\{[\sinh(t/s)]/[\sinh(t/2s)]\}} \tag{1.12}$$

for a *non-conducting* bottom wafer surface boundary, where t is the wafer or layer thickness. For a *conducting* bottom surface the correction factor becomes

$$F_{12} = \frac{t/s}{2 \ln\{[\cosh(t/s)]/[\cosh(t/2s)]\}} \tag{1.13}$$

F_{11} and F_{12} are plotted in Fig. 1.3. For thin samples Eq. (1.12) reduces to

$$F_{11} = \frac{t/s}{2 \ln(2)} \tag{1.14}$$

Equation (1.14) is valid for $t \leq s/2$. For very thin samples for which $F_2 = \pi/\ln(2)$ and $F_3 = 1$, we find from Eqs. (1.10), (1.11), and (1.14)

$$\rho = \frac{\pi t}{\ln(2)} \frac{V}{I} = 4.532 t (V/I) \tag{1.15}$$

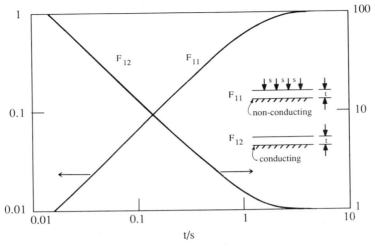

Fig. 1.3 Wafer thickness correction factors as a function of normalized wafer thickness.

Thin layers are often characterized by their *sheet resistance* ρ_s expressed in units of ohms per square. The sheet resistance is given by

$$\rho_s = \frac{\rho}{t} = \frac{\pi}{\ln(2)} \frac{V}{I} = 4.532(V/I) \tag{1.16}$$

subject to the constraint $t \leq s/2$. The sheet resistance is frequently used to characterize thin semiconductor sheets or layers, such as diffused or ion-implanted layers, polycrystalline silicon, and metallic conductors.

The sheet resistance measurement is subject to a further correction factor. This is concerned with the size of the sample. The sheet resistance expression of Eq. (1.16) holds for a sample of infinite lateral extent. The more general expression is

$$\rho_s = F_2(V/I) \tag{1.17}$$

For circular wafers of diameter d, the correction factor F_2 is given by[15,16]

$$F_2 = \frac{\pi}{[\ln(2) + \ln\{[(d/s)^2 + 3]/[(d/s)^2 - 3]\}]} \tag{1.18}$$

F_2 is plotted in Fig. 1.4 for circular wafers. It is interesting to note that the sample must have a diameter $d \geq 40s$ for Eq. (1.16) to be valid. For a probe spacing of 62.5 mils, this implies that the wafer must be at least 2.5 inches in diameter. This is often satisfied experimentally. Also shown in Fig. 1.4 is the correction factor for rectangular samples.[5]

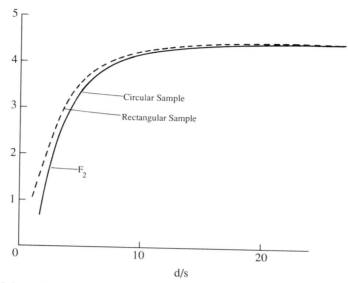

Fig. 1.4 Wafer diameter correction factors as a function of normalized wafer thickness. For rectangular samples: d = width.

The correction factor 4.532 in Eq. (1.16) is for the current and voltage probes shown in Fig. 1.2 in which the current flows into probe 1, out of probe 4, and the voltage is sensed across probes 2 and 3. For the current applied to and the voltage sensed across other probes, appropriate correction factors are given in Table 1.1.

For probes *perpendicular* to and at a distance d from a *non-conducting boundary*, as indicated in the insets of Fig. 1.5, the correction factor is given as a function of the normalized distance d/s by[2]

$$F_{31} = \cfrac{1}{1 + \cfrac{1}{1 + 2d/s} - \cfrac{1}{2 + 2d/s} - \cfrac{1}{4 + 2d/s} + \cfrac{1}{5 + 2d/s}} \tag{1.19}$$

TABLE 1.1 Current Probes, Voltage Probes and the Corresponding Correction Factors for Collinear Four-Point Probes

Current Probes	Voltage Probes	Correction Factor
1, 4	2, 3	4.532
2, 3	1, 4	4.532
1, 2	3, 4	21.84
3, 4	1, 2	21.84
1, 3	2, 4	15.50
2, 4	1, 3	15.50

Source: Rymaszewski.[17]

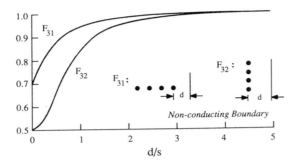

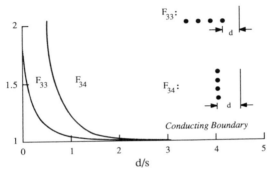

Fig. 1.5 Boundary proximity correction factors as a function of normalized distance from the boundary.

The correction factor for the probes *parallel* to a *non-conducting boundary* is

$$F_{32} = \frac{1}{1 + 2[1 + (2d/s)^2]^{-1/2} - [1 + (d/s)^2]^{-1/2}} \qquad (1.20)$$

For probes *perpendicular* to a *conducting boundary*,

$$F_{33} = \frac{1}{\left[1 - \dfrac{1}{1 + 2d/s} + \dfrac{1}{2 + 2d/s} + \dfrac{1}{4 + 2d/s} - \dfrac{1}{5 + 2d/s}\right]} \qquad (1.21)$$

and for probes *parallel* to a *conducting boundary*,

$$F_{34} = \frac{1}{1 - 2[1 + (2d/s)^2]^{-1/2} + [1 + (d/s)^2]^{-1/2}} \qquad (1.22)$$

Equations (1.19)–(1.22) are plotted in Fig. 1.5. These correction factors apply to infinitely thick samples. It is obvious from the figures that as long as the probe distance from the wafer boundary is three to four probe spacings or more, the correction factors F_{31} to F_{34} reduce to unity. For most four-point measurements this condition is easily satisfied. Correction factors F_{31} to F_{34} only become important for small samples in which the probe is, of necessity, close to a sample boundary.

Other corrections must be applied when the probe is not centered even in a wafer of substantial diameter.[16] For rectangular samples it has been found that the sensitivity of the geometrical correction factor to positional error is minimized by orienting the probe with its electrodes within about 10% of the center.[11] For square arrays the error is minimized by orienting the probe array with its electrodes equidistant from the midpoints of the sides. There is also an angular dependence of the placement of a square array on the rectangular sample.[9,11] We should mention that if the probe spacings are not exactly identical, there is a further correction.[18] This correction is small, however. The key to high precision four-point measurements, including reducing geometric effects associated with proximity of the probe to a non-conducting boundary, is the use of two measurement configurations at each probe location.[19–21] The first configuration is usually with current into probe 1 and out of probe 4 and with the voltage detected across probes 2 and 3. The second measurement is made with current driven through probes 1 and 3 and voltage measured across probes 2 and 4. The noise level is considerably reduced in this dual configuration, and the accuracy is considerably improved.

The resistivity of semiconductor ingots is also measured with the four-point probe. The simple equation

$$\rho = 2\pi s(V/I) \qquad (1.23)$$

is valid only if the diameter of the ingot d is related to the probe spacing by $d \geq 10s$.[10,22,23]

1.2.2 Resistivity of Arbitrarily Shaped Samples

The collinear probe configuration is the most common four-point probe arrangement. Arrangement of the points in a square has the advantage of occupying a smaller area since the spacing between points is only s, whereas in a collinear configuration the spacing between the outer two probes is $3s$. However, the square arrangement is more commonly used, not as an array of four mechanical probes but rather as contacts to square semiconductor samples. Occasionally it is difficult to provide a sample in a square format. In fact sometimes the sample is irregularly shaped. The theoretical foundation of measurements on irregularly shaped samples is based on conformal mapping developed by van der Pauw.[24,25] He showed how the specific resistivity of a flat sample of arbitrary shape can be measured without knowing the current pattern, if the following conditions are met: (1) the contacts are at the circumference of the sample, (2) the contacts are sufficiently small, (3) the sample is uniformly thick, and (4) the surface of the sample is singly connected, i.e., the sample does not contain any isolated holes.

Consider the flat sample of a conducting material of arbitrary shape, with contacts 1, 2, 3, and 4 along the periphery as shown in Fig. 1.6 to satisfy the conditions above. The resistance $R_{12,34}$ is defined as

$$R_{12,34} = V_{34}/I_{12} \qquad (1.24)$$

where the current I_{12} enters the sample through contact 1 and leaves through contact 2 and $V_{34} = V_3 - V_4$ is the voltage difference between the contacts 3 and 4. $R_{23,41}$ is defined similarly. The resistivity is given by[24]

$$\rho = \frac{\pi t}{\ln(2)} \frac{(R_{12,34} + R_{23,41})}{2} F \qquad (1.25)$$

where F is a function only of the ratio $R_r = R_{12,34}/R_{23,41}$, satisfying the relation

$$\frac{R_r - 1}{R_r + 1} = \frac{F}{\ln(2)} \operatorname{arcosh}\left(\frac{\exp[\ln(2)/F]}{2}\right) \qquad (1.26)$$

The function F on the right side of Eq. (1.26) depends only on R_r. The dependence of F on R_r calculated from Eq. (1.26) is shown in Fig. 1.7.

For a symmetrical sample such as a circle or a square, $R_r = 1$ and $F = 1$. This allows Eq. (1.25) to be simplified to give the resistivity as

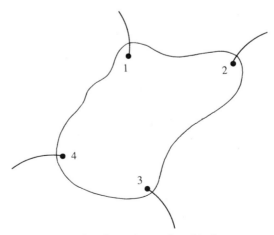

Fig. 1.6 Arbitrarily shaped sample with four contacts.

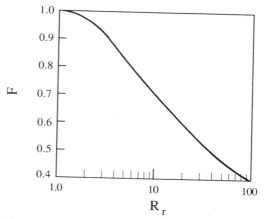

Fig. 1.7   The van der Pauw correction factor F as a function of R_r.

$$\rho = \frac{\pi t}{\ln(2)} R_{12,34} = 4.532 t R_{12,34}$$ (1.27)

The sheet resistance is given by

$$\rho_s = \frac{\pi}{\ln(2)} R_{12,34} = 4.532 R_{12,34}$$ (1.28)

similar to the four-point probe expression in Eq. (1.16).

The van der Pauw equations are based on the assumption of negligibly small contacts located on the sample periphery. Real contacts have finite dimensions and may not be exactly on the periphery of the sample. The influence of non-ideal contacts has been calculated.[24] The error so introduced can be eliminated by the use of the clover-leaf configuration of Fig. 1.8(b). Such configurations make sample preparation more complicated and are therefore undesirable, so square samples are generally used. One of the advantages of the van der Pauw structure is the small sample size compared with the area required for four-point probe measurements. Van der Pauw structures are therefore preferred for integrated circuit technology. For simple processing it is preferable to use the circular or square sample geometries shown in Fig. 1.8. For such structures it is not always possible to align the contacts exactly.

The placement of contacts for square samples is better at the midpoint of the sides as in Fig. 1.8(d), even if the sample is square, than at the corners as in Fig. 1.8(c).[11] This, coupled with the rather large sample size in conventional van der Pauw structures has led to other sample shapes. The *Greek cross* is shown in Fig. 1.9. Using photolithographic techniques, it is

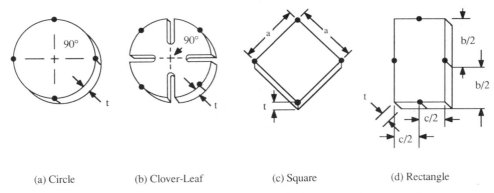

(a) Circle   (b) Clover-Leaf   (c) Square   (d) Rectangle

Fig. 1.8 Typical symmetrical circular and square sample geometries.

possible to make such structures very small and place many of them on a wafer for uniformity characterization. It has been shown that L should be slightly larger than W with $L \geq 1.02W$ being sufficient.[26] Surface leakage can introduce errors if L is too large.[27] A variety of cross sheet resistor structures have been investigated and their performance compares very well with conventional bridge-type structures.[28] The measured voltages in cross and van der Pauw structures are much lower than those in conventional bridge structures discussed in Chapter 5.

The cross and the bridge structures have been combined into the cross-bridge structure shown in Fig. 1.10. This structure allows not only the sheet resistance to be determined but the line width as well. The sheet resistance is given by

$$\rho_s = \frac{\pi}{\ln(2)} \frac{V_{12}}{I_{12}} \tag{1.29}$$

where $V_{12} = V_1 - V_2$ and I_{12} is the current flowing into contact I_1 and out of contact I_2. This part of the cross-bridge structure is a conventional cross.

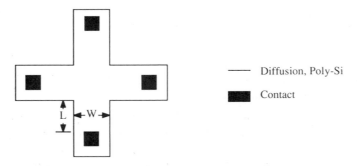

—— Diffusion, Poly-Si

■ Contact

Fig. 1.9 A cross sheet resistance test structure.

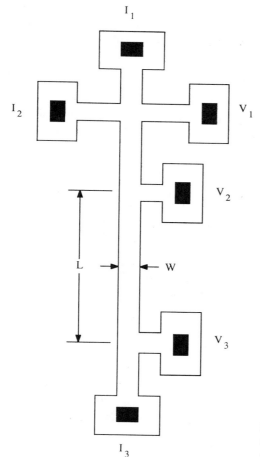

Fig. 1.10 A cross bridge sheet resistance and line width test structure.

The lower part of Fig. 1.10 is a bridge resistor that is used to determine the line width W. We mention the line width measurement feature only briefly here since it is more fully discussed in Chapter 9. The sheet resistance of the bridge resistor is

$$\rho_s = \frac{W}{L}\frac{V_{23}}{I_{13}} \tag{1.30}$$

where $V_{23} = V_2 - V_3$ and I_{13} is the current flowing into contact I_1 and out of contact I_3. From Eq. (1.30) we find the line width to be

$$W = \rho_s L \frac{I_{13}}{V_{23}} \tag{1.31}$$

with ρ_s determined from the cross structure and Eq. (1.29). Comparison sheet resistance measurements between cross and bridge structures agree within 1%. The bridge structure is found to be sensitive to line-width variations of ± 0.1 µm.[28]

1.2.3 Measurement Circuits

The basic four-point probe measurement circuit is simple. One implementation recommended by ASTM is shown in Fig. 1.11(a).[18] The double pole switch allows voltage measurements for both current directions. The resis-

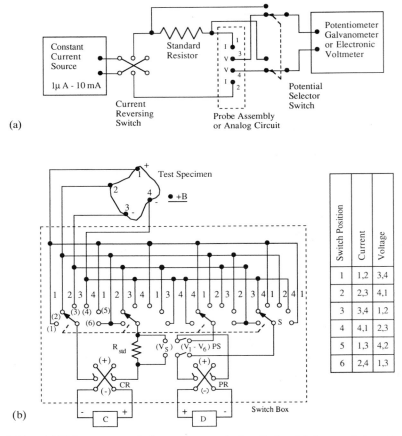

Fig. 1.11   ASTM recommended measurement circuits for (a) four-point probe, and (b) van der Pauw structure. For (b): C = constant current supply, CR = current reversing switch, D = electrometer or potentiometer/galvanometer, PR = potential reversing switch, PS = potential selector switch, R_{std} = standard resistor, S = contact selector switch. Copyright ASTM; reprinted with permission after ASTM F84[18] and ASTM F76.[41]

tances for the two current directions are measured, and the average resistance is calculated. Recent implementations are supplied with a microcomputer that makes the appropriate corrections. A circuit for van der Pauw measurements is shown in Fig. 1.11(b).

1.2.4 Measurement Errors and Precautions

For a four-point probe measurement to be successful, a number of precautions must be taken and appropriate correction factors must be applied for the interpretation of the measured data.

Sample Size   As mentioned earlier a number of corrections must be applied, depending on the location of the probe and the sample thickness. For those cases where the wafer is uniformly doped in the lateral direction and its diameter is appreciably larger than the probe spacing, the wafer thickness is the chief correction. If the wafer or the layer to be measured is appreciably thinner than the probe spacing, which is usually the case, the calculated resistivity varies directly with thickness. It is therefore very important to determine the thickness accurately.

Minority Carrier Injection   Although the probes constitute a metal-semiconductor contact, there is some minority carrier injection at the current probes. This is generally a small effect, but under high current conditions it may not be negligible. To reduce this effect, the surface should have a high recombination rate for minority carriers. This is best achieved by using lapped surfaces. For a highly polished wafer it may not be possible to achieve the necessary high surface recombination. Minority carrier injection causes conductivity modulation. If the voltage probes are sufficiently far from the injecting current probe, even if minority carriers are injected, they will have decayed by recombination and cause very little error. However, for high lifetime material the diffusion length may be longer than the probe spacing, and the measured resistivity will be in error.

Minority carrier injection may be important for high resistivity materials. For silicon this applies for $\rho \geq 100$ ohm-cm. An error of less than 2% is introduced by minority carrier injection if the voltage across the two voltage-sensing probes is held to less than 100 mV for 1 mm probe spacings for samples with lapped surfaces.[29]

Probe Spacing   A mechanical four-point probe exhibits small random probe spacing variations. Such variations give erroneous values of resistivity or sheet resistance, especially when evaluating highly uniformly doped wafers. In such cases it is very important to know whether any non-uniformities are due to the wafer, due to process variations, or due to measurement errors. An example is the evaluation of ion-implanted layers. It is known that ion-implanted layers can have sheet resistance uniformities better than 1%.[30] For small probe spacing variations the correction factor[18]

$$F_s \approx 1 + 1.082(1 - s_2/s_m) \tag{1.32}$$

must be applied. The spacing s_2 is shown in Fig. 1.2, and s_m is the mean value of the probe spacings. Errors due to probe wander can be reduced by averaging several independent readings.

Current Effects   Additional sources of error are the current and surface leakage current. The current can affect the measured resistivity in two ways: by an apparent resistivity increase produced by wafer heating and by an apparent resistivity decrease due to minority carrier injection. The suggested current for silicon wafers as well as ingots is shown in Fig. 1.12 as a function of resistivity. The data for the solid lines were obtained by measuring the four-point probe resistivity as a function of current for a given sample. Such resistivity-current curves show typically a flat region bounded by non-linearities at both low and high currents.[31] The flat region gives the appropriate current. The dashed line is the recommended ASTM current.[18] Surface leakage is reduced or eliminated by enclosing the probe in a shielded enclosure held at a potential equal to the inner probe potential.[29]

Temperature Effects   It is important that the temperature along the sample be uniform in order not to introduce thermoelectric voltages.

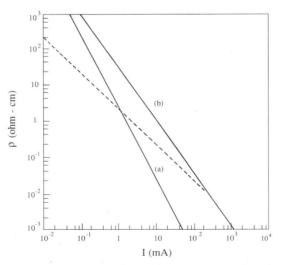

Fig. 1.12   The recommended four-point probe current as a function of Si resistivity. (a) Si wafers, (b) Si ingots, ---- ASTM recommended.

Temperature gradients can be caused by ambient effects but are more likely due to sample heating by the probe current. Current heating is most likely to occur in low resistivity samples where large currents are required to obtain readily measurable voltages.

Even if temperature variations are not caused by the measurement apparatus and there are no temperature gradients, there may still be temperature variations due to temperature fluctuations in the measurement room. Since semiconductors have relatively large temperature coefficients of resistivity, errors are easily introduced by failing to compensate for such temperature variations. Figure 1.13 gives the temperature coefficients of resistivity for n- and p-Si.[32] For resistivities of 10 ohm-cm or higher, the coefficient is on the order of 1%/°C. Temperature corrections are made according to[32]

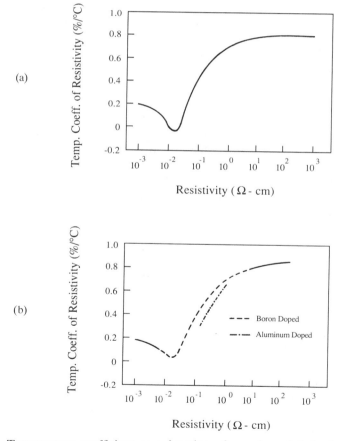

Fig. 1.13 Temperature coefficient as a function of sample resistivity for (a) n-Si, and (b) p-Si. Reprinted with permission after Bullis et al.[32]

$$\rho = \frac{\rho_{unc}}{[1 + C_T(T - T_{ref})]} \tag{1.33}$$

where ρ is the corrected resistivity, ρ_{unc} is the uncorrected resistivity, C_T is the temperature coefficient of resistivity, T is the measurement temperature, and T_{ref} is the temperature of the reference standard.

High Resistivity Materials   Materials of very high resistivity, such as semi-insulating GaAs with $\rho \approx 10^7 - 10^9$ ohm-cm, are difficult to measure by conventional four-point probe or van der Pauw methods. Moderately doped wafers can become highly resistive at low temperatures and are similarly difficult to measure. Special measurement precautions must be observed. The simplest measurement method relies on providing the wafer with a large contact on one side and small contacts on the other side of the wafer. A current is passed through the contacts and the voltage is measured. This arrangement, by itself, is not sufficient because surface leakage currents can introduce measurement errors. By surrounding the small contacts with a guard ring and holding the guard ring at the same or nearly the same potential as the small contact, surface currents are essentially suppressed.[33] It is of course necessary to ensure that the contacts are ohmic or as close to ohmic as possible so that the bulk resistivity and not the contact resistance is measured.

Two-terminal measurements are notorious for being complicated by contact effects and the true sample resistivity is not easy to determine as indicated by Eq. (1.2). Conventional van der Pauw measurements suitable for moderate or low resistivity materials are suspect for high resistance samples unless care is taken to eliminate current leakage paths and sample loading by the voltmeter. One approach around this problem is the "guarded" approach using high input impedance, unity gain amplifiers between each probe on the sample, and the external circuitry.[34] The unity gain amplifiers drive the shields on the leads between the amplifer and the sample, thereby effectively eliminating the stray capacitance in the leads. This reduces leakage currents and the system time constant. Measurements of resistances up to 10^{12} ohms have been made with such a system. The "guarded" approach has also been automated.[35]

1.2.5   Wafer Mapping

Four-point probe resistivity measurements have been made for many years. Manual measurements are suitable for only relatively few measurements per wafer. Wafer mapping techniques consisting of a four-point probe that is stepped across the wafer in two dimensions under computer control have been developed to generate many data points.[21,36,37] Data are gathered in a two-dimensional array, and graphical outputs are generated in the form of contour maps.

This four-point probe mapping implementation is well suited for rapid data generation and for display in a format that allows very effective nonuniformity observation. It also allows for rapid comparison between samples. It has been used, for example, in the evaluation of the performance of ion implanters and planar channeling studies as well as for sheet resistance measurements for diffused layers, polycrystalline silicon films, and metal layers.[38-40] An example is shown in Fig. 1.14. Wafer mapping has become an important process monitor. Displaying data in a graphical, two-dimensional display is a much more powerful indicator of process uniformity than displaying the same data in tabular form. A well-designed uniformity map gives instant information about implant uniformity, flow patterns during diffusion, epitaxial reactor nonuniformities, etc.

High Sheet Resistance Wafer Mapping   The sheet resistance of high resistance layers can be determined by capacitance-voltage, spreading resistance, van der Pauw, MOSFET threshold voltage shift, thermal wave, optical density changes in photoresist, and four-point probe measurements. When the sheet resistance is measured by the four-point probe technique, experimental difficulties are experienced for high sheet resistances. An example is a low-dose, ion-implanted layer typically used for threshold voltage control of MOS field-effect transistors. The reasons for the measurement difficulties are: (1) it is difficult to make good electrical contact from

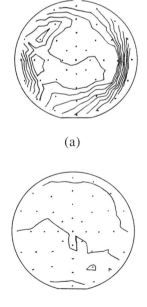

(a)

(b)

Fig. 1.14 Sheet resistance contour maps for 10 mA P$^+$ implants into (100) Si. (a) without, and (b) with the use of a "flood gun" source of secondary electrons. The lines are 1% contour lines. Reprinted with permission after Current.[30]

the probe to the semiconductor, (2) low doses give low carrier concentrations and therefore poor conductivity, and (3) the surface leakage current can be comparable to the measurement current. The conventional four-point probe method has been used provided the starting wafers are of high resistivity, and they are oxidized before the implant to stabilize the surface resistance and to prevent ion channeling. The wafer is annealed, the oxide is stripped, and a surface stabilization is performed using a hot sulfuric acid and hydrogen peroxide solution.

A modified four-point probe method, the *double implant technique*, has proved to be useful for the sheet resistance measurement of such layers.[20,42] It is implemented in the following manner. A *p*-type (*n*-type) impurity is implanted into an *n*-type (*p*-type) substrate at a dose Φ_1 and energy E_1. For example, boron is implanted at a dose of $\Phi_1 = 10^{14}$ cm^{-2} and energy $E_1 = 120$ keV. The wafer is annealed to remove the implant damage and to activate the implanted ions electrically. The sheet resistance ρ_{s1} is then measured. Next the desired low-dose impurity is implanted at dose Φ_2 and energy E_2, with $\Phi_2 < \Phi_1$. E_2 should be less than E_1 to prevent penetration through the first implanted layer. The first implant energy is typically at least 10–20% higher, and the first implant dose is at least two orders of magnitude higher than the second implant. The second implant conditions might be $\Phi_2 = 10^{11}$ cm^{-2} and $E_2 = 100$ keV. The sheet resistance ρ_{s2} after the second implant is measured and compared to ρ_{s1} without annealing the second implant.

The second sheet resistance measurement relies on the implant damage of the second implant being proportional to the implant dose, and therefore $\rho_{s2} > \rho_{s1}$. The impurity atomic mass of the first implant should be approximately the same mass as the second implant. It has also been found that (111)-oriented Si wafers are preferred over (100)-oriented wafers to reduce channeling effects. The double-implant method allows measurements immediately after the second implant. Implant doses as low as 10^{10} cm^{-2} can be measured by this technique. Test wafers can be annealed and reused, provided the anneal temperature is kept sufficiently low to prevent impurity redistribution. The method is also applicable for electrically inactive species, such as oxygen, argon, or nitrogen implants. A more detailed discussion is given in Smith et al.[42]

The double-implant technique suffers from several problems. Any sheet resistance nonuniformities resulting from the first implant and its activation cycle alter the low-dose measurement. Additionally, since this method derives its low-dose sensitivity from ion-implant *damage*, it is sensitive to post-implant relaxation. This is an effect where implant damage decreases over a period of hours to days following the implant. If the measurement is made immediately after the second implant, damage relaxation has little effect. However, if the measurement is made several hours or days after the implant, damage relaxation can reduce the measured resistance by 10–20% for the types of implant dose and energy typical for low-dose implants.

1.3 RESISTIVITY PROFILING

Conventional four-point probe measurements give an average resistivity. This is suitable for uniformly doped substrates but is not sufficient for non-uniformly doped samples in which resistivity *profiles* need to be determined. For example, the resistivity profile of a diffused or ion-implanted layer cannot be determined from a singular four-point probe measurement. Furthermore it is usually the dopant concentration profile that is desired, not the resistivity profile.

Techniques suitable for determining dopant concentration profiles include anodic oxidation–four-point probe, spreading resistance, capacitance-voltage, secondary ion mass spectrometry, and thermal waves. We will discuss the first two methods in this chapter and defer discussion of the others to Chapter 2.

1.3.1 Anodic Oxidation—Four-Point Probe

To be able to measure a resistivity depth profile, it is obvious that depth information must be provided. It is possible to measure the resistivity profile of a diffused or ion-implanted layer by removing thin layers of the sample, measuring the resistivity, removing, measuring, etc. The sheet resistance of a layer of thickness t is given by

$$\rho_s(x) = \frac{1}{q \int_x^t [n(x)\mu_n(x) + p(x)\mu_p(x)] \, dx} \tag{1.34}$$

where x is the coordinate measured from the surface into the diffused layer. The sheet resistance of a uniformly doped layer with constant carrier concentrations and mobilities becomes

$$\rho_s = \frac{1}{q[n\mu_n + p\mu_p]t} \tag{1.35}$$

The sheet resistance is a meaningful descriptor not only for uniformly doped layers but also for non-uniformly doped layers, where both carrier concentrations and mobilities are depth dependent. In Eq. (1.34) $\rho_s(x)$ represents an averaged value.

The sheet resistance is measured as a function of depth by incremental layer removal and $1/\rho_s$ is plotted versus depth. The slope of such a plot is[43]

$$\frac{d[1/\rho_s(x)]}{dx} = -q[n(x)\mu_n(x) + p(x)\mu_p(x)] = -\sigma(x) \tag{1.36}$$

where $\sigma(x)$ is the layer conductivity. The resistivity is determined from Eq. (1.36) and from the identity $\rho(x) = 1/\sigma(x)$ as

$$\rho(x) = \left[\frac{1}{\rho_s^2(x)} \frac{d\rho_s(x)}{dx} \right]^{-1} = \frac{0.4343\rho_s(x)}{d[\log\rho_s(x)]/dx} \qquad (1.37)$$

A plot of $\log\rho_s(x)$ versus x is shown in Fig. 1.15. Once the resistivity variation is derived from such a plot, the dopant profile is obtained from published N_A, N_D versus ρ curves. Such curves are given in Appendix 1.1.

Repeated removing of well-controlled thin layers from a heavily doped semiconductor is difficult to do by chemical etching. It is, however, not difficult with *anodic oxidation*. During anodic oxidation a semiconductor is immersed in a suitable electrolyte in an anodization cell. A current is passed from an electrode to the semiconductor sample through the electrolyte, causing an oxide to grow. The oxidation proceeds at room temperature, and the oxide grows by consuming a portion of the semiconductor sample.[44] By subsequently etching the oxide, that portion of the semiconductor consumed during the oxidation is removed as well. This can be done very reproducibly.

Two anodization methods are possible. In the first method a constant voltage is applied, and the anodization current is allowed to fall from an initial to a final predetermined value. In the second method a constant current is forced to flow, and the voltage is allowed to rise until a preset value is attained. The oxide thickness is directly proportional to the net forming voltage in the constant current anodization method, where the net forming voltage is the final cell voltage minus the initial cell voltage.

A variety of anodization solutions have been used. The non-aqueous solutions N-methylacetamide, tetrahydrofurfuryl alcohol and ethylene glycol is suitable for silicon.[45] Ethylene glycol containing $0.04N$ KNO_3 and 1–5% water produces uniform, reproducible oxides at current densities of 2 to $10\,mA/cm^2$. For the ethylene glycol mixture $2.2\,\text{Å}$ of Si are removed per volt.[45] A forming voltage of 100 V removes $220\,\text{Å}$ of Si. Ge,[46] InSb,[47] and GaAs[48–49] have all been anodically oxidized.

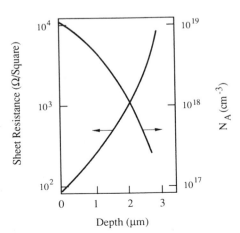

Fig. 1.15 Doping profile of a boron-diffused layer using the log(sheet resistance) method. Reprinted with permission after Evans and Donovan.[43]

The laborious nature of the differential conductivity profiling technique limits its applicability if the entire process is done manually. The measurement time can be substantially reduced by automating the method. Computer-controlled experimental methods have been developed in which the sample is anodized, etched and then the resistivity and the mobility are measured in situ.[50–51]

1.3.2 Spreading Resistance

Spreading resistance profiling (SRP) is a technique to generate a resistivity and a dopant profile. It consists of two carefully aligned probes that are stepped along the semiconductor surface and the resistance between the probes is measured at each location as shown in Fig. 1.16.[52–53] The sample is prepared by mounting it on a bevel block with melted wax. Bevel angles of 15' to 5° are typical. The bevel block is inserted into a well-fitting cylinder, and the sample is lapped using 0.25 μm diamond paste. Next the

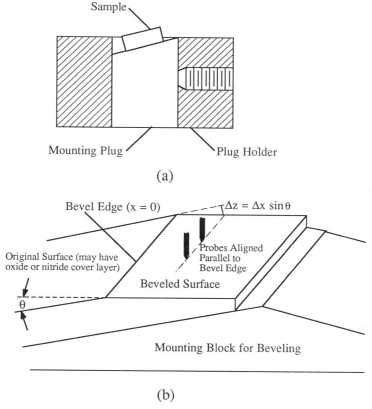

(a)

(b)

Fig. 1.16 Spreading resistance bevel block in (a), and the bevelled sample with the probes in (b). Copyright ASTM; reprinted with permission after ASTM F672.[67]

sample is positioned in the measurement apparatus with the bevel edge perpendicular to the probe stepping direction. It is very useful to provide the sample with an oxide coating. The oxide provides a sharp corner at the bevel and also clearly defines the start of the beveled surface because the spreading resistance of the oxide is very high. Spreading resistance measurements are primarily used for silicon.

To understand spreading resistance, consider a mechanical probe contacting a semiconductor surface as in Fig. 1.17. The current I flows from the probe with diameter $2r$ into a semiconductor of resistivity ρ. The concentric circles represent equipotential lines and the dashed lines represent the electric field lines perpendicular to the equipotential lines. The current, flowing in the direction of the electric field along the dashed lines, is concentrated at the probe tip and *spreads* out radially from the tip. Hence the name *spreading resistance*. Most of the voltage drop or most of the resistance is near the probe and decreases rapidly away from the probe into the semiconductor.

For a non-indenting, cylindrical contact with a planar, circular interface and a highly conductive probe with negligible resistance, the spreading resistance for a semi-infinite sample is[54]

$$R_{sp} = \frac{\rho}{4r} \tag{1.38}$$

For a hemispherical, indenting probe tip of radius r, the spreading resistance is

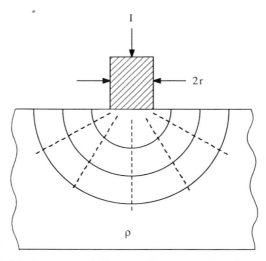

Fig. 1.17 A cylindrical contact of diameter $2r$ to a semiconductor. The equipotential lines are indicated by the solid lines and the electric field lines by the dashed lines.

$$R_{sp} = \frac{\rho}{2\pi r} \tag{1.39}$$

The spreading resistance is measured in units of ohms. Equation (1.38) has been verified by comparing spreading resistance with four-point probe measurements.[55]

About 80% of the potential drop due to the spreading phenomenon occurs within a distance of about five times the contact radius for both types of contacts. The probe penetration is about 100 Å for probe loads of 10 to 12 g.[56] Hence the contact damage in silicon is typically several hundred ångströms. With a probe radius of several microns, it is customary to use the contact type of Eq. (1.38) as a starting point. For a contact radius of 2.5 μm, Eq. (1.38) gives $R_{sp} \approx 1000\rho$. Measured values are $R_{sp} \sim (3000 \text{ to } 4000)\rho$.[57] The idealized spreading resistance expression of Eq. (1.38) should be modified to $R_{sp} = k\rho/4r$, where k is an empirical constant which is found to be $k \approx 1$–3 for n-Si and $k \approx 1$ for p-Si.[52] The higher value for n-Si is believed to be due to the higher barrier height of n-Si contacts compared to p-Si contacts.

The spreading resistance measurement method is characterized by four major features: (1) specially prepared probes and the apparatus to raise, lower and step the probes, (2) low applied voltages during the measurement, (3) calibration of the measured resistance against known standards, and (4) a multilayer correction procedure to correct for boundary effects.[58]

The probes, consisting of tungsten-osmium alloy, are mounted in gravity-loaded probe arms. The probe tips are shaped so that they can be positioned very close together, often with less than 20 μm spacing. The probe arms are supported by a kinematic bearing system with five contacts giving the arms only one degree of freedom, which is a rotation around the horizontal axis. This virtually eliminates lateral probe motion during contact to the sample minimizing probe wear and damage to the semiconductor. The probes deform only slightly elastically upon contacting the semiconductor, thus making very reproducible contacts. The contact pressure is on the order of 1000 kg/mm^2, plastically deforming the semiconductor and creating some damage. The probes are "conditioned" using the "Gorey-Schneider technique"[59] for the contact area of the probe to consist of a large number of microscopic protrusions. These protrusions are sufficiently small that they penetrate the thin oxide layer that generally exists on silicon surfaces.

The voltage between the probes during measurement is kept at around 5 mV to reduce the effect of contact resistance. The probe-semiconductor contact is a metal-semiconductor contact that generally has a nonlinear current-voltage characteristic. However, for applied voltages less than $kT/q \approx 25$ mV, the contact current-voltage characteristic is linear. The resistance between the two contacts consists of the contact resistance, the spreading resistance, and the semiconductor resistance as given by Eq. (1.2). The spreading resistance technique is successful because it is a comparative technique. Calibration curves are generated for a particular set of probes at a particular time using samples of known resistivity. Such calibration samples are available from the National Institute of Standards and Technology for (100) n-Si, (111) n-Si, (100) p-Si, and (111) p-Si. Comparison of the spreading resistance data to the calibration samples is necessary and sufficient for uniformly doped samples. An example of spreading resistance profiles along "uniformly" doped silicon is shown in Fig. 1.18. Note the uniformity in the neutron-doped wafer compared to a conventional float-zone grown sample. For samples containing pn or high-low junctions, additional corrections are necessary. These multilayer corrections have evolved over the years where today very sophisticated correction schemes are used.[57,60-65]

Shallow layers are measured by angle-lapping the sample, as shown in Fig. 1.16. The bevel angle θ is typically 1°–5° for junction depths of 1–2 μm and $\theta \leq 0.5°$ for junction depths less than 0.5 μm. The equivalent depth, Δz, for each Δx step along the surface beveled at angle θ, is

$$\Delta z = \Delta x \sin \theta \qquad (1.40)$$

For a step of 5 μm and an angle of 1°, the equivalent step height or measurement resolution is 0.087 μm or 870 Å. A very high resolution spreading resistance plot is shown in Fig. 1.19 for which the bevel angle is

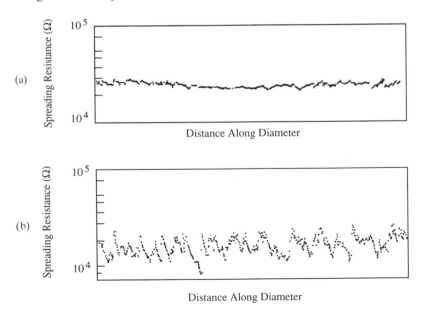

Fig. 1.18 Spreading resistance profiles along two (111) n-Si wafers of 65 ohm-cm resistivity. (a) Neutron-transmutation doped wafer, (b) float zone grown wafer. Probe step = 25 μm. Reprinted with permission after ref. 57.

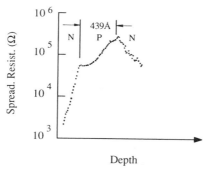

Fig. 1.19 Spreading resistance profiles of a narrow-base *npn* Si bipolar transistor. Bevel angle 3.5 min, $\Delta x = 1\ \mu$m, $\Delta z = 10.2$ Å, probe load 5 g. Reprinted with permission from *Microelectronics Processing*: *Inorganic Materials Characterization*, Fig. 6, p. 40. Copyright 1986 American Chemical Society, after Mazur.[58]

3.5 min and the step increment is 1 μm. The equivalent step height is 10.2 Å. Such small angles are determined by measuring a small slit of light that is reflected from the beveled and the unbeveled surfaces so that two images are detected. When the slit is rotated, the two images rotate also, and the rotation angle is measured and related to the bevel angle.[66]

Almost all spreading measurements are done with two probes, but three-probe arrangements have been used.[60] In the three-probe configuration one probe serves as the common point to both voltage and current circuits and is the only proble contributing to the measured resistance. The three-probe system is more difficult to keep aligned. Since probe alignment parallel to the bevel intersection with the top surface is crucial for depth profiling, the three-point spreading resistance probe is rarely used.

1.4 CONTACTLESS METHODS

Contactless resistivity measurement techniques have become popular during the past few years in line with the general trend toward contactless semiconductor measurements wherever possible. The measurements methods fall into two broad categories: electrical and optical measurements. Commercial products are available for electrical characterization, but optical characterization is also used. Among the various electrical methods, only the *eddy current* method has found application to date.

1.4.1 Eddy Current

Electrical contactless measurement techniques fall into several categories.[68] (1) The sample is placed into a microwave circuit and perturbs the transmission or reflection characteristics of a waveguide or cavity,[69–72] (2) the sample is capacitively coupled to the measuring apparatus,[73-74] and (3) the sample is inductively coupled to the apparatus.[75-77] The concepts for all of these contactless characterization methods have been known for many years, but have not found applications until recently. With the trend toward contactless measurements wherever practical, contactless resistivity apparatus have become commercially available.

To be a viable commercial instrument, the apparatus should be simple with no special sample requirements. This ruled out special sample configurations to fit microwave cavities, for example, and led to a variation of the inductively coupled approach. The measurement technique is based on the parallel resonant tank circuit of Fig. 1.20(a). The Q of such a circuit is reduced when a conducting material is brought close to the coil due to the power absorbed by the conducting material. A practical implementation of this concept is shown in Fig. 1.20(b), where the LC circuit is replaced by dual coils on ferrite cores separated to provide a gap for wafer insertion. A semiconductor wafer is coupled to the circuit via the high permeability ferrite cores. The oscillating magnetic field sets up eddy currents in the semiconductor leading to Joule heating of the material.

The absorbed power P_a is[78]

$$P_a = K(V_T/n)^2 \sigma t \qquad (1.41)$$

where K is a constant involving the coupling parameters of the core, V_T the rms primary rf voltage, n the number of primary turns of the coil, σ the semiconductor conductivity, t the thickness. Equation (1.41) is valid provided the sample thickness is less than the skin depth. This can always be satisfied by choosing the appropriate frequency. With power given by $P_a = V_T I_T$, where I_T is the in-phase drive current, we find

$$I_T = K(V_T/n^2)\sigma t = K(V_T/n^2)/\rho_s \qquad (1.42)$$

If V_T is held constant through a feedback circuit, the current is proportional to the sample conductivity-thickness product, or it is inversely proportional the sample sheet resistance. Eddy current and other contactless techniques are discussed further in Section 8.4 in reference to lifetime measurements.

To determine the wafer resistivity, the wafer thickness must be known. In contactless measurements provision must be made to measure the thickness without wafer contact. Two methods are used: differential capacitance probe, and ultrasound.[79] The capacitance probe determines the thickness by measuring the capacitance change between a probe and a conductive base plane when the wafer is placed on the base. The resulting capacitance change is proportional to the distance between the probe head and wafer. In the ultrasound method sound waves are reflected from the upper and lower wafer surfaces located between the two probes shown in Fig. 1.20(b). The phase shift of the reflected sound caused by the impedance variation of the

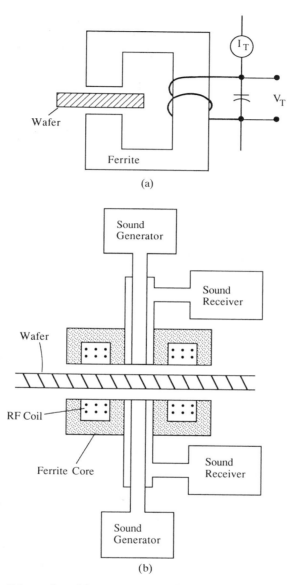

Fig. 1.20 (a) Schematic eddy current experimental arrangement, and (b) the schematic of the Tencor commercial apparatus showing the eddy current coils and the sonic thickness arrangement.

air gap is detected by the sonic receiver. The phase shift is proportional to the distance from each probe to each surface. With the spacing between probes known, the wafer thickness can be determined.

Resistivity measurements based on the eddy current technique are useful for uniformly doped wafers. The eddy current technique has also found use for the measurement of highly conductive layers on less conductive substrates. The sheet resistance of the layer should be at least a hundred times lower than the sheet resistance of the substrate to measure the layer and not the substrate. This rules out the use of this technique for measurements of diffused or ion-implemented layers, which generally do not satisfy this rule. For example, sheet resistances of diffused or ion-implanted layers are typically 10 to 100 ohms/square, and the sheet resistance of a 10 ohm-cm, 500 μm thick Si wafer is 200 ohms/square. However, the sheet resistance of metal layers on semiconductor substrates can be measured. The sheet resistance of a 5000 Å Al layer is typically 0.06 to 0.1 ohms/square, making such layers 2000 times less resistive than the Si substrate. From a measurement of the layer sheet resistance, one can determine the layer thickness, knowing the layer resistivity from an independent measurement. Contactless resistance measurements are primarily used to determine sheet resistances and thicknesses of conducting layers.

1.5 CONDUCTIVITY TYPE

Several techniques are used to determine the semiconductor conductivity type. They are wafer flat location, thermal emf, rectification, and Hall effect. The Hall effect is discussed in Chapter 2. The simplest method utilizes the shape of the wafer flats for wafers following a standard pattern. Silicon wafers are almost always circular with characteristic flats, illustrated in Fig. 1.21, provided for alignment and identification purposes. The primary flat is usually along the $\langle 110 \rangle$ direction, and secondary flats are used to identify the conductivity type and orientation. Many wafers use the standard flats of Fig. 1.21.

In the *hot* or *thermoelectric probe* method the type is determined by the sign of the thermal emf or Seebeck voltage generated by a temperature gradient. Two probes contact the sample surface: one is hot; the other is cold as illustrated in Fig. 1.22(a). Thermal gradients produce currents in a semiconductor; the majority carrier currents for *n*- and *p*-type materials are[83]

$$J_n = -qn\mu_n \mathscr{P}_n \, dT/dx \tag{1.43a}$$

$$J_p = -qp\mu_p \mathscr{P}_p \, dT/dx \tag{1.43b}$$

where $\mathscr{P}_n < 0$ and $\mathscr{P}_p > 0$ are the differential thermoelectric power.

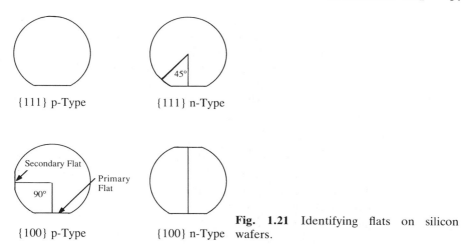

{111} p-Type

{111} n-Type

{100} p-Type

{100} n-Type

Fig. 1.21 Identifying flats on silicon wafers.

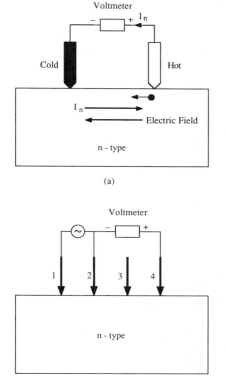

(a)

(b)

Fig. 1.22 (a) Hot probe, (b) rectifying probe for conductivity type measurements.

Consider the experimental arrangement of Fig. 1.22(a). The right probe is hot; the left probe is cold. $dT/dx > 0$, and the electron current in an n-type sample flows from left to right. The thermoelectric power can be thought of as a current generator, with some of that current flowing through the voltmeter causing the hot probe to have a positive potential with respect to the cold probe.[84–85] There is a simple alternative view. Electrons diffuse from the hot to the cold region setting up an electric field that opposes the diffusion. The electric field produces a potential detected by the voltmeter. Analogous reasoning leads to the opposite potential for p-type samples. Hot probes are effective over the 10^{-3} to 10^3 ohm-cm resistivity range. The voltmeter tends to indicate n-type for high resistivity material even if the sample is weakly p-type because the method actually determines the $n\mu_n$ or the $p\mu_p$ product. With $\mu_n > \mu_p$ intrinsic or high resistivity material is measured n-type if $n \approx p$. In semiconductors with $n_i > n$ or $n_i > p$ at room temperature (narrow band gap semiconductors), it may be necessary to cool one of the probes and let the room temperature probe be the "hot" one.

In the *rectification* method the sign of the conductivity is determined by the polarity of a rectified ac signal at a point contact to the semiconductor.[84–85] When two probes are used, one should be rectifying and the other should be ohmic. Current flows through a rectifying contact to n-type material if the metal is positive and for p-type if it is negative. Rectifying and ohmic contacts are difficult to implement with two-point contacts. Fortunately four-point probes can be used with appropriate connections not requiring ohmic contacts. An ac voltage is applied between probes 1 and 2, and the resulting potential is measured between probes 4 and 2 as shown in Fig. 1.22(b). The voltage drop V_{42} is small when the ac voltage at probe 2 is positive because the metal-semiconductor junction is forward biased. But for negative voltage at probe 2, the junction is reverse biased; V_{42} is large and positive. The large positive and small negative ac V_{42} result in a dc component with the polarity of the semiconductor-metal junction voltage necessary to reverse bias the junction. For n-type $V_{42} > 0$, and for p-type $V_{42} < 0$.[84] Probe 3 can also be used as the voltage-sensing probe. This method of conductivity type measurement is built into some commercial four-point probe systems.

1.6 STRENGTHS AND WEAKNESSES

- *Four-Point Probe* The weakness of the four-point probe technique is the damage it introduces when the probes contact the surface. The damage is not very large but sufficient not to make measurements on wafers to be used for device fabrication. The probe also samples a relatively large volume of the wafer, preventing high-resolution measurements.

The method's strength lies in its established use. It has been used for many years in the semiconductor industry and is well understood. With the recent advent of wafer mapping, the four-point probe has become a very powerful process-monitoring tool. This is where its major strength lies today.

- *Differential Resistivity* The weakness of this method is its tediousness. The layer removal by anodic oxidation is very controlled, but it is also slow, limiting the method to relatively few data points per profile. Repeated four-point probe measurements on the same area will create damage, rendering the measurements questionable. The method is also destructive.

  The method's strength lies in its inexpensive equipment. For those dopant profiles that cannot be profiled by capacitance-voltage measurements, only secondary ion mass spectroscopy and spreading resistance methods are the alternatives. Equipment for those measurements is significantly more expensive, leaving anodic oxidation/four-point probe as a viable, inexpensive alternative.

- *Spreading Resistance* The weakness of the spreading resistance profiling technique is the necessity of a skilled operator to obtain reliable profiles. The system must be periodically calibrated against known standards, and the probes must be periodically reconditioned. It does not work well for semiconductors other than Si and Ge, but advances for use on GaAs are being made. The sample preparation is not trivial, and the measurement is destructive. The conversion of the measured spreading resistance data to doping concentration profiles depends very much on the algorithm. Several algorithms are in use, and others are being developed.

  The strengths of SRP lie in the ability to profile practically any combination of layers with very high resolution and no depth limitation and no real doping concentration limitations. Very high resistivity material must be carefully measured and interpreted. Although the equipment is not inexpensive, it is commercially available and it is used extensively. Hence there is large background of knowledge related to this method which has been developed over the past 20 years.

- *Eddy Current* The weakness of the eddy current technique is its inability to determine the sheet resistance of thin diffused or ion-implanted layers. In order to detect such sheet resistances, it is necessary for the sheet resistance of the layer to be on the order of a hundred times lower than the sheet resistance of the substrate. This is only attainable when the sheet consists of a metal and the eddy current technique is predominantly utilized to measure the resistance of metal layers on semiconductor substrates to determine its thickness.

  The strength of the method lies in its non-contacting nature and the availability of commercial equipment. This is ideal for measuring the resistivity of semiconductor wafers. However, the wafer resistivity is not measured all that frequently anymore. The user relies predominantly on the wafer specifications being supplied by the vendor.

APPENDIX 1.1   RESISTIVITY AS A FUNCTION OF DOPING CONCENTRATION

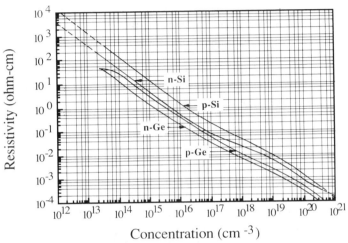

Fig. A1.1   Resistivity versus doping concentration at 23°C for *p*-type (boron-doped) and *n*-type (phosphorus-doped) silicon and for *n*- and *p*-type germanium. Data adapted from ASTM F723[80] and Cuttriss.[81]

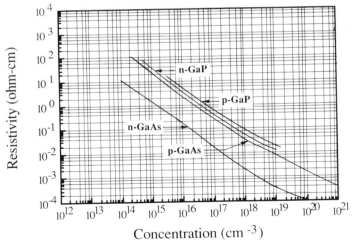

Fig. A1.2   Resistivity versus doping concentration for GaAs and GaP. Data adapted from Sze and Irvin.[82,83]

REFERENCES

[1]   F. Wenner, "A Method of Measuring Earth Resistivity," *Bulletin of the Bureau of Standards* **12**, 469–478, 1915.

[2]   L. B. Valdes, "Resistivity Measurements on Germanium for Transistors," *Proc. IRE* **42**, 420–427, Feb. 1954.

[3]   H. H. Wieder, "Four Terminal Nondestructive Electrical and Galvanomagnetic Measurements," in *Nondestructive Evaluation of Semiconductor Materials and Devices* (J.N. Zemel, ed.), Plenum Press, New York, NY, 1979, pp. 67–104.

[4]   R. Hall, "Minimizing Errors of Four-Point Probe Measurements on Circular Wafers," *J. Sci. Instrum.* **44**, 53–54, Jan. 1967.

[5]   F. M. Smits, "Measurement of Sheet Resistivities with the Four-Point Probe," *Bell Syst. Tech. J.* **37**, 711–718, May 1958.

[6]   A. Uhlir, Jr., "The Potentials of Infinite Systems of Sources and Numerical Solutions of Problems in Semiconductor Engineering," *Bell Syst. Tech. J.* **34**, 105–128, Jan. 1955.

[7]   M. G. Buehler, "A Hall Four-Point Probe on Thin Plates," *Solid-State Electron.* **10**, 801–812, Aug. 1967.

[8]   M. G. Buehler, "Measurement of the Resistivity of a Thin Square Sample with a Square Four-Probe Array," *Solid-State Electron.* **20**, 403–406, May 1977.

[9]   M. Yamashita, "Geometrical Correction Factor for Resistivity of Semiconductors by the Square Four-Point Probe Method," *Japan. J. Appl. Phys.* **25**, 563–567, April 1986.

[10] S. Murashima and F. Ishibashi, "Correction Devisors for the Four-Point Probe Resistivity Measurement on Cylindrical Semiconductors II," *Japan. J. Appl. Phys.* **9**, 1340–1346, *Nov.* 1970.

[11] D. S. Perloff, "Four-Point Probe Sheet Resistance Correction Factors for Thin Rectangular Samples," *Solid-State Electron.* **20**, 681–687, Aug. 1977.

[12] D. S. Perloff, "Four-Point Probe Correction Factors for Use in Measuring Large Diameter Doped Semiconductor Wafers," *J. Electrochem. Soc.* **123**, 1745–1750, Nov. 1976.

[13] M. Yamashita and M. Agu, "Geometrical Correction Factor for Semiconductor Resistivity Measurements by Four-Point Probe Method," *Japan. J. Appl. Phys.* **23**, 1499–1504, Nov. 1984.

[14] J. Albers and H.L. Berkowitz, "An Alternative Approach to the Calculation of Four-Probe Resistances on Nonuniform Structures," *J. Electrochem. Soc.* **132**, 2453–2456, Oct. 1985.

[15] J. F. Combs and M. P. Albert, "Diameter Correction Factors for the Resistivity Measurement of Semiconductor Slices," *Semic. Prod./Solid State Techn.* **6**, 26–27, Feb. 1963.

[16] M. P. Albert and J. F. Combs, "Correction Factors for Radial Resistivity Gradient Evaluation of Semiconductor Slices," *IEEE Trans. Electron Dev.* **ED-11**, 148–151, April 1964.

[17] R. Rymaszewski, "Relationship Between the Correction Factor of the Four-Point Probe Value and the Selection of Potential and Current Electrodes," *J. Sci. Instrum.* **2**, 170–174, Feb. 1969.

[18] ASTM Standard F84, "Standard Method for Measuring Resistivity of Silicon Slices With a Collinear Four-Point Probe," *1988 Annual Book of ASTM Standards*, Am. Soc. Test., Philadelphia, 1988.

[19] D. S. Perloff, J. N. Gan, and F. E. Wahl, "Dose Accuracy and Doping Uniformity of Ion Implantation Equipment," *Solid State Technol.* **24**, 112–120, Feb. 1981.

[20] A. K. Smith, D. S. Perloff, R. Edwards, R. Kleppinger, and M. D. Rigik, "The Use of Four-Point Probe Sheet Resistance Measurements for Characterizing Low Dose Ion Implantation," *Nucl. Instrum. and Meth.* **B6**, 382–388, Jan. 1985.

[21] M. I. Current and M. J. Markert, "Mapping of Ion Implanted Wafers," in *Ion Implantation: Science and Technology* (J. F. Ziegler, ed.), Academic Press, Orlando, FL, 1984, pp. 487–536.

[22] H. H. Gegenwarth, "Correction Factors for the Four-Point Probe Resistivity Measurement on Cylindrical Semiconductors," *Solid-State Electron.* **11**, 787–789, Aug. 1968.

[23] S. Murashima, H. Kanamori, and F. Ishibashi, "Correction Devisors for the Four-Point Probe Resistivity Measurement on Cylindrical Semiconductors," *Japan. J. Appl. Phys.* **9**, 58–67, Jan. 1970.

[24] L. J. van der Pauw, "A Method of Measuring Specific Resistivity and Hall Effect of Discs of Arbitrary Shape," *Phil. Res. Rep.* **13**, 1–9, Feb. 1958.

[25] L. J. van der Pauw, "A Method of Measuring the Resistivity and Hall Coefficient on Lamellae of Arbitrary Shape," *Phil. Tech. Rev.* **20**, 220–224, Aug. 1958; R. Chwang, B. J. Smith, and C. R. Crowell, "Contact Size Effects on the van der Pauw Method for Resistivity and Hall Coefficient Measurement," *Solid-State Electron.* **17**, 1217–1227, Dec. 1974.

[26] J. M. David and M. G. Buehler, "A Numerical Analysis of Various Cross Sheet Resistor Test Structures," *Solid-State Electron.* **20**, 539–543, June 1977.

[27] M. G. Buehler and W. R. Thurber, "An Experimental Study of Various Cross Sheet Resistor Test Structures," *J. Electrochem. Soc.* **125**, 645–650, April 1978.

[28] M. G. Buehler, S. D. Grant, and W. R. Thurber, "Bridge and van der Pauw Sheet for Characterizing the Line Width of Conducting Layers," *J. Electrochem. Soc.* **125**, 650–654, April 1978.

[29] J. K. Hargreaves and D. Millard, "The Accuracy of Four-Probe Resistivity Measurements on Silicon," *Brit. J. Appl. Phys.* **13**, 231–234, May 1962.

[30] M. I. Current, "Current Status of Ion Implantation Equipment and Techniques for Semiconductor IC Fabrication," *Nucl. Instrum. and Meth.* **B6**, 9–15, Jan. 1985.

[31] L. H. Garrison, "Proper Current Input for True Resistivity Measurements of Single Crystal Silicon Both *n*-and *p*-Type and Single Crystal Slices," *Semic. Prod./Solid State Techn.* **9**, 47–49, May 1966.

[32] W. M. Bullis, F. H. Brewer, C. D. Kolstad, and L. J. Swartzendruber, "Temperature Coefficient of Resistivity of Silicon and Germanium Near Room Temperature," *Solid-State Electron.* **11**, 639–646, July 1968.

[33] T. Matsumara, T. Obokata, and T. Fukuda, "Two-Dimensional Microscopic

Uniformity of Resistivity in Semi-Insulating GaAs," *J. Appl. Phys.* **57**, 1182–1185, Feb. 1985.

[34] P. M. Hemenger, "Measurement of High Resistivity Semiconductors Using the van der Pauw Method," *Rev. Sci. Instrum.* **44**, 698–700, June 1973.

[35] L. Forbes, J. Tillinghast, B. Hughes, and C. Li, "Automated System for the Characterization of High Resistivity Semiconductors by the van der Pauw Method," *Rev. Sci. Instrum.* **52**, 1047–1050, July 1981.

[36] P. A. Crossley and W. E. Ham, "Use of Test Structures and Results of Electrical Test for Silicon-On-Sapphire Integrated Circuit Processes," *J. Electron. Mat.* **2**, 465–483, Aug. 1973.

[37] D. S. Perloff, F. E. Wahl, and J. Conragan, "Four-Point Sheet Resistance Measurements of Semiconductor Doping Uniformity," *J. Electrochem. Soc.* **124**, 582–590, April 1977.

[38] J. N. Gan and D. S. Perloff, "Post-Implant Methods for Characterizing the Doping Uniformity and Dose Accuracy of Ion Implantation Equipment," *Nucl. Instrum. and Meth.* **189**, 265–274, Nov. 1981.

[39] M. I. Current and W. A. Keenan, "Ion Implant Round Robbins," *Nucl. Instrum. and Meth.* **B6**, 418–426, Jan. 1985.

[40] M. I. Current, N. L. Turner, T. C. Smith, and D. Crane, "Planar Channelling Effects in Si (100)," *Nucl. Instrum. and Meth.* **B6**, 336–348, Jan. 1985.

[41] ASTM Standard F76, "Standard Method for Measuring Hall Mobility and Hall Coefficient in Extrinsic Semiconductor Single Crystals," *1988 Annual Book of ASTM Standards*, Am. Soc. Test. Mat., Philadelphia, 1988.

[42] A. K. Smith, W. H. Johnson, W. A. Keenan, M. Rigik, and R. Kleppinger, "Sheet Resistance Monitoring of Low Dose Ion Implants Using the Double Implant Technique," *Nucl. Instrum. and Meth.* **B21**, 529–536, March 1987.

[43] R. A. Evans and R. P. Donovan, "Alternative Relationship for Converting Incremental Sheet Resistivity Measurements into Profiles of Impurity Concentration," *Solid-State Electron.* **10**, 155–157, Febr. 1967.

[44] E. Tannenbaum, "Detailed Analysis of Thin Phosphorus-Diffused Layers in *p*-Type Silicon," *Solid-State Electron.* **2**, 123–132, March 1961.

[45] H. D. Barber, H. B. Lo, and J. E. Jones, "Repeated Removal of Thin Layers of Silicon by Anodic Oxidation," *J. Electrochem. Soc.* **123**, 1404–1409, Sept. 1976 and references therein.

[46] S. Zwerdling and S. Sheff, "The Growth of Anodic Oxide Films on Germanium," *J. Electrochem. Soc.* **107**, 338–342, April 1960.

[47] J. F. Dewald, "The Kinetics and Mechanism of Formation of Anode Films on Single-Crystal InSb," *J. Electrochem. Soc.* **104**, 244–251, April 1957.

[48] B. Bayraktaroglu and H. L. Hartnagel, "Anodic Oxides on GaAs: I Anodic Native Oxides on GaAs," *Int. J. Electron.* **45**, 337–352, Oct. 1978; "II Anodic Al_2O_3 and Composite Oxides on GaAs," *Int. J. Electron.* **45**, 449–463, Nov. 1978; "III Electrical Properties," *Int. J. Electron.* **45**, 561–571, Dec. 1978; "IV Thin Anodic Oxides on GaAs," *Int. J. Electron.* **46**, 1–11, Jan. 1979.

[49] H. Müller, F. H. Eisen, and J. W. Mayer, "Anodic Oxidation of GaAs as a Technique to Evaluate Electrical Carrier Concentration Profiles," *J. Electrochem. Soc.* **122**, 651–655, May 1975.

[50]   R. Galloni and A. Sardo, "Fully Automatic Apparatus for the Determination of Doping Profiles in Si by Electrical Measurements and Anodic Stripping," *Rev. Sci. Instrum.* **54**, 369–373, March 1983.

[51]   L. Bouro and D. Tsoukalas, "Determination of Doping and Mobility Profiles by Automatic Electrical Measurements and Anodic Stripping," *J. Phys. E: Sci. Instrum.* **20**, 541–544, May 1987.

[52]   R. G. Mazur and D. H. Dickey, "A Spreading Resistance Technique for Resistivity Measurements in Si," *J. Electrochem. Soc.* **113**, 255–259, March 1966.

[53]   J. R. Ehrstein, "Spreading Resistance Measurements—An Overview," in *Emerging Semiconductor Technology* (D. C. Gupta and R. P. Langer, eds.), STP 960, Am. Soc. Test. Mat., Philadelphia, 1987, pp. 453–479.

[54]   R. Holm, *Electric Contacts Theory and Application*, Springer Verlag, New York, NY, 1967.

[55]   G. P. Carver, S. S. Kang, J. R. Ehrstein, and D. B. Novotny, "Well-Defined Contacts Produce Accurate Spreading Resistance Measurements," *J. Electrochem. Soc.* **134**, 2878–2882, Nov. 1987.

[56]   W. B. Vandervorst and H. E. Maes, "Probe Penetration in Spreading Resistance Measurements," *J. Appl. Phys.* **56**, 1583–1590, Sept. 1984.

[57]   J. R. Ehrstein, "Two-Probe (Spreading Resistance) Measurements for Evaluation of Semiconductor Materials and Devices," in *Nondestructive Evaluation of Semiconductor Materials and Devices* (J. N. Zemel, ed.), Plenum, New York, 1979, pp. 1–66.

[58]   R. G. Mazur, "Doping Profiles by the Spreading Resistance Technique," in *Microelectronics Processing: Inorganic Materials Characterization* (L. A. Casper, ed.), Am. Chem. Soc., Washington, DC, ACS Symp. Ser. 295, 1986, pp. 34–48.

[59]   R. G. Mazur and G. A. Gruber, "Dopant Profiling on Thin Layer Silicon Structures with the Spreading Resistance Tecnhique," *Solid State Technol.* **24**, 64–70, Nov. 1981.

[60]   P. A. Schumann, Jr. and E. E. Gardner, "Application of Multilayer Potential Distribution to Spreading Resistance Correction Factors," *J. Electrochem. Soc.* **116**, 87–91, Jan. 1969.

[61]   S. C. Choo, M. S. Leong, H. L. Hong, L. Li, and L. S. Tan, "Spreading Resistance Calculations by the Use of Gauss-Laguerre Quadrature," *Solid-State Electron.* **21**, 769–774, May 1978.

[62]   H. L. Berkowitz and R. A. Lux, "An Efficient Integration Technique for Use in the Multilayer Analysis of Spreading Resistance Profiles," *J. Electrochem. Soc.* **128**, 1137–1141, May 1981.

[63]   R. Piessens, W. B. Vandervorst, and H. E. Maes, "Incorporation of a Resistivity-Dependent Contact Radius in an Accurate Integration Algorithm for Spreading Resistance Calculations," *J. Electrochem. Soc.* **130**, 468–474, Feb. 1983.

[64]   S. C. Choo, M. S. Leong, and J. H. Sim, "An Efficient Numerical Scheme for Spreading Resistance Calculations Based on the Variational Method," *Solid-State Electron.* **26**, 723–730, Aug. 1983.

[65] M. Pawlik, "Spreading Resistance: A Comparison of Sampling Volume Correction Factors in High Resolution Quantitative Spreading Resistance," in *Emerging Semiconductor Technology* (D. C. Gupta and R. P. Langer, eds.), STP 960, Am. Soc. Test. Mat., Philadelphia, 1987, pp. 502–520.

[66] A. H. Tong, E.F. Gorey, and C. P. Schneider, "Apparatus for the Measurement of Small Angles," *Rev. Sci. Instrum.* **43**, 320–323, Feb. 1972.

[67] ASTM Standard F672, "Standard Method for Measuring Resistivity Profile Perpendicular to the Surface of a Silicon Wafer Using a Spreading Resistance Probe," *1988 Annual Book of ASTM Standards*, Am. Soc. Test. Mat., Philadelphia, 1988.

[68] W. R. Runyan, *Semiconductor Measurements and Instrumentaion*, McGraw-Hill, New York, 1975.

[69] M. R. E. Bichara and J. P. R. Poitevin, "Resistivity Measurement of Semiconductor Epitaxial Layers by the Reflection of a Hyperfrequency Electromagnetic Wave," *IEEE Trans. Instrum. Meas.* **IM-13**, 323–328, Dec. 1964.

[70] M. E. Brodwin and P. S. Lu, "A Precise Cavity Technique for Measuring Low Resistivity Semiconductors," *Proc. IEEE* **53**, 1742–1743, Nov. 1965.

[71] J. A. Naber and D. P. Snowden, "Application of Microwave Reflection Technique to the Measurement of Transient and Quiescent Electrical Conductivity of Silicon," *Rev. Sci. Instrum.* **40**, 1137–1141, Sept. 1969.

[72] G. P. Srivastava and A. K. Jain, "Conductivity Measurements of Semiconductors by Microwave Transmission Technique," *Rev. Sci. Instrum.* **42**, 1793–1796, *Dec. 1971.*

[73] C. A. Bryant and J. B. Gunn, "Noncontact Technique for the Local Measurement of Semiconductor Resistivity," *Rev. Sci. Instrum.* **36**, 1614–1617, Nov. 1965.

[74] N. Miyamoto and J. I. Nishizawa, "Contactless Measurement of Resistivity of Slices of Semiconductor Materials," *Rev. Sci. Instrum.* **38**, 360–367, March 1967.

[75] H. K. Henisch and J. Zucker, "Contactless Method for the Estimation of Resistivity and Lifetime of Semiconductors," *Rev. Sci. Instrum.* **27**, 409–410, June 1956.

[76] J. C. Brice and P. Moore, "Contactless Resistivity Meter for Semiconductors," *J. Sci. Instrum.* **38**, 307, July 1961.

[77] P. J. Olshefski, "A Contactless Method for Measuring Resistivity of Silicon," *Semic. Prod.* **4**, 34–36, Dec. 1961.

[78] G. L. Miller, D. A. H. Robinson, and J. D. Wiley, "Contactless Measurement of Semiconductor Conductivity by Radio Frequency-Free Carrier Power Absorption," *Rev. Sci. Instrum.* **47**, 799–805, July 1976.

[79] P. S. Burggraaf, "Resistivity Measurement Systems," *Semicond. Int.* **3**, 37–44, June 1980.

[80] ASTM Standard F723, "Standard Practice for Conversion Between Resistivity and Dopant Density for Boron-Doped and Phosphorus-Doped Silicon," *1988 Annual Book of ASTM Standards*, Am. Soc. Test. Mat., Philadelphia, 1988.

[81] D. B. Cuttriss, "Relation Between Surface Concentration and Average Con-

ductivity in Diffused Layers in Germanium," *Bell Syst. Tech. J.* **40**, 509–521, March 1961.

[82] S. M. Sze and J. C. Irvin, "Resistivity, Mobility, and Impurity Levels in GaAs, Ge, and Si at 300 K," *Solid-State Electron.* **11**, 599–602, June 1968.

[83] S. M. Sze, *Physics of Semiconductor Devices*, 2d ed., Wiley, New York, 1981.

[84] W. A. Keenan, C. P. Schneider, and C. A. Pillus, "Type-All System for Determining Semiconductor Conductivity Type," *Solid State Technol.* **14**, 51–56, March 1971.

[85] ASTM Standard F42, "Standard Test Methods for Conductivity Type of Extrinsic Semiconducting Materials," *1988 Annual Book of ASTM Standards*, Am. Soc. Test. Mat., Philadelphia, 1988.

CHAPTER 2

CARRIER AND DOPING CONCENTRATION

2.1 INTRODUCTION

A knowledge of the carrier or dopant concentration is important for many semiconductor devices. Although the carrier concentration is related to the resistivity, it is usually not derived from resistivity measurements but is measured independently.

We discuss in this chapter several methods for determining the carrier or the doping concentration. We should stress that these two concentrations are not necessarily identical. They are identical only for uniformly doped substrates but not in regions of the wafer in which there are substantial doping gradients. Electrical characterization methods are generally preferred over optical methods, although both have been used. Among the electrical methods the capacitance-voltage, the spreading resistance, and the Hall effect techniques are most commonly used. More recently, secondary ion mass spectrometry has also found wide application. Optical methods are sparingly employed, but they have the advantage of very high sensitivity and the ability to *identify* the doping impurities.

2.2 CAPACITANCE MEASUREMENTS

2.2.1 Differential Capacitance

The *capacitance-voltage* (*C-V*) technique relies on the fact that the width of a reverse-biased space-charge region (scr) of a semiconductor junction device depends on the applied voltage. This scr width dependence on

41

voltage lies at the heart of the *C-V* technique. The *C-V* profiling method has been used with Schottky barrier diodes using metal and liquid electrolyte contacts, *pn* junctions, MOS capacitors, and MOSFETs.

We consider the Schottky barrier diode of Fig. 2.1(a). The semiconductor is *p*-type with doping concentration N_A. A dc bias V is applied to the metal contact. The reverse bias produces a space-charge region of width W. The capacitance is defined by

$$C = -\frac{dQ_s}{dV} \tag{2.1}$$

where Q_s is the semiconductor charge. The negative sign accounts for more *negative* charge in the semiconductor scr (negatively charged ionized acceptors) for increased *positive* voltages on the metal. The capacitance is determined by superimposing a small-amplitude ac voltage v on the dc voltage V. The ac voltage typically varies at a frequency of 1 MHz with an amplitude of 10 to 20 mV, but other frequencies and other voltages can be used.

Let us consider the diode to be biased to dc voltage V plus a sinusoidal ac voltage. Imagine the ac voltage increasing from zero to a small positive voltage adding a charge increment dQ_m to the metal contact. The charge

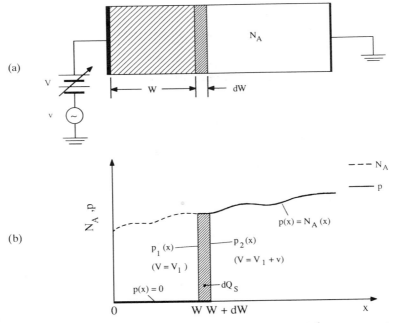

Fig. 2.1   (a) A reverse-biased Schottky diode and (b) the doping concentration and majority carrier profiles in the depletion approximation.

increment dQ_m must be balanced by an equal semiconductor charge increment dQ_s for overall charge neutrality, where dQ_s is given by

$$dQ_s = -qAN_A(W)\, dW \qquad (2.2)$$

The charge increment dQ_s, shown in Fig. 2.1(*b*), comes about through a slight increase in the scr width. From Eqs. (2.1) and (2.2) we find

$$C = -\frac{dQ_s}{dV} = qAN_A(W)\frac{dW}{dV} \qquad (2.3)$$

The capacitance of a reverse-biased junction, when considered as a parallel plate capacitor, is expressed as

$$C = \frac{K_s\varepsilon_0 A}{W} \qquad (2.4) \;\leftarrow$$

Differentiating Eq. (2.4) with respect to voltage and substituting dW/dV into Eq. (2.3) gives

$$N_A(W) = -\frac{C^3}{qK_s\varepsilon_0 A^2(dC/dV)} \qquad (2.5a)$$

which can also be written as

$$N_A(W) = \frac{2}{qK_s\varepsilon_0 A^2[d(1/C^2)/dV]} \qquad (2.5b)$$

using the identity $d(1/C^2)/dV = -(2/C^3)\,dC/dV$. Note the *area dependence* in these expressions. Since the area appears as A^2, it is very important that the device area be precisely determined for accurate doping profiling. From Eq. (2.4) we find the scr width dependence on capacitance as

$$W = \frac{K_s\varepsilon_0 A}{C} \qquad (2.6)$$

Equations (2.5) and (2.6) are the key equations for doping profiling.[1-3] The doping concentration is obtained from a *C-V* curve by taking the slope dC/dV or by plotting $1/C^2$ versus V and taking the slope $d(1/C^2)/dV$. The depth at which the doping concentration is evaluated is obtained from Eq. (2.6). For a Schottky barrier diode there is no ambiguity in the scr width since it can only spread into the substrate. Space-charge region spreading into the metal is totally negligible.

It is worthwhile to say a few words about the *C-V* interpretation of Eq. (2.5). A *C-V* curve is usually measured with a capacitance meter. Both dC/dV and $d(1/C^2)/dV$ methods are used. Sometimes one is preferred over the other. We demonstrate this in Fig. 2.2. The *C-V* curves of two Schottky devices are shown in Fig. 2.2(a). These curves give little indication of doping concentration uniformity. When the *C-V* curves are converted to $1/C^2$-*V* curves, shown in Fig. 2.2(b), it is immediately obvious which device has the more uniform doping concentration profile because the $1/C^2$-*V* data follow a straight line for uniformly doped wafers. It is clearly much easier to take the slopes from Fig. 2.2(b) than it is from Fig. 2.2(a). The dopant profiles for both devices are shown in Fig. 2.2(c). Before digital data acquisition and computerized measurement equipment became widespread, a special slide rule was used for doping profile evaluations.[4]

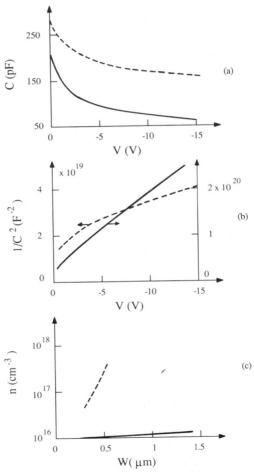

Fig. 2.2 (a) Capacitance-voltage curves for two Si p^+n diodes, (b) $1/C^2$ versus *V* curves, (c) *n* versus *W* profiles. $A = 7.9 \times 10^{-3}$ cm^2, $T = 300$ K.

The doping profile theory is equally well applicable for asymmetrical pn junctions with one side of the junction more highly doped than the other side. Such junctions are generally referred to as p^+n or n^+p junctions. If the doping concentration of the heavily doped side is 100 or more times higher than that of the lowly doped side, then the scr spreading into the heavily doped region can be neglected, and Eqs. (2.5) and (2.6) hold. If that condition is not met, the equations must be modified or both doping concentration and depth will be in error.[5] The correction, however, is fraught with difficulty. It has been proposed that no unique doping profile can be derived from *C-V* measurements under those conditions.[6] If the doping profile of one side of the junction is known, then the profile on the other side can be derived from the measurements.[7] Fortunately, most *pn* junctions utilized for doping concentration profiling, are of the p^+n or n^+p type, and corrections due to doping asymmetries are not required.

MOS capacitors (MOS-C) can also be used for profiling.[8] The measurement is more complicated because the device must remain in deep depletion during the measurement. This is accomplished with a rapidly varying ramp voltage or by using pulsed gate voltages. Two device parameters influence MOS-C dopant profile measurements that are absent in *pn* junctions: interface traps and minority carrier generation. Both are discussed in more detail in Section 2.5. Since a capacitance measurement is a measure of the change of charge with voltage, charges in addition to the bulk charge contribute additional capacitance. Equation (2.5) applies directly to MOS-C's when both interface states and minority carriers can be neglected, but the scr width is given by[9-10]

$$W = K_s \varepsilon_0 A \left[\frac{1}{C} - \frac{1}{C_{\text{ox}}} \right] \qquad (2.7)$$

Equation (2.7) differs from Eq. (2.6) by the oxide capacitance C_{ox}. The MOS-C profiling technique has also been implemented by driving the device into deep depletion and measuring the current instead of the capacitance.[11-12]

The interference of minority carrier generation with differential capacitance profile measurements can be avoided by providing a *pn* junction adjacent to the MOS-C to collect minority carriers. A MOSFET provides such minority carrier collecting junctions. Minority carriers are drained from the gate region of the MOSFET with sufficiently high drain and source voltages. The *C-V* curve of an *n*-type implant into an *n*-epitaxial film on an n^+ substrate and the doping profile derived from this curve are shown in Fig. 2.3.

For the derivation of Eq. (2.5) we used the *depletion approximation*, which completely neglects minority carriers and assumes total depletion of majority carriers in the space-charge region to a depth W and perfect charge neutrality beyond W. This is generally a reasonably good approximation

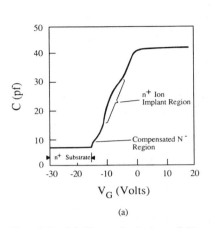

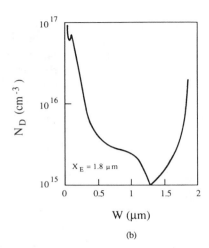

(a)   (b)

Fig. 2.3 (a) Deep-depletion *C-V* curve of an ion-implanted sample, (b) doping profile obtained from (a). Note the bulge in the *C-V* curve due to the ion-implanted layer, the rapid drop due to the compensated, low doping concentration N^- region, and the flattened capacitance due to the N^+ substrate. X_E = epitaxial layer thickness. Implant: arsenic, 1.2×10^{12} cm^{-2}, 300 keV. Reprinted after Brown et al.[13] by permission of the publisher, the Electrochemical Society, Inc.

when the scr is reverse biased and when the substrate is uniformly doped. Furthermore we used as the incremental charge variation the acceptor ion concentration at the edge of the space-charge region. The ac probe voltage exposes more or less ionized acceptors at the scr edge, as shown in Fig. 2.1. The charges that actually move in response to the ac voltage are the mobile holes, not the acceptor ions. From that point of view, the differential capacitance-voltage profiling technique determines the *majority carrier concentration* not the *doping concentration* and the relevant equations become

$$p(W) = -\frac{C^3}{qK_s\varepsilon_0 A^2 (dC/dV)} \qquad (2.8a)$$

$$p(W) = \frac{2}{qK_s\varepsilon_0 A^2 [d(1/C^2)/dV]} \qquad (2.8b)$$

$$W = \frac{K_s\varepsilon_0 A}{C} \qquad (2.9)$$

$$W = K_s\varepsilon_0 A\left[\frac{1}{C} - \frac{1}{C_{ox}}\right] \qquad (2.10)$$

The equations for the majority carrier concentration rather than the doping concentration can be derived from either majority carrier current flow in diodes[14] or from surface potential considerations in MOS capacitors.[15]

The use of the majority carrier concentration rather than the doping concentration in the profile equations is an important point and has been the subject of much discussion.[14-30] We demonstrate the concept for a non-uniform acceptor dopant profile shown by the dashed curve in Fig. 2.4(a). The majority hole concentration profile shown by the solid line, differs from the doping concentration profile even in thermal equilibrium. Some of the holes diffuse from the highly doped region to the more lowly doped region and an equilibrium profile is established as a result of both diffusion and drift. The steeper the doping gradient, the more p and N_A differ from one another. The majority carrier deviation from the dopant concentration is governed by the extrinsic Debye length L_D, more generally called the Debye length,

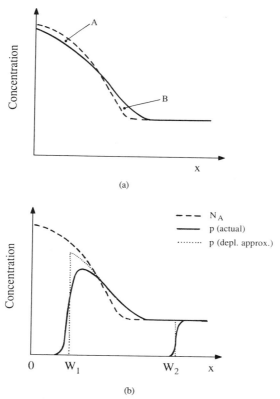

(a)

(b)

Fig. 2.4 A schematic representation of the doping and majority carrier concentration profiles of a non-uniformly doped layer. (a) Zero-biased junction, (b) reverse-biased junction showing the doping profile, the majority carrier profiles in the depletion approximation, and the actual majority carrier profiles for two reverse-bias voltages.

$$L_D = \sqrt{\frac{kTK_s\varepsilon_0}{q^2(p + n)}} \tag{2.11}$$

L_D is a measure of the distance over which a charge imbalance is neutralized by majority carriers under steady-state or equilibrium conditions.

When a scr is formed as a result of a reverse bias on a Schottky diode, for example, the carrier distribution becomes that in Fig. 2.4(b). We show by the dotted curve the majority carrier distribution expected from the depletion approximation for scr widths W_1 and W_2, corresponding to two different reverse-bias voltages. The actual majority carrier distribution is that shown by the solid lines. The two differ appreciably and it is quite obvious from these curves that the doping profile is not what is measured by differential capacitance profiling. It is also not clear that it is the majority carrier distribution that is measured. It has been shown by detailed computer calculations that what is actually measured is an *effective* or *apparent* majority carrier profile.[19] This effective profile is closer to the true majority carrier profile than it is to the doping profile. The doping profile, the majority carrier profile, and the effective majority carrier profile are identical for uniformly doped substrates. It is only for non-uniformly doped substrates where the problem arises.

The Debye length sets a limit to the *spatial resolution* of the measured profile. This Debye length problem arises because the capacitance is determined by the movement of majority carriers and the majority carrier distribution cannot follow abrupt spatial changes in dopant concentration profiles. The result of detailed calculations is shown in Fig. 2.5. For a doping concentration step occurring within one Debye length, the majority carrier and the apparent concentrations agree fairly well with one another [Fig. 2.5(a)]. But both differ appreciably from the true doping concentration profile. For a more gradual transition, as in Fig. 2.5(b), the majority carrier concentration agrees quite well with the apparent concentrations with depletion occurring either from the lowly doped or from the highly doped side. The agreement with the doping concentration profile is also quite reasonable.

The majority carrier concentration is equal to the doping concentration for uniformly doped substrates, and both Eqs. (2.5) and (2.8) are correct. However, for non-uniformly doped substrates Eq. (2.8) should be used. A relationship between the measured majority carrier concentration and the doping concentration is[16]

$$N_A(x) = p(x) - \frac{kT}{q}\frac{K_s\varepsilon_0}{q}\frac{d}{dx}\left[\frac{1}{p(x)}\frac{dp(x)}{dx}\right] \tag{2.12}$$

Extensive computer simulations have shown that Eq. (2.12) is too much of a simplification.[18-19,28] These simulations also found for low-high junctions, such as the p-p^+ junction, that the results depend on whether the junction is

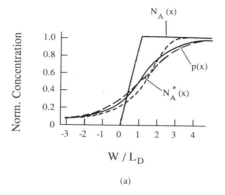

(a)

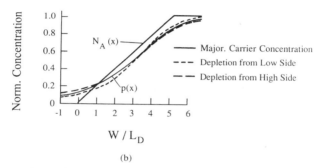

(b)

Fig. 2.5 Comparison of an assumed dopant profile with the majority carrier profile and the apparent dopant profiles deduced using Eq. (2.5). (a) 100:1 change of dopant concentration in one Debye length, (b) 100:1 change of dopant concentration in three Debye lengths. Reprinted after Johnson and Panousis[19] by permission of IEEE (© 1971, IEEE).

profiled from the p-side or from the p^+-side. The simulations show that a step profile cannot be resolved accurately to less than about 2 to 3 Debye lengths. The Debye length is determined by the carrier concentration on the high side of the junction. A doping concentration ramp profile cannot be distinguished accurately from a step unless its width is appreciably larger than a Debye length.

Equations (2.4) to (2.9) are derived in the *depletion approximation*, which assumes zero mobile carrier concentration in the space-charge region. This is a reasonably good aproximation when the scr is reverse biased. However, for zero- or forward-biased Schottky and *pn* junctions, the approximation loses its validity, and doping profiling becomes inaccurate under such bias conditions. Under forward bias an additional capacitance due to excess minority carrier storage in the quasi-neutral regions is

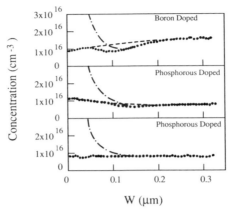

Fig. 2.6 Doping concentration profiles for three samples. The filled circles are experimental data. The dashed lines indicate the profiles in the absence of interface states. The dot-dash lines show the profiles when the depletion approximation is used. Reprinted with permission after Ziegler et al.[32]

introduced, rendering the method still less accurate. The concept of a zero- or forward-biased junction does not apply for an MOS-C. However, the role of mobile carriers is clearly just as important as it is for junction devices.

Neglect of *majority carriers* has been shown to lead to errors in pulsed MOS-C doping profile determinations for surface potentials ≤ 0.1 V.[9,20,31] This corresponds to a distance from the SiO_2-Si interface of approximately 2 to 3 Debye lengths. It has been suggested that profiling below this limit is possible by accounting for majority carriers.[32] Fairly complex equations are necessary for this correction, but they apply only to uniformly-doped substrates. Nevertheless, they are useful, and results of such a modified analysis are shown in Fig. 2.6 where the dash-dot lines show the profile under the usual Debye length limitation and the corrected experimental data points show the profile all the way to the surface. This majority carrier correction has been incorporated into an automated C-V profiler.[33]

2.2.2 Maximum–Minimum MOS-C Capacitance

Equations (2.8) and (2.10) hold for the depletion portion of the equilibrium and the deep-depletion portion of the nonequilibrium MOS-C C-V curve but not for strong inversion. A simple method to determine the doping concentration of an equilibrium MOS-C is to measure the maximum capacitance of an MOS-C in strong accumulation C_{ox} and the minimum capacitance in strong inversion C_f.[34–36] Interface traps play no role in this measurement if the gate voltage is sufficiently high for the device to be in strong inversion. Minority carrier generation does not exist with the device in equilibrium. The max-min capacitance method is not a doping profile measurement, however, and only a single, average doping concentration value is extracted from the measurement which is the average doping concentration over the scr width with the device in strong inversion.

Such a measurement is quite sufficient for uniformly doped substrates but not accurate for non-uniform doping concentrations. Recently it has been shown that some information about non-uniformly doped substrates can be extracted from such *equilibrium* MOS-C C-V curves.[37] This variation on the basic technique is based on linearizing a non-uniformly doped layer on a uniformly doped substrate. The measurement requires a knowledge of the substrate doping concentration and extracts the surface concentration and layer depth from the measured capacitance-voltage curves by iteration.

The *maximum-minimum capacitance* technique is extensively used because of its simplicity. The measurement relies on the dependence of the scr width of a strongly inverted MOS capacitor on the substrate doping concentration. The general MOS-C capacitance is given by

$$C = \frac{C_{ox} C_s}{C_{ox} + C_s} \qquad (2.13)$$

where $C_s = K_s \varepsilon_0 A / W$ is the semiconductor capacitance. The capacitance C_f is the strong inversion or *final* capacitance which obtains for $W = W_f$ $[W_f = (2 K_s \varepsilon_0 \phi_{s,inv} / q N_A)^{1/2}]$. The surface potential $\phi_{s,inv}$ is frequently approximated by $\phi_{s,inv} = 2\phi_F$.[38] But $\phi_{s,inv}$ is actually slightly higher than $2\phi_F$.[39] This should be considered in the interpretation of the measurement. For the simple case of $\phi_{s,inv} \approx 2\phi_F = 2(kT/q) \ln(N_A/n_i)$, we find

$$N_A = \frac{4\phi_F C_{sf}^2}{q K_s \varepsilon_0 A^2} \qquad (2.14)$$

where $C_{sf} = R C_{ox}/(1 - R)$ and $R = C_f / C_{ox}$. C_f and C_{ox} are shown in the inset of Fig. 2.7.

An empirical relationship between C_{sf} in F and N_A in cm^{-3} and A in cm^2 for silicon at room temperature is[40]

$$\log(N_A) = 30.38759 + 1.68278 \log(C_{sf}/A) - 0.03177[\log(C_{sf}/A)]^2 \qquad (2.15)$$

The equation is identical for *n*-type substrates with N_D substituted for N_A. We show in Fig. 2.7 curves calculated from Eq. (2.15) giving the doping concentration as a function of C_f/C_{ox}. These curves are useful for a first-order estimate of the doping concentration, but they may hide depth-dependent features for spatially varying doping profiles. Depth-dependent doping profiles may be measured by gradually immersing the wafer in an etch bath so that the surface becomes a slightly sloped plane along which the

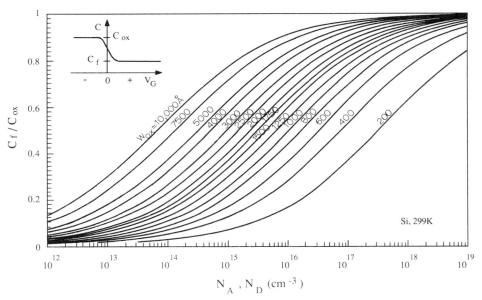

Fig. 2.7   Curves of C_f/C_{ox} versus doping concentration as a function of oxide thickness for silicon at $T = 299$ K.

impurity gradient is gradually changing. MOS capacitors formed on the etched and oxidized surface can be used to determine the doping concentration under each MOS-C as determined from its C_f/C_{ox} ratio. Combining the concentration data with the local etch depth for each capacitor gives the doping profile.[41] This method has been used to determine the doping profile and the band gap profile for HgCdTe.[42]

2.2.3   Integral Capacitance

The differential capacitance technique has been shown to have limitations when used as a process monitor where accuracy and measurement time are important.[43] In particular, the required differentiation often results in noisy profiles, especially for low-dose, ion-implanted samples. It is shown in Section 1.2.5 that low-dose, ion-implanted layers are difficult to characterize even with a four-point probe. Furthermore the dependence of the doping concentration on area ($N_A \sim A^{-2}$) renders measurements inaccurate when the gates are defined by evaporation through a metal mask where the device area may not be precisely known. Gates are simpler to form by mask evaporation, but their areas are not as well defined as they are with photolithographically defined gates. The effect of these inaccuracies is that measurement repeatability using differential capacitance was found to be 5 to 10%.[43] This makes for a poor process monitor when the process itself is capable of repeatability of 1% or better.

The integral capacitance characterization method is based on integrating the C-V curve of a pulsed MOS-C. The pulsed, deep-depletion C-V curves of two MOS capacitors are shown in Fig. 2.8(a). One curve is for a uniformly doped sample and the other for an ion-implanted sample. The doping profile obtained by the differential capacitance method is shown in Fig. 2.8(b). Note that the rather ragged curve with sharp features is due to data reduction and not to variations in the actual profile. The important parameters for an implanted layer are the implant dose and implant depth. Both are somewhat ambiguous when extracted from Fig. 2.8(b).

The integral capacitance technique is based on integrating a portion of the C-V curve to obtain a *partial dose* P_Φ. The partial dose is proportional to the implanted dose and is shown by the shaded area in Fig. 2.8(b). The dose chosen includes the doping concentration between $x = x_1$ and $x = x_2$ and contains most of the implanted layer, but does not extend into the region

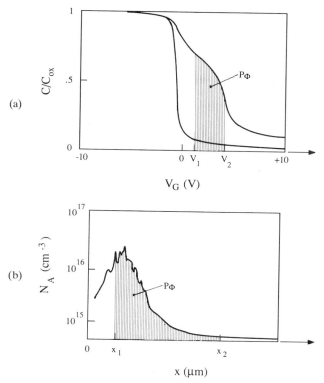

Fig. 2.8   (a) Deep-depletion C-V curves for a conventional and an ion-implanted Si sample. The area under the "implanted" curve is used to find P_Φ. (b) The concentration profile is obtained from the differential capacitance method. The shaded area is the partial dose P_Φ. Implant condition: boron, 50 keV, 10^{12} cm^{-2}. Reprinted with permission of Solid State Technology.

where the doping concentration equals the uniform background doping concentration nor into the region within 2 to 3 Debye lengths from the surface. The partial dose is given by[43]

$$P_\Phi = \int_{x_1}^{x_2} N_A(x)\,dx = \frac{1}{qA} \int_{V_1}^{V_2} C\,dV \qquad (2.16)$$

The second parameter that is measured is related to the projected range or implant depth R at the concentration peak. It is defined by[43]

$$R = W_{ox} + \frac{1}{P_\Phi} \int_{x_1}^{x_2} x N_A(x)\,dx = \frac{K_s \epsilon_0}{q P_\Phi} (V_2 - V_1) - \frac{(K_s - K_{ox})}{K_{ox}} W_{ox} \qquad (2.17)$$

This expression for R incorporates P_Φ, requiring only one integration which reduces the analysis time considerably. The time to acquire the data is about ten seconds using a local computer in which the capacitance-voltage curve is integrated as it is measured. The repeatability of a given device was accurate to 0.1%, and the authors claim that the repeatability in partial dose measurement has been improved by over a factor of ten by going to the integral capacitance technique.[43]

Mercury Probe Contacts   Every differential capacitance profiling technique requires some kind of junction device. At times it is desirable to use a device whose junction can be fabricated without subjecting the material to high temperature treatments. Conventional Schottky barrier device fabrication is done near room temperature, but a metal must be deposited on the wafer. When a temporary contact is needed, as in evaluation of epitaxial layers, a mercury probe is frequently used. The mercury contact is formed by mercury rising through a well-defined orifice, in which case the orifice opening defines the contact area. The mercury is forced against the semiconductor wafer by air pressure[44] or it can be drawn to the wafer surface by a vacuum between the orifice and the vacuum groove as shown in Fig. 2.9.[45] The contact area is sufficiently well defined to be useful for profile measurements. A mercury probe with a probe diameter of 7 μm has been employed to make conventional C-V measurements. This probe was also used to make lateral capacitance profiles by continuously dragging the probe across the wafer.[46]

The mercury contact leaves a mercury residue on the surface.[45] The stray capacitance, indicated in Fig. 2.9 as C_s for stray capacitance from the probe to the wafer and C_c for coupling capacitance between the two Hg columns, is on the order of 2 or 3 pF and should be compensated for precise measurements. The semiconductor surface should be treated before the Hg contacts the surface for reproducible measurements. For n-type Si, the following procedure has been found satisfactory.[44-45,47] Dip the wafer in

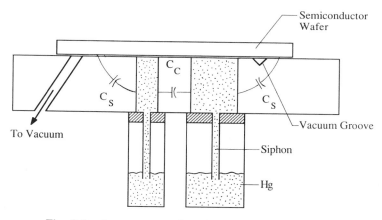

Fig. 2.9 A mercury probe with two contacting areas.

HF for 2 min to remove any residual oxide, rinse in flowing deionized (DI) water for 2 min.; place wafer in boiling HNO_3 for 5 min (alternately immerse wafer in 90°C H_2O_2 for at least 10 min); rinse in flowing DI water and dry the wafer. The purpose of this process is to achieve a stable surface by growing a thin oxide film on the wafer, which is necessary for a low leakage current Schottky diode. For p-type Si:[44,47] dip wafer in HF for 30 s, rinse in flowing DI water and dry the wafer. This gives an oxide-free surface which is desirable for most reproducible results. The mercury should be very pure, so twice-a-month mercury changes are recommended for cases of heavy usage.[45] Mercury probes for differential capacitance probing are commercially available.

2.2.4 Electrochemical Profiler

The electrochemical profiling technique is based on making a liquid electrolyte-semiconductor Schottky contact and measuring the capacitance at a *constant* dc voltage. Depth profiling is achieved by electrolytically etching the semiconductor between capacitance measurements. There is thus no depth limitation. However, the method is destructive because of the hole it etches into the semiconductor. Early measurements divided the measurement and etch processes; later they were combined into one operation.[48–49] The present technique uses a combined process in which both etching and measurement are performed with the same apparatus. An excellent review is given by Blood.[50]

The electrochemical method is schematically shown in Fig. 2.10. The semiconductor wafer is pressed against a well-defined sealing ring in the electrochemical cell containing an electrolyte. The opening defines the

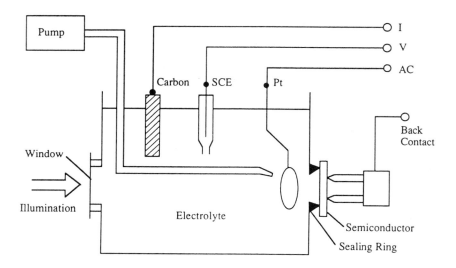

Fig. 2.10 Schematic diagram of the electrochemical cell showing the Pt, saturated calumel and carbon electrodes and the pump used to agitate the electrolyte and disperse bubbles on the semiconductor surface. Reprinted with permission after Blood.[50]

contact area by means of spring-loaded back contacts pressing the wafer against the sealing ring. The etching and measuring conditions are controlled by the potential across the cell. This is established by passing a dc current between the semiconductor and the carbon electrode to maintain the required overpotential measured with respect to the saturated calomel electrode. To reduce series resistance, the ac voltages are measured with a platinum electrode located near the sample.

With a small reverse dc bias applied between the electrolyte and the semiconductor sample, two low-voltage signals of different frequencies are applied to the electrolyte. The carrier concentration measurement is based on the relationship

$$p(W) = \frac{2K_s \varepsilon_0}{q} \frac{\Delta V}{\Delta(W^2)} \tag{2.18}$$

where ΔV is the modulation component of the applied ac voltage (100 mV at 30 Hz) and $\Delta(W^2)$ is the resulting scr width modulation. The scr width is obtained from the relationship

$$W = \frac{K_s \varepsilon_0 A}{C} \tag{2.19}$$

W is determined by measuring the imaginary component of the current with a phase-sensitive amplifier using a 50 mV, 3 kHz signal. W and $p(W)$ are obtained through appropriate electronic circuits.[49] The 3 kHz frequency is significantly lower than the 1 MHz frequency often used for conventional differential capacitance profiling to reduce the r_sC time constant, where r_s is the series resistance of the electrolyte and C the device capacitance. The resistance-capacitance product must meet certain criteria for the measurements to be valid as discussed in Section 2.5 on *Series Resistance*.

Equations (2.18) and (2.19) provide the concentration at depth W. Depth profiling is achieved by dissolving the semiconductor electrolytically. Dissolution of the semiconductor depends on the presence of holes. For p-type semiconductors, holes are plentiful and dissolution is readily achieved by forward biasing the electrolyte-semiconductor junction. For n-type material, holes are generated by illuminating and reverse biasing the junction. The depth to which the semiconductor is etched W_R depends on the dissolution current I_{dis} according to the relationship[49]

$$W_R = \frac{M}{zF\rho A} \int_0^t I_{dis} \, dt \tag{2.20}$$

where M is the semiconductor molecular weight, z the dissolution valency (number of charge carriers required to dissolve one molecule of the semiconductor), F the Faraday constant (9.64×10^4 C), ρ the semiconductor density, and A the effective contact area. W_R is determined by integrating the dissolution current electronically. The measurement depth of the carrier concentration is

$$x = W + W_R \tag{2.21}$$

There is no depth limitation, since the semiconductor can be etched to any desirable depth. The electrolyte must be chosen appropriately for each semiconductor and suitable electrolytes are HCl for InP,[51] Tiron (dihydroxybenzene-3,5-disulphonic acid disodium salt) and EDTA/$0.2M$ NaOH for GaAs,[50,52–53] and NaF/H_2SO_4 for Si.[54–56] The technique is eminently suitable for III–V materials because the dissolution valency, $z = 6$, is well defined and the electrolyte etches the semiconductor very controllably. The dissolution valency is not well defined for Si where it can vary between 2 and 5. Furthermore hydrogen bubbles generated during the dissolution process hinder the uniformity and degrade the depth resolution. The hydrogen bubble problem has been overcome by using a pulsed jet of the electrolyte.[55–56] It is for these reasons that electrochemical silicon profiling is limited primarily to thin layers grown by molecular beam epitaxy (MBE).

The etch rate is typically a few microns/hour and depths to 20 µm are readily obtained in III–V materials. The etch rate for Si is on the order of 1 µm/hr. The etch crater may develop irregular topography for large depths and a correction for the crater sides may be necessary. A profile of a Zn-diffused layer in GaAs with 35 Å resolution is shown in Fig. 2.11(a). An example of a Si profile is shown in Fig. 2.11(b), where the electrochemical

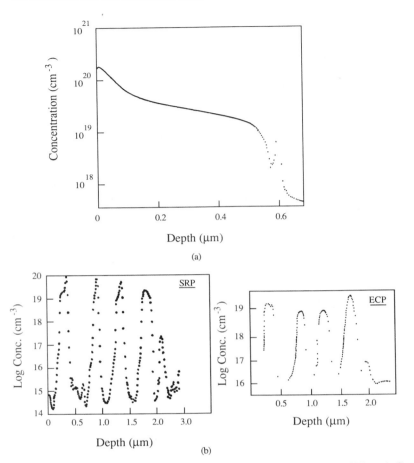

Fig. 2.11 Profiles obtained with the electrolytic profiler. (a) Zinc-diffused GaAs, courtesy of R. J. Roedel, Arizona State University, and (b) MBE-grown Si where the electrochemical profile is compared with a spreading resistance profile. Copyright ASTM; reprinted with permission after Pawlik et al.[56]

profile is compared to a spreading resistance profile showing good agreement. The equipment is commercially available and is routinely used for III–V materials but has not yet found extensive application in silicon.

2.3 CURRENT-VOLTAGE MEASUREMENTS

2.3.1 Second Harmonic

The second harmonic profiling technique relies on the nonlinear properties of a semiconductor junction, consisting of a nonlinear conductance in parallel with a nonlinear capacitance. When a sinusoidally varying current of frequency f is applied to such a junction, the fundamental frequency f as well as higher order components $2f$, $3f$, etc., are generated in the output signal. The capacitance component dominates at sufficiently high frequencies. It has been shown that the fundamental frequency is related to the scr width, the second-order harmonic is related to the carrier concentration, and the third-order harmonic is related to the doping concentration gradient.[57] For an ac current given by $I \sin(\omega t)$ the incremental voltage obtained by changing the dc bias is[58]

$$\Delta V = \frac{I \cos(\omega t)}{K_s \varepsilon_0 \omega A} W + \frac{I^2 [1 + \cos(2\omega t)]}{4 q K_s \varepsilon_0 \omega^2 A^2} \frac{1}{p(W)} \qquad (2.22)$$

The first term is the incremental voltage component with the same frequency as the driving current. It has an amplitude proportional to the scr width. The second term is the incremental voltage component at twice the frequency of the driving current. This term is proportional to the reciprocal of the carrier concentration.

The second harmonic profiling concept has been implemented using a fundamental frequency of $f = 5$ MHz.[59] A circuit utilizing a capacitance compensating scheme to overcome the stray capacitance problem has also been implemented.[60] The method itself has not found wide application because it is a high-frequency measurement, and any stray capacitance in the measurement circuit and in the connecting leads is very important. Such stray capacitance can lead to appreciable error especially for low-capacitance devices. The second harmonic component is small and it is therefore not easy to detect unambiguously. With the second harmonic component inversely proportional to A^2, it suggests small-area, low-capacitance diodes for good signal/noise detection. But low-capacitance diodes are more prone to stray capacitance problems. Lastly, there are no commercial instruments.

2.3.2 MOSFET Substrate Voltage–Gate Voltage

Capacitance-doping profile measurements are typically made at a frequency of 1 MHz. In order to reduce stray capacitances and increase the signal/noise ratio, the measurements are usually made on large-diameter devices. Typically diodes or MOS capacitors of 0.25 to 1 mm diameter are used. These constraints make measurements on small-geometry MOSFET's very difficult because the capacitance of such MOSFET's is extremely small and difficult to measure. To overcome this limitation, a method has been developed that allows the doping profile to be extracted from voltage measurements of a MOSFET.

The method is based on the following principle. The MOSFET is biased in its *linear* region by a small drain-source voltage V_{DS} and a gate-source voltage V_{GS}. A source-substrate potential V_{SB} forces the space-charge region under the gate to extend into the substrate, allowing the dopant profile to be obtained provided the inversion charge density is held constant. Constant inversion charge is approximated by a constant drain current and the drain current can be held constant by adjusting V_{GS} whenever V_{SB} is changed. The relevant equations are[61-64]

$$p(W) = \frac{K_{ox}^2 \varepsilon_0}{q K_s W_{ox}^2} \left[\frac{d^2 V_{SB}}{dV_{GS}^2} \right]^{-1} \tag{2.23}$$

$$W = \frac{K_s W_{ox}}{K_{ox}} \frac{dV_{SB}}{dV_{GS}} \tag{2.24}$$

This measurement is performed at *constant* drain current. The assumption that constant drain current corresponds to constant inversion charge is only true to a first approximation. It is known that in a MOSFET the effective mobility varies with gate voltage (see Chapter 5), requiring a correction in the analysis; otherwise, the measured profile gives higher concentrations than the true profile.[63,65] The technique is also subject to the Debye limit where profiles cannot be reliably obtained closer than about

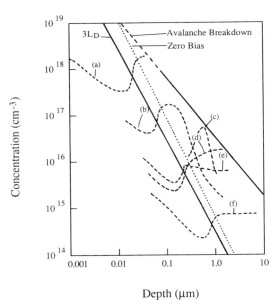

Depth (μm)

Fig. 2.12 Curves of (a) boron diffusion, (b) phosphorus implant, (c) boron implant, (d) boron-doped bulk, (e) phosphorus-doped bulk, and (f) boron-doped bulk. Also shown are the curves for the 3 Debye length limit, the zero-bias limit, and the avalanche breakdown limit. Profiles below the $3L_D$ curve are meaningless. Reprinted after Buehler[63] by permission of the publisher, The Electrochemical Society, Inc.

three Debye lengths from the oxide-semiconductor interface. Experimentally, a dip in the profile observed for $W < 3L_D$ is an artifact of this limitation and not a reflection of the true doping profile. As with differential capacitance profiling there is also an upper limit to the profile depth determined by the breakdown voltage. In addition, the drain-source voltage should be maintained below about 100 mV, and the profile is affected by short-channel effects.[63–64,66] Profiles obtained by this technique are shown in Fig. 2.12.

2.3.3 MOSFET Threshold Voltage

In the MOSFET threshold voltage profiling technique, the *threshold voltage* is measured as a function of substrate bias.[67–70] It is a dc technique requiring only voltage measurements, and it lends itself to measurements on small geometry MOSFET's. The threshold voltage of a MOSFET is given by[10]

$$V_T = V_{FB} + 2\phi_F + \frac{\sqrt{2qK_s\varepsilon_0 N_A(2\phi_F - V_{BS})}}{C_{ox}} \qquad (2.25)$$

where the substrate bias $V_{BS} = V_B - V_S$ is negative for *n*-channel devices. The doping profile is obtained by plotting the threshold voltage against $(2\phi_F - V_{BS})^{1/2}$ and measuring the slope $m = \Delta V_T / \Delta(2\phi_F - V_{BS})^{1/2}$ of this plot. The doping concentration is from Eq. (2.25)

$$N_A = \frac{m^2 C_{ox}^2}{2qK_s\varepsilon_0} \qquad (2.26a)$$

and the profile depth is

$$W = \sqrt{\frac{2K_s\varepsilon_0(2\phi_F - V_{BS})}{qN_A}} \qquad (2.26b)$$

In Eq. (2.26b) ϕ_F depends on N_A which is not known *a priori*. A suitable approach is to plot V_T versus $(2\phi_F - V_{BS})^{1/2}$ using $2\phi_F = 0.6$ V, and then to take the slope and find N_A. With this value of N_A one can find a new $\phi_F = (kT/q)\ln(N_A/n_i)$, replot V_T versus $(2\phi_F - V_{BS})^{1/2}$, repeating the procedure until a profile is obtained. One or two iterations usually suffice. In Fig. 2.13 we show doping profiles obtained from MOSFET threshold voltage, spreading resistance, and pulsed MOS-C *C-V* measurements. The pulsed MOS-C measurements were made on a test MOS-C structure processed identically to the MOSFET. The data are compared to a SUPREM3 calculated profile. Note the good agreement between all four methods. The threshold voltage technique can also be used for depletion-mode devices.[68–70]

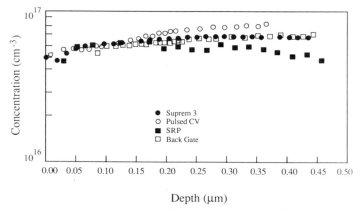

Fig. 2.13   Dopant profiles determined by MOSFET threshold voltage, SRP, pulsed *C-V*, and SUPREM3. Courtesy of D. Feldbaumer, Motorola.

2.3.4   Spreading Resistance

Spreading resistance profiling is commonly used for Si. The sample is beveled, and two spreading resistance probes are stepped along the beveled surface. The spreading resistance is measured as a function of sample depth, and the doping profile is calculated from the measured resistance profile. A detailed discussion of this technique is given in Section 1.3.2 Very high resolution profiles can be generated by using shallow bevel angles. A recent application of SRP to MBE Si layers is given by Jorke and Herzog who also discuss carrier spilling and low-high and high-low transitions.[71]

2.4   MEASUREMENT CIRCUITS

Many profile measurements are made with automated equipment. Typically the capacitance-voltage curve is measured with a capacitance meter which has provision to apply the necessary reverse-bias voltage. The capacitance meter has an output voltage proportional to the capacitance. This output is digitized; then the appropriate calculations to compute the doping concentration and the depth are made with a digital computer, and the profile is plotted. For *pn* junctions and Schottky diodes it is sufficient to apply a slowly varying bias voltage. For MOS capacitors the bias voltage is provided as a series of pulses of successively higher amplitude to eliminate the problem of minority carrier generation.

When *C-V* curves are analyzed manually, it is advantageous to use the $1/C^2$-*V* rather than the *C-V* format, since the former has less curvature and is easier to differentiate. Early instrumentation used a bias voltage modulation across the diode and extracted the appropriate signals from the resulting current flow. The computation was done by an analog technique.[72-73] An interesting analog approach, proposed by Miller, became a commercial product.[74] It is based on the equation

$$\Delta \mathscr{E} = \frac{\Delta V}{W} = \frac{q}{K_s \varepsilon_0} N(W) \Delta W \tag{2.27}$$

where $\mathscr{E}$ is the electric field. For $\Delta W =$ constant, we find $\Delta \mathscr{E} \sim N(W)$ while for $\Delta \mathscr{E} =$ constant, we find $\Delta W \sim 1/N(W)$. It is necessary to design a circuit that holds either $\Delta \mathscr{E}$ or ΔW constant.

The circuit used for the second harmonic profiler is described in Spiwak[59] and is not reproduced here since the method is not commonly used. A feedback approach is used in the MOSFET substrate/gate voltage technique as shown in Fig. 2.14. V_{DS} is held constant, and V_{GS} is varied. With the operational amplifier differential input voltage and input current nearly zero, the source current is $I_S = I_D = V_1/R_1$. When V_{GS} is changed, the op amp adjusts the voltage V_{SB} between the source and substrate to maintain a constant current.[62]

Other considerations to be observed during profile measurements are discussed in the ASTM standard.[75] As with all ASTM methods this is a good source of practical information and precautions to observe during measurement.

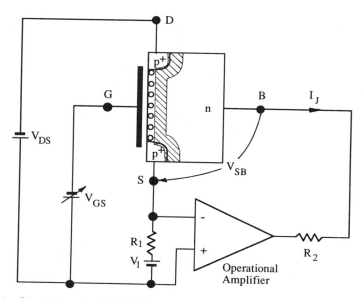

Fig. 2.14 Circuit for the MOSFET substrate/gate voltage method. Reprinted with permission after Buehler.[62]

2.5   MEASUREMENT ERRORS AND PRECAUTIONS

Many *C-V* measurements are made with no corrections of any kind because corrections often only produce small changes in the measured doping profile. Sometimes corrections are not made because the experimenter is unaware of possible corrections or they are too difficult to make. Nevertheless, the reader should be aware of possible measurement errors.

2.5.1   Debye Length and Voltage Breakdown

The Debye length limitation is discussed in Section 2.2.1 and in numerous papers.[14–32,76] To summarize briefly, mobile majority carriers do not follow the profile of the dopant atoms if the dopant atom profile varies spatially over distances less than the Debye length. The majority carriers are more smeared out than the dopant atoms and a measured profile of steep dopant gradients will result in neither the doping nor the majority carrier profile. Instead an effective or apparent carrier profile is obtained, which is closer to the majority carrier profile than it is to the dopant profile. This applies to abrupt high-low junctions and to steep-gradient ion implants. It is possible to correct the measured profile by iterative calculations, but due to the mathematical complexity this is rarely done.

   Another consequence of the Debye length limitation is the inability to profile closer than about $3L_D$ from the surface using MOS capacitors. Although corrections are possible to calculate the profile to the surface, it is not routinely done. Even considering the Debye length limitation, it is possible to profile closer to the surface with MOS-C's and MOSFET's than it is with Schottky barrier diodes or *pn* junctions. For MOS devices the limit is approximately $3L_D$, for Schottky diodes it is the zero-bias scr width, and for *pn* junction it is the junction depth plus the zero-biased scr width. The $3L_D$ limit is shown as the lower profile depth limit in Fig. 2.15.

   When the profile is generated by sweeping a reverse-bias voltage, the upper profile depth limit is determined by the semiconductor voltage breakdown. The electric field in the semiconductor is related to the semiconductor charge by

$$\mathscr{E} = \frac{Q_s}{K_s \varepsilon_0} \qquad (2.28)$$

With $Q_s = qN_AW$, we find the maximum N_AW product to be $(N_AW)_{max} = (K_s\varepsilon_0/q)\mathscr{E}_{BD}$, where $\mathscr{E}_{BD}$ is the semiconductor breakdown electric field. For Si with $\mathscr{E}_{BD} \approx 3 \times 10^5$ V/cm, this gives $(N_AW)_{max} \approx 2 \times 10^{12}$ cm^{-2}. The scr width at breakdown, W_{BD}, is related to $\mathscr{E}_{BD}$ by

$$W_{BD} = \frac{K_s \varepsilon_0 \mathscr{E}_{BD}}{qN_A} \qquad (2.29)$$

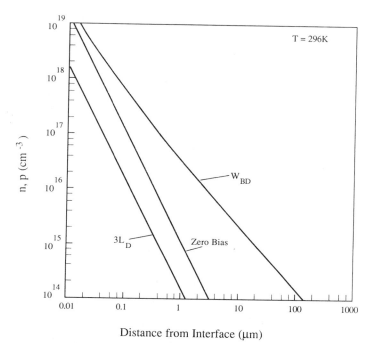

Fig. 2.15 Spatial profiling limits. The "$3L_D$" line is the lower limit for conventional MOS-C profiling, the "zero-bias" line is the lower limit for *pn* and Schottky diode profiling, and the "W_{BD}" line is the upper profile limit governed by bulk voltage breakdown.

For silicon $W_{BD} \approx 2 \times 10^{16}/N_A$ μm. The maximum depth to which Si can be profiled is also shown on Fig. 2.15. For this figure the more exact breakdown electric field dependence on doping concentration was used.[77] Breakdown considerations do not apply to the electrochemical profiler and that technique has no depth limit.

A theoretical study incorporating the Debye length limitation, the breakdown limitation, and the diffusion of majority carriers in steep-gradient profiles gives the dose and energy limits of Si and GaAs ion-implanted layers that can be profiled by differential capacitance techniques to 5% accuracy.[30]

2.5.2 Series Resistance

A *pn* or Schottky diode consists of a junction capacitance C, a junction conductance G, and a series resistance r_s as shown in Fig. 2.16(a). The conductance determines the junction leakage current and can be varied by processing conditions. The series resistance depends on the bulk wafer resistivity and on the contact resistances. Capacitance meters assume the

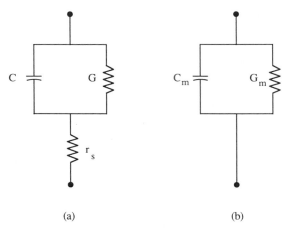

(a)   (b)

Fig. 2.16   (a) Actual circuit, and (b) capacitance meter or bridge equivalent circuit for a *pn* or Schottky diode.

device to be represented by the parallel circuit in Fig. 2.16(b), where C_m and G_m are the *measured* capacitance and conductance. Comparing the two circuits allows C_m and G_m to be written as[78]

$$C_m = \frac{C}{(1 + r_s G)^2 + (2\pi f r_s C)^2} \tag{2.30}$$

$$G_m = \frac{G(1 + r_s G) + r_s(2\pi f C)^2}{(1 + r_s G)^2 + (2\pi f r_s C)^2} \tag{2.31}$$

We only address the capacitance here. For the measured capacitance to be approximately equal to the true capacitance, it is necessary that $r_s G \ll 1$ and $(2\pi f r_s C)^2 \ll 1$. The "$r_s G \ll 1$" requirement is usually met by junctions with reasonably low leakage currents. The "$(2\pi f r_s C)^2 \ll 1$" requirement is not always met. For example, for 1% accuracy in capacitance measurements using a capacitance meter operating at $f = 1\,\text{MHz}$, r_s should be less than $1.6 \times 10^4 / C(\text{pf})$. With $C = 50\,\text{pF}$ the series resistance should be less than 320 ohms.

Care must be exercised when preparing samples for capacitance measurements, especially if the device is at the wafer stage and measurements are made on a probe station. If the wafer is provided with a metallic back contact, there is usually no problem, provided the wafer resistivity itself does not contribute significant series resistance. However, wafers placed on a probe station without any back metallization can have significant contact resistance and their capacitance measurements can be quite inaccurate. If the wafer has an oxide on the back surface, it is preferable to leave the oxide intact and place the wafer on the probe station making a large-area *capacitive* back contact. This is quite sufficient for capacitance measurements but of course cannot be used for dc current-voltage measurements. It is important that a vacuum be pulled for all probe capacitive measurements to reduce the resistance between the wafer and the probe chuck.

Series resistance also interferes with dopant profile measurements. For a wafer with negligible series resistance, there is zero phase shift between the rf voltage applied to the device and the rf current flowing through it when the *conductance* is measured. For the *capacitance* measurement there is a 90° phase shift, which is the basis of phase-sensitive capacitance measurements. When series resistance is not negligible, an additional phase shift ϕ is introduced into the measurement. This must be taken into account or the measured dopant profile determined from Eqs. (2.5) and (2.6) will be incorrect. These equations can be corrected to yield[79]

$$N_A(W) = -\frac{C_m^3}{qK_s\varepsilon_0 A^2}\left[\sin^4(\phi)\left(\frac{dC_m}{dV} - 2C_m\cot(\phi)\frac{d\phi}{dV}\right)\right]^{-1} \quad (2.32)$$

$$W = \frac{K_s\varepsilon_0 A\sin^2(\phi)}{C_m} \quad (2.33)$$

Another way to consider series resistance is from Eqs. (2.5a), (2.6), and (2.30) with $r_sG \ll 1$, where it can be shown that the measured concentration, $N_{A,meas}(W)$, and depth, W_{meas}, are related to N_A and W by the relationships

$$N_{A,meas} = \frac{N_A}{1 - (2\pi fr_sC)^4} \quad (2.34)$$

$$W_{meas} = W[1 + (2\pi fr_sC)^2] \quad (2.35)$$

Clearly both the concentration and depth are affected by series resistance.

The effect of series resistance on a dopant profile of an epitaxial GaAs layer grown on a semi-insulating substrate is illustrated in Fig. 2.17. The

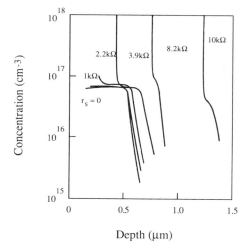

Fig. 2.17 Measured dopant profiles for a GaAs epitaxial layer on a semi-insulating substrate. The series resistance was obtained by placing resistors in series with the device. Reprinted after Wiley and Miller[79] by permission of IEEE (© 1975, IEEE).

correct profile is the one labeled $r_s = 0$. To obtain the other curves, external resistors were placed in series with the device to demonstrate the effect. Conducting layers on insulating or semi-insulating substrates are particularly prone to series resistance effects since both contacts are made on the top surface and lateral series resistance can be substantial. This problem is addressed in Wiley.[80]

2.5.3 Minority Carriers and Interface Traps

When a Schottky barrier or *pn* junction diode is reverse biased, the scr width remains constant as a function of time because thermally generated electron-hole pairs are swept out of the scr and leave through the ohmic contacts of the device. Thermally generated minority carriers in an MOS-C drift to the SiO$_2$-Si interface to form an inversion layer. The device is unable to remain in deep depletion when that is allowed to happen and doping profile measurements will be in error. For a more complete discussion of the behavior of MOS capacitors in their non-equilibrium or deep-depletion state see Schroder[10] or Section 8.6.2. The role of minority carriers can be neglected when the MOS-C is rapidly driven into deep depletion by applying a ramp gate voltage with a sufficiently high ramp rate for minority carrier generation to be negligible during the measurement time. Alternately, a pulse train of successively higher gate voltage can be applied with the device being cycled between accumulation and deep depletion.

The effect of minority carrier generation is shown on Fig. 2.18. The equilibrium *C-V* curve of an MOS-C is shown in Fig. 2.18(a). When the device is driven into deep depletion by a rapidly varying ramp voltage, curve (i) in Fig. 2.18(b) results. For negligible minority carrier generation the curve is identical for the gate voltage being swept from left to right or from right to left as indicated by the arrows. The doping concentration profile obtained from this curve is shown in Fig. 2.18(c) by (i). If the curve is swept extremely slowly, then the equilibrium high-frequency curve is obtained. For an intermediate sweep rate curve (ii) results. This curve lies above curve (i) and the extracted doping profile, shown in Fig. 2.18(c) by (ii), is in error because dC/dV for (ii) is lower than dC/dV for (i). If curve (ii) is swept from right to left, resulting in curve (iii), its doping profile is lower for similar reasons, as shown in Fig. 2.18(c) by curve (iii). It is possible to correct for these effects but corrections are not necessary for high sweep rates.[81]

It is interesting to use the simple *max−min MOS-C capacitance* method to determine N_A. From Fig. 2.18(a) and (b) we find $C_f/C_{ox} = 0.19$. Coupled with $W_{ox} = 1200$ Å, we find $N_A = 3.5 \times 10^{14}$ cm^{-3} from Fig. 2.7. This value is very close to curve (i) in Fig. 2.18(c). Of course the C_f/C_{ox} approach does not give a doping profile, but considering its simplicity, it yields a concentration that compares favorably with the differential capacitance derived value.

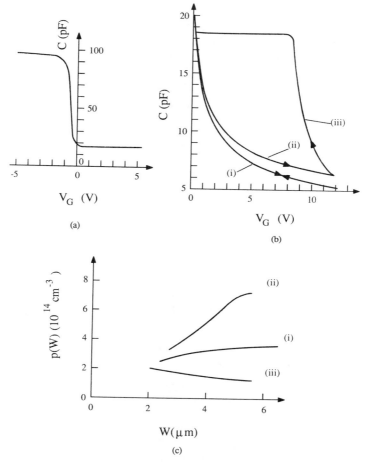

Fig. 2.18   (a) Equilibrium C-V_G curve of an MOS-C, (b) deep-depletion curves for (i) 5 V/s and (ii), (iii) 0.1 V/s sweep rates, (c) the carrier concentration profiles determined from (b). $C_{ox} = 98$ pF, $W_{ox} = 1200$ Å. Courtesy of J. S. Kang, Arizona State University.

The effects of *minority carrier* generation are a problem for high carrier generation rates found in devices with low generation lifetimes because gate voltage sweep rates necessary to prevent minority carrier buildup must be extremely high. An example is silicon-on-sapphire in which minority carriers are rapidly generated due to the low generation lifetime. It is more difficult to drive the MOS-C into deep depletion under those conditions. Cooling the device reduces the generation rate substantially, allowing doping profile measurements to be made. Cooling to liquid nitrogen temperatures for high generation rates works well to reduce the effects of minority carrier generation.[13,82] Providing a collection junction is another way to reduce the effect of minority carriers. As soon as minority carriers are generated, they are collected by the reverse-biased junction instead of the MOS-C. A MOSFET with source and drain reverse biased is such an implementation. A gate-controlled diode is another.

A further complication is introduced by interface traps invariable present in all MOS capacitors. The interface trap density is usually negligibly low for properly annealed, high quality SiO_2-Si interfaces. When interface states do play a role, their effect on doping profiling can be corrected by measuring the high-frequency capacitance C_{HF} and the low-frequency capacitance C_{LF} according to[83–84]

$$N_{A,corr} = \frac{(1 - C_{LF}/C_{ox})}{(1 - C_{HF}/C_{ox})} N_{A,uncorr} \qquad (2.36)$$

The effects of interface traps are considerably reduced in the pulsed MOS-C doping profile technique when the modulation frequency is increased. Modulation frequencies of 30 MHz have been suggested,[20] but most measurements are made at 1 MHz. Interface trap effects are also reduced when the device is cooled.

Interface traps or interfacial layers can also give errors in Schottky barrier capacitance profiling. It has been found that if the ideality factor n is larger than 1.1, erroneous profiles are obtained.[85] Ideality factors $n \leq 1.1$ are satisfactory for profiling.

2.5.4 Diode Edge and Stray Capacitance

C-V profiling relies on an accurate measure of the capacitance and of the device area. While the capacitance can be accurately measured, the area cannot always be accurately determined. Furthermore the capacitance may contain stray capacitance components. The device contact area can be measured but the effective area differs from the contact area due to lateral space-charge region spreading. This problem was originally addressed by Goodman.[78] The effective capacitance is[86–87]

$$C_{eff} = C(1 + bW/r) \qquad (2.37)$$

where $C = K_s \varepsilon_0 A/W$, $A = \pi r^2$, r is the contact radius, $b \approx 1.5$ for Si and GaAs, and $b \approx 1.46$ for Ge. Equation (2.37) shows the lateral scr effect to diminish as the contact radius increases and that $r \geq 100bW$ ensures for the second term in the bracket to contribute no more than 1% to the effective capacitance. For $W = 1\,\mu m$, $r \geq 150\,\mu m$ whereas for $W = 10\,\mu m$, $r \geq 1500\,\mu m$. This is not a particularly severe limitation. It should be considered, however, because the effective doping concentration is related to the actual doping concentration by

$$N_{A,eff} = (1 + bW/r)^3 N_A \qquad (2.38)$$

Equation (2.37) shows the edge capacitance to be a constant, and it can therefore be nulled out prior to differential profile measurements by using a dummy capacitor of an appropriate value. For mercury-probe profiling it has been proposed to make the contact sufficiently large that the edge capacitance effects can be neglected. The minimum recommended contact radius depends on the substrate doping concentration and should be[47]

$$r_{min} = 3.7 \times 10^{-2}(N/10^{16})^{-0.35} \text{ cm} \qquad (2.39)$$

where N is the doping concentration. Equation (2.39) is valid for the doping range of 10^{13} to 10^{16} cm^{-3}. The minimum radius is about 8.3×10^{-2} cm = 30 mils for $N = 10^{15}$ cm^{-3}.

Stray capacitance is more difficult to determine. It includes cable and probe capacitances, bonding pads, and gate protection diodes in MOSFET's. Cable and probe capacitances can be eliminated by nulling the capacitance meter without contact to the diode. Bonding pad capacitance can often be calculated. Since the diode, MOS-C, or MOSFET can be made extremely small, but the bonding pad is approximately 100 μm × 100 μm, it becomes important to know the bonding pad capacitance contribution accurately.

2.5.5 Excess Leakage Current

Junction devices occasionally show excessively high reverse-biased leakage currents. This is especially true of Schottky barrier devices. Errors are introduced into the doping profiles when that occurs. The assumption in the conventional profile equations is that the voltage is measured across the reverse-biased space-charge region only. For most devices that is a good assumption since the impedance of the reverse-biased scr is much higher than the semiconductor quasi-neutral region resistance. If, however, excessive leakage currents flow, then an appreciable voltage can be developed across the quasi-neutral regions. This voltage is automatically included in the recorded voltage introducing errors in the measured profiles.[88]

2.5.6 Deep Traps

Capacitance measurements, being a measure of charge responding to an applied time-varying voltage, will determine any charge that can respond to the applied voltage. We have already considered the contribution of interface traps to the capacitance. Deep level impurities or traps in the semiconductor bulk can also produce errors in capacitance profiles.[84,89–91] The contribution of traps is a complicated function of the density and energy level of the traps as well as the sample temperature and the frequency of the ac voltage. The ac voltage frequency is often assumed to be sufficiently high for the traps to be unable to follow it. Even if that is true, there is still cause for concern because the reverse bias dc voltage usually changes sufficiently slowly for the traps to be able to respond. This can give rise to profile errors that are both time and depth dependent. Fortunately for trap concentrations much less than the doping concentration, say, 1% or less, the contribution of traps is usually negligible.

A good discussion of the various considerations of the influence of traps is given by Kimerling.[90] The necessary considerations, too detailed to be reproduced here, are given there. Capacitance measurements are also used to determine the properties of the traps themselves as discussed in Chapter 7.

2.5.7   Semi-Insulating Substrates

Epitaxial or implanted layers on semi-insulating (SI) substrates present some unique problems when these layers are profiled. Due to the semi-insulating nature of the substrate, both contacts must be made to the top surface. This introduces series resistance problems, especially if the reverse-biased scr extends close to the SI substrate when the remaining neutral region of the layer is very thin, and appreciable series resistance results. Similar problems occur when an n-type (p-type) layer is formed on a p-type (n-type) substrate. The measured concentration profiles sometimes exhibit minima in the upper layer near the interface between the two. Such minima are frequently not real, but are artificially introduced by the sample geometry.[92–93]

2.5.8   Instrumental Limitations

Capacitance meters determine the accuracy with which $p(x)$ and W are measured. The depth resolution should be limited by the Debye length rather than by the instrument. According to Amron the overriding influence on the precision of $p(x)$ is the precision with which ΔC is measured.[94] There is a temptation to make ΔC large, but this introduces errors in the determination of the local value of $\Delta C/\Delta V$ because the C-V curve is not linear. It also degrades the depth resolution by increasing the modulation of W. It is common practice in analog profilers to keep ΔV constant by using a modulation voltage of constant amplitude. According to Eq. (2.18) and (2.19), we find

$$\Delta V = \frac{qW\,p(W)\,\Delta W}{K_s \varepsilon_0} \qquad \text{and} \qquad \frac{\Delta W}{W} = -\frac{\Delta C}{C} \qquad (2.40)$$

so that

$$\Delta C = -\frac{K_s \varepsilon_0 C\,\Delta V}{qW^2 p(W)} \qquad (2.41)$$

For constant $p(W)$ and constant ΔV, ΔC decreases because W increases and C decreases, and profiles become noisier as the profile is measured deeper into the sample. Constant electric field increment feedback profilers allevi-ate this problem somewhat. An excellent discussion of instrumental limita-tions is given by Blood.[50]

2.6   HALL EFFECT

Those aspects of the *Hall effect* pertaining to carrier concentration measure-ments are discussed here. A more complete treatment of the Hall effect, including a derivation of the appropriate equations, can be found in Chapter 5. The key feature of Hall measurements is the ability to determine the carrier concentration, the carrier type, the resistivity, and the mobility with a relatively simple measurement.

The Hall theory predicts the Hall coefficient R_H as[95]

$$R_H = \frac{r(p - b^2 n)}{q(p + bn)^2} \tag{2.42}$$

where $b = \mu_n/\mu_p$ and r is the scattering factor that lies between 1 and 2, depending on the scattering mechanism in the semiconductor. For lattice scattering $r = 3\pi/8 = 1.18$, for ionized impurity scattering $r = 315\pi/512 = 1.93$, and for neutral impurity scattering $r = 1$.[95-96] The scattering factor is also a function of the magnetic field and the temperature. In the high magnetic field limit $r \rightarrow 1$. The scattering factor can be determined by measuring R_H in the high magnetic field limit, i.e., $r = R_H(B)/R_H(B = \infty)$ where B is the magnetic field. The scattering factor has been measured in *n*-type GaAs as a function of magnetic field and was found to vary from 1.17 at $B = 0.1\,kG$ to 1.006 at $B = 83\,kG$.[9] The high fields necessary for r to approach unity are not achievable in most laboratories. Typical magnetic fields are 0.5 to 10 kG, making $r > 1$ for typical Hall measurements.

For extrinsic *p*-type material, where $p \gg n$, Eq. (2.42) reduces to

$$R_H = \frac{r}{qp} \tag{2.43}$$

and for extrinsic *n*-type it becomes

$$R_H = -\frac{r}{qn} \tag{2.44}$$

Equations (2.43) and (2.44) show that a knowledge of the Hall coefficient leads to a determination of the *carrier type* as well as the *carrier concen-tration*. Usually r is assumed to be unity—an assumption generally intro-ducing an error of less than 30%.[99]

The Hall effect is used to measure the carrier concentration, resistivity and mobility at a given temperature, and the carrier concentration as a function of temperature to extract additional information. Consider a p-type semiconductor of doping concentration N_A compensated with donors of concentration N_D. The hole concentration is given by[99]

$$\frac{p(p + N_D) - n_i^2}{(N_A - N_D - p + n_i^2/p)} = \frac{N_v}{g} e^{-E_A/kT} \tag{2.45}$$

where N_v is the effective density of states in the valence band, g is the degeneracy factor for acceptors (usually taken as 4), and E_A is the energy level of the acceptors above the valence band with the top of the valence band being the reference energy. Equation (2.45) can be simplified for certain conditions.

1. At low temperatures where $p \ll N_D$, $p \ll (N_A - N_D)$, and $n = n_i^2/p \approx 0$

$$p \approx \frac{(N_A - N_D)N_v}{gN_D} e^{-E_A/kT} \tag{2.46}$$

2. When N_D is negligibly small,

$$p \approx \sqrt{\frac{(N_A - N_D)N_v}{g}} e^{-E_A/2kT} \tag{2.47}$$

3. At higher temperatures where $p \gg n_i$,

$$p \approx N_A - N_D \tag{2.48}$$

According to Eqs. (2.46) and (2.47), the slope of a $\log(p)$ versus $1/T$ plot gives an activation energy of either E_A or $E_A/2$, depending on whether there is an appreciable donor concentration in the material or not. At higher temperatures, typically room temperature, the net majority carrier concentration is obtained.

Instead of merely measuring the slopes of a $\log(p)$ versus $1/T$ plot, the experimental data can be fitted with an appropriate model, and a wealth of information can be extracted. An example is given in Fig. 2.19.[100] This figure shows the Hall carrier concentration data for two indium-doped silicon samples. In addition to In, the samples contain Al, B, and P. For the acceptors (B, Al, and In) both the concentrations and the energy levels were extracted from the data. For the donor only the concentration can be determined from a p-type sample. This figure demonstrates the extremely powerful nature of Hall measurements if the data are properly analyzed.

The Hall coefficient is given by

$$R_H = \frac{tV_H}{BI} \tag{2.49}$$

where V_H is the Hall voltage, t is the sample thickness, I is the current, and

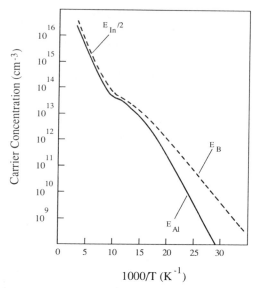

Fig. 2.19 Carrier concentration versus reciprocal temperature for Si:In with Al and B contamination. Solid line: $N_{In} = 7.9 \times 10^{16}$ cm^{-3}, $E_{In} = 0.165$ eV, $N_{Al} = 7 \times 10^{13}$ cm^{-3}, $E_{Al} = 0.07$ eV, $N_B = 1.5 \times 10^{13}$ cm^{-3}, $E_B = 0.044$ eV, $N_D = 1.3 \times 10^{13}$ cm^{-3}; dashed line: $N_{In} = 4.5 \times 10^{16}$ cm^{-3}, $E_{In} = 0.164$ eV, $N_{Al} = 6.4 \times 10^{13}$ cm^{-3}, $E_{Al} = 0.07$ eV, $N_B = 1.6 \times 10^{13}$ cm^{-3}, $N_D = 2 \times 10^{13}$ cm^{-3}. Reprinted after Braggins et al.[100] by permission of IEEE (© 1980, IEEE).

B is the magnetic field. The thickness is well defined for uniformly doped wafers. The active layer thickness is not necessarily the total layer thickness for thin epitaxial or implanted layers on substrates of opposite conductivity or on semi-insulating substrates. If depletion effects caused by Fermi level pinned band bending at the surface and by band bending at the layer-substrate interface are not considered, the Hall coefficient will be in error and so will those semiconductor parameters derived from it.[101] Even the temperature dependence of the surface and interface space-charge regions should be considered for unambiguous measurements.[102]

Hall measurements are generally made on samples from which an average carrier concentration is derived. For uniformly doped samples the correct concentration is obtained, but for non-uniformly doped samples an average value is determined. Occasionally one wants to measure spatially varying carrier concentration profiles. The Hall technique is suitable through the use of alternating layer stripping and Hall measurements.[103] Layers can be stripped reliably by anodic oxidation and subsequent oxide etch. Anodic oxidation consumes a certain fraction of the semiconductor which is removed during the oxide etch. Although anodic oxidation/etch is time consuming, it has been used to remove a 600 Å BF$_2$ implanted layer in Si in

25 Å increments.[104] Anodic oxidation, oxide etch, and Hall measurement were done in situ in an electrolytic cell located between the pole pieces of a magnet. The technique has also been applied to GaAs.[105] For a further discussion of anodic oxidation etching, see Section 1.3.1.

The interpretation of the Hall data becomes more complex when successive measurements are made. In order to generate a carrier concentration profile, the *sheet* Hall coefficient R_{Hs} given by $R_{Hs} = R_H/t$, and the sheet conductance σ_s must be measured repeatedly. The carrier concentration profile is obtained from Hall coefficient versus depth and from sheet conductance versus depth curves according to the relationship[103]

$$p(x) = \frac{r(d\sigma_s/dx)^2}{q\, d(R_{Hs}\sigma_s^2)/dx} \tag{2.50}$$

where $\sigma_s = 1/\rho_s$ and ρ_s is the sheet resistance (see Section 1.3.1).

Occasionally the Hall sample consists of an *n*, or *p*-film on a *p*- or *n*-substrate. For film and substrate of opposite conductivity, the *pn* junction between them is usually assumed to be an insulating boundary. If that is not true, then the Hall data must be corrected.[106] This correction must also be made if the junction is a good insulator, but the ohmic contact to the Hall sample is alloyed through the top layer, shorting it to the substrate. This can happen if the upper layer is an unintentional type conversion as has been observed in materials like HgCdTe where the doping concentration and type is determined by stoichiometry.[107]

The theory for a simple two-layer structure with an upper layer having thickness t_1 and conductivity σ_1 and a substrate of thickness t_2 and conductivity σ_2 gives the Hall constant as[107–108]

$$R_H = R_{H1}\,\frac{t_1}{t}\left(\frac{\sigma_1}{\sigma}\right)^2 + R_{H2}\,\frac{t_2}{t}\left(\frac{\sigma_2}{\sigma}\right)^2 \tag{2.51}$$

where R_{H1} is the Hall constant of layer 1, R_{H2} is the Hall constant of substrate 2, $t = t_1 + t_2$, and σ is given by

$$\sigma = \frac{t_1}{t}\,\sigma_1 + \frac{t_2}{t}\,\sigma_2 \tag{2.52}$$

For $t_1 = 0$ we have $t = t_2$, $\sigma = \sigma_2$, and $R_H = R_{H2}$, with the substrate being characterized. If the upper layer is more heavily doped than the substrate or is formed by inversion through surface states, for example, and the carriers in the substrate freeze out at low temperatures making σ_2 very small, then

$$\sigma \approx \frac{t_1}{t}\,\sigma_1 \qquad R_H \approx R_{H1}\,\frac{t}{t_1} \tag{2.53}$$

and the Hall measurement characterizes the surface layer. This can be especially serious if the existence of the upper layer is not suspected and it is believed that the substrate is being characterized. Examples of such measurements are given in Lou and Frye.[107]

2.7 OPTICAL TECHNIQUES

2.7.1 Plasma Resonance

The optical reflection coefficient of a semiconductor is given by

$$R = \frac{(n-1)^2 + k^2}{(n+1)^2 + k^2} \qquad (2.54)$$

where n is the refractive index and $k = \alpha\lambda/4\pi$ is the extinction coefficient, α is the absorption coefficient and λ is the photon wavelength. The reflection coefficient of semiconductors, shown in the inset of Fig. 2.20, is high at short wavelengths, tends to a constant, and then shows an anomaly at higher wavelengths. First, it decreases toward a minimum; then it rises rapidly toward unity. When the photon frequency v, related to the wavelength through the relation $v = c/\lambda$, approaches the *plasma resonance frequency R* tends to unity. The *plasma resonance wavelength* is given by[109]

$$\lambda_p = \frac{2\pi c}{q} \sqrt{\frac{K_s \varepsilon_0 m^*}{p}} \qquad (2.55)$$

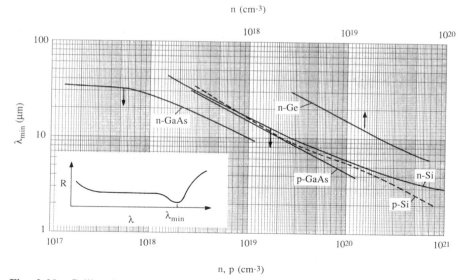

Fig. 2.20 Calibration curves of λ_{min} versus carrier concentration for Si, Ge, and GaAs. The inset shows a schematic R vs. λ curve. Copyright ASTM; reprinted with permission after ASTM F398.[113]

TABLE 2.1   Constants for Calculating n and p from λ_{min}

Material	Type	Applicable λ (μm)	A	B	C
Si	n	2.8–42.5	3.039×10^{-12}	-1.835	-5.516×10^{-12}
Si	p	2.5–5.4	4.097×10^{-11}	-2.071	0
Si	p	5.4–32.4	8.247×10^{-16}	-1.357	-2.626×10^{-15}
Ge	n	7–25	1.294×10^{-13}	-1.611	-2.934×10^{-13}
GaAs	n	9.4–18.5	5.803×10^{-11}	-2.051	0
GaAs	n	18.5–30.4	2.405×10^{-8}	-2.898	0
GaAs	n	30.4–33.9	1.188×10^{-3}	-12.308	0
GaAs	p	3.7–30	5.566×10^{-12}	-1.884	0

Source: ASTM Standard F398-77.[113]

where p is the free carrier concentration in the semiconductor and m^* the effective mass. It is, in principle, possible to determine p from λ_p. Plasma resonance has also been used for the determination of optical constants, conductivity effective mass, and carrier concentration.[110]

The plasma resonance wavelength is difficult to determine because it is not well defined. It is for this reason that the carrier concentration is determined not from the plasma resonance wavelength but from the wavelength λ_{min} at the *reflectivity minimum*, where $\lambda_{min} < \lambda_p$. The minimum wavelength is related to the carrier concentration through an empirical relationship given by the equation[111–113]

$$p = (A\lambda_{min} + C)^B \qquad (2.56)$$

where the constants A, B, and C are listed in Table 2.1. In Eq. (2.56) p is in cm^{-3} and λ_{min} in μm. The technique is useful only for carrier concentrations higher than 10^{18} to 10^{19} cm^{-3}.

The carrier concentrations derived from Fig. 2.20 or from Table 2.1 are for uniformly doped substrates or for uniformly doped layers with layer thicknesses at least equal to $1/\alpha$. For example, for n-type Si with $n = 10^{19}$ cm^{-3}, $\alpha = 5000$ cm^{-1}. Hence the layer should be at least $1/\alpha \approx 2$ μm thick. For a diffused or implanted layer with varying carrier concentration profile, a determination of the surface concentration is only possible if the shape of the profile and the junction depth are known.[114] A further complication for thin epitaxial layers is introduced by the phase shift at the epitaxial layer—substrate surface. This adds an oscillatory component to the R-λ curve, making it more difficult to extract λ_{min}.[115]

2.7.2 Infrared Spectroscopy

Infrared spectroscopy relies on optical excitation of electrons (holes) from their respective donors (acceptors) into excited states. Consider the n-type semiconductor containing only donors shown in Fig. 2.21(a). At low temperatures most of the electrons are "frozen" onto the donors, and the free carrier concentration in the conduction band is very low. The electrons are mainly located in the lowest energy level or donor ground state illustrated in Fig. 2.21(b). When photons of energy $h\nu \leq (E_C - E_D)$ are incident on the sample, two optical absorption processes can occur: electrons are excited from the ground state to the conduction band giving a broad absorption continuum, and electrons are excited from the ground state to one of several excited states producing sharp absorption lines in the transmission spectrum, characteristic of the shallow-level impurities.[116-118] Such a transmittance curve is shown in Fig. 2.22(a) for phosphorus- and arsenic-containing silicon.[119] Additional information can be obtained by splitting the energy levels with a magnetic field.[120]

Through the use of the Fourier transform techniques (Fourier transform infrared spectroscopy is discussed in Chapter 9), high sensitivity is obtained, and the detection limits are extremely low. Impurity concentrations as low 5×10^{11} cm^{-3} have been measured in Si.[119] Such low carrier concentrations can also be determined by Hall measurements, but optical transmission

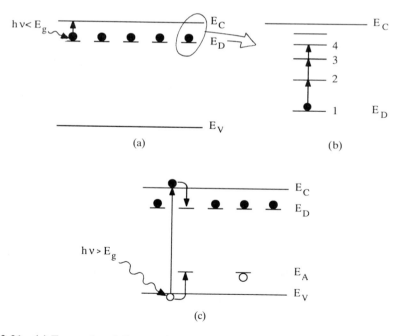

Fig. 2.21   (a) Energy band diagram for a semiconductor containing donors at low temperature, (b) energy band diagram showing the donor energy levels, (c) band diagram when both donors and acceptors are present. The above-band-gap light fills donors and acceptors.

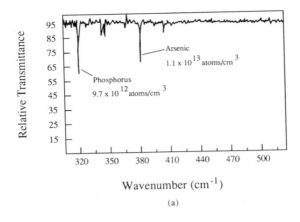

(a)

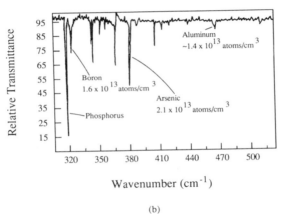

(b)

Fig. 2.22   (a) Donor impurity spectrum for 265 Ω-cm n-Si at $T \approx 12$ K, (b) spectrum for the sample in (a) with above-band-gap illumination. Reprinted with permission after Baber.[119]

techniques are simpler to implement because no contacts are required. The appropriate optical equipment suitable for long wavelength generation and detection is of course required.

Most electrical carrier concentration measurement methods determine the net carrier concentration. The infrared spectroscopy technique as discussed so far also measures $N_D - N_A$, because there are only $n = N_D - N_A$ electrons frozen onto the donors at the low measurement temperatures. Any compensating acceptors are empty because the holes are compensated by electrons. However, it is possible to determine the acceptor concentration also. To measure N_D *and* N_A, the sample is illuminated with background light of energy $h\nu > E_g$.[119,121–122] The excess electron-hole pairs generated by the background light neutralize the remaining ionized donors and acceptors. Virtually all donors and acceptors are neutralized, as shown in Fig. 2.21(c) if sufficient ehp's are created. The long wavelength infrared radiation now can excite electrons into excited donor states *and* holes into excited acceptor states.

A spectrum for a Si sample with background light is shown in Fig. 2.22(b). Two features distinguish Fig. 2.22(a) from 2.22(b): (i) the P and As signals are increased, and (ii) the compensating B and Al impurities appear in the spectrum. It is possible to determine the *concentration* of all impurities and to *identify* them because each impurity has unique absorption peaks. The strongest absorption lines for Si are given in Baber.[119]

The infrared spectroscopy technique is very quantitative in identifying the *impurity type* but is qualitative in determining the *impurity concentration*. In order to determine the relationship between the absorption peak height and the impurity concentration, calibration data must be established using samples whose doping concentration is determined from electrical measurements. For uncompensated material this is fairly unambiguous. For compensated samples the procedure is more complex, and Baber gives a good discussion of the procedure.[119]

The optical transmittance through a semiconductor wafer is approximately

$$T \approx (1 - R)^2 e^{-\alpha t} \qquad (2.57)$$

where t is the sample thickness. For reasonable measurement sensitivity, αt should be on the order of unity or $t \approx 1/\alpha$. For $\alpha = 1$ to $10 \, \text{cm}^{-1}$, applicable for shallow impurity absorption at low concentrations, the sample must be 1 to 10 mm thick. Samples of this thickness can be provided for bulk wafers but not for epitaxial layers, making IR spectroscopy impractical for thin layers.

A variation of this technique is known as *photothermal ionization spectroscopy* (PTIS) or *photoelectric spectroscopy*. Bound donor electrons are optically excited from the ground state to one of the excited states. At $T \approx 5$ to 10 K the sample phonon population is sufficiently high for carriers in the excited state to be transferred into the conduction band where they lead to a change in the sample conductivity. It is this photoconductivity change that is detected as a function of wavelength.[123–125] Concentrations as low as $10^9 \, \text{cm}^{-3}$ boron and gallium acceptors in Ge have been measured by the technique.[126] A disadvantage of PTIS is the need for ohmic contacts, but the advantage is its sensitivity for thin films. PTIS can be combined with magnetic fields for easier identification of impurities for GaAs and InP.[123]

2.7.3 Photoluminescence

Photoluminescence (PL) is used to detect and identify impurities in semiconductor materials. The technique is described in more detail in Chapter 9. PL relies on the creation of electron-hole pairs by incident radiation and subsequent *radiative* recombination photon emission. Only radiative recombination events emit light. Any non-radiative recombination events are not detected and detract from the PL signal. We discuss here briefly the application of PL to the measurement of the doping concentrations in semiconductors.

Impurity identification is very precise because the energy resolution of photoluminescence is very high. It is the concentration measurement that is more difficult because it is not easy to draw a correlation between the intensity of a given impurity spectral line and the concentration of that impurity. This is due to differential nonradiative effects of non-radiative recombination through deep-level bulk or surface recombination centers.[127] Since the concentration of such recombination centers can vary from sample to sample, even for constant shallow level concentrations, it is obvious that the photoluminescence signal can vary greatly.

This problem has been overcome by measuring both the intrinsic and the extrinsic PL peaks and using their ratio. It has been determined that the ratio $X_{TO}(BE)/I_{TO}(FE)$ is proportional to the doping concentration.[128] $X_{TO}(BE)$ is the transverse optical phonon PL intensity peak of bound excitons for element $X = B$ or P, and $I_{TO}(FE)$ is the transverse optical phonon intrinsic PL intensity peak of free excitons. Good agreement is found between the resistivity measured electrically and the resistivity determined from photoluminescence for Si as shown in Fig. 2.23. The technique has also been applied to InP, where the donor concentration as well as the compensation ratio was determined.[130]

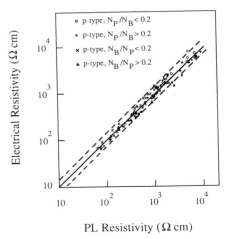

PL Resistivity (Ω cm)

Fig. 2.23   Comparison between the PL resistivity for B and P in Si. Reprinted with permission after Tajima et al.[129] This paper was originally presented at the Spring 1981 Meeting of the Electrochemical Society, Inc., held in Minneapolis, Minnesota.

2.7.4 Free Carrier Absorption

When photons of energy $hv > E_g$ are incident on a semiconductor, electron-hole pairs are generated. We saw in Section 2.7.2 that photons of energy $hv < E_g$ can excite electrons from the ground state of shallow-level impurities onto excited states. It is also possible that photons of energy $hv < E_g$ excite free electrons (holes) in the conduction (valence) band to higher energy states in the band, i.e., photons are absorbed by free carriers. This is the basis of *free carrier absorption*. The free carrier absorption coefficient for holes is given by[95]

$$\alpha_{fc} = \frac{q^3\lambda^2 p}{4\pi^2\varepsilon_0 c^3 n m^{*2}\mu_p} = 5.27 \times 10^{-17} \frac{\lambda^2 p}{n(m^*/m)^2\mu_p} \qquad (2.58)$$

where λ is the wavelength, c the velocity of light, n the refractive index, m^* the effective mass, and μ_p the hole mobility. The free carrier absorption coefficient depends on the free carrier concentration and on the square of the wavelength. Care should be taken during the measurement not to use wavelengths that coincide with impurity or lattice absorption lines. For example, there is an absorption line in silicon due to interstitial oxygen at $\lambda = 9.05\mu m$, and substitution carbon at $\lambda = 16.47 \mu m$, and lattice absorption lines are found near $16 \mu m$.

By fitting curves to experimental Si data[131] fairly good agreement is observed for[132]

$$\alpha_{fc,n} \approx 1 \times 10^{-18}\lambda^2 n \qquad (2.59a)$$

$$\alpha_{fc,p} \approx 2.7 \times 10^{-18}\lambda^2 p \qquad (2.59b)$$

where n and p are the free carrier concentrations in cm^{-3} for n-Si and p-Si, respectively, and the wavelength is given in units of μm. Carrier concentrations of $10^{17} cm^{-3}$ or higher can be measured by this technique. The measurement becomes difficult for lower concentrations because the absorption coefficient is too low.

Free carrier absorption also lends itself to sheet resistance ρ_s measurements. Good agreement with experiment has been found for the transmission using the relationships[132]

$$T \approx (1 - R)^2 \exp(-k\lambda^2/\rho_s) \qquad (2.60)$$

with $k = 0.15$ for n-type Si and $k = 0.3375$ for p-type Si layers. In Eq. (2.60) λ is in μm and ρ_s in ohms/square. Free carrier concentration maps have been generated by scanning the infrared light beam. Carrier concentrations as low as $10^{16} cm^{-3}$ have been determined with a 1 mm resolution using $\lambda = 10.6 \mu m$.[133]

2.7.5 Optical Dosimetry

In *optical dosimetry* the doping concentration is determined by a technique entirely different from any of the methods discussed in this chapter. The method was developed for ion-implant uniformity and dose monitoring and does not even use semiconductor wafers. A transparent glass wafer is provided with a thin coating of positive photoresist by conventional spin-coating methods. The photoresist is then decolored by UV annealing to render it optically transparent. When this photoresist-coated glass wafer is ion implanted, a complex chemical process occurs that darkens the photoresist. The amount of darkening depends on the implant energy and dose.

The optical dosimeter, using a sensitive microdensitometer, detects the UV transparency of the entire wafer before and after implant and compares the final-to-initial difference in optical transparency with internal calibration tables.[134] The optical transparency is measured on the entire implanted wafer and then displayed as a contour map. Calibration curves of optical density as a function of implant dose have been developed for implant doses of 10^{12} to 10^{15} cm^{-2}.

The method requires no implantation activation anneal and the results can be displayed within a few minutes of the implantation. However, for best results one should wait some time for post-implant photoresist stabilization. The optical density is measured on 3 mm spot resolution and lends itself well to low-dose implants. As discussed earlier in this chapter, the doping concentration of low-dose implants is not easy to measure electrically, and this optical method is a viable alternate technique.

2.7.6 Thermal Waves

Thermal and plasma waves are generated whenever a periodic heat stimulus is applied to a semiconductor wafer. The source of heat can be a modulated light or an electron beam. In the *thermal wave* method an Ar^{+} ion laser beam is modulated at a frequency of 0.1 to 10 MHz. A periodic temperature variation is established in the semiconductor in response to this periodic heat stimulus. The amplitude of this temperature variation is around 10°C in silicon. The thermal wave diffusion length at a 1 MHz modulation frequency is 2 to 3 μm.[135] The small temperature variations cause small volume changes of the wafer near the surface. These changes include both thermoelastic and optical effects,[136] and they are detected with a second laser—the probe beam—by measuring a reflectivity change.

The ability to determine ion-implant concentrations by thermal waves depends on two sources. First, the conversion of the single crystal substrate to a partially disordered layer by the implant process and, second, the perturbation of the lattice by the implanted ions themselves. The thermal wave-induced thermoelastic and optical effects are changed in proportion to the number of implanted ions in both cases. Thermal wave implant monitoring is subject to post-implant damage relaxation. However, the laser detection scheme accelerates the damage relaxation process, and the sample stabilizes within a few minutes.

Both laser beams are focused to approximately $1\,\mu m$ diameter spots, giving the technique a spatial resolution of around $1\,\mu m$. This allows measurements to be made not only on uniformly implanted wafers but also on patterned wafers with implant patterns of $2.5\,\mu m$ or larger. The technique is contactless and non-destructive and has been used to measure implant doses from 10^{11} to $10^{15}\,cm^{-2}$.[137] Measurements can be made on bare and on oxidized wafers having the advantage of allowing measurements of implants through oxides, so oxides do not have to be removed. The technique can discriminate between implant species since the lattice damage increases with implant atom size and the thermal wave signal depends on the lattice damage. It has been used for ion implantation monitoring, wafer polish damage, and reactive and plasma etch damage studies. Its chief strength lies in the ability to detect low-dose implants and to display the information as contour maps. The equipment is commercially available.

2.8 SECONDARY ION MASS SPECTROMETRY

Secondary ion mass spectrometry (SIMS) has become a very powerful technique for the analysis of impurities in solids. SIMS is discussed in more detail in Chapter 10. In this section we briefly discuss the application of SIMS to semiconductor dopant profiling. The technique relies on removal of material from a solid by sputtering and on analysis of the sputtered ionized species. Most of the sputtered material consists of neutral atoms and cannot be analyzed. Only the *ionized* atoms can be analyzed by passing them through an energy filter and a mass spectrometer.

SIMS has good detection sensitivity for many elements, but its sensitivity is not as high as electrical or optical methods. It allows simultaneous detection of different elements, has a depth resolution of 50 to 100 Å, and can give lateral surface characterization on a scale of several microns. It is a destructive method since the very act of removing material by sputtering leaves a crater in the sample.

A SIMS doping concentration is produced by sputtering the sample and monitoring the secondary ion signal of a given element as a function of time. Such an "ion signal versus time" plot contains the necessary information for a dopant concentration profile, but the axes must be converted. The *time axis* is converted to *a depth axis* by measuring the depth of the crater at the end of the measurement. This should be done for each sample, since the sputter rate varies with spot focus and ion current.[138] The *secondary ion signal* is converted to *impurity concentration* through standards of known dopant profile. The proportionality between ion signal and concentration is strictly true only if the matrix in which the impurity is contained is uniform.

The ion yield of a given element is highly dependent on the matrix, and the conversion of the secondary ion signal to an impurity concentration requires the use of a standard. For example, to determine an unknown boron profile in silicon, boron is implanted into Si at a given energy and dose. The secondary ion signal is calibrated by assuming the total amount of boron in the sample to be equal to the implanted boron. Errors can be introduced if the impurity is implanted through an oxide layer since a fraction of the impurity remains in the oxide.

SIMS determines the *total* impurity concentration, not the *electrically active* impurity concentration. For example, implanted, non-annealed samples give SIMS profiles very close to the predicted Gaussian distribution. Electrical measurements give very different results, with the ions not yet electrically activated. SIMS and electrical measurement agree quite well for activated samples as shown in Fig. 2.24. In this figure the profiles for B-implanted into Si were determined by SIMS, spreading resistance profiling (SRP), and neutron depth profiling (NDP). Note the good agreement over most of the profile. The sample with the highest implant dose shows a small hump in both SIMS and NDP determined profiles. Both of these methods measure the total concentration. The SRP profile does not show this feature, indicating that the atoms in that hump are not electrically active.

Comparisons of SIMS dopant profiles with profiles measured by spreading resistance sometimes show a discrepancy in the lowly doped portions of the profile giving shallower junctions for SRP compared to SIMS determined profiles as in Fig. 2.24 and also observed by others.[140–144] The discrepancy in the low concentration tail is not fully understood. The deeper

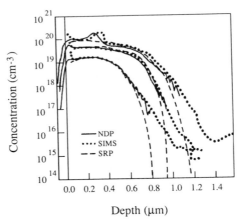

Fig. 2.24   Boron-concentration profiles determined by SIMS, spreading resistance, and neutron depth profiling. The implant conditions are: 70 keV and fluences of 10^{15}, 4×10^{15}, and 10^{16} cm^{-2}. Copyright ASTM; reprinted with permission after Ehrstein et al.[139]

SIMS junctions have been attributed to atomic mixing and permanent displacement of atoms changing their original distribution contributing to slightly deeper junctions.[144] In addition the beveled SRP geometry is believed to influence the carrier distribution through surface charges and through the bevel itself.[145]

SIMS is not applicable when the dopant species is not a foreign impurity. For the usual group IV and III–V semiconductors, the dopant atoms are foreign impurities, and they can be determined by SIMS. For semiconductors like HgCdTe the doping is frequently due to stoichiometric defects. For example, *p*-type HgCdTe is believed to be the result of Hg vacancies.[146] These can of course not be determined by SIMS, and electrical or optical methods must be employed to determine the doping or carrier concentration.

2.9 STRENGTHS AND WEAKNESSES

- *Differential Capacitance*   The major weakness of the differential capacitance profiling method is its limited profile depth, limited at the surface by the zero-bias space-charge region width and in depth by voltage breakdown. The latter limitation is particularly serious for heavily doped regions. Further limitations are due to the Debye limit, which applies to all carrier profiling techniques.

  The method's strength lies in the ability to give the carrier concentration profile with little data processing. A simple differentiation of the *C-V* data is all that is required. It is an ideal method for moderately doped materials and is nondestructive through the use of a mercury probe. Its depth profiling capability is extended significantly for the electrolytic method.

- *Max-Min MOS-C Capacitance*   The weakness of this technique lies in its inability to provide a profile. It determines merely an average doping concentration in the space-charge region width of an MOS-C in equilibrium. It is, however, a very simple method if an MOS capacitor is available as the test structure and is routinely used for that reason to give a first order value for the concentration.

- *Integral Capacitance*   The integral capacitance technique also does not provide a profile, which limits its usefulness. It does, however, provide, a value for an implant dose and depth, and its major strength lies in its accuracy. This is very important when monitoring ion implants that can have uniformities of 1%.

- *MOSFET Current-Voltage*   The substrate/gate voltage technique requires two differentiations and has not found wide application. The threshold voltage method utilizes threshold voltage measurements and needs a proper definition of threshold voltage in its interpretation. The advantage of both methods is the fact that a MOSFET is measured directly. No special, large-area test structures are required. This is especially important when no test structures are available.

- *Spreading Resistance*   The weakness of SRP is the complexity of sample preparation and the required spreading resistance apparatus as well as the interpretation of the measured spreading resistance profile. The measured data must be deconvolved, and either the mobility must be known or well calibrated standards must be used to extract the dopant profile.

  Its strength lies in it being a well-known method that is routinely used by the semiconductor industry for Si profiling. It has no depth limit and can profile through an arbitrary number of *pn* junctions; it spans a very large concentration range from about 10^{13} cm^{-3} to 10^{21} cm^{-3}.

- *Hall Effect*   The Hall effect is limited in its profiling ability through the inconvenience of providing repeated layer removal. Although it is utilized for profiling, it is not a routine method for generating profiles. Its advantage lies in providing average values of carrier concentration and mobility. For that it is used a great deal, as discussed in Chapter 5.

- *Optical Techniques*   Optical techniques require specialized equipment with quantitative doping measurements requiring known standards. Profiling is generally not possible, and only average values are obtained. The major advantage of optical methods is their unprecedented sensitivity and accuracy in impurity identification. These methods are generally employed in specialized laboratories. However, the optical dosimetry and thermal wave techniques have become commercially available methods. They are mainly used for ion-implant monitoring by displaying contour maps of the implanted wafers. Their strength lies in their ability to measure the implants non-destructively and rapidly and in displaying the information in the form of contour plots.

- *Secondary Ion Mass Spectrometry*   The weakness of SIMS lies in the complexity of the equipment. It does not have the sensitivity of electrical and optical techniques. It is most sensitive for B in Si, for all other impurities it has reduced sensitivity and is unlikely to yield a complete profile. It is useless for semiconductors with stoichiometric dopant species. Reference standards must be used for quantitative interpretation of the raw SIMS data, and matrix effects can render measurement interpretation difficult.

  The strength of SIMS lies in its accepted use for dopant profiling. It measures the dopant profile not the carrier profile and can be used for implanted samples before any activation anneals. That is not possible with electrical methods. It has high spatial resolution and can be used for any semiconductor.

REFERENCES

[1]   W. Schottky, "Simplified and Expanded Theory of Boundary Layer Rectifiers (in German)," *Z. Phys.* **118**, 539–592, Febr. 1942.

[2]   J. Hilibrand and R. D. Gold, "Determination of the Impurity Distribution in Junction Diodes from Capacitance-Voltage Measurements," *RCA Rev.* **21**, 245–252, June 1960.

[3]   C. O. Thomas, D. Kahng and R. C. Manz, "Impurity Distribution in Epitaxial Silicon Films," *J. Electrochem. Soc.* **109**, 1055–1061, Nov. 1962.

[4]   I. Amron, "A Slide Rule for Computing Dopant Profiles in Epitaxial Semiconductor Films," *Electrochem. Technol.* **2**, 327–328, Nov/Dec. 1964.

[5]   R. Decker, "Measurement of Epitaxial Doping Density vs. Depth," *J. Electrochem. Soc.* **115**, 1085–1089, Oct. 1968.

[6]   L. E. Coerver, "Note on the Interpretation of *C-V* Data in Semiconductor Junctions," *IEEE Trans. Electron Dev.* **ED-17**, 436, May 1970.

[7]   H. J. J. DeMan, "On the Calculation of Doping Profiles from $C(V)$ Measurements on Two-Sided Junctions," *IEEE Trans. Electron Dev.* **ED-17**, 1087–1088, Dec. 1970.

[8]   W. van Gelder and E. H. Nicollian, "Silicon Impurity Distribution as Revealed by Pulsed MOS *C-V* Measurements," *J. Electrochem. Soc.* **118**, 138–141, Jan. 1971.

[9]   Y. Zohta, "Rapid Determination of Semiconductor Doping Profiles in MOS Structures," *Solid-State Electron.* **16**, 124–126, Jan. 1973.

[10] D. K. Schroder, *Advanced MOS Devices*, Addison-Wesley, Reading, MA, 1987, pp. 64–71.

[11] C. D. Bulucea, "Investigation of Deep-Depletion Regime of MOS Structures Using Ramp-Response Method" *Electron. Lett.* **6**, 479–481, July 1970.

[12] G. Baccarani, S. Solmi, and G. Soncini, "The Silicon Impurity as Revealed by High-Frequency Non Equilibrium MOS *C-V* Characteristics," *Alta Frequ.* **16**, 113–115, Feb. 1972.

[13] D. M. Brown, R. J. Connery and P. V. Gray, "Doping Profiles by MOSFET Deep Depletion $C(V)$," *J. Electrochem. Soc.* **122**, 121–127, Jan. 1973.

[14] D. P. Kennedy, P. C. Murley, and W. Kleinfelder, "On the Measurement of Impurity Atom Distributions in Silicon by the Differential Capacitance Technique," *IBM J. Res. Develop.* **12**, 399–409, Sept. 1968.

[15] J. R. Brews, "Threshold Shifts Due to Nonuniform Doping Profiles in Surface Channel MOSFET's," *IEEE Trans. Electron Dev.* **ED-26,** 1696–1710, Nov. 1979.

[16] D. P. Kennedy and R. R. O'Brien, "On the Measurement of Impurity Atom Distributions by the Differential Capacitance Technique," *IBM J. Res. Develop.* **13**, 212–214, March 1969.

[17] R. A. Moline, "Ion Implanted Phosphorus in Silicon: Profiles Using *C-V* Analysis," *J. Appl. Phys.* **42**, 3553–3558, Aug. 1971.

[18] W. E. Carter, H. K. Gummel, and B. R. Chawla, "Interpretation of

Capacitance vs. Voltage Measurements of *PN* Junctions," *Solid-State Electron.* **15**, 195–201, Feb. 1972.

[19] W. C. Johnson and P. T. Panousis, "The Influence of Debye Length on the *C-V* Measurement of Doping Profiles," *IEEE Trans. Electron Dev.* **ED-18**, 965–973, Oct. 1971.

[20] E. H. Nicollian, M. H. Hanes, and J. R. Brews, "Using the MIS Capacitor for Doping Profile Measurements with Minimal Interface State Error," *IEEE Trans. Electron Dev.* **ED-20**, 380–389, April 1973.

[21] C. P. Wu, E. C. Douglas, and C. W. Mueller, "Limitations of the *CV* Technique for Ion-Implanted Profiles," *IEEE Trans. Electron Dev.* **ED-22**, 319–329, June 1975.

[22] M. Nishida, "Depletion Approximation Analysis of the Differential Capacitance-Voltage Characteristics of an MOS Structure With Nonuniformly Doped Semiconductors," *IEEE Trans. Electron Dev.* **ED-26**, 1081–1085, July 1979.

[23] G. Baccarani, M. Rudan, G. Spadini, H. Maes, W. Vandervorst, and R. Van Overstraeten, "Interpretation of *C-V* measurements for Determining the Doping Profile in Semiconductors," *Solid-State Electron.* **23**, 65–71, Jan. 1980.

[24] M. Nishida and M. Aoyama, "An Improved Definition for the Onset of Heavy Inversion in an MOS Structure with Nonuniformly Doped Semiconductors," *IEEE Trans. Electron Dev.* **ED-27**, 1222–1230, July 1980.

[25] C. L. Wilson, "Correction of Differential Capacitance Profiles for Debye-Length Effects," *IEEE Trans. Electron Dev.* **ED-27**, 2262–2267, Dec. 1980.

[26] D. J. Bartelink, "Limits of Applicability of the Depletion Approximation and Its Recent Augmentation," *Appl. Phys. Lett.* **38**, 461–463, March 1981.

[27] H. Kroemer and W. Y. Chien, "On the Theory of Debye Averaging in the *C-V* Profiling of Semiconductors," *Solid-State Electron.* **24**, 655–660, July 1981.

[28] S. Sikorski, "Theory of *C-U* Profiling of a Nonuniformly Doped Semiconductor," *Phys. Stat. Sol.* (*a*) **82**, 265–274, March 1984.

[29] K. Lehovec, "*C-V* Profiling of Steep Dopant Distributions," *Solid-State Electron.* **27**, 1097–1105, Dec. 1984.

[30] J. Voves, V. Rybka, and V. Trestikova, "*C-V* Technique on Schottky Contacts—Limitation of Implanted Profiles," *Appl. Phys.* **A37**, 225–229, Aug. 1985.

[31] A. R. LeBlanc, D. D. Kleppinger, and J. P. Walsh, "A Limitation of the Pulsed Capacitance Technique of Measuring Impurity Profiles," *J. Electrochem. Soc.* **119**, 1068–1071, Aug. 1972.

[32] K. Ziegler, E. Klausmann, and S. Kar, "Determination of the Semiconductor Doping Profile Right Up to its Surface Using the MIS Capacitor," *Solid-State Electron.* **18**, 189–198, Feb. 1975.

[33] B. J. Gordon, "On-Line Capacitance-Voltage Doping Profile Measurement of Low-Dose Ion Implants, "*IEEE Trans. Electron Dev.* **ED-27**, 2268–2272, Dec. 1980.

[34] B. E. Deal, A. S. Grove, E. H. Snow and C. T. Sah, "Observation of

Impurity Redistribution During Thermal Oxidation of Silicon Using the MOS Structure," *J. Electrochem. Soc.* **112**, 308–314, March 1965.

[35] C. Jund and R. Poirier, "Carrier Concentration and Minority Carrier Lifetime Measurement in Semiconductor Epitaxial Layers by the MOS Capacitance Method," *Solid-State Electron.* **9**, 315–319, April 1966.

[36] D. C. Gupta and N. G. Anantha, "Measurement of Epitaxial Layer Resistivity Using MOS Capacitance Method," *Proc. IEEE* **55**, 1108, June 1967.

[37] K. Iniewski and A. Jakubowski, "Procedure for Determination of a Linear Approximation Doping Profile in a MOS Structure," *Solid-State Electron.* **30**, 295–298, March 1987.

[38] A. S. Grove, *Physics and Technology of Semiconductor Devices*, Wiley, New York, 1967.

[39] E. H. Nicollian and J. R. Brews, *MOS Physics and Technology*, Wiley, New York, 1982.

[40] W. E. Beadle, J. C. C. Tsai, and R. D. Plummer, *Quick Reference Manual for Silicon Integrated Circuit Technology*, Wiley-Interscience, New York, 1985, Ch. 14.

[41] J. Shappir, A. Kolodny, and Y. Shacham-Diamand, "Diffusion Profiling Using the Graded $C(V)$ Method," *IEEE Trans. Electron Dev.* **ED-27**, 993–995, May 1980.

[42] J. P. Rosbeck and M. E. Harper, "Doping and Composition Profiling in $Hg_{1-x}Cd_xTe$ by the Graded Capacitance-Voltage Method," *J. Appl. Phys.* **62**, 1717–1722, Sept. 1987.

[43] R. O. Deming and W. A. Keenan, "*C-V* Uniformity Measurements," *Nucl. Instrum. and Meth.* **B6**, 349–356, Jan. 1985; "Low Dose Ion Implant Monitoring," *Solid State Technol.* **28**, 163–167, Sept. 1985.

[44] P. J. Severin and G. J. Poodt, "Capacitance-Voltage Measurements With a Mercury-Silicon Diode," *J. Electrochem. Soc.* **119**, 1384–1389, Oct. 1972.

[45] P. L. Jones and J. W. Corbett, "Investigation of the Electrical Degradation of Silicon Schottky Contacts due to Mercury Contamination," *Appl. Phys. Lett.* **55**, 2331–2332, Nov. 1989.

[46] R. S. Nakhmanson and S. B. Sevastianov, "Investigations of Metal-Insulator-Semiconductor Structure Inhomogeneities Using a Small-Size Mercury Probe," *Solid-State Electron.* **27**, 881–891, Oct. 1984.

[47] P. S. Schaffer and T. R. Lally, "Silicon Epitaxial Wafer Profiling Using the Mercury-Silicon Schottky Diode Differential Capacitance Method," *Solid State Technol.* **26**, 229–233, April 1983.

[48] A. Yamashita, T. Aoki, and M. Yamaguchi, "A Method for Determining a GaAs Epitaxial Layer Impurity Profile," *Japan. J. Appl. Phys.* **14**, 991–997, July 1975.

[49] T. Ambridge and M. M. Faktor, "An Automatic Carrier Concentration Profile Plotter Using an Electrochemical Technique," *J. Appl. Electrochem.* **5**, 319–328, Nov. 1975.

[50] P. Blood, "Capacitance-Voltage Profiling and the Characterisation of III–V Semiconductors Using Electrolyte Barriers," *Semicond. Sci. Technol.* **1**, 7–27, 1986.

[51] T. Ambridge and D. J. Ashen, "Automatic Electrochemical Profiling of Carrier Concentration in Indium Phosphide," *Electron. Lett.* **15**, 647–648, Sept. 1979.

[52] T. Ambridge, J. L. Stevenson, and R. M. Restall, "Application of Electrochemical Methods for Semiconductor Characterization: I. Highly Reproducible Carrier Concentration Profiling of VPE "Hi-Lo" n-GaAs," *J. Electrochem. Soc.* **127**, 222–228, Jan. 1980.

[53] A. C. Seabaugh, W. R. Frensley, R. J. Matyi, and G. E. Cabaniss, "Electrochemical C-V Profiling of Heterojunction Device Structures," *IEEE Trans. Electron Dev.* **ED-36,** 309–313, Feb. 1989.

[54] C. D. Sharpe and P. Lilley, "The Electrolyte-Silicon Interface; Anodic Dissolution and Carrier Concentration Profiling," *J. Electrochem. Soc.* **127**, 1918–1922, Sept. 1980.

[55] W. Y. Leong, R. A. A. Kubiak, and E. H. C. Parker, "Dopant Profiling of Si-MBE Material Using the Electrochemical CV Technique," in *Proc. of the First Int. Symp. on Silicon MBE*, Electrochem. Soc., Pennington, NJ, 1985, pp. 140–148.

[56] M. Pawlik, R. D. Groves, R. A. Kubiak, W. Y. Leong, and E. H. C. Parker, "A Comparative Study of Carrrier Concentration Profiling Techniques in Silicon: Spreading Resistance and Electrochemical CV," in *Emerging Semiconductor Technology* (D. C. Gupta and R. P. Langer, eds.), STP 960, Am. Soc. Test. Mat., Philadelphia, 1987, pp. 558–572.

[57] N. I. Meyer and T. Guldbrandsen, "Method for Measuring Impurity Distributions in Semiconductor Crystals," *Proc. IEEE* **51**, 1631–1636, Nov. 1963.

[58] J. Copeland, "A Technique for Directly Plotting the Inverse Doping Profile of Semiconductor Wafers," *IEEE Trans. Electron. Dev.* **ED-16**, 445–449, May 1969.

[59] R. R. Spiwack, "Design and Construction of a Direct-Plotting Capacitance Inverse-Doping Profiler for Semiconductor Evaluation," *IEEE Trans. Instrum. Meas.* **IM-18**, 197–202, Sept. 1969.

[60] D. C. Gupta and J. Y. Chan, "Direct Measurement of Impurity Distribution in Semiconducting Materials," *J. Appl. Phys.* **43**, 515–522, Feb. 1972.

[61] J. M. Shannon, "DC Measurement of the Space Charge Capacitance and Impurity Profile Beneath the Gate of an MOST," *Solid-State Electron.* **14**, 1099–1106, Nov. 1971.

[62] M. G. Buehler, "Dopant Profiles Determined from Enhancement-Mode MOSFET DC Characteristics," *Appl. Phys. Lett.* **31**, 848–850, Dec. 1977.

[63] M. G. Buehler, "The D-C MOSFET Dopant Profile Method," *J. Electrochem. Soc.* **127**, 701–704, March 1980.

[64] M. G. Buehler, "Effect of the Drain-Source Voltage on Dopant Profiles Obtained from the DC MOSFET Profile Method," *IEEE Trans. Electron Dev.* **ED-27**, 2273–2277, Dec. 1980.

[65] M. Chi and C. Hu, "Errors in Threshold-Voltage Measurements of MOS Transistors for Dopant-Profile Determinations," *Solid-State Electron.* **24**, 313–316, April 1981.

[66] G. P. Carver, "Influence of Short-Channel Effects on Dopant Profiles Obtained from the DC MOSFET Profile Method," *IEEE Trans. Electron Dev.* **ED-30**, 948–954, Aug. 1983.

[67] N. D. Arora, "Semi-Empirical Model for the Threshold Voltage of a Double Implanted MOSFET and Its Temperature Dependence," *Solid-State Electron.* **30**, 559–569, May 1987.

[68] R. A. Haken, "Analysis of the Deep Depletion MOSFET and the Use of DC Characteristics for Determining Bulk-Channel Charge-Coupled Device Parameters," *Solid-State Electron.* **21**, 753–761, May 1978.

[69] D. S. Wu, "Extraction of Average Doping Density and Junction Depth in an Ion-Implanted Deep-Depletion Transistor," *IEEE Trans. Electron Dev.* **ED-27**, 995–997, May 1980.

[70] R. A. Burghard and Y. A. El-Mansy, "Depletion Transistor Threshold Voltage as a Process Monitor," *IEEE Trans. Electron Dev.* **ED-34**, 940–942, April 1987.

[71] H. Jorke and H. J. Herzog, "Carrier Spilling in Spreading Resistance Analysis of Si Layers Grown by Molecular-Beam Epitaxy," *J. Appl. Phys.* **60**, 1735–1739, Sept. 1986.

[72] A. Ambrozy, "A Simple dC/dV Mesurement Method and Its Applications," *Solid-State Electron.* **13**, 347–353, March 1970.

[73] P. J. Baxandall, D. J. Colliver, and A. F. Fray, "An Instrument for the Rapid Determination of Semiconductor Impurity Profiles," *J. Sci. Instrum.* **4**, 213–221, March 1971.

[74] G. L. Miller, "A Feedback Method for Investigating Carrier Distributions in Semiconductors," *IEEE Trans. Electron Dev.* **ED-19**, 1103–1108, Oct. 1972.

[75] ASTM Standard F419, "Standard Test Method for Net Carrier Density in Silicon Epitaxial Layers by Voltage-Capacitance of Gated and Ungated Diodes," *1985 Annual Book of ASTM Standards*, Am. Soc. Test. Mat., Philadelphia, 1985.

[76] H. Maes, W. Vandervorst, and R. Van Overstraeten, "Impurity Profile of Implanted Ions in Silicon," in *Impurity Doping Processes in Silicon* (F. F. Y. Wang, ed.) North-Holland, Amsterdam, 1981, pp. 443–638.

[77] S. M. Sze, *Physics of Semiconductor Devices*, 2nd ed., Wiley, New York, 1981, p. 103.

[78] A. M. Goodman, "Metal-Semiconductor Barrier Height Measurement by the Differential Capacitance Method—One Carrier System," *J. Appl. Phys.* **34**, 329–338, Feb. 1963.

[79] J. D. Wiley and G. L. Miller, "Series Resistance Effects in Semiconductor CV Profiling," *IEEE Trans. Electron Dev.* **ED-22**, 265–272, May 1975.

[80] J. D. Wiley, "*C-V* Profiling of GaAs FET Films," *IEEE Trans. Electron Dev.* **ED-25**, 1317–1324, Nov. 1978.

[81] S. T. Lin and J. Reuter, "The Complete Doping Profile Using MOS *CV* Technique," *Solid-State Electron.* **26**, 343–351, April 1983.

[82] D. K. Schroder and P. Rai Choudhury, "Silicon-on-Saphire with Microsecond Carrier Lifetimes," *Appl. Phys. Lett.* **22**, 455–457, May 1973.

[83] J. R. Brews, "Correcting Interface-State Errors in MOS Doping Profile Determinations," *J. Appl. Phys.* **44**, 3228–3231, July 1973.

[84] Y. Zohta, "Frequency Dependence of $\Delta V/\Delta(C^{-2})$ of MOS Capacitors," *Solid-State Electron.* **17**, 1299–1309, Dec, 1974.

[85] B. L. Smith and E. H. Rhoderick, "Possible Sources of Error in the Deduction of Semiconductor Impurity Concentrations from Schottky-Barrier (C, V) Characteristics," *Brit. J. Appl. Phys.* **D2**, 465–467, March 1969.

[86] J. A. Copeland, "Diode Edge Effect on Doping-Profile Measurements," *IEEE Trans. Electron. Dev.* **ED-17**, 404–407, May 1970.

[87] W. Tantraporn and G. H. Glover, "Extension of the *C-V* Doping Profile Technique to Study the Movements of Alloyed Junction and Substrate Out-Diffusion, the Separation of Junctions, and Device Area Trimming," *IEEE Trans. Electron Dev.* **ED-35**, 525–529, April 1988.

[88] P. Kramer, C. de Vries, and L. J. van Ruyven, "The Influence of Leakage Current on Concentration Profile Measurements," *J. Electrochem. Soc.* **122**, 314–316, Feb. 1975.

[89] G. H. Glover, "Determination of Deep Levels in Semiconductors from *C-V* Measurements," *IEEE Trans. Electron Dev.* **ED-19**, 138–143, Feb. 1972.

[90] L. C. Kimerling, "Influence of Deep Traps on the Measurement of Free-Carrier Distributions in Semiconductors by Junction Capacitance Techniques," *J. Appl. Phys.* **45**, 1839–1845, April 1974.

[91] G. Goto, S. Yanagisawa, O. Wada, and H. Takanashi, "An Improved Method of Determining Deep Impurity Levels and Profiles in Semiconductors," *Japan. J. Appl. Phys.* **13**, 1127–1133, July 1974.

[92] B. L. Smith, "The Majority Carrier Profile in Thin Epitaxial Layers of GaAs from Schottky Barrier (*C.V.*) Characteristics," *Brit. J. Appl. Phys.* **D3**, 1179–1182, Aug. 1970.

[93] K. Lehovec, "*C-V* Analysis of a Partially Depleted Semiconducting Channel," *Appl. Phys. Lett.* **26**, 82–84, Feb. 1975.

[94] I. Amron, "Errors in Dopant Concentration Profiles Determined by Differential Capacitance Measurements," *Electrochem. Technol.* **5**, 94–97, March/April 1967.

[95] R. A. Smith, *Semiconductors*, Cambridge University Press, Cambridge, 1959, Ch. 5.

[96] A. C. Beer, *Galvanomagnetic Effects in Semiconductors*, Academic Press, New York, 1963, p. 308.

[97] D. L. Rode, C. M. Wolfe, and G. E. Stillman, "Magnetic-Field Dependence of the Hall Factor of Gallium Arsenide," in *GaAs and Related Compounds* (G. E. Stillman, ed.) Conf. Ser. No. 65, Inst. Phys., Bristol, 1983, pp. 569–572.

[98] E. H. Putley, *The Hall Effect and Related Phenomena*, Butterworths, London, 1960, p. 106.

[99] G. E. Stillman and C. M. Wolfe, "Electrical Characterization of Epitaxial Layers," *Thin Solid Films* **31**, 69–88, Jan. 1976.

[100] T. T. Braggins, H. M. Hobgood, J. C. Swartz and R. N. Thomas, "High

Infrared Responsivity Indium-Doped Silicon Detector Material Compensated by Neutron Transmutation," *IEEE Trans. Electron Dev.* **ED-27**, 2–10, Jan. 1980.

[101] A. Chandra, C. E. C. Wood, D. W. Woodard, and L. F. Eastman, "Surface and Interface Depletion Corrections to Free Carrier-Density Determinations by Hall Measurements," *Solid-State Electron.* **22**, 645–650, July 1979.

[102] T. R. Lepkowski, R. Y. DeJule, N. C. Tien, M. H. Kim, and G. E. Stillman, "Depletion Corrections in Variable Temperature Hall Measurements," *J. Appl. Phys.* **61**, 4808–4811, May 1987.

[103] R. Baron, G. A. Shifrin, O. J. Marsh, and J. W. Mayer, "Electrical Behavior of Group III and V Implanted Dopants in Silicon," *J. Appl. Phys.* **40**, 3702–3719, Aug. 1969.

[104] N. D. Young and M. J. Hight, "Automated Hall Effect Profiler for Electrical Characterisation of Semiconductors," *Electron. Lett.* **21**, 1044–1046, Oct. 1985.

[105] H. Müller, F. H. Eisen, and J. W. Mayer, "Anodic Oxidation of GaAs as a Technique to Evaluate Electrical Carrier Concentration Profiles," *J. Electrochem. Soc.* **122**, 651–655, May 1975.

[106] R. D. Larrabee and W. R. Thurber, "Theory and Application of a Two-Layer Hall Technique," *IEEE Trans. Electron Dev.* **ED-27**, 32–36, Jan. 1980.

[107] L. F. Lou and W. H. Frye, "Hall Effect and Resistivity in Liquid-Phase-Epitaxial Layers of HgCdTe," *J. Appl. Phys.* **56**, 2253–2267, Oct. 1984.

[108] R. L. Petritz, "Theory of an Experiment for Measuring the Mobility and Density of Carriers in the Space-Charge Region of a Semiconductor Surface," *Phys. Rev.* **110**, 1254–1262, June 1958.

[109] T. S. Moss, G. J. Burrell, and B. Ellis, *Semiconductor Opto-Electronics*, Wiley, New York, 1973, pp. 42–46.

[110] P. A. Schumann, Jr., "Current Problems in the Electrical Characterization of Semiconductor Materials," in *Semiconductor Silicon/1969* (R. Haberecht and E. L. Kern, eds.), Electrochem. Soc., New York, 1969, pp. 662–692.

[111] P. A. Schumann, Jr., "Plasma Resonance Calibration Curves for Silicon, Germanium and Gallium Arsenide," *Solid State Technol.* **13**, 50–52, Jan. 1970.

[112] J. W. Philbrick, C. A. Pillus, and C. P. Schneider, "Plasma Resonance on *P*-Type Gallium Arsenide," *Solid State Technol.* **16**, 66–68, April 1973.

[113] ASTM Standard F398-77, "Standard Method for Majority Carrier Concentration in Semiconductors by Measurement of Wavelength of the Plasma Resonance Minimum," 1985 *Annual Book of ASTM Standards*, Am. Soc. Test. Mat., Philadelphia, 1985.

[114] T. Abe and Y. Nishi, "Non-Destructive Measurement of Surface Concentrations and Junction Depths of Diffused Semiconductor Layers," *Japan. J. Appl. Phys.* **7**, 397–403, April 1968.

[115] A. H. Tong, P. A. Schumann, Jr., and W. A. Keenan, "Epitaxial Substrate Carrier Concentration Measurement by the Infrared Interference Envelope (IRIE) Technique," *J. Electrochem. Soc.* **119**, 1381–1384, Oct. 1972.

[116] E. Burstein, G. Picus, B. Henvis, and R. Wallis, "Absorption Spectra of Impurities in Silicon; I. Group III Acceptors," *J. Phys. Chem Solids* **1**, 65–74, Sept./Oct. 1956.

[117] G. Picus, E. Burstein, and B. Henvis, "Absorption Spectra of Impurities in Silicon; II. Group V Donors," *J. Phys. Chem Solids* **1**, 75–81, Sept./Oct. 1956.

[118] H. J. Hrostowski and R. H. Kaiser, "Infrared Spectra of Group III Acceptors in Silicon," *J. Phys. Chem Solids* **4**, 148–153, 1958.

[119] S. C. Baber, "Net and Total Shallow Impurity Analysis of Silicon by Low Temperature Fourier Transform Infrared Spectroscopy," *Thin Solid Films* **72**, 201–210, Sept. 1980.

[120] T. S. Low, M. H. Kim, B. Lee, B. J. Skromme, T. R. Lepkowski, and G. E. Stillman, "Neutron Transmutation Doping of High Purity GaAs," *J. Electron. Mat.* **14**, 477–511, Sept. 1985.

[121] J. J. White, "Effects of External and Internal Electric Fields on the Boron Acceptor States in Silicon," *Can. J. Phys.* **45**, 2695–2718, Aug. 1967; "Absorption-Line Broadening in Boron-Doped Silicon," *Can. J. Phys.* **45**, 2797–2804, Aug. 1967.

[122] B. O. Kolbesen, "Simultaneous Determination of the Total Content of Boron and Phosphorus in High-Resistivity Silicon by IR Spectroscopy at Low Temperatures," *Appl. Phys. Lett.* **27**, 353–355, Sept. 1975.

[123] G. E. Stillman, C. M. Wolfe, and J. O. Dimmock, "Far Infrared Photoconductivity in High Purity GaAs," in *Semiconductors and Semimetals*, Vol. 12 (R. K. Willardson and A. C. Beer, eds.) Academic Press, New York, 1977, pp. 169–290.

[124] M. J. H. van de Steeg, H. W. H. M. Jongbloets, J. W. Gerritsen, and P. Wyder, "Far Infrared Photothermal Ionization Spectroscopy of Semiconductors in the Presence of Intrinsic Light," *J. Appl. Phys.* **54**, 3464–3474, June 1983.

[125] E. E. Haller, "Semiconductor Physics in Ultra-Pure Germanium," in *Festkörperprobleme*, Vol. 26 (P. Grosse, ed.), Vieweg, Braunschweig, 1986, pp. 203–229.

[126] S. M. Kogan and T. M. Lifshits, "Photoelectric Spectroscopy–A New Method of Analysis of Impurities in Semiconductors," *Phys. Stat. Sol. (a)* **39**, 11–39, Jan. 1977.

[127] K. K. Smith, "Photoluminescence of Semiconductor Materials," *Thin Solid Films* **84**, 171–182, Oct. 1981.

[128] M. Tajima, "Determination of Boron and Phosphorus Concentration in Silicon by Photoluminescence Analysis," *Appl. Phys. Lett.* **32**, 719–721, June 1978.

[129] M. Tajima, T. Masui, T. Abe, and T. Iizuka, "Photoluminescence Analysis of Silicon Crystals," in *Semiconductor Silicon* 1981 (H. R. Huff, R. J. Kriegler, and Y. Takeishi, eds.) Electrochem. Soc., Pennington, NJ, 1981, pp. 72–89.

[130] G. Pickering, P. R. Tapster, P. J. Dean, and D. J. Ashen, "Determination of

Impurity Concentration in *n*-Type InP by a Photoluminescence Technique," in *GaAs and Related Compounds* (G. E. Stillman, ed.) Conf. Ser. No. 65, Inst. Phys., Bristol, 1983, pp. 469–476.

[131] P. A. Schumann, Jr., W. A. Keenan, A. H. Tong, H. H. Geganwarth, and C. P. Schneider, "Silicon Optical Constants in the Infrared," *J. Electrochem. Soc.* **118**, 145–148, Jan. 1971, and references therein.

[132] D. K. Schroder, R. N. Thomas, and J. C. Swartz, "Free Carrier Absorption in Silicon," *IEEE Trans. Electron Dev.* **ED-25**, 254–261, Feb. 1978.

[133] J. L. Boone, M. D. Shaw, G. Cantwell, and W. C. Harsh, "Free Carrier Density Profiling by Scanning Infrared Absorption," *Rev. Sci. Instrum.* **59**, 591–595, April 1988.

[134] J. R. Golin and J. A. Glaze, "High Resolution Dose Uniformity Monitoring of Ion Implanters," *Solid State Technol.* **27**, 137–141, Aug. 1984; **27**, 289–295, Sept. 1984.

[135] A. Rosencwaig, "Thermal-wave Imaging," *Science* **218**, 223–228, Oct. 1982.

[136] A. Rosencwaig, J. Opsal, W. L. Smith, and D. L. Willenborg, "Detection of Thermal Waves Through Optical Reflectance," *Appl. Phys. Lett.* **46**, 1013–1015, June 1985.

[137] W. L. Smith, A. Rosencwaig, and D. L. Willenborg, "Ion Implant Monitoring with Thermal Wave Technology," *Appl. Phys. Lett.* **47**, 584–586, Sept. 1985; W. L. Smith, A. Rosencwaig, D. L. Willenborg, J. Opsal, and M. W. Taylor, "Ion Implant Monitoring with Thermal Wave Technology," *Solid State Technol.* **29**, 85–92, Jan. 1986.

[138] M. Pawlik, "Dopant Profiling in Silicon," in *Semiconductor Processing*, ASTM STP 850 (D. C. Gupta, ed.), Am. Soc. Test. Mat., Philadelphia, PA, 1984, pp. 391–408.

[139] J. R. Ehrstein, R. G. Downing, B. R. Stallard, D. S. Simons, and R. F. Fleming, "Comparison of Depth Profiling ^{10}B in Silicon Using Spreading Resistance Profiling, Secondary Ion Mass Spectrometry, and Neutron Depth Profiling," in *Semiconductor Processing*, ASTM STP 850 (D. C. Gupta, ed.) Am. Soc. Test. Mat., Philadelphia, PA, 1984, pp. 409–425.

[140] J. A. Albers, P. Roitman, and C. L. Wilson, "Verification of Models for Fabrication of Arsenic Source-Drains in VLSI MOSFET's," *IEEE Trans. Electron Dev.* **ED-30**, 1453–1462, Nov. 1983.

[141] D. J. Godfrey, R. D. Groves, M. G. Dowsett, and A. F. W. Willoughby, "A Comparison Between SIMS and Spreading Resistance Profiles for Ion Implanted Arsenic and Boron After Heat Treatments in an Inert Ambient," *Physica B&C* **129**, 181–186, March 1985.

[142] G. G. Sweeney and T. R. Alvarez, "Comparison of Impurity Profiles Generated by Spreading Resistance Probe and Secondary Ion Mass Spectrometry," in *Emerging Semiconductor Technology*, ASTM STP 960 (D. C. Gupta and P. H. Langer, ed.), Am. Soc. Test. Mat. Philadelphia, PA, 1987, pp. 521–534.

[143] G. W. Banke, Jr., K. Varahramyan, and G. J. Slusser, "Analysis of Boron Profiles as Determined by Secondary Ion Mass Spectrometry, Spreading

Resistance, and Process Modeling," in *Emerging Semiconductor Technology*, ASTM STP 960 (D. C. Gupta and P. H. Langer, eds.), Am. Soc. Test. Mat., Philadelphia, PA, 1987, pp. 573–585.

[144] S. Clayton, L. Springer, B. Offord, T. Sedgwick, R. Reedy, A. Michel, and G. Scilla, "Formation and Analysis of Shallow Arsenic Profiles," *Electron. Lett.* **24**, 831–833, July 1988.

[145] S M. Hu, "Between Carrier Distributions and Dopant Atomic Distribution in Beveled Silicon Substrates," *J. Appl. Phys.* **53**, 1499–1510, March 1982.

[146] C. L. Jones, M. J. T. Quelch, P. Capper, and J. J. Gosney, "Effects of Annealing on the Electrical Properties of $Cd_x Hg_{1-x} Te$," *J. Appl. Phys.* **53**, 9080–9092, Dec. 1982.

CHAPTER 3

CONTACT RESISTANCE AND SCHOTTKY BARRIER HEIGHT

3.1 INTRODUCTION

All semiconductor devices have contacts, all contacts exhibit contact resistance, and it is important to measure the contact resistance. Contacts consist generally of metal-semiconductor contacts, but they may be semiconductor-semiconductor contacts, where both semiconductors can be single crystal, polycrystalline, or amorphous. In the conceptual discussion of ohmic contacts and contact resistance we will be mainly concerned with the metal-semiconductor contact because it is the most common one. For the discussion of the measurement techniques the type of contact is unimportant, but the resistance of the contact material is important.

The metal-semiconductor contact, discovered by Braun in 1874, forms the basis of one of the oldest semiconductor devices.[1] The device was used for a long time without a thorough understanding. The first acceptable theory was developed by Schottky in the 1930s.[2] In his honor metal-semiconductor devices are frequently referred to as *Schottky barrier devices*. Usually this name denotes the use of these devices as rectifiers with distinctly nonlinear current-voltage characteristics. The use of metal-semiconductor devices as *ohmic contacts* is the subject of this chapter.

Ohmic contacts have linear or quasi-linear current-voltage characteristics. It is not necessary that ohmic contacts have linear *I-V* characteristics. The contacts must be able to supply the necessary device current, and the voltage drop across the contact should be small compared to the voltage drops across the active device regions. An ohmic contact should not degrade the device to any significant extent, and it should not inject minority carriers. In addition one should be able to make such contacts in a reproducible manner.

The first comprehensive publication on ohmic contacts was the result of a conference devoted to this topic.[3] The theory of metal-semiconductor contacts with emphasis on ohmic contacts was presented by Rideout.[4] Ohmic contacts to III–V devices were reviewed by Yoder,[5] Braslau,[6] and Piotrowska et al.,[7] and ohmic contacts to solar cells were discussed by Schroder and Meier.[8] Yu and Cohen have presented discussions of contact resistance.[9–10] Additional information can be found in the books by Milnes and Feucht,[11] Sharma and Purohit,[12] Rhoderick,[13] and Cohen and Gildenblat.[14]

3.2 METAL-SEMICONDUCTOR CONTACTS

The Schottky model of the metal-semiconductor barrier is shown in Fig. 3.1. The energy bands are shown before contact in the upper part of the figure and after contact in the lower part. We assume intimate contact between the metal and the semiconductor with no interfacial layer. The metal work

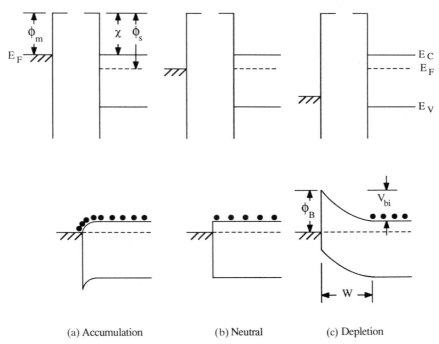

(a) Accumulation   (b) Neutral   (c) Depletion

Fig. 3.1 Metal-semiconductor contacts according to the simple Schottky model. The upper and lower parts of the figure show the metal-semiconductor system before and after contact, respectively.

function ϕ_M less than the semiconductor work function ϕ_S is shown in Fig. 3.1(a). In Fig. 3.1(b) $\phi_M = \phi_S$, and in Fig. 3.1(c) $\phi_M > \phi_S$. The work function is the potential difference between the Fermi level and the vacuum level. The barrier height after contact for this model is given by

$$\phi_B = \phi_M - \chi \qquad (3.1)$$

χ is the electron affinity of the semiconductor, defined as the potential difference between the bottom of the conduction band and the vacuum level. The barrier height is independent of the semiconductor doping concentration. According to the Schottky theory the barrier height depends only on the metal work function and on the semiconductor electron affinity. This should make it easy to vary the barrier height by merely using metals of the appropriate work function to implement any one of the three barrier types of Fig. 3.1. We have for convenience named them *accumulation*, *neutral*, and *depletion contacts* because the majority carriers are accumulated, unchanged (neutral), or depleted compared to their concentration in the quasi-neutral substrate.

As is evident from Fig. 3.1 an accumulation-type contact is the preferred ohmic contact because electrons in the metal and in the semiconductor encounter the least barrier to their flow into or out of the semiconductor. In practice it is very difficult to alter the barrier height by using metals of varying work functions. It is experimentally observed that the barrier height for the common semiconductors Ge, Si, GaAs, and other III–V materials is relatively independent of the work function of the metal.[15–17] A *depletion* contact is generally formed on both n-type and p-type substrates, as shown in Fig. 3.2. For n-type substrates $\phi_B \approx 2E_g/3$ and for p-substrates $\phi_B \approx E_g/3$.[18] The relative constancy of the barrier height with work function of metals is sometimes called *Fermi level pinning*, referring to the fact that the Fermi level in the semiconductor is pinned at some energy in the band gap to create a depletion-type contact. The details of Schottky barrier formation are not yet fully understood. It appears, however, that defects at the semiconductor surface play an important role during contact formation. Bardeen was the first to point out the importance of surface states in determining the barrier height.[19] Such surface states may be dangling bonds at the surface or some other types of defects.[15–17,20] There is, however, still disagreement between the various proposed mechanisms causing Fermi level pinning.[21–23]

Whatever the mechanisms that cause barrier heights to be relatively independent of the metal work function, they make it extremely difficult to engineer an accumulation-type contact. With barrier height engineering being impractical, we must look to other means of implementing ohmic contacts. Ohmic contacts are frequently defined as regions of high recombination rates. This implies that highly damaged regions should serve as good ohmic contacts. Such fabrication methods are not practical because damage

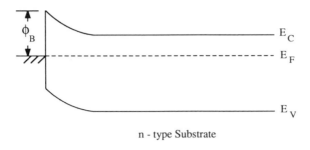

n - type Substrate

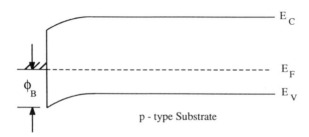

p - type Substrate

Fig. 3.2   Depletion-type contacts on *n*- and *p*-substrates.

is usually the last thing one wants in a semiconductor device. Damage-induced ohmic contacts are also not very reproducible leaving the semiconductor doping concentration as the only alternative.[24] As stated earlier, the *barrier height* is independent of the doping concentration, but the *barrier width* does depend on the doping concentration. The barrier height does actually depend weakly on doping concentration through image force barrier lowering.[13] We neglect that effect here.

With the space-charge region (scr) width W being proportional to $N_D^{-1/2}$, it is obvious that highly doped semiconductors have narrow scr widths. For metal-semiconductor contacts with narrow scr widths, electrons can tunnel from the metal to the semiconductor and from the semiconductor to the metal. For *p*-type semiconductors holes tunnel. Some readers may be uncomfortable with the concept of holes tunneling from a metal to a semiconductor. It may be helpful to think of hole tunneling from the metal to the semiconductor as electron tunneling from the semiconductor valence band to the metal.

The conduction mechanisms for a metal-n-type semiconductor are illustrated in Fig. 3.3. For lowly-doped semiconductors ($N_D \leq 10^{17}$ cm^{-3}) the current flows as a result of *thermionic emission* (TE) as shown in Fig. 3.3(a)

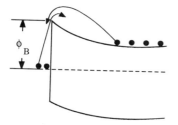

(a) Low N_D - Thermionic Emission

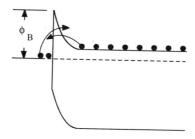

(b) Intermediate N_D - Thermionic/Field Emission

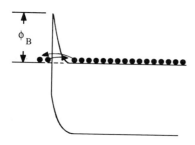

(c) High N_D - Field Emission

Fig. 3.3 Depletion-type contacts to n-type substrates with increasing doping concentrations. The electron flow is schematically indicated by the electrons and their arrows.

with electrons thermally excited over the barrier.[25] In the intermediate doping range $(10^{17} < N_D < 10^{19}$ cm$^{-3})$ *thermionic-field emission* (TFE) dominates.[26–27] The carriers are thermally excited to an energy where the barrier is sufficiently narrow for tunneling to take place. For $N_D > 10^{19}$ cm^{-3} the barrier is sufficiently narrow at or near the bottom of the conduction band for the electrons to tunnel directly, known as *field emission* (FE).[4,26]

The simple structure of Fig. 3.3(c) is not realized in most real contacts. Generally only the semiconductor directly under the contact is heavily doped; the region farther from the contact being more lightly doped as illustrated in Fig. 3.4. The contact resistance becomes the sum of the

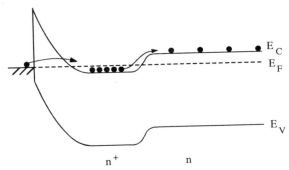

Fig. 3.4 A metal-n^+-n semiconductor contact band diagram.

metal-semiconductor contact resistance and the n^+n junction resistance. Such a structure has a contact resistance similar to a uniformly doped structure if the metal-semiconductor junction dominates.[28] However, the contact resistance dependence on doping concentration is expected to be different when the n^+n junction dominates over the metal-semiconductor junction. The inverse dependence of contact resistance on doping concentration has been attributed to the resistance of the high-low junction.[29-30]

3.3 CONTACT RESISTANCE

Metal-semiconductor contacts fall into two basic categories illustrated by the bipolar transistor structure in Fig. 3.5. The emitter current flows nearly vertically into the emitter contact. The base current flows decidedly horizon-

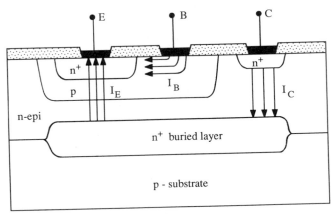

Fig. 3.5 A bipolar transistor with the arrows indicating the main directions of current flow. Both vertical and horizontal or lateral contacts are shown.

tally or laterally, and the collector current flows vertically from the contact to the buried layer. This device incorporates both *vertical* as well as *horizontal* current flow into and out of the contacts. Vertical and horizontal or lateral contacts can behave quite differently.

Let us consider the resistance between points A and B of a sample having metallic conductors lying on an insulator and making ohmic contacts to an *n*-type layer diffused into a *p*-type substrate as shown in Fig. 3.6. We divide the total resistance R_T between points A and B into three components: (1) the resistance of the metallic conductor R_m, (2) the contact resistances R_c, and (3) the semiconductor resistance R_s. The total resistance is

$$R_T = 2R_m + 2R_c + R_s \qquad (3.2)$$

The metal resistance is the resistance of the conductor and the semiconductor resistance is determined by the sheet resistance of the diffused layer. The contact resistance is less clearly defined. It certainly includes the resistance of the metal-semiconductor contact, sometimes called the *specific interfacial resistance* ρ_i.[10] But it also includes a portion of the metal immediately above the metal-semiconductor interface, a part of the semiconductor below that interface, current crowding effects, spreading resistance under the contact, and any interfacial oxide that may be present between the metal and the semiconductor.

The current density J of a metal-semiconductor contact depends on the applied voltage V and the barrier height ϕ_B in a manner that varies for each of the three conduction mechanisms in Section 3.2. We write that dependence as

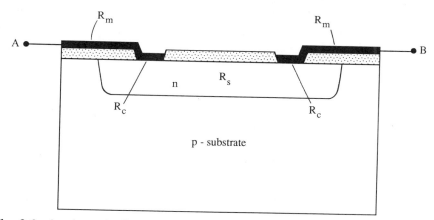

Fig. 3.6 A schematic diagram showing two contacts to a diffused semiconductor layer, with the metal resistance, the contact resistances, and the semiconductor resistance indicated.

$$J = f(V, \phi_{\mathrm{B}}) \tag{3.3}$$

The contact resistance is characterized by two quantities: the *contact resistance* (ohms) and the *specific contact resistance*, ρ_c (ohm-cm^2), sometimes referred to as *contact resistivity* or *specific contact resistivity*. The specific contact resistance includes not only the actual interface but the regions immediately above and below the interface.

We define a *specific interfacial resistance* ρ_i(ohm-cm^2) by

$$\rho_i = \left. \frac{\partial V}{\partial J} \right|_{V=0} \tag{3.4}$$

This specific interfacial resistance is a theoretical quantity referring to the metal-semiconductor interface only. It is not actually measurable because of the effects referred to above. The parameter that is calculated from measured contact resistance is the specific contact resistance. It is a very useful term for ohmic contacts because it is independent of contact area and is a convenient parameter when comparing contacts of various sizes. We will use ρ_i only when deriving theoretical expressions of metal-semiconductor contacts. Thereafter we use ρ_c when we discuss real contacts, their measurements, and measurement interpretations.

The current density of a metal-semiconductor contact for which thermionic emission is dominant is given in its simplest form by[13]

$$J = A^{**}T^2 e^{-q\phi_{\mathrm{B}}/kT}(e^{qV/kT} - 1) \tag{3.5}$$

where $A^{**} = 4\pi q k^2 m^*/h^3 = 120(m^*/m)$ A/cm^{2}K^2 is Richardson's constant, m is the electron mass, m^* is the effective electron mass, and T is the temperature. With Eq. (3.4) we find the specific interfacial resistance for *thermionic emission* to be

$$\rho_i(\mathrm{TE}) = \rho_1 e^{q\phi_{\mathrm{B}}/kT} \tag{3.6}$$

where

$$\rho_1 = \frac{k}{qA^{**}T} \tag{3.7}$$

For *thermionic-field emission* ρ_i is given by[9]

$$\rho_i(\mathrm{TFE}) = C_1 \rho_1 e^{q\phi_{\mathrm{B}}/E_0} \tag{3.8}$$

and for *field emission* it is[9]

$$\rho_i(\mathrm{FE}) = C_2 \rho_1 e^{q\phi_{\mathrm{B}}/E_{00}} \tag{3.9}$$

C_1 and C_2 are functions of N_D, T, and ϕ_B. The energy E_{00} is a characteristic energy that characterizes the tunneling process. It is given by[26]

$$E_{00} = \frac{qh}{4\pi} \sqrt{\frac{N_D}{K_s \varepsilon_0 m^*}}$$

The effective mass m^* is the tunneling effective mass in this expression.[31] E_0 in Eq. (3.8) is related to E_{00} by[26]

$$E_0 = E_{00} \coth(E_{00}/kT)$$

Substituting for E_{00} in Eq. (3.9) leads to

$$\rho_i(FE) = C_2 \rho_1 \exp(C_3/\sqrt{N_D}) \tag{3.10}$$

where C_3 is another constant. Taking the logarithm of Eq. (3.10) gives the dependence of the specific interfacial resistance as

$$\log[\rho_i(FE)] \sim 1/\sqrt{N_D} \tag{3.11}$$

$\rho_i(FE)$ is very sensitive to the doping concentration under the contact. N_D should be as high as possible for lowest interfacial resistance.

The ratio kT/E_{00} is a measure of the ratio of thermionic emission current to field emission or tunnel current. For lightly doped semiconductors, $kT/E_{00} \gg 1$ and thermionic emission dominates. For heavily doped semiconductors, $kT/E_{00} \ll 1$ and tunneling is dominant. For thermionic-field emission, $kT/E_{00} \approx 1$.

We have given the specific interfacial resistance by these simple expressions in order not to obscure the main points in this discussion. More complex relations have been derived.[32–34] The detailed expressions for the various conduction mechanisms are rather complicated and a calculation of the specific interfacial resistance for each of the three regions is difficult. Various approximations have been proposed and theoretical curves of ρ_i versus N_A or N_D have been generated.[32–34] These curves depend of course on the effective masses, the barrier height, and various other parameters. The barrier height depends also on the contact metal, and it is therefore impossible to derive "universal" ρ_i versus N_A or ρ_i versus N_D curves.

We show in Figs. 3.7 and 3.8 experimental ρ_c versus N_D and N_A data for Si and GaAs. There is considerable scatter, but a definite trend of lower specific contact resistance with higher doping concentrations, predicted by Eq. (3.11), is obvious in the data. The Si data show a tighter distribution because for Si the surface is generally heavily doped *before* the metal is deposited, and the doping concentration can be accurately measured. For GaAs there is considerably more scatter probably because the doping concentration under the contact is not well known. Most GaAs ohmic

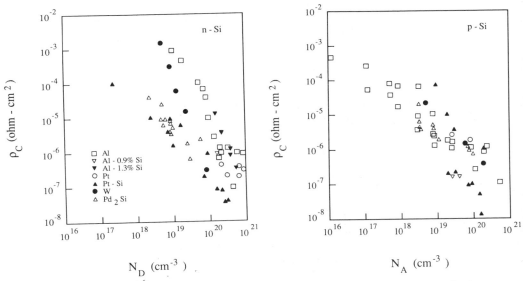

Fig. 3.7 The specific contact resistance as a function of donor and acceptor doping concentrations for Si. The references for the data points are given for *n*-Si in [36] and for *p*-Si in [37].

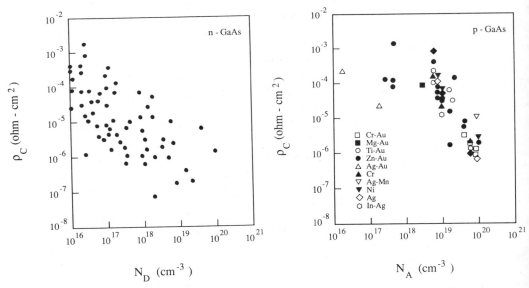

Fig. 3.8 The specific contact resistance as a function of donor and acceptor doping concentrations for GaAs. The references for the data points are given for *n*-GaAs in [38] and for *p*-GaAs in [39].

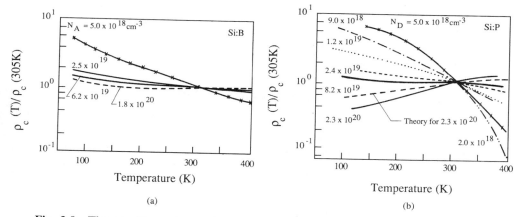

Fig. 3.9 The specific contact resistance, normalized to $T = 305$ K, as a function of temperature for (a) p-Si and (b) n-Si. The data for $N_D = 2 \times 10^{18}$ cm^{-3} extend from $T = 305$ to 400 K only. The metal is tungsten. Reprinted after Swirhun and Swanson[40] by permission of IEEE (© 1986, IEEE).

contacts are made by alloying. A popular technique is the deposition of a AuGe-containing metal. When alloyed with the GaAs, the Ge occupies Ga sites, converting a thin layer near the surface to n^+-GaAs. However, the doping concentration in this thin layer, which determines the contact resistance, is not well known and is difficult to measure.

The temperature dependence of the specific contact resistance for tungsten contacts to n-Si and p-Si, normalized at a temperature of 305 K, is shown in Fig. 3.9.[40] These figures show that there is not a simple ρ_c-T relationship. The temperature behavior of ρ_c is very much dependent on the doping concentration. For surface concentrations around 10^{20} cm^{-3}, there is almost no temperature dependence whereas for surface concentrations above and below that value, there are significant variations of ρ_c with temperature.

3.4 MEASUREMENT TECHNIQUES

Contact resistance measurement techniques fall into four main categories: two-terminal, three-terminal, four-terminal, and six-terminal methods. None of these methods is capable of determining the specific interfacial resistance. Instead, they determine the specific contact resistance that is not the resistance of the metal-semiconductor interface alone but is a practical quantity describing the real contact. It is therefore difficult to compare theory with experiment because theory cannot predict ρ_c accurately and experiment cannot determine ρ_i accurately. As we will see in this chapter, it is frequently even difficult to measure ρ_c unambiguously. We limit ourselves to discussions of measurement techniques. Contact formation and the impact of contact resistance on device behavior can be found in numerous references of which [7, 12, 14, and 41] are a few.

3.4.1 Two-Terminal Contact Resistance Methods

The two-terminal contact resistance measurement method is the simplest and earliest method.[42] It is also of questionable accuracy. The simplest implementation is shown in Fig. 3.10(a). Originally large contacts of ~ 1 cm^2 area were used. For a homogeneous semiconductor of resistivity ρ with two identical contacts of area A_c, the total resistance $R_T = V/I$ is measured by passing a current I through the sample and measuring the voltage V across the two contacts. The total resistance is given by

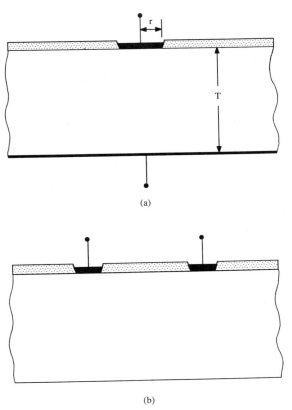

(a)

(b)

Fig. 3.10 (a) A vertical two-terminal contact resistance structure, (b) a lateral two-terminal contact resistance structure.

$$R_T = 2R_c + \rho T/A_c \qquad (3.12)$$

The contact resistance R_c is obtained as

$$R_c = \frac{R_T - \rho T/A_c}{2} \qquad (3.13)$$

For uniform current flow through area A_c, the specific contact resistance can be calculated from R_c through the relationship

$$\rho_c = R_c A_c \qquad (3.14)$$

A more refined analysis was introduced by Cox and Strack in 1967.[43] They employed a small top contact and a large bottom contact, giving the total resistance as

$$R_T = R_{sp} + R_c + R_0 \qquad (3.15)$$

where R_{sp} is the spreading resistance under the top contact, R_c is the contact resistance of the top contact to be determined, and R_0 accounts for the resistance of the bottom contact. The bottom contact usually has a very large contact area with a concomitant small resistance. Consequently R_0 is frequently neglected.

The spreading resistance accounts for the bulk sample resistance and is given by[43]

$$R_{sp} = \frac{\rho}{2\pi r} \arctan\left(\frac{2T}{r}\right) \qquad (3.16)$$

where r is the circular top contact radius and T is the sample thickness. Equation (3.16) is a reasonable approximation to the resistance of a semiconductor slab with a contact of radius r on the top and a large contact on the bottom. More exact expressions for the spreading resistance are given in Brookes and Mathes.[44] With the current flowing vertically into the top contact the contact resistance can be written as

$$R_c = \frac{\rho_c}{A_c} = \frac{\rho_c}{\pi r^2} \qquad (3.17)$$

For small R_0 it is obvious from Eq. (3.15) that the contact resistance is the difference between the total resistance and the spreading resistance—it is the difference of two large numbers. The spreading resistance cannot be measured and it is difficult to calculate accurately. But small errors in R_{sp} can lead to large errors in R_c. Hence the two-terminal method works best when $R_{sp} < R_c$. This can be approximated by using small-area contacts. For small r we find the requirement of $R_{sp} < R_c$ leading to

$$r < \frac{4\rho_c}{\pi\rho} \qquad (3.18)$$

The vertical two-terminal method has been used to measure the contact resistance of Si and GaAs.[43,45–50] Some authors claim questionable accuracy even with contacts of diameter as small as 5 μm.[28] The specific contact resistance is difficult to extract due to the difficulty of calculating the bulk resistance contribution accurately. Sometimes negative ρ_c values are obtained.

A variation of this basic theme is the use of top contacts of varying diameters. Then one plots R_c, calculated from Eq. (3.15) using experimental R_T data, as a function of $1/A_c$ and determines ρ_c from the slope of this linear plot.[45] Alternately, the total resistance can be plotted against $1/r$, and Eq. (3.15) is fitted to this curve.[50] By using various diameters, one can see from the shape of the curve whether there are any anomalies in the data.

The two-terminal method has also been implemented on lateral structures as shown in Fig. 3.10(b). The analysis is more difficult due to lateral current flow, current crowding at the contact, and sample geometry.[51] For the geometry of Fig. 3.11 the total resistance is given by

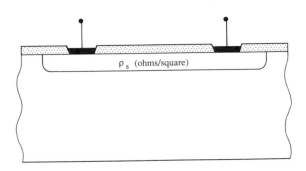

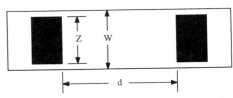

Fig. 3.11 A lateral two-terminal contact resistance structure in cross section and top view.

$$R_T = \rho_s d/W + R_d + R_w + 2R_c \tag{3.19}$$

where R_d is the resistance due to current crowding under the contact, and R_w is a contact width correction if $Z < W$. Expressions for these resistances are given in Ting and Chen.[51] A recent version of the two-terminal method uses two small (5 μm diameter) top contacts.[52] High accuracy is achieved by carefully calculating the spreading resistance under the contacts and calculating the specific contact resistance from the measured resistance.

The *contact chain* is sometimes used for contact resistance measurements. It incorporates many contacts of the type shown in Fig. 3.11 in series as illustrated in Fig. 3.12(a). The total resistance is the sum of the semiconductor resistance and the contact resistance. The semiconductor resistance is calculated from the sheet resistance and the string geometry. By subtracting the semiconductor resistance from the total resistance, one obtains the total contact resistance. The contact resistance for each contact is obtained by

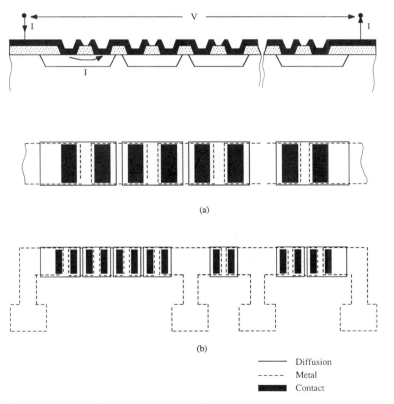

(a)

(b)

——— Diffusion
----- Metal
▉▉ Contact

Fig. 3.12 A contact string test structure. (a) Cross section and top view, and (b) a contact string laid out in three separate sections as discussed in the text.

dividing by the number of contacts. A refined contact string divides the string into sections with intermediate contact pads, as shown in Fig. 3.12(b). By measuring between any two pads, it is possible to vary the number of contacts between 2 and 14, with 2, 4, 6, 8, 10, or 14 contacts being accessible.[53]

The contact string technique is considered to be a coarse measurement method that is useful for a first order contact resistance determination but not for detailed evaluations of contact resistance. It is used extensively as a process monitor in which resistances are compared from day to day and where detailed values are not required. If the measured resistance is higher than the norm, it is difficult to know whether all contacts are poor, or whether one particular contact is poor, unless intermediate contact pads are provided. Frequently they are not provided.

3.4.2 Three-Terminal Contact Resistance Methods

The three-terminal contact resistance measurement technique was developed to overcome the deficiencies of the two-terminal method. A simple three-terminal structure is shown in Fig. 3.13. Three identical contacts are made to the diffused or ion-implanted sheet with contact spacings d_1 and d_2. Assuming identical contact resistances for all three contacts allows the total resistance to be written as

$$R_{Ti} = \rho_s d_i / Z + 2R_c \qquad (3.20)$$

where $i = 1$ or 2. Solving for R_c gives

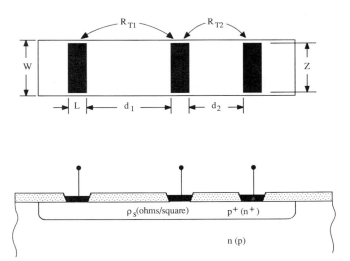

Fig. 3.13 A three-terminal contact resistance test structure.

$$R_c = \frac{(R_{T2}d_1 - R_{T1}d_2)}{2(d_1 - d_2)}$$

(3.21)

This structure does not have the ambiguities of the two-terminal structures because neither the bulk resistance nor the layer sheet resistance needs be known. The assumption of identical contact resistance for all three contacts is somewhat questionable but reasonable for a sample that is not too large. A difficulty with this method is the fact that the contact resistance is obtained by taking the difference of two large numbers. This is especially troublesome for contacts with low contact resistances. The lengths d_1 and d_2 are a further source of inaccuracy. Occasionally negative contact resistances are obtained by this method.

The structure of Fig. 3.13 only allows the contact resistance to be determined. The specific contact resistance cannot be directly extracted from the two resistance measurements. To find ρ_c one has to make a more detailed evaluation of the nature of the current flow into and out the lateral contacts. An early two-dimensional current flow analysis by Kennedy and Murley in diffused semiconductor resistors revealed current crowding at the contacts.[54] The analysis, based on zero contact resistance, showed that only a fraction of the total contact length was active during the transfer of current from the metal to the semiconductor and from the semiconductor to the metal. This fraction was found to be approximately equal to the thickness of the diffused semiconductor sheet. Clearly for lateral contacts a more detailed theory was required for a proper interpretation of the experimental data.

To take current crowding into account and to be able to extract the specific contact resistance, a detailed theoretical investigation was undertaken.[55-60] Murrmann and Widmann used a simple *transmission line model* (TLM) for both the semiconductor sheet resistance and the contact resistance.[55] They also described a structure to determine the contact resistance using linear and concentric contacts.[56] Berger extended the transmission line method.[57-58] In contrast to the Kenndey-Murley, in which the contact resistance is assumed to be zero, in the TLM the contact resistance is considered to be nonzero. However, the semiconductor sheet thickness is assumed to be zero in the TLM, with the layer retaining its sheet resistance ρ_s. This assumption allows one-dimensional current flow only. The "zero sheet thickness" restriction was somewhat relaxed by Berger in his extended TLM where he allowed nonzero sheet thickness, but with the current still restricted to one-dimensional flow.[57] The extended TLM was further modified by Schuldt using conformal mapping techniques.[59] The TLM model was later extended to two dimensions by the dual-level transmission line model in which the current is allowed to flow perpendicular to the contact interface. A comparison between the simple and the revised TLM shows a maximum contact resistance deviation of 12%.[61]

For the contact configuration of Fig. 3.14 with the contact width Z equal

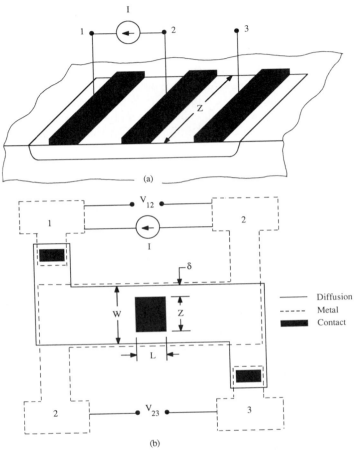

Fig. 3.14 (a) Ideal and (b) actual three-terminal contact resistance test structure used for front and end contact resistance measurements. The contact width and length are Z and L and the diffusion width is W. δ is the gap between the contact and the diffusion window.

to the width of the diffused sheet W the contact resistance is given as[57]

$$R_c = \frac{\sqrt{\rho_s \rho_c}}{Z} \coth\left(\frac{L}{L_T}\right) = \frac{\rho_c}{L_T Z} \coth\left(\frac{L}{L_T}\right) \qquad (3.22)$$

The contact resistance R_c of Eq. (3.22) is sometimes referred to as the contact *front* resistance. $L_T = \sqrt{\rho_c / \rho_s}$ is defined as the *transfer length*. It can be thought of as the length where the voltage due to the current transferring from the semiconductor to the metal or from the metal to the semiconductor has dropped to $1/e$ of its maximum value. The transfer length is plotted in Fig. 3.15 against the specific contact resistance as a function of the sheet resistance. The transfer length is on the order of $1\ \mu m$ or less for $\rho_c \leq 10^{-6}\ \Omega\text{-cm}^2$. For contacts appreciably longer than $1\ \mu m$, most of the contact is inactive during current transfer. Equation (3.22) is only an approximation when the sheet is wider than Z because the model does not consider the current flow around the contacts.

Two cases lead to simplifications of Eq. (3.22): First, for $L \leq 0.5L_T$,

$$R_c \approx \frac{\rho_c}{LZ} \tag{3.23}$$

and second for $L \geq 1.5L_T$,

$$R_c \approx \frac{\rho_c}{L_T Z} \tag{3.24}$$

The effective contact area is the actual contact area $A = LZ$ for the first case. But in the second case the effective contact area is the contact width times the transfer length. In other words, the *effective* contact area can be much smaller than the *actual* contact area. This can have important consequences. For example, consider a structure with $\rho_s = 20\ \Omega/\text{square}$ and $\rho_c = 10^{-6}\ \Omega\text{-cm}^2$. The transfer length $L_T = 2.2\ \mu m$. For a contact length of $L = 10\ \mu m$ and width $Z = 50\ \mu m$, the actual contact area is $LZ = 5 \times 10^{-6}\ \text{cm}^2$. However, the effective contact area is only $L_T Z = 1.1 \times 10^{-6}\ \text{cm}^2$. The current density flowing across the contact is $5 \times 10^{-6}/1.1 \times 10^{-6} = 4.5$ times higher than if the entire contact were active. This higher current density can cause reliability problems by degrading the contact. The reduced contact area can burn out in extreme cases, shifting the effective area along the contact until the entire contact is destroyed.

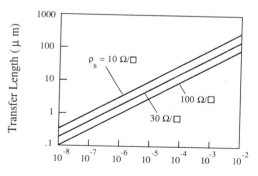

Fig. 3.15 The transfer length as a function of specific contact resistance and semiconductor sheet resistance.

The effect of contact length on contact resistance is illustrated in Fig. 3.16(a). It shows plots of the contact resistance given by Eq. (3.22) multiplied by the contact width Z, for normalization purposes, against the contact length as a function of the specific contact resistance. Note the initial R_c decrease with contact length. However, R_cZ reaches its minimum value at $L \approx L_T$ from which it departs no further, no matter how long the contact becomes.

When the voltage in Fig. 3.14 is measured between contacts 2 and 3 with the current flowing from 2 to 1 as shown, a contact *end* resistance is defined by[57]

$$R_e = \frac{V_{23}}{I} = \frac{\sqrt{\rho_s\rho_c}}{Z}\frac{1}{\sinh(L/L_T)} = \frac{\rho_c}{L_TZ}\frac{1}{\sinh(L/L_T)} \qquad (3.25)$$

The contact end resistance measurement can be used to determine the specific contact resistance by measuring R_e and using an iteration of Eq.

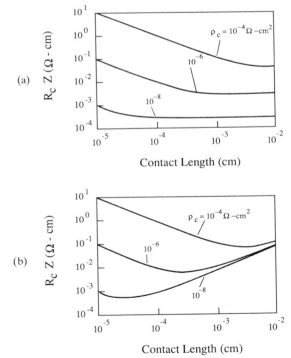

Fig. 3.16 Contact resistance–contact width product as a function of contact length and specific contact resistance for (a) $\rho_s = 10\ \Omega/\text{square}$ and $\rho_m = 0$, and (b) $\rho_{sc} = 10\ \Omega/\text{square}$ and $\rho_m = 30\ \Omega/\text{square}$.

(3.25).[62] R_e was found to be sensitive to contact length variations for short contacts, with the error in determining L limiting the accuracy of the method. For long contacts R_e becomes very small, and the measurement accuracy is limited by instrumentation.

Equation (3.25) assumes the contact width Z to be identical to the sheet width W. This is rarely realized in practice. A more realistic structure is that of Fig. 3.14(b), in which $Z < W$. Experiments with $Z = 5$ μm and W ranging from 10 μm to 60 μm showed the contact end resistance to give erroneously high ρ_c. The error increased as ρ_c decreased or as ρ_s increased. Appreciable error resulted even for $\delta = W - Z$ as small as 5 μm.[63] The error arises from the potential difference between the front edge and the rear edge of the contact, allowing current to leak around the contact edges. The measured resistance is proportional to the sheet resistance and insensitive to the contact resistance for very large δ. For the simple one-dimensional theory to hold, the test structure should meet the conditions: $L \leq L_T$, $Z \gg L$, and $\delta \ll Z$. The one-dimensional analysis is not valid if these conditions are not met. Accurate extraction of ρ_c, however, is possible by fitting numerical simulations to measured data. The problem of $W \neq Z$ is avoided by use of circular test structures. The theory of such a structure using a central contact surrounded by two concentric circular contacts has been derived and the method has been used to determine ρ_c.[64]

Equations (3.22) and (3.25) are derived under the assumption that $\rho_c > 0.2\rho_s t^2$, where t is the diffused layer thickness. For $\rho_s = 20$ ohms/square and $t = 1$ μm, this constraint leads to $\rho_c > 4 \times 10^{-8}$ ohm-cm^2. The TLM method must be modified if that condition is not satisfied, as verified by experiments and by modeling.[65] Fortunately, most specific contact resistances are above 4×10^{-8} ohm-cm^2 and the TLM method is valid.

The difficulty of deciding where to measure the voltage in the configuration of Fig. 3.14 has led to a slightly more complicated test structure and a measurement technique known as the *transfer length method*, originally proposed by Shockley.[66] Unfortunately, it is also abbreviated as TLM, but it should be distinguished from the transmission line model. The technique is very much like that of Fig. 3.13, but it consists of more than three contacts with unequal spacing between contacts. For contacts with $L \geq 1.5 L_T$ and for a front contact resistance measurement, we find the total resistance between any two contacts to be given by

$$R_T = \frac{\rho_s d}{Z} + 2R_c \approx \frac{\rho_s d}{Z} + 2\frac{\rho_s L_T}{Z} \qquad (3.26)$$

where we have used the approximation leading from Eq. (3.22) to Eq. (3.24).

The total resistance is measured for various contact spacings d and plotting R_T as function of d as illustrated in Fig. 3.17. Three parameters can be extracted from such a plot. The slope $\Delta(R_T)/\Delta(d) = \rho_s/Z$ leads to the

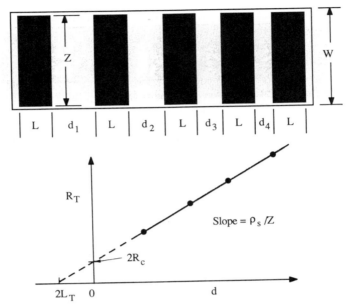

Fig. 3.17 A transfer length method test structure and a plot of total resistance as a function of contact spacing d. Typical values might be: $L = 50\ \mu m$, $W = 100\ \mu m$, $Z - W = 5\ \mu m$ (should be as small as possible), $d \approx 5$ to $50\ \mu m$.

sheet resistance, with the contact width Z independently measured. The intercept at $d = 0$ is $R_T = 2R_c$, giving the *contact resistance*. The intercept at $R_T = 0$ gives $d = 2L_T$, which in turn can be used to calculate the *specific contact resistance* with ρ_s known from the slope of the plot. The transfer length method gives a complete characterization of the contact by providing the sheet resistance, the contact resistance, and the specific contact resistance.

The transfer length method is used a great deal, but it has its own problems. The intercept at $R_T = 0$ giving L_T is sometimes not very distinct leading to incorrect ρ_c values. Perhaps a more serious problem is the uncertainty of the sheet resistance under the contacts. Equation (3.26) assumes ρ_s to be uniform. But the sheet resistance under the contacts may be different from the sheet resistance between contacts due to the effects of contact formation itself. This leads to a modified expression for the total resistance[67]

$$R_T = \frac{\rho_s d}{Z} + 2R_c \approx \frac{\rho_s d}{Z} + 2\frac{\rho_{sc} L_{Tc}}{Z} \tag{3.27}$$

where ρ_{sc} is the sheet resistance *under* the contact and $L_{Tc} = \sqrt{\rho_c/\rho_{sc}}$. The slope of the R_T versus d plot still gives ρ_s/Z, and the intercept at $d = 0$ gives $2R_c$. However, the intercept at $R_T = 0$ now yields $2L_{Tc}(\rho_{sc}/\rho_s)$, and it is no longer possible to determine ρ_c since ρ_{sc} is unknown. However, by determining R_c from the transfer length method and R_e from the end resistance method, it can be shown that[67]

$$R_c/R_e = \cosh(L/L_{Tc}) \tag{3.28}$$

Thus L_{Tc} can be found and then combined with

$$R_e = \frac{\sqrt{\rho_{sc}\rho_c}}{Z}\frac{1}{\sinh(L/L_{Tc})} = \frac{\rho_c}{L_{Tc}Z}\frac{1}{\sinh(L/L_{Tc})} \tag{3.29}$$

to give ρ_c. In this way it is possible to find the contact resistance and the specific contact resistance in addition to the sheet resistance *between* and *under* the contacts. Another method to separate ρ_s from ρ_{sc} involves a slight reduction in the sheet thickness between the contacts by etching the semiconductor.[35]

We have so far considered the specific contact and sheet resistance of the semiconductor, but have neglected the resistance of the metal. This generally introduces little error, although at times the metal resistance increases with aging and can no longer be neglected.[68] The resistance of silicides is higher than that of pure metals and may not always be negligible. But a more serious limitation arises when polysilicon conductors are used instead of metals. Their resistance is significantly higher than that of metals and must surely be considered for proper interpretation of the experimental results. The contact resistance expression given in Eq. (3.22) has been modified to[69-70]

$$R_c = \frac{L_T'}{Z}\left\{\frac{(\rho_m^2 + \rho_{sc}^2)}{(\rho_m + \rho_{sc})}\coth(L/L_T') + \frac{\rho_m\rho_{sc}}{(\rho_m + \rho_{sc})}\left[\frac{2}{\sinh(L/L_T')} + \frac{L}{L_T'}\right]\right\} \tag{3.30}$$

were ρ_m is the metal or polysilicon sheet resistance and $L_T' = \sqrt{\rho_c/(\rho_m + \rho_{sc})}$. Equation (3.30) reduces to Eq. (3.22) for $\rho_{sc} = \rho_s$ and $\rho_m = 0$. The contact resistance from Eq. (3.30), normalized by multiplying by Z, is plotted in Fig. 3.16(b) against the contact length as a function of the specific contact resistance. The main difference between Fig. 3.16(a) and 3.16(b) is the minimum in the lower figure. For each combination of ρ_c, ρ_{sc}, and ρ_m, there is an optimum contact length for minimum contact resistance. For lengths above and below this optimum value, the contact resistance increases. Further discussions of the effects of finite-resistance metal conductors can be found in Kovacs and Mojzes.[71]

3.4.3 Four-Terminal Contact Resistance Method

All the specific resistance measurement techniques discussed so far require a knowledge of the semiconductor bulk resistivity or the semiconductor sheet resistance. It is desirable to measure R_c and ρ_c by minimizing or eliminating, if possible, the parasitic contribution from bulk or sheet resistance. The measurement technique that comes closest to this goal is the four-terminal Kelvin test structure also known as the cross-bridge Kelvin resistor (CBKR) structure. It appears to have been first used for evaluating metal-semiconductor contacts by Shih and Blum[72] in 1972. Anderson and Reith considered it again a few years later.[73] But it was only in the early 1980s that it was evaluated seriously.[74-77] In principle, this method allows the specific contact resistance to be measured without being affected by the underlying semiconductor or the contacting metal or polysilicon conductors.

The principle of the method is illustrated in Fig. 3.18. Current is forced between contact pads 1 and 2, and the voltage is measured between contact

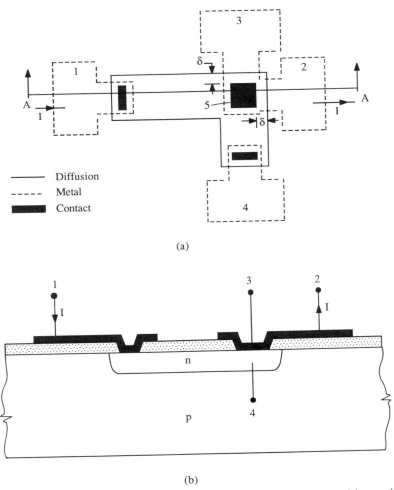

(a)

(b)

Fig. 3.18 A four-terminal or Kelvin contact resistance test structure. (a) top view of the structure, (b) cross section through section A-A.

pads 3 and 4. There are three voltage drops between pad 1 and pad 2. The first is between pad 1 and the semiconductor sheet, the second along the semiconductor sheet, and the third between the semiconductor sheet and pad 2/3. A high impedance voltmeter, used to measure the voltage $V_{34} = V_3 - V_4$, allows very little current flow between pads 3 and 4. Hence the potential at pad 4 is essentially identical to that in the semiconductor directly under contact 5 and V_{34} is solely due to the voltage drop across the contact metal-semiconductor interface. The name "Kelvin test structure" refers to the fact that a voltage is measured with no current flow, similar to four-point probe resistance measurements that can also be thought of as Kelvin measurements.

In principle we have

$$R_c = \frac{V_{34}}{I}$$

(3.31)

where the contact resistance is simply the ratio of the voltage to the current. The specific contact resistance is calculated from R_c through the relation

$$\rho_c = R_c A_c$$

(3.32)

where A_c is the contact area.

Equation (3.32) does not agree with experiment for most exerimental conditions. The specific contact resistance calculated according to Eq. (3.32) should be thought of as an *apparent* specific contact resistance that differs from the true specific contact resistance by lateral current crowding for contact windows smaller than the diffusion tap, shown as $\delta \neq 0$ in Fig. 3.18.[78] Contact window to diffused layer misalignment and lateral dopant diffusion account for $\delta \neq 0$. Part of the current flowing from the sheet into the contact, indicated by the arrows in Fig. 3.19, flows *around* the diffusion tap area. The error introduced by this geometrical factor is highest for low ρ_c and/or high ρ_s and lowest for high ρ_c and/or low ρ_s. $R_c A_c$ tends to overestimate the true ρ_c.[79]

Figure 3.19 shows eperimental data and calculated curves. For the ideal case of $W = L$ or $\delta = 0$ the lower dashed curve is calculated according to $R_c = \rho_c / A_c$ with $\rho_c = 4.5 \times 10^{-8}$ Ω-cm^2. The condition $\delta = 0$ is virtually impossible to attain experimentally. The experimental contact resistance data are higher by over a factor of 10 compared to the values predicted by the simple one-dimensional theory with $\delta \neq 0$ being the main reason for this discrepancy. If ρ_c were calculated from the experimental data points directly, a specific contact resistance much higher than $\rho_c = 4.5 \times 10^{-8}$ Ω-cm^2 would be obtained. This higher value is the *apparent* specific contact resistance, which in this case is approximately ten times the true value. Furthermore the predicted inverse behavior with contact area is not ob-

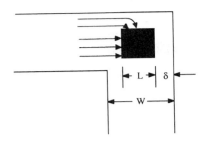

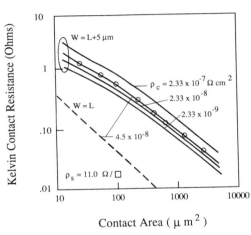

Fig. 3.19   Contact resistance of a four-terminal test structure as a function of contact area for square contacts. The dashed line is the ideal case for $\delta = 0$. The upper curves are for $\delta = 5\ \mu m$ with ρ_c varied from 2.33×10^{-9} to 2.33×10^{-7} Ω-cm^2. The solid lines are simulations and the points are experimental data. Reprinted after Loh et al.[80] by permission of IEEE (© 1985, IEEE).

served. Figure 3.19 clearly shows simple one-dimensional theory to give erroneous results. However, two-dimensional modeling gives reasonably good agreement with experiment.

A simplified approach to this problem of two-dimensional current flow approximates the measured contact resistance R_k by the actual contact resistance R_c and a geometry-dependent resistance R_{geom} as[81]

$$R_k = R_c + R_{geom} = \frac{\rho_c}{A_c} + \frac{4\rho_s\delta^2}{3W_xW_y}\left[1 + \frac{\delta}{(W_x - \delta)}\right] \qquad (3.33)$$

with the various dimensions shown on Fig. 3.20. R_k is plotted in Fig. 3.20 as a function of δ, the gap between the contact window and the diffusion tap. R_k at $\delta = 0$ is the true contact resistance given by ρ_c/A_c. As δ increases, so

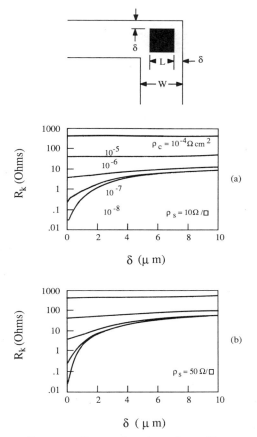

Fig. 3.20 Resistance R_k versus δ as a function of specific contact resistance. The geometry of the square contact is shown in the upper part of the figure: (a) $\rho_s = 10\ \Omega/$square, (b) $\rho_s = 50\ \Omega/$square and $L = 5\ \mu m$.

does R_k despite the fact that the contact itself has not been altered. Lateral current flow around the contact accounts for the additional resistance. The resistance increase gets worse, the lower the specific contact resistance. Note that for $\rho_c \geq 10^{-5}\ \Omega\text{-cm}^2$, there is little change with δ. But for high quality contacts with $\rho_c \leq 10^{-6}\ \Omega\text{-cm}^2$, a large R_k increase is predicted. This increase is further aggravated for higher sheet resistances as seen by comparing Fig. 3.20(a) with Fig. 3.20(b). Unfortunately, the trend in the technology of today's high density integrated circuits is toward lower ρ_c and higher ρ_s due to shallower junctions. Both are in the direction of complicating the interpretation of four-terminal contact resistance test structure measurements. Simple one-dimensional interpretations must be carefully evaluated for their accuracy.

Two-dimensional models of the transmission line (TLM), the contact end resistor (CER), and the cross-bridge Kelvin resistor (CBKR) structures were used to calculate and plot the contact resistance normalized by the sheet resistance against the contact length normalized by δ. These plots are shown in Fig. 3.21 for all three geometries. Significant deviations from the simple one-dimensional analysis is predicted for all three cases. Little deviation from one-dimensionality is indicated by the straight lines in Fig. 3.21(c) for high L_T/δ. The TLM has the least sensitivity to δ because it detects the front contact potential, which is only weakly perturbed by peripheral current flow. However, the TLM method relies on extrapolation

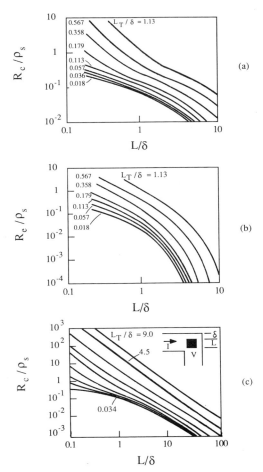

Fig. 3.21   Universal curves of R_c/ρ_s versus L/δ as a function of L_T/δ for (a) the front resistance of the TLM structure, (b) the end resistance of the TLM structure, and (c) the contact resistance of the Kelvin structure. In (a) experimental data are also shown. Reprinted after Loh et al.[82] by permission of IEEE (© 1987, IEEE).

of experimental data to determine ρ_c. That has a potential error especially if the data points do not lie on a well-defined straight line. Both the CER and the CBKR structures show significant deviations due to peripheral current flow. The contact resistances determined by the CER method are generally low, $R_c(CER) < R_c(CBKR)$, making the measurement more difficult. The models used for Fig. 3.21 assume the contacts to be aligned within the diffused region which may not be true. Contact misalignment introduces further departures from one-dimensional behavior.[83] Self-aligned contacts solve the misalignment problem but not the lateral diffusion problem.[84] Other models of contact resistance calculations are given in Lieneweg and Hannaman[85] and Chalmers and Streetman.[86]

Silicided contacts are difficult to implement with the CBKR structure because the break in the self-aligned silicide between contacts 2 and 4 in Fig. 3.18 cannot be made after the silicide is formed without an additional etch step. A solution to this problem is a modified MOSFET consisting of three diffusions and two gates.[87] The "sheet" between contacts 1 and 2 and between contacts 3 and 4 is replaced with an inversion layer formed by biasing the two MOSFET sections into conduction. This structure is compatible with standard silicide processes.

A *vertical* Kelvin test structure, shown in Fig. 3.22, has been developed to overcome the lateral current flow problems of the conventional Kelvin structure.[88] The device requires one additional mask level during its fabrication compared to conventional Kelvin structures. The metal contact is made to a diffused or ion-implanted layer. Current I, confined to the contact area by the isolation pn junction, is forced between contact 1 and the substrate B. Voltage V_{23} is measured between contacts 2 and 3. V_2 is the voltage of the metal, and V_3 is the voltage of the semiconductor layer just below the metal, even though V_3 is measured at some distance from the contact. Just as in a conventional Kelvin structure, there is very little lateral voltage drop along the layer during the voltage measurement because essentially no current is drawn. The contact resistance and the specific contact resistance are given by $R_c = V_{23}/I$ and $\rho_c = R_c A_c$.

Lateral effects, so important in all methods that rely on lateral current flow, also play a role in this structure. This comes about not because the current flows laterally to reach a collecting contact but because of current spreading. The current does not flow strictly vertically. It has a small lateral, spreading component, making the voltage at the sensing contact (contact 3) not exactly equal to the voltage under the metal. The additional spreading causes the measured contact resistance to be higher than the true contact resistance.[89]

Additional contacts are provided. V_{24}, V_{25}, and V_{26} can be used to average the voltage readings, with V_{23} to reduce experimental errors. Furthermore, conventional lateral six-terminal measurements can be made to obtain the end resistance R_e, the front resistance R_f, and the sheet resistance ρ_s.

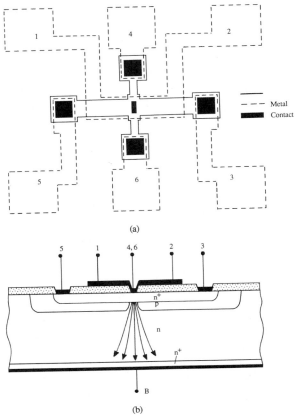

Fig. 3.22 (a) Top view and (b) cross section of the vertical Kelvin structure. The top view shows additional contacts used for six-terminal measurements. Reprinted after Lei et al.[88] by permission of IEEE (© 1987, IEEE).

3.4.4 Six-Terminal Contact Resistance Method

The six-terminal contact resistance structure, shown in Fig. 3.23, is related to the four-terminal Kelvin structure.[76] However, two more contacts provide additional measurement options and additional information not available with the conventional Kelvin structure. The structure allows the *contact resistance*, the *specific contact resistance*, the *end resistance*, the *front resistance*, and the *sheet resistance* under the contact to be determined. For the conventional Kelvin structure contact resistance measurement, the current is forced between contacts 1 and 3 in Fig. 3.23, and the voltage is measured between contacts 2 and 4. The analysis is that of Eqs. (3.31) and (3.32) in the one-dimensional case, where $R_c = V_{24}/I$ and $\rho_c = R_c A_c$. All the two-dimensional complications, not reflected in Eqs. (3.31) and (3.32), manifest themselves in the six-terminal structure also.

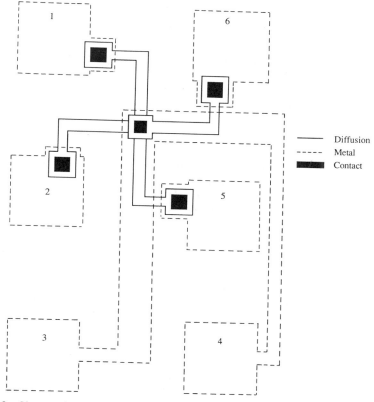

Fig. 3.23 Six-terminal Kelvin structure for the determination of R_c, R_e, R_f, and ρ_{sc}.

To measure the end resistance, $R_e = V_{54}/I$, current is forced between contacts 1 and 3, and the voltage is sensed across contacts 5 and 4. With the contact resistance and the specific contact resistance determined from the Kelvin part of this structure, the sheet resistance under the contact can be determined from the end resistance using Eq. (3.29). Then the front resistance, given by R_c in Eqs. (3.22) and (3.28), can be calculated using Eq. (3.28).

Non-Planar Contacts   Thus far we have only concerned ourselves with deviations from simple theory due to two-dimensional current flow. We have assumed the contact itself to be a smooth, intimate contact between the metal and the semiconductor. Real contacts are not this perfect and this imperfection introduces additional complications. When aluminum contacts silicon, there is a tendency for the silicon to migrate into the aluminum, leaving voids in the silicon.[90] Aluminum can subsequently migrate into these voids creating *spiking*, a phenomenon in which the metal migrates into the semiconductor. Under extreme conditions this can lead to junction shorts. Additions of 1 to 3 wt% Si to the Al reduces spiking considerably but creates other problems. For example, it is possible for the Si to precipitate and to grow epitaxially between the original Si surface and the Al film. The epitaxially regrown layer is *p*-type because it contains a high concentration of aluminum creating a *pn* junction at the regrown epi/n^+ interface. It has been observed that the propensity for such epitaxial films to form is much higher for (100) than for (111)-oriented substrates.[91] This can be a severe problem for small contact areas (1 μm × 1 μm) where the contact resistance for (100)-oriented substrates increases tenfold over similar (111) surfaces.[91]

Unless the semiconductor is carefully cleaned, there can be interfacial layers between the metal and the semiconductor. These can consist of oxides that are allowed to form prior to metal deposition. But interfacial layers can also be due to poor substrate cleaning or even due to poor vacuum during metal deposition.[92]

Contacts to GaAs are typically formed through alloying. A Ge-containing alloy is deposited on the device and heated until alloying occurs. The metal-semiconductor interface after contact formation can be very non-planar. It has been suggested that the current in such alloyed contacts flows through Ge-rich islands with the contact resistance largely determined by the spreading resistance under the Ge-rich regions.[93] The effective contact area is likely to be very different from the actual contact area for that model. Very smooth metal-GaAs interfaces can be formed by evaporating Ge, Au, and Cr layers separately and keeping the annealing temperature below the AuGe eutectic temperature.[94] All of these "technological" imperfections make contact resistance measurement interpretation yet more difficult.

3.5 SCHOTTKY BARRIER HEIGHT

The band diagram of a Schottky barrier diode on an *n*-type substrate is shown in Fig. 3.24. The ideal barrier height of ϕ_{B0} is approached only when the diode is strongly forward biased. The actual barrier height ϕ_B is less than ϕ_{B0} due to image force barrier lowering and other factors. V_{bi} is the built-in potential and V_0 is the potential of the Fermi level with respect to the conduction band. The thermionic current-voltage expression of a Schottky barrier diode is given by

$$I = I_s(e^{qV/nkT} - 1) \tag{3.34}$$

where I_s is the saturation current

$$I_s = AA^{**}T^2 e^{-q\phi_B/kT} = I_{s1}e^{-q\phi_B/kT} \tag{3.35}$$

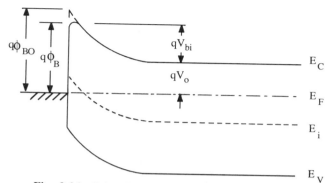

Fig. 3.24 Schottky barrier energy band diagram.

A is the diode area, A^{**} is A^* multiplied by a factor that takes into account optical phonon scattering and quantum mechanical reflection,[95] $A^* = 4\pi q k^2 m^*/h^3 = 120(m^*/m)$ A/cm^{2}K^2 is Richardson's constant, ϕ_B is the effective barrier height, and n is the ideality factor. Equation (3.34) is sometimes expressed as (see Appendix 4.1)

$$I = I_s e^{qV/nkT}(1 - e^{-qV/kT}) \qquad (3.36)$$

Data plotted according to Eq. (3.34) are linear only for $V \gg kT/q$. When plotting $\log[I/(1 - \exp(-qV/kT))]$ versus V using Eq. (3.36), the data are linear all the way to $V = 0$.

3.5.1 Current-Voltage

Among the current-voltage methods the barrier height is most commonly calculated from the current I_s which is determined by an extrapolation of the $\log(I)$ versus V curve to $V = 0$. The current axis intercept for the straight-line portion of this semi-log plot at $V = 0$ is given by I_s. The barrier height ϕ_B is calculated from I_s in Eq. (3.35) according to

$$\phi_B = \frac{kT}{q} \ln\left(\frac{AA^{**}T^2}{I_s}\right) \qquad (3.37)$$

The barrier height so determined is ϕ_B for zero bias. The most uncertain of the parameters in Eq. (3.37) is A^{**} rendering this method only as accurate as a knowledge of A^{**}. Fortunately A^{**} appears in the "ln" term and an error of two in A^{**} gives rise to an error of less than kT/q in ϕ_B. Nonetheless, appreciable errors can occur due to this uncertainty.

An experimental $\log(J)$ versus V plot for an Al/p-type InP Schottky diode is shown in Fig. 3.25. From the slope $n = 1.12$ and from the $V = 0$

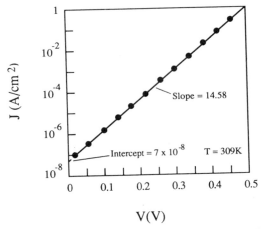

Fig. 3.25   Log(J) versus V for an Al/n-InP Schottky barrier diode. $J = I/A$. Data from Song et al.[96]

intercept $J_s = I_s/A = 7 \times 10^{-8}$ A/cm^2. The barrier height, calculated from Eq. (3.37), is $\phi_B(I - V) = 0.81$ V for $A^{**} = 10.7$ A/cm^{2}K^2 for InP.[96] Other methods based on current-voltage measurements are discussed in Section 4.3.1.

3.5.2   Current-Temperature

For $kT/q \ll V$ Eq. (3.34) can be written as

$$\ln(I/T^2) = \ln(AA^{**}) - q(\phi_B - V/n)/kT \qquad (3.38)$$

A plot of $\ln(I/T^2)$ versus $1/T$ at a constant forward bias voltage V_1, sometimes called a *Richardson plot*, has a slope of $-q(\phi_B - V_1/n)/k$ and an intercept $\ln(AA^{**})$ on the vertical axis. A Richardson plot for the diode of Fig. 3.25 is shown in Fig. 3.26. The slope is well defined, but the extraction of A^{**} from the intercept is prone to error. Generally the $1/T$ axis ranges only over a narrow range, 2.8 to about 3.5 in this example. Extrapolating the data from that narrow range to $1/T = 0$ involves extrapolation over a long distance and any uncertainty in the data produces a large uncertainty in A^{**}. In Fig. 3.26 the current density J is plotted instead of the current I, and hence the intercept is given by $\log(A^{**})$ from which $A^{**} = 10.7$ A/cm^{2}K^2.[96]
    The barrier height is given by

$$\phi_B = \frac{V_1}{n} - \frac{k}{q}\frac{\Delta(\ln(I/T^2))}{\Delta(1/T)} \qquad (3.39)$$

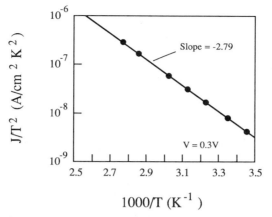

Fig. 3.26   Richardson plot of the diode of Fig. 3.25 for $V = 0.3$ V. Data from Song et al.[96]

The barrier height is obtained from the slope for a known forward bias voltage, but n must be determined independently. For the data of Fig. 3.26 with $n = 1.12$ determined from Fig. 3.25, $V_1 = 0.3$, and the slope $\Delta \log(J/T^2)/\Delta(1000/T) = 2.79$ we find $\phi_B(I - 1/T) = 0.82$ V, very close to $\phi_B(I - V) = 0.81$ V from the $\log I$ versus V plot. Sometimes $\ln(I_s/T^2)$ is plotted against $1/T$, with I_s obtained from the intercept of $\ln I$ versus V plots. The current I in Eq. (3.39) should then be replaced by I_s and $V_1 = 0$.

An implicit assumption in the barrier height determination by the Richardson plot method is a temperature-independent barrier height. Should the barrier height be temperature dependent, as has been observed for Al on InP barriers,[96] we can write ϕ_B as

$$\phi_B(T) = \phi_B(0) - \xi T \tag{3.40}$$

With this temperature dependence, Eq. (3.38) becomes

$$\ln(I/T^2) = \ln(AA^{**}) + q\xi/k - q(\phi_B(0) - V/n)/kT \tag{3.41}$$

A Richardson plot now gives the "zero Kelvin" barrier height, and the intercept contains the barrier height temperature dependence factor ξ. Nonlinearities are sometimes observed in Richardson plots at low temperatures. These may be due to current mechanisms other than thermionic emission current usually manifesting themselves as $n > 1.1$. Nonlinear Richardson plots are observed when both the barrier height and the ideality factor are temperature dependent. Accurate extraction of ϕ_B and A^{**} becomes impossible then, but linearity can be restored if $n \ln(I/T^2)$ is plotted against $1/T$.[97]

3.5.3   Capacitance-Voltage

The capacitance per unit area of a Schottky diode is given by[98]

$$\frac{C}{A} = \sqrt{\frac{qK_s\varepsilon_0|N_D - N_A|}{2(V_{bi} + |V| - kT/q)}} \tag{3.42}$$

where V is the reverse-bias voltage. For n-type substrates $N_D > N_A$ and $V < 0$, whereas for p-type substrates $N_D < N_A$ and $V > 0$. The kT/q in the denominator accounts for the majority carrier tail in the space-charge region which is omitted in the depletion approximation. The built-in potential is related to the barrier height by the relationship

$$\phi_B = V_{bi} + V_0 \tag{3.43}$$

as seen in Fig. 3.24. $V_0 = (kT/q) \ln(N_c/N_D)$, where N_c is the effective density of states in the conduction band. Plotting $1/(C/A)^2$ versus V gives a curve with slope $2/(qK_s\varepsilon_0|N_D - N_A|)$, and with intercept on the V-axis, $V_i = -V_{bi} + kT/q$. The barrier height is determined from the intercept voltage by

$$\phi_B = -V_i + V_0 + kT/q \tag{3.44}$$

The doping concentration can be determined from the slope as discussed in Chapter 2. $\phi_B(C\text{-}V)$ is approximately the flat-band barrier height.

A $(C/A)^{-2}$ versus V plot of the diode of Fig. 3.25 is shown in Fig. 3.27. From the slope we find $N_A = 3.8 \times 10^{17} \text{ cm}^{-3}$, and from Eq. (3.44) the barrier height is $\phi_B(C\text{-}V) = 0.82$ V using the intercept voltage $V_i = -0.8$ V and $n_i = 1.5 \times 10^7 \text{ cm}^{-3}$ for InP.

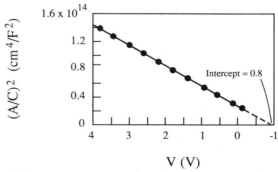

Fig. 3.27 $(C/A)^{-2} - V$ plot of the diode of Fig. 3.25. $N_A = 3.8 \times 10^{17} \text{ cm}^{-3}$. Data from Song et al.[96]

3.5.4 Photocurrent

When a Schottky diode is irradiated with photons of sub-band gap energy, it is possible to excite carriers from the metal into the semiconductor as shown in Fig. 3.28(a). Electron emission is not possible for $h\nu < \phi_B$. However, for $h\nu > \phi_B$ electrons are excited from the metal over the barrier into the semiconductor where they are detected as photocurrent I_{ph}. The yield Y, defined as the ratio of the photocurrent to the absorbed photon flux, is given by[99]

$$Y = B(h\nu - q\phi_B)^2 \qquad (3.45)$$

where B is a constant. $Y^{1/2}$ is plotted versus $h\nu$, and an extrapolation of the linear portion of this curve, sometimes called a *Fowler plot*, to $Y^{1/2} = 0$ gives the barrier height. An example of a Fowler plot is shown in Fig. 3.28(b) by the solid points.

A Fowler plot is not always linear as predicted by the theory. When it is nonlinear it is difficult to determine ϕ_B, as illustrated in Fig. 3.28(b). By differentiating Eq. (3.45) with respect to $h\nu$, one obtains

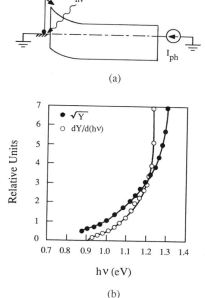

Fig. 3.28 (a) Photoresponse schematic, (b) Fowler plot and $dY/d(h\nu)$ plot of a Pt/GaAs Schottky barrier diode. Data from Fontaine et al.[101]

$$\frac{dY}{d(hv)} = 2B(hv - q\phi_B) \tag{3.46}$$

In the first derivative plot the deviation from linearity is much smaller than it is in the conventional Fowler plot.[100] This is because the extended tail of the Fowler plot in the vicinity of the barrier height has been removed by the differentiation procedure as seen in Fig. 3.28(b) where the barrier height from the differentiated curve is $\phi_B = 0.905$ V. Moreover, the derivative plot is much more sensitive to contact non-uniformities and has been used to detect such contact non-uniformities.[100] The photocurrent technique relies only on photo-excited current flow and is little influenced by tunnel currents, especially if ϕ_B is obtained by extrapolating from $hv \gg \phi_B$, where only those electrons well above the barrier height contribute to the photocurrent.

Comparison of Methods A number of studies have been undertaken to compare barrier heights determined by the current-voltage (I-V), current-temperature (I-T), capacitance-voltage (C-V), and photocurrent (PC) techniques. In one study the barrier height of evaporated Pt films on GaAs substrates was determined by the I-V, C-V, and PC methods giving $\phi_B(I$-$V) = 0.81$ V, $\phi_B(C$-$V) = 0.98$ V, and $\phi_B(PC) = 0.905$ V.[101] Which is the most reliable value? Any damage at the interface affects the I-V behavior because defects may act as recombination centers or as intermediate states for trap-assisted tunnel currents. Either one of these mechanisms raises n and lowers ϕ_B. C-V measurements are less prone to such defects. However, defects can alter the space-charge region width and hence the intercept voltage. Photocurrent measurements are less sensitive to such defects, and this method is judged to be most reliable. Nevertheless, Fowler plots are not always linear. The first derivative plot usually does have a straight-line portion, making ϕ_B extraction more reliable.

The sequence of $\phi_B(I$-$V) < \phi_B(PC) < \phi_B(C$-$V)$ was also observed for a variety of metals deposited on n-GaAs and p-GaAs.[102] Result of barrier height measurements of Schottky barriers on p-type InP gave $\phi_B(I$-$T) < \phi_B(C$-$V)$.[96] The difference was attributed to patchiness of barrier heights across the contact. When two Schottky diodes of different barrier height are connected in parallel, the *lower barrier height* dominates the I-V behaviour, but the barrier height with the largest contact area dominates the C-V behavior.[103] This concept was extended theoretically to mixed-phase contacts of varying dimensions but fixed area ratios, predicting that generally $\phi_B(C$-$V) > \phi_B(I$-$V)$.[104] For large contact regions results similar to those in [103] were obtained. For smaller contact regions, however, the low barrier height regions were found to be pinched off by the high barrier height regions.

Barrier height measurements on atomically clean Si gave $\phi_B(C$-$V) >$

$\phi_B(I\text{-}V) \approx \phi_B(PC)$.[105] The equality of the barrier heights for the *I-V* and the PC methods is because both measure the energy necessary to excite electrons from the metal into the semiconductor and barrier height lowering is almost equally effective. In the *C-V* method the data are extrapolated to $1/C^2 = 0$ corresponding to near flatband conditions in the semiconductor and barrier height lowering is close to zero. In another study, Richardson plots were compared with photocurrent measurements for Mo-AlGaAs junctions. Good agreement between the two methods was obtained for various Al mole fractions.[106]

The barrier height patchiness invoked to explain the differing barrier heights also predicts varying Richardson constants. It is frequently observed that A^{**} varies greatly with processing conditions such as annealing. It may well be that annealing causes the patchiness to vary and therefore A^{**} to change. This would rule against using those methods that rely on a knowledge of A^{**} for ϕ_B determination. Consequently *C-V* and photocurrent measurements are preferred over *I-V* and *I-T* measurements. PC probes the device from outside the semiconductor, that is, photoemission is from the metal to the semiconductor, *I-V* and *C-V* methods probe the device from the semiconductor side. It is for this reason that these two methods are more sensitive to spatial inhomogeneities, insulating layers between the metal and the semiconductor, doping inhomogeneities, surface damage, and tunneling. The PC technique is least influenced by these parameters and is therefore likely to yield the most reliable value of barrier height. For well-behaved contacts with few of these degradation factors, all methods give values that agree reasonably well with one another.

3.6 STRENGTHS AND WEAKNESSES

- *Two-Terminal Methods* The two-terminal contact resistance measurement technique is simple but the least reliable. The contact resistance data are corrupted by either the semiconductor bulk or sheet resistance. The method is only infrequently used today. The two-terminal *contact string* is used mainly as a process monitor. It does not give detailed contact resistance information nor can specific contact resistance be reliably extracted.

- *Three-Terminal Methods* The three-terminal technique is usually employed in its transfer length method implementation, where the effect of the semiconductor sheet resistance is separated from the contact resistance and both contact resistance as well as specific contact resistance can be determined. The three-terminal method allows both front and end contact resistance measurements to be made. Complications in the interpretation of the experimental data arise due to three main effects: (1) the extrapolation of experimental data to obtain intercepts, (2) lateral current flow around the contact, and (3) the sheet resistance under the contact which is not necessarily indentical to the sheet resistance outside the contact window. Current flows laterally around the contact window whenever the contact window is narrower than the diffusion tap, leading to erroneous contact resistances if the experimental data are analyzed by the conventional one-dimensional theory. For the most reliable measurements the test structure should be configured to satisfy the following requirements: $L > L_T$, $Z \gg L$, and $\delta \ll W$ as defined in Fig. 3.14.

- *Four-Terminal Method*   The four-terminal or Kelvin structure is preferred over the two- and three-terminal structures. There are several reasons for this. (1) There is only one metal-semiconductor contact, and the contact resistance is measured directly as the ratio of a voltage to a current. R_c can therefore be very small. (2) Neither metal nor semiconductor sheet resistance enters into the R_c determination. Hence there is no practical limit to the value of R_c that can be measured. (3) The contact area can be made small to be consistent with contact areas used in high density ICs. This makes the method very simple and attractive. However, any lateral current flow obscures the interpretation. Modeling has shown two-dimensional effects to be very important, especially for appreciable gaps between the contact window and the diffusion edge. Even for gaps as narrow as 5 μm appreciable error is introduced, with the contact resistance becoming relatively independent of the true contact resistance.

- *Six-Terminal Method*   The six-terminal method is very similar to the four-terminal technique. It incorporates the Kelvin structure but additionally allows measurements of the front and end contact resistance as well as the contact sheet resistance. It is only slightly more complex than the four-terminal structure but does not require additional masking operations.

For any of the contact resistance measurement methods it is difficult to determine absolute values of ρ_c. Simple one-dimensional interpretations of the experimental data frequently give incorrect values of specific contact resistance. Proper interpretation of the experimental data requires more exact modeling. This makes many of the data, determined in the past by simple one-dimensional interpretation, suspect. Nevertheless, ρ_c can be used as a figure of merit but the experimental conditions under which they were obtained should be carefully specified. The *contact resistance* can be measured directly, but again the measured resistance may not be the true contact resistance.

APPENDIX 3.1   ALLOYS FOR CONTACTS TO SEMICONDUCTOR MATERIALS

Material	Alloy	Contact Type
n-Si	Au-Sb	ohmic
p-Si	Au-Ga	ohmic
p-Si	Al	ohmic
n-GaAs	Au-Ge	ohmic
n-GaAs	Sn	ohmic
p-GaAs	Au-Zn	ohmic
p-GaAs	In	ohmic
n-GaInP	Au-Sn	ohmic
n-InP	Ni/Au-Ge/Ni	ohmic
n-InP	Au-Sn	ohmic
p-InP	Au-Zn	ohmic
n-AlGaAs*	Ni/Au-Ge/Ni	ohmic
p-AlGaAs*	In-Sn	ohmic
GaAs (n or p type)	Ni	Schottky
GaAs (n or p type)	Al	Schottky
GaAs (n or p type)	Au-Ti	Schottky
InP (n or p type)	Au	Schottky
InP (n or p type)	Au-Ti	Schottky

Source: Bio-Rad.[107]

* with GaAs capping layer.

REFERENCES

[1]   F. Braun, "On the Conduction Through Sulfurmetals (in German)," *Annal. Phys. Chem.* **153**, 556–563, 1874.

[2]   W. Schottky, "Semiconductor Theory of the Blocking Layer (in German)," *Naturwissenschaften* **26**, 843, Dec. 1938; "On the Semiconductor Theory of Blocking and Point Contact Rectifiers (in German)," *Z. Phys.* **113**, 367–414, July 1939; "Simplified and Expanded Theory of Boundary Layer Rectifiers (in German)," *Z. Phys.* **118**, 539–592, Feb. 1942.

[3]   B. Schwartz (ed.), *Ohmic Contacts to Semiconductors*, Electrochem. Soc., New York, 1969.

[4]   V. L. Rideout, "A Review of the Theory and Technology for Ohmic Contacts to Group III–V Compound Semiconductors," *Solid-State Electron.* **18**, 541–550, June 1975.

[5]   M. N. Yoder, "Ohmic Contacts in GaAs," *Solid-State Electron.* **23**, 117–119, Feb. 1980.

[6]   N. Braslau, "Alloyed Ohmic Contacts to GaAs," *J. Vac. Sci. Technol.* **19**, 803–807, Sept./Oct. 1981.

[7]   A. Piotrowska, A. Guivarch, and G. Pelous, "Ohmic Contacts to III–V Compound Semiconductors: A Review of Fabrication Techniques," *Solid-State Electron.* **26**, 179–197, March 1983.

[8]   D. K. Schroder and D. L. Meier, "Solar Cell Contact Resistance—A Review," *IEEE Trans. Electron Dev.* **ED-31**, 637–647, May 1984.

[9]   A. Y. C. Yu, "Electron Tunneling and Contact Resistance of Metal-Silicon Contact Barriers," *Solid-State Electron.* **13**, 239–247, Feb. 1970.

[10] S. S. Cohen, "Contact Resistance and Methods for Its Determination," *Thin Solid Films* **104**, 361–379, June 1983.

[11] A. G. Milnes and D. L. Feucht, *Heterojunction and Metal-Semiconductor Junctions*, Academic Press, New York, 1972.

[12] B. L. Sharma and R. K. Purohit, *Semiconductor Heterojunctions*, Pergamon, London, 1974; B. L. Sharma, "Ohmic Contacts to III–V Compound Semiconductors," in *Semiconductors and Semimetals*, Vol. 15 (R. K. Willardson and A. C. Beer, eds.), Academic Press, New York, 1981, pp. 1–38.

[13] E. H. Rhoderick and R. H. Williams, *Metal-Semiconductor Contacts*, 2nd ed., Clarendon, Oxford, 1988.

[14] S. S. Cohen and G. S. Gildenblat, *Metal-Semiconductor Contacts and Devices*, Academic Press, Orlando, FL, 1986.

[15] W. E. Spicer, I. Lindau, P. R. Skeath, and C. Y. Su, "The Unified Model for Schottky Barrier Formation and MOS Interface States in 3–5 Compounds," *Appl. Surf. Sci.* **9**, 83–91, Sept. 1981.

[16] W. E. Spicer, S. Eglash, I. Lindau, C. Y. Su, and P. R. Skeath, "Development and Confirmation of the Unified Model for Schottky Barrier Formation and MOS Interface States on III–V Compounds," *Thin Solid Films* **89**, 447–460, March 1982.

[17] A. M. Cowley and S. M. Sze, "Surface States and Barrier Height of Metal-Semiconductor Systems," *J. Appl. Phys.* **36**, 3212–3220, Oct. 1965.

[18] C. A. Mead, "Physics of Interfaces," in *Ohmic Contacts to Semiconductors* (B. Schwartz, ed.). Electrochem. Soc., New York, 1969, pp. 3–16.

[19] J. Bardeen, "Surface States and Rectification at Metal-Semiconductor Contact," *Phys. Rev.* **71**, 717–727, May 1947.

[20] R. H. Williams, "The Schottky Barrier Problem," *Contemp. Phys.* **23**, 329–351, July/Aug. 1982.

[21] L. J. Brillson, "Surface Photovoltage Measurements and Fermi Level Pinning: Comment on 'Development and Confirmation of the Unified Model for Schottky Barrier Formation and MOS Interface States on III–V Compounds'," *Thin Solid Films* **89**, L27–L33, March 1982.

[22] J. Tersoff, "Recent Models of Schottky Barrier Formation," *J. Vac. Sci. Technol.* **B3**, 1157–1161, July/Aug. 1985.

[23] I. Lindau and T. Kendelewicz, "Schottky Barrier Formation on III–V Semiconductor Surfaces: A Critical Evaluation," *CRC Crit. Rev. in Solid State and Mat. Sci.* **13**, 27–55, Jan. 1986.

[24] F. A. Kroger, G. Diemer, and H. A. Klasens, "Nature of Ohmic Metal-Semiconductor Contacts," *Phys. Rev.* **103**, 279, July 1956.

[25] S. M. Sze, *Physics of Semiconductor Devices*, 2d ed., Wiley, New York, 1981, pp. 255–258.

[26] F. A. Padovani and R. Stratton, "Field and Thermionic-Field Emission in Schottky Barriers," *Solid-State Electron.* **9**, 695–707, July 1966.

[27] C. R. Crowell and V. L. Rideout, "Normalized Thermionic-Field (TF) Emission in Metal-Semiconductor (Schottky) Barriers," *Solid-State Electron.* **12**, 89–105, Feb. 1969; "Thermionic-Field Resistance Maxima in Metal-Semiconductor (Schottky) Barriers," *Appl. Phys. Lett.* **14**, 85–88, 1 Feb. 1969.

[28] R. S. Popovic, "Metal-N-Type Semiconductor Ohmic Contact with a Shallow N^+ Surface Layer," *Solid-State Electron.* **21**, 1133–1138, Sept. 1978.

[29] D. F. Wu, D. Wang, and K. Heime, "An Improved Model to Explain Ohmic Contact Resistance of n-GaAs and Other Semiconductors," *Solid-State Electron.* **29**, 489–494, May 1986.

[30] G. Brezeanu, C. Cabuz, D. Dascalu, and P. A. Dan, "A Computer Method for the Characterization of Surface-Layer Ohmic Contacts," *Solid-State Electron.* **30**, 527–532, May 1987.

[31] F. A. Padovani, "The Current-Voltage Characteristics of Metal-Semiconductor Contacts," in *Semiconductors and Semimetals* (R. K. Willardson and A. C. Beer, eds.), Academic Press, New York, **7A**, 75–146, 1971.

[32] C. Y. Chang and S. M. Sze, "Carrier Transport Across Metal-Semiconductor Barriers," *Solid-State Electron.* **13**, 727–740, June 1970.

[33] C. Y. Chang, Y. K. Fang, and S. M. Sze, "Specific Contact Resistance of Metal-Semiconductor Barriers," *Solid-State Electron.* **14**, 541–550, July 1971.

[34] W. J. Boudville and T. C. McGill, "Resistance Fluctuations in Ohmic Contacts due to Discreteness of Dopants," *Appl. Phys. Lett.* **48**, 791–793, 24 March 1986.

[35] M. Heiblum, M. I. Nathan, and C. A. Chang, "Characteristics of AuGeNi Ohmic Contacts to GaAs," *Solid-State Electron.* **25**, 185–195, March 1982.

[36] Data for n-Si were taken from: [9], [32], [39], [43], [44], [45], [49], [55], [58], [69]; S. S. Cohen, P. A. Piacente, G. Gildenblat, and D. M. Brown, "Platinum Silicide Ohmic Contacts to Shallow Junctions in Silicon," *J. Appl. Phys.* **53**, 8856–8862, Dec. 1982; S. Swirhun, K. C. Saraswat, and R. M. Swanson, "Contact Resistance of LPCVD W/Al and PtSi/Al Metallization," *IEEE Electron Dev. Lett.* **EDL-5**, 209–211, June 1984; S. S. Cohen and G. S. Gildenblat, "Mo/Al Metallization for VLSI Applications," *IEEE Trans. Electron Dev.* **ED-34**, 746–752, April 1987.

[37] Data for p-Si were taken from: [39], [43], [44], [49], [55], [69]; S. S. Cohen, P. A. Piacente, G. Gildenblat, and D. M. Brown, "Platinum Silicide Ohmic Contacts to Shallow Junctions in Silicon," *J. Appl. Phys.* **53**, 8856–8862, Dec. 1982; S. Swirhun, K. C. Saraswat, and R. M. Swanson, "Contact Resistance of LPCVD W/Al and PtSi/Al Metallization," *IEEE Electron Dev. Lett.* **EDL-5**, 209–211, June 1984; S. S. Cohen and G. S. Gildenblat, "Mo/Al Metallization for VLSI Applications," *IEEE Trans. Electron Dev.* **ED-34**, 746–752, April 1987.

[38] Data for n-GaAs were taken from [6] and references therein.

[39]   Data for *p*-GaAs were taken from: [6], C. J. Nuese and J. J. Gannon, "Silver-Manganese Evaporated Ohmic Contacts to *p*-type GaAs," *J. Electrochem. Soc.* **115**, 327–328, March 1968; K. L. Klohn and L. Wandinger, "Variation of Contact Resistance of Metal-GaAs Contacts with Impurity Concentration and Its Device Implications," *J. Electrochem. Soc.* **116**, 507–508, April 1969; H. Matino and M. Tokunaga, "Contact Resistances of Several Metals and Alloys to GaAs," *J. Electrochem. Soc.* **116**, 709–711, May 1969; H. J. Gopen and A. Y. C. Yu, "Ohmic Contacts to Epitaxial GaAs," *Solid-State Electron.* **14**, 515–517, June 1971; O. Ishihara, K. Nishitani, H. Sawano, and S. Mitsue, "Ohmic Contacts to *P*-Type GaAs," *Japan. J. Appl. Phys.* **15**, 1411–1412, July 1976; C. Y. Su and C. Stolte, "Low Contact Resistance Non Alloyed Ohmic Contacts to Zn Implanted GaAs," *Electron. Lett.* **19**, 891–892, Oct. 1983; R. C. Brooks, C. L. Chen, A. Chu, L. J. Mahoney, J. G. Mavroides, M. J. Manfra, and M. C. Finn, "Low-Resistance Ohmic Contacts to *p*-Type GaAs Using Zn/Pd/Au Metallization", *IEEE Electron Dev. Lett.* **EDL-6**, 525–527, Oct. 1985.

[40]   S. E. Swirhun and R. M. Swanson, "Temperature Dependence of Specific Contact Resistivity," *IEEE Electron Dev. Lett.* **EDL-7**, 155–157, March 1986.

[41]   D. M. Brown, M. Ghezzo, and J. M. Pimbley, "Trends in Advanced Process Technology—Submicrometer CMOS Device Design and Process Requirements," *Proc. IEEE* **74**, 1678–1702, Dec. 1986.

[42]   M. V. Sullivan and J. H. Eigler, "Five Metal Hydrides as Alloying Agents on Silicon," *J. Electrochem. Soc.* **103**, 218–220, April 1956.

[43]   R. H. Cox and H. Strack, "Ohmic Contacts for GaAs Devices," *Solid-State Electron.* **10**, 1213–1218, Dec. 1967.

[44]   R. D. Brookes and H. G. Mathes, "Spreading Resistance Between Constant Potential Surfaces," *Bell Syst. Tech. J.* **50**, 775–784, March 1971.

[45]   A. Shepela, "The Specific Contact Resistance of Pd$_2$Si Contacts on *n*- and *p*-Si," *Solid-State Electron.* **16**, 477–481, April 1973.

[46]   A. K. Sinha, "Electrical Characteristics and Thermal Stability of Platinum Silicide-to-Silicon Ohmic Contacts Metallized with Tungsten," *J. Electrochem. Soc.* **120**, 1767–1771, Dec. 1973.

[47]   H. Muta, "Electrical Properties of Platinum-Silicon Contact Annealed in an H$_2$ Ambient," *Japan. J. Appl. Phys.* **17**, 1089–1098, June 1978.

[48]   W. D. Edwards, W. A. Hartman, and A. B. Torrens, "Specific Contact Resistance of Ohmic Contacts to Gallium Arsenide," *Solid-State Electron.* **15**, 387–392, April 1972.

[49]   K. Heime, U. König, E. Kohn, and A. Wortman, "Very Low Resistance Ni-AuGe-Ni Contacts to *n*-GaAs," *Solid-State Electron.* **17**, 835–837, Aug. 1974.

[50]   G. Y. Robinson, "Metallurgical and Electrical Properties of Alloyed Ni/Au-Ge Films on *n*-Type GaAs," *Solid-State Electron.* **18**, 331–342, April 1975.

[51]   C. Y. Ting and C. Y. Chen, "A Study of the Contacts of a Diffused Resistor," *Solid State Electron.* **14**, 433–438, June 1971.

[52]   G. P. Carver, J. I. Kopanski, D. B. Novotny, and R. A. Forman, "Specific Contact Resistivity of Metal-Semiconductor Contacts—A New, Accurate

Method Linked to Spreading Resistance," *IEEE Trans. Electron Dev.* **ED-35**, 489–497, April 1988.

[53] J. M. Andrews, "A Lithographic Mask System for MOS Fine-Line Process Development," *Bell Syst. Tech. J.* **62**, 1107–1160, April 1983.

[54] D. P. Kennedy and P. C. Murley, "A Two-Dimensional Mathematical Analysis of the Diffused Semiconductor Resistor," *IBM J. Res. Dev.* **12**, 242–250, May 1968.

[55] H. Murrmann and D. Widmann, "Current Crowding on Metal Contacts to Planar Devices," *IEEE Trans. Electron Dev.* **ED-16**, 1022–1024, Dec. 1969.

[56] H. Murrmann and D. Widmann, "Measurement of the Contact Resistance Between Metal and Diffused Layer in Si Planar Devices (in German)," *Solid-State Electron.* **12**, 879–886, Dec. 1969.

[57] H. H. Berger, "Models for Contacts to Planar Devices," *Solid-State Electron.* **15**, 145–158, Feb. 1972.

[58] H. H. Berger, "Contact Resistance and Contact Resistivity," *J. Electrochem. Soc.* **119**, 507–514, April 1972.

[59] S. B. Schuldt, "An Exact Derivation of Contact Resistance to Planar Devices," *Solid-State Electron.* **21**, 715–719, May 1978.

[60] I. F. Chang, "Contact Resistance in Diffused Resistors," *J. Electrochem. Soc.* **117**, 368–372, March 1970.

[61] J. M. Pimbley, "Dual-Level Transmission Line Model for Current Flow in Metal-Semiconductor Contacts," *IEEE Trans. Electron Dev.* **ED-33**, 1795–1800, Nov. 1986.

[62] J. G. J. Chern and W. G. Oldham, "Determining Specific Contact Resistivity from Contact End Resistance Measurements," *IEEE Electron Dev. Lett.* **EDL-5**, 178–180, May 1984.
Comments on this Paper are: J. A. Mazer and L. W. Linholm, "Comments on 'Determining Specific Contact Resistivity from Contact End Resistance Measurements'," *IEEE Electron Dev. Lett.* **EDL-5**, 347–348, Sept. 1984; J. Chern and W. G. Oldham, "Reply to 'Comments on Determining Specific Contact Resistivity from Contact End Resistance Measurements'," *IEEE Electron Dev. Lett.* **EDL-5**, 349, Sept. 1984; M. Finetti, A. Scorzoni, and G. Soncini, "A Further Comment on 'Determining Specific Contact Resistivity from Contact End Resistance Measurements'," *IEEE Electron Dev. Lett.* **EDL-6**, 184–185, April 1985.

[63] S. E. Swirhun, W. M. Loh, R. M. Swanson, and K. C. Saraswat, "Current Crowding Effects and Determination of Specific Contact Resistivity from Contact End Resistance (CER) Measurements," *IEEE Electron Dev. Lett.* **EDL-6**, 639–641, Dec. 1985.

[64] G. K. Reeves, "Specific Contact Resistance Using a Circular Transmission Line Model," *Solid-State Electron.* **23**, 487–490, May 1980; A. J. Willis and A. P. Botha, "Investigation of Ring Structures for Metal-Semiconductor Contact Resistance Determination," *Thin Solid Films* **146**, 15–20, 2 Jan. 1987.

[65] E. G. Woelk, H. Kräutle, and H. Beneking, "Measurement of Low Resistive

Ohmic Contacts on Semiconductors," *IEEE Trans. Electron Dev.* **ED-33**, 19–22, Jan. 1986.

[66] W. Shockley, A. Goetzberger, and R. M. Scarlett, "Research and Investigation of Inverse Epitaxial UHF Power Transistors," Rep. No. AFAL-TDR-64-207, Air Force Avionics Lab., Wright-Patterson Air Force Base, OH, Sept. 1964.

[67] G. K. Reeves and H. B. Harrison, "Obtaining the Specific Contact Resistance from Transmission Line Model Measurements," *IEEE Electron Dev. Lett.* **EDL-3**, 111–113, May 1982.

[68] G. S. Marlow and M. B. Das, "The Effects of Contact Size and Non-Zero Metal Resistance on the Determination of Specific Contact Resistance," *Solid-State Electron.* **25**, 91–94, Feb. 1982.

[69] D. B. Scott, W. R. Hunter, and H. Shichijo, "A Transmission Line Model for Silicided Diffusions: Impact on the Performance of VLSI Circuits," *IEEE Trans. Electron Dev.* **ED-29**, 651–661, April 1982.

[70] G. K. Reeves and H. B. Harrison, "Contact Resistance of Polysilicon-Silicon Interconnections," *Electron. Lett.* **18**, 1083–1085, Dec. 1982; G. Reeves and H. B. Harrison, "Determination of Contact Parameters of Interconnecting Layers in VLSI Circuits," *IEEE Trans. Electron Dev.* **ED-33**, 328–334, March 1986.

[71] B. Kovacs and I. Mojzes, "Influence of Finite Metal Overlayer Resistance on the Evaluation of Contact Resistivity," *IEEE Trans. Electron Dev.* **ED-33**, 1401–1403, Sept. 1986.

[72] K. K. Shih and J. M. Blum, "Contact Resistances of Au-Ge-Ni, Au-Zn and Al to III–V Compounds," *Solid-State Electron.* **15**, 1177–1180, Nov. 1972.

[73] R. M. Anderson and T. M. Reith, "Microstructural and Electrical Properties of Thin PtSi Films and Their Relationships to Deposition Parameters," *J. Electrochem. Soc.* **122**, 1337–1347, Oct. 1975.

[74] S. S. Cohen, G. Gildenblat, M. Ghezzo, and D. M. Brown, "Al–0.9%Si/Si Ohmic Contacts to Shallow Junctions," *J. Electrochem. Soc.* **129**, 1335–1338, June 1982.

[75] S. J. Proctor and L. W. Linholm, "A Direct Measurement of Interfacial Contact Resistance," *IEEE Electron Dev. Lett.* **EDL-3**, 294–296, Oct. 1982.

[76] S. J. Proctor, L. W. Linholm, and J. A. Mazer, "Direct Measurements of Interfacial Contact Resistance, End Resistance, and Interfacial Contact Layer Uniformity," *IEEE Trans. Electron Dev.* **ED-30**, 1535–1542, Nov. 1983.

[77] J. A. Mazer, L. W. Linholm, and A. N. Saxena, "An Improved Test Structure and Kelvin-Measurement Method for the Determination of Integrated Circuit Front Contact Resistance," *J. Electrochem. Soc.* **132**, 440–443, Feb. 1985.

[78] A. A. Naem and D. A. Smith, "Accuracy of the Four-Terminal Measurement Techniques for Determining Contact Resistance," *J. Electrochem. Soc.* **133**, 2377–2380, Nov. 1986.

[79] M. Finetti, A. Scorzoni, and G. Soncini, "Lateral Current Crowding Effects on Contact Resistance Measurements in Four Terminal Resistor Test Patterns," *IEEE Electron Dev. Lett.* **EDL-5**, 524–526, Dec. 1984.

[80] W. M. Loh, S. E. Swirhun, E. Crabbe, K. Saraswat, and R. M. Swanson, "An Accurate Method to Extract Specific Contact Resistivity Using Cross-Bridge Kelvin Resistors," *IEEE Electron Dev. Lett.* **EDL-6**, 441–443, Sept. 1985. Comments on this paper are: H. B. Harrison and G. K. Reeves, "Comment on 'An Accurate Method to Extract Specific Contact Resistivity Using Cross-Bridge Kelvin Resistors'," *IEEE Electron Dev. Lett.* **EDL-7**, 142, Feb. 1986; S. E. Swirhun, W. M. Loh, E. Crabbe, R. M. Swanson, and K. C. Saraswat, "Reply to Comment on 'An Accurate Method to Extract Specific Contact Resistivity Using Cross-Bridge Kelvin Resistors," *IEEE Electron Dev. Lett.* **EDL-7**, 142–144, Feb. 1986.

[81] T. A. Schreyer and K. C. Saraswat, "A Two-Dimensional Analytical Model of the Cross-Bridge Kelvin Resistor," *IEEE Electron Dev. Lett.* **EDL-7**, 661–663, Dec. 1986.

[82] W. M. Loh, S. E. Swirhun, T. A. Schreyer, R. M. Swanson, and K. C. Saraswat, "Modeling and Measurement of Contact Resistances," *IEEE Trans. Electron Dev.* **ED-34**, 512–524, March 1987.

[83] A. Scorzoni, M. Finetti, K. Grahn, I. Suni, and P. Cappelletti, "Current Crowding and Misalignment Effects as Sources of Error in Contact Resistivity Measurements—Part I: Computer Simulation of Conventional CER and CKR Structures," *IEEE Trans. Electron Dev.* **ED-34**, 525–531, March 1987.

[84] P. Cappelletti, M. Finetti, A. Scorzoni, I. Suni, N. Cirelli, and G. D. Libera, "Current Crowding and Misalignment Effects as Sources of Error in Contact Resistivity Measurements—Part II: Experimental Results and Computer Simulation of Self-Aligned Test Structures," *IEEE Trans. Electron Dev.* **ED-34**, 532–536, March 1987.

[85] U. Lieneweg and D. J. Hannaman, "New Flange Correction Formula Applied to Interfacial Resistance Measurements of Ohmic Contacts to GaAs," *IEEE Electron Dev. Lett.* **EDL-8**, 202–204, May 1987.

[86] S. A. Chalmers and B. G. Streetman, "Lateral Diffusion Contributions to Contact Mismatch in Kelvin Resistor Structures," *IEEE Trans. Electron Dev.* **ED-34**, 2023–2024, Sept. 1987.

[87] W. T. Lynch and K. K. Ng, "A Tester for the Contact Resistivity of Self-Aligned Silicides," *IEEE Int. Electron Dev. Meet. Digest*, San Francisco, CA, 1988, pp. 352–355.

[88] T. F. Lei, L. Y. Leu, and C. L. Lee, "Specific Contact Resistivity Measurement by a Vertical Kelvin Test Structure," *IEEE Trans. Electron Dev.* **ED-34**, 1390–1395, June 1987.

[89] C. L. Lee, W. L. Yang, and T. F. Lei, "The Spreading Resistance Error in the Vertical Kelvin Test Resistor Structure for the Specific Contact Resistivity," *IEEE Trans. Electron Dev.* **ED-35**, 521–523, April 1988.

[90] J. G. J. Chern, W. G. Oldham, and N. Cheung, "Contact-Electromigration-Induced Leakage Failure in Aluminum-Silicon to Silicon Contacts," *IEEE Trans. Electron Dev.* **ED-32**, 1341–1346, July 1985.

[91] H. Onoda, "Dependence of Al-Si/Al Contact Resistance on Substrate Surface Orientation," *IEEE Electron Dev. Lett.* **EDL-9**, 613–615, Nov. 1988.

[92] T. J. Faith, R. S. Iven, L. H. Reed, J. J. O'Neill Jr., M. C. Jones, and B. B. Levin, "Contact Resistance Monitor for Si ICs," *J. Vac. Sci. Technol.* **B2**, 54–57, Jan./March 1984.

[93] N. Braslau, "Alloyed Ohmic Contacts to GaAs," *J. Vac. Sci. Technol.* **19**, 803–807, Sept./Oct. 1981; "Ohmic Contacts to GaAs," *Thin Solid Films* **104**, 391–397, June 1983.

[94] J. Willer, D. Ristow, W. Kellner, and H. Oppolzer, "Very Stable Ge/Au/Cr/Au Ohmic Contacts to GaAs," *J. Electrochem. Soc.* **135**, 179–181, Jan. 1988.

[95] S. M. Sze, *Physics of Semiconductor Devices*, 2d ed., Wiley, New York, 1981, pp. 256–263.

[96] Y. P. Song, R. L. Van Meirhaeghe, W. H. Laflère, and F. Cardon, "On the Difference in Apparent Barrier Height as Obtained from Capacitance-Voltage and Current-Voltage-Temperature Measurements on Al/p-InP Schottky Barriers," *Solid-State Electron.* **29**, 633–638, June 1986.

[97] A. S. Bhuiyan, A. Martinez, and D. Esteve, "A New Richardson Plot for Non-Ideal Schottky Diodes," *Thin Solid Films* **161**, 93–100, July 1988.

[98] A. M. Goodman, "Metal-Semiconductor Barrier Height Measurement by the Differential Capacitance Method—One Carrier System," *J. Appl. Phys.* **34**, 329–338, Feb. 1963.

[99] R. H. Fowler, "The Analysis of Photoelectric Sensitivity Curves for Clean Metals at Various Temperatures," *Phys. Rev.* **38**, 45–56, July 1931.

[100] T. Okumura and K. N. Tu, "Analysis of Parallel Schottky Contacts by Differential Internal Photoemission Spectroscopy," *J. Appl. Phys.* **54**, 922–927, Feb. 1983.

[101] C. Fontaine, T. Okumura, and K. N. Tu, "Interfacial Reaction and Schottky Barrier Between Pt and GaAs," *J. Appl. Phys.* **54**, 1404–1412, March 1983.

[102] T. Okumura and K. N. Tu, "Electrical Characterization of Schottky Contacts of Au, Al, Gd and Pt on *n*-Type and *p*-Type GaAs," *J. Appl. Phys.* **61**, 2955–2961, April 1987.

[103] I. Ohdomari and K. N. Tu, "Parallel Silicide Contacts," *J. Appl. Phys.* **51**, 3735–3739, July 1980.

[104] J. L. Freeouf, T. N. Jackson, S. E. Laux, and J. M. Woodall, "Size Dependence of 'Effective' Barrier Heights of Mixed-Phase Contacts," *J. Vac. Sci. Technol.* **21**, 570–574, July/Aug. 1982.

[105] A. Thanailakis, "Contacts Between Simple Metals and Atomically Clean Silicon," *J. Phys. C: Solid State Phys.* **8**, 655–668, March 1975.

[106] M. Eizenberg, M. Heiblum, M. I. Nathan, N. Braslau, and P. M. Mooney, "Barrier Heights and Electrical Properties of Intimate Metal-AlGaAs Junctions," *J. Appl. Phys.* **61**, 1516–1522, Feb. 1987.

[107] Bio-Rad, *Semiconductor Newsletter*, Winter 1988.

CHAPTER 4

SERIES RESISTANCE, CHANNEL LENGTH, THRESHOLD VOLTAGE

4.1 INTRODUCTION

Semiconductor device performance is generally degraded by series resistance. The extent of performance degradation depends on the series resistance, on the device, on the operating current flowing through the device, and on a number of other parameters. The series resistance r_s depends on the semiconductor resistivity, on the contact resistance, and sometimes on geometrical factors. Series resistance may be very large before it causes device degradation for some devices that are much less sensitive to it than others. For example, in a reverse-biased photodiode series resistance is a minor consideration. However, series resistances of a few ohms are detrimental for a solar cells. The effect of r_s on capacitance and carrier concentration profiling measurements is discussed in Chapter 2. The aim of the device designer should be a design in which series resistance is negligibly small. However, since r_s cannot be zero it is important to be able to measure it.

4.2 *PN* JUNCTION DIODES

4.2.1 Current-Voltage

The diode current of a *pn* junction is usually written as a function of the diode voltage V_d as

$$I = I_0(e^{qV_d/nkT} - 1) \qquad (4.1)$$

147

where I_0 is the prefactor and n is the ideality factor. The diode voltage in Eq. (4.1) does not include any resistance effects. I_0 and n generally vary over the I-V range of a diode. At low currents, where space-charge region (scr) recombination/generation dominates, I_0 is the scr current and $n = 1.5$ to 2 for Si diodes. At higher currents, I_0 accounts for quasi-neutral region recombination/generation and $n = 1$. The I-V behavior may be different for diodes made on other semiconductor materials. For the purposes of this chapter we assume Eq. (4.1) to hold although the parameters in the equation may differ with current.

If both I_0 and n are constant over some current range, then a plot of $\log(I)$ versus V_d yields a straight line. When series resistance contributes to device behavior the diode voltage becomes $V_d = V - Ir_s$, where V is the measured voltage across the entire diode including substrate and contact resistance as well as other series resistance components. The current-voltage expression becomes

$$I = I_0(e^{q(V - Ir_s)/nkT} - 1) \qquad (4.2)$$

A plot of $\log(I)$ versus V, shown in Fig. 4.1, gives a straight line only over that portion of the curve where $kT/q \ll V$ and $Ir_s \ll V$. The measured current deviates from the straight line at low currents due to the -1 in the bracket of Eq. (4.2), and it deviates from the straight line at high currents due to series resistance. The straight-line portion of the plot yields I_0 by extrapolation to $V = 0$, and it gives n from the slope $S = d \log(I)/dV$. Knowing the sample temperature allows the ideality factor to be obtained from the measured slope, according to

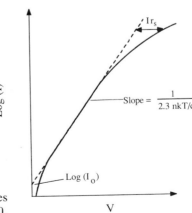

Fig. 4.1 $\log(I)$ versus V for a diode with series resistance. The upper dashed line is for $r_s = 0$.

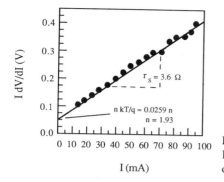

Fig. 4.2 $I\,dV/dI$ versus I for a GaAs LED. Reprinted after Escher et al.[1] by permission of IEEE (© 1982, IEEE).

$$n = \frac{1}{2.3SkT/q} \qquad (4.3)$$

where the 2.3 accounts for the conversion from $\ln(I)$ (base e) to $\log(I)$ (base 10). We will generally use the logarithm to base 10, written as "log," instead of the logarithm to base e, written as "ln," because experimental data are plotted on "log," not "ln," paper.

One technique to determine r_s is to measure the deviation of the experimental current-voltage curve from the extrapolated straight line as illustrated on Fig. 4.1. A more accurate series resistance is obtained by expressing Eq. (4.2) as[1]

$$I\,\frac{dV}{dI} = Ir_s + \frac{nkT}{q} \qquad (4.4)$$

A plot of $I(dV/dI)$ versus I gives nkT/q as the intercept and r_s as the slope, as shown in Fig. 4.2. Equation (4.4) can be slightly rewritten. Instead of $I(dV/dI)$, one can use $dV/d[\ln(I)]$ by using the identity $d[\ln(I)] = dI/I$. $d[\ln(I)]/dV$ can be obtained directly from the $\log(I)$ versus V curve. In a slight variation of Eq. (4.4), dV/dI is plotted versus $1/I$. The slope of this plot is nkT/q, and the intercept on the dV/dI axis is r_s.

4.2.2 Open-Circuit Voltage Decay

Open-circuit voltage decay is a method to determine the minority carrier lifetime of pn junctions as discussed in detail in Chapter 8. The same method also lends itself to a determination of the diode series resistance, as illustrated in Fig. 4.3.[2] The diode is forward biased. At $t = 0$ switch is opened, and the open-circuit diode voltage is monitored as a function of time. The *lifetime* is determined from the time-varying open-circuit voltage. The *series resistance* is obtained from the voltage discontinuity at $t = 0$.

The voltage drop across the diode just before opening the switch $V_{oc}(0^-)$

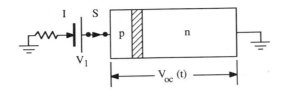

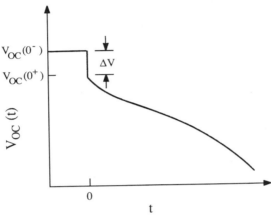

Fig. 4.3 Open-circuit voltage decay of a *pn* junction showing the voltage discontinuity at $t = 0$.

consists of the diode voltage V_d and the voltage drop across any device resistances

$$V_{oc}(0^-) = V_d + Ir_s \qquad (4.5)$$

At $t = 0$ switch S is opened, and the voltage drops abruptly because the current becomes zero and $V_{oc}(0^+) = V_d$. With the measured voltage drop given by $V_{oc}(0^-) - V_{oc}(0^+) = Ir_s$ and I measured independently, it is a simple matter to calculate the series resistance $r_s = [V_{oc}(0^-) - V_{oc}(0^+)]/I$. This is an absolute measure that does not rely on slopes or intercepts and is suitable for low r_s measurements. Solar cell series resistances as low as 10 to 20 mΩ have been determined this way.

4.2.3 Capacitance-Voltage

We showed in Chapter 2 the degradation in capacitance caused by series resistance effects. The measured capacitance C_m of a junction device is related to the true capacitance C by

$$C_m = \frac{C}{(1 + r_s G)^2 + (2\pi f r_s C)^2} \qquad (4.6)$$

where G is the conductance and f the measurement frequency. For reasonably good junction devices, the condition $r_s G \ll 1$ is generally satisfied, and Eq. (4.6) can be simplified to

$$C_m \approx \frac{C}{1 + (2\pi f r_s C)^2} \qquad (4.7)$$

By lowering the frequency, it is usually possible to reduce the second term in the denominator so that it is less than unity and the true capacitance is determined. Then the frequency is raised until the second term dominates, and r_s can be calculated with all other quantities known. This method is only effective when $r_s \gg 1/2\pi f C$ and can also be used when dc current techniques are unable to determine the series resistance. For example, for an MOS capacitor with no dc current flow through the device, r_s can be determined from capacitance measurements.

4.3 SCHOTTKY BARRIER DIODES

4.3.1 Series Resistance

The current-voltage characteristic of a Schottky barrier diode without series resistance is discussed in Section 3.5. The thermionic current-voltage expression of a Schottky barrier diode with series resistance is given by

$$I = I_s(e^{q(V - Ir_s)/nkT} - 1) \qquad (4.8)$$

where I_s is the saturation current

$$I_s = A A^{**} T^2 e^{-q\phi_B/kT} = I_{s1} e^{-q\phi_B/kT} \qquad (4.9)$$

A is the diode area, A^{**} is A^* multiplied by a factor taking into account optical phonon scattering and quantum mechanical reflection,[3] $A^* = 4\pi q k^2 m^*/h^3 = 120(m^*/m)$ A/cm^{2}K^2 is Richardson's constant, ϕ_B is the effective barrier height, and n is the ideality factor. Equation (4.8) is sometimes expressed as (see Appendix 4.1)

$$I = I_s e^{q(V - Ir_s)/nkT}(1 - e^{-qV/kT}) \qquad (4.10)$$

Data plotted according to Eq. (4.8) are linear only for $V \gg kT/q$. When plotting $\log[I/(1 - \exp(-qV/kT))]$ versus V, using Eq. (4.10), the data are linear all the way to $V = 0$.

The method of extracting r_s from a plot of $I\,dV/dI$ versus I, given in Eq. (4.4), can also be used for Schottky diodes.[4] Another method, proposed by Norde, defines a function F as[5]

$$F = \frac{V}{2} - \frac{kT}{q}\ln\left(\frac{I}{I_{s1}}\right)$$ (4.11)

Using Eqs. (4.8) and (4.9), Eq. (4.11) can be written as

$$F = \left(\frac{1}{2} - \frac{1}{n}\right)V + \frac{Ir_s}{n} + \phi_B$$ (4.12)

Why is this rather peculiarly defined F function used? When F is plotted against V, it exhibits a minimum which is used to determine r_s and ϕ_B. To see the dependence of F on V, we consider the low and high voltage limits. At low applied voltages, where $Ir_s \ll V$, Eq. (4.12) gives $dF/dV = \frac{1}{2} - 1/n \approx -\frac{1}{2}$ for $n \approx 1$. At high voltages, where $Ir_s \gg V$, $dF/dV = \frac{1}{2}$. Hence, F has a minimum lying between these two limits. The voltage at the minimum is designated $V = V_m$. The current corresponding to this voltage is $I = I_m$. From $dF/dV = 0$ at the minimum, we find the series resistance as

$$r_s = \frac{(2 - n)}{I_m}\frac{kT}{q}$$ (4.13)

The minimum F-value, found by substituting Eq. (4.13) into Eq. (4.12) is

$$F_m = \left(\frac{1}{2} - \frac{1}{n}\right)V_m + \frac{(2 - n)}{n}\frac{kT}{q} + \phi_B$$ (4.14)

The series resistance of the Schottky diode is calculated from the ideality factor n and from I_m. The ideality factor is obtained from the slope of the $\log(I)$ versus V plot, and I_m is the current on a $\log(I)$ versus V plot at $V = V_m$. I_{s1}, and therefore A^{**}, must be known for this method. This is a disadvantage of this technique since A^{**} is not necessarily known. In the absence of an experimentally determined A^{**}, one must assume that published values for A^{**} apply. That is not always a good assumption since it has been shown that A^{**} depends on the contact preparation, including the surface cleaning procedure[6] and sample annealing temperature; it even depends on the metal thickness and the method of metal deposition.[7]

It has been pointed out that the original Norde method of plotting F versus V assumes the ideality factor $n = 1$, and the statistical error is increased by using only a few data points near the minimum of the F versus V curve. A modification of the Norde plot has been proposed, increasing the accuracy and allowing r_s, n, and ϕ_B to be extracted from an experimental

$\log(I)$ versus V plot.[8] Alternately, R_s, n, and ϕ_B can be determined by measuring the I-V curves at two different temperatures.[9]

Barrier height measurements in the absence of series resistance are discussed in Section 3.5. The barrier height is commonly calculated from the saturation current I_s determined by an extrapolation of the $\log(I)$ versus V curve to $V = 0$. Series resistance is not important in this extrapolation because the current I_s is very low. The barrier height ϕ_B is calculated from I_s in Eq. (4.9) according to

$$\phi_B = \frac{kT}{q} \ln\left(\frac{AA^{**}T^2}{I_s}\right) \tag{4.15}$$

The barrier height so determined is the value of ϕ_B for zero bias. The most uncertain of the parameters in Eq. (4.15) is A^{**}, rendering this method only as accurate as A^{**} is known. Fortunately A^{**} appears in the "ln" term, and an error of two in A^{**} gives rise to an error of less than kT/q in ϕ_B. Nonetheless, appreciable errors can result due to this uncertainty if A^{**} varies significantly from sample to sample.

Another method to find ϕ_B is based on Eq. (4.14), where ϕ_B is calculated from F_m, V_m, T, and n. In the original plot of F versus V [Eq. (4.11)] A^{**} is assumed to be known. A slight variation on the F-function theme is the H-function, defined as[4]

$$H = V - \frac{nkT}{q} \ln\left(\frac{I}{I_{s1}}\right) = Ir_s + n\phi_B \tag{4.16}$$

A plot of H versus I has a slope given by r_s and an H-axis intercept of $n\phi_B$. Like the F plot, the H plot also requires a knowledge of A^{**}.

An approach which does *not* require a knowledge of A^{**} is the modified Norde plot, given by[10]

$$F1 = \frac{V}{2kT/q} - \ln\left(\frac{I}{T^2}\right) \tag{4.17}$$

The function $F1$ is plotted versus V for several different temperatures. Each of these plots exhibits a minimum and each minimum defines an $F1_m$, a voltage V_m, and a current I_m. Using Eqs. (4.8), (4.9), and (4.13) and for $V \gg kT/q$,

$$2F1_m + (2-n)\ln\left(\frac{I_m}{T^2}\right) = 2 - n[\ln(AA^{**}) + 1] + \frac{n\phi_b}{kT/q} \tag{4.18}$$

When the left side of Eq. (4.18) is plotted against q/kT, a straight line results whose slope is $n\phi_B$ and whose y-axis intercept is $\{2 - n[\ln(AA^{**}) + 1]\}$. With n independently determined from the slope of a $\log(I)$ versus V

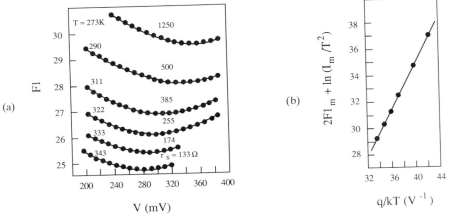

Fig. 4.4 (a) F_{m1} versus V, (b) $[2F1_m + (2 - n)\ln(I/T^2)]$ versus q/kT for a Au/n-Si Schottky diode. Reprinted with permission after Chot.[10]

plot, it is possible to extract both ϕ_B and A^{**}, provided the area A is known.

Plots of $F1$ versus V at several different temperatures are shown in Fig. 4.4(a). Each curve exhibits a minimum, leading to $F1_m$ and I_m. From I_m the series resistance, shown on Fig. 4.4(a), is determined according to Eq. (4.13). Next $[2F1_m + (2 - n)\ln(1/T^2)]$ is plotted against q/kT, shown in Fig. 4.4(b). In this example, the barrier height is found to $(0.867/n)$ V from the slope of this plot. The intercept yields $\{2 - n[\ln(AA^{**}) + 1]\} = 0.18$. The intercept is prone to error because the data points cover only a limited temperature range and must be extrapolated to $q/kT = 0$.

A very different method of determining n, I_s, and r_s is the fitting of a curve to the experimental I-V data. This has been done over the temperature range of 180 K to 340 K using a least square curve fitting routine.[11]

4.4 SOLAR CELLS

Solar cells are particularly prone to series resistance. Series resistance reduces the short-circuit current, but more important, it reduces the maximum power available from the device. The series resistance should be approximately $r_s < (0.8/X)\,\Omega$ for 1 cm^2 area cells, where X is the solar concentration.[12] $X = 1$ for conventional non-concentrator cells, whereas for concentrator cells, X can be several hundred. For $X = 100$, $r_s < 8 \times 10^{-3}\,\Omega$. Under "one-sun" conditions 10–20% of the maximum power available from a solar cell can be lost due to a series resistance of 1 ohm.[13] Although solar cells are *pn* junction diodes, their I-V characteristics are often not suitable for the types of measurements for conventional diodes. Furthermore the operation of solar cells in the presence of sunlight may alter the series resistance, and r_s should be determined under actual operating conditions. Shunt resistance is also important during solar cell operation.

Several methods have been used to determine r_s. They are generally neither simple to implement nor to interpret. A solar cell may be represented by the equivalent circuit of Fig. 4.5(a) consisting of a photon or light-induced current generator I_{ph}, a diode, a series resistor r_s, and a shunt resistor r_{sh}. Frequently r_s and r_{sh} are assumed to be constant, but they may depend on the cell current. The current I flows through the load resistor R_L and develops a voltage V across it. The current is given by

$$I = I_{ph} - I_0(e^{q(V + Ir_s)/nkT} - 1) - \frac{(V + Ir_s)}{r_{sh}} \qquad (4.19)$$

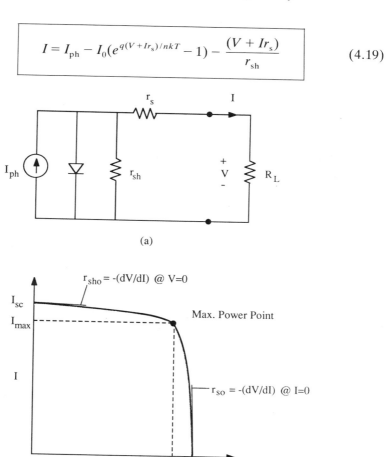

(a)

(b)

Fig. 4.5 (a) Equivalent circuit of a solar cell, (b) current-voltage curve of a solar cell.

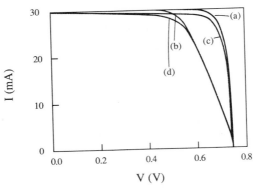

Fig. 4.6 Current-voltage curve of a solar cell with $I_{ph} = 30$ mA, $I_0 = 10^{-14}$ A, $n = 1$, $T = 300$ K. The series and shunt resistances are (a) $r_s = 0$, $r_{sh} = \infty$, (b) $r_s = 5 \, \Omega$, $r_{sh} = \infty$, (c) $r_s = 0$, $r_{sh} = 500 \, \Omega$, and (d) $r_s = 5 \, \Omega$, $r_{sh} = 500 \, \Omega$.

This equation does not take into account that both I_0, and n are not constant over the entire I-V curve. At low voltage, space-charge region (scr) recombination generally dominates, but at higher voltage quasi-neutral region (qnr) recombination is dominant. Equation (4.19) is used for most solar cell analyses in spite of its simplifications. However, both scr and qnr recombination are occasionally considered separately.

A general current-voltage curve of a solar cell is shown in Fig. 4.5(b). The open-circuit voltage V_{oc}, the short-circuit current I_{sc}, and the maximum power point voltage V_{max} and current I_{max} are also shown. The quantities r_{so} and r_{sho} are the resistances defined by the slopes of the I-V curve at $I = 0$ and at $V = 0$, respectively. The effects of series and shunt resistances are shown on the I-V characteristics in Fig. 4.6 calculated from Eq. (4.19). Series resistances of a few ohms degrade the device performance, as do shunt resistances of several hundred ohms. Small I-V degradations have a significant effect on cell efficiency since every percent efficiency decrease in a solar cell is very important.

4.4.1 Multiple Light Intensities

An early method to determine r_s is based on the measurement of the I-V curves at two different light intensities giving the short-circuit currents I_{sc1} and I_{sc2}, respectively. A current δI below I_{sc}, $I = I_{sc} - \delta I$, is picked on both I-V curves. The currents $I_1 = I_{sc1} - \delta I$ and $I_2 = I_{sc2} - \delta I$ correspond to voltages V_1 and V_2. The series resistance is then given by[14]

$$r_s = \frac{V_1 - V_2}{I_2 - I_1} = \frac{V_1 - V_2}{I_{sc2} - I_{sc1}} \tag{4.20}$$

This technique was later modified. By using more than two light intensities, more than two points are generated. Drawing a line through all of the points gives the series resistance by the slope of this line, $\Delta I/\Delta V$, as[15]

$$r_s = \frac{\Delta V}{\Delta I} \tag{4.21}$$

The method is demonstrated in Fig. 4.7.

The slope method lends itself to r_s determination at any current with no limiting approximations and is generally considered to give good results. It is also independent of I_0, n, and r_{sh}, provided they do not change with operating point. This is an important consideration. Those techniques that require a knowledge of I_0, n, and r_{sh}, and even I_{ph} in some cases, are at a disadvantage because these parameters may not be accurately known. It is important that the temperature of the cell be constant during the measurements at different light intensities, as temperature variations can alter the series resistance.

Comparison of experimental *I-V* curves with a theoretical curves ($r_s = 0$) has also been used to determine r_s. The shift of the maximum power point from its theoretical value, $\Delta V_{max} = V_{max}(\text{theory}) - V_{max}(\text{exp})$, is given by Smirnov and Mahan[16]

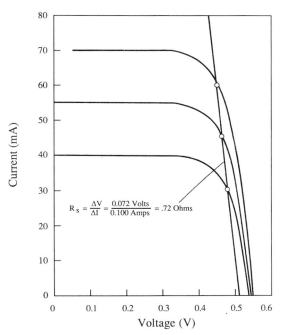

Fig. 4.7 Series resistance determination of an n^+p solar cell. Reprinted with permission after Handy.[15]

$$r_s = \frac{\Delta V_{max}}{I_{max}} \tag{4.22}$$

A weakness of this method is the assumption that parameters like I_0 and n be known. If unknown, they must be determined by other means, for they are required to calculate the theoretical *I-V* curve.

Under short-circuit conditions, where $I = I_{sc}$ and $V = 0$, Eq. (4.19) becomes

$$\ln\left(\frac{I_{ph} - I_{sc}}{I_0}\right) = \frac{qI_{sc}r_s}{nkT} \tag{4.23}$$

A plot of $\ln[(I_{ph} - I_{sc})/I_0]$ versus I_{sc} has a slope given by qr_s/nkT.[17] The series resistance is calculated from the slope, provided n and I_{ph} are known.

Another method relies on a *dark I-V* curve, the open-circuit voltage, and the short-circuit current. From Eq. (4.19) with r_{sh} very large, the *dark* voltage is

$$V_{dk} = \frac{nkT}{q} \ln\left(\frac{I_{dk}}{I_0}\right) - I_{dk}r_s \tag{4.24}$$

The open-circuit voltage is given by

$$V_{oc} = \frac{nkT}{q} \ln\left(\frac{I_{ph}}{I_0}\right) \tag{4.25}$$

V_{oc} is independent of r_s since there is no current flow during an open circuit voltage measurement. Hence, by comparing V_{oc} with V_{dk} at a given current I_{dk}, it is possible to determine r_s at that current. To reduce any error, one should choose that point on the $I_{dk} - V_{dk}$ curve where the diode parameters are the same as those of the open-circuit condition.[18] That corresponds to $I_{dk} = I_{ph}$ and since generally $I_{ph} \approx I_{sc}$, we can write

$$r_s \approx \frac{V_{dk}(I_{sc}) - V_{oc}}{I_{sc}} \tag{4.26}$$

Choosing $I_{dk} = I_{sc}$ assures that the upper limit of the series resistance for a given light intensity is obtained.[18]

4.4.2 Constant Light Intensity

The series resistance can be determined by a method based on the area under the *I-V* curve.[19] The area is given by the power P_1,

$$P_1 = \int_0^{I_{sc}} V(I)\, dI \tag{4.27}$$

The series resistance, obtained from Eqs. (4.19) and (4.27), is[19]

$$r_s = 2\left(\frac{V_{oc}}{I_{sc}} - \frac{P_1}{I_{sc}^2} - \frac{nkT}{qI_{sc}}\right) \qquad (4.28)$$

This method has been used to measure the very low resistances of concentrator solar cells of $r_s = 5$ to 6×10^{-3} Ω. Such cells, because they are operated under solar concentrations with high photocurrents, are particularly prone to series resistance degradation.

Series resistances determined by the "area" method have been compared to values determined by the "slope" methods.[14-15] Such comparisons have shown the "area" method to overestimate r_s at "one-sun" and lower illuminations.[20] This is because n must be known accurately in Eq. (4.28) and r_{sh} may not be negligible. The values determined by the two methods at *high* illumination are in reasonably good agreement.

Various analytical techniques have also been used to determine r_s. Some are based on complete curve fitting of the solar equation to experimental I-V curves. Others use several points on the experimental I-V curve to determine the key parameters. In the *five-point method* the parameters I_{ph}, I_0, n, r_s, and r_{sh} are calculated from the experimental V_{oc}, I_{sc}, V_{max}, r_{so}, and r_{sho} shown in Fig. 4.5.[21] Later simplifications in the equations make the analysis more tractable.[22] A comparison of the five parameters determined by the exact five point, by the approximate five point, and by numerical techniques gave very good agreement for I_{ph}, I_0, and n. The main differences were found for r_s and r_{sh} at low light intensities. The five point method has been simplified to the *three-point method* in which I_{ph}, I_0, n, r_s, and r_{sh} are determined from the open-circuit voltage, the short-circuit current, and the maximum power point. Both five-point and three-point methods give comparable results.[23-24]

Considering that both scr and qnr recombination takes place in a solar cell, parameters that describe both of these processes should be determined for complete solar cell modeling. By applying small current steps to a solar cell in both the forward and reverse current directions and measuring the resulting voltage, it is possible to determine $I_0(scr)$, $I_0(qnr)$, $n(scr)$, $n(qnr)$, r_s, and r_{sh}.[25]

A technique especially suitable for concentrator solar cells with low series resistances is based on high intensity flash illumination.[26] Neglecting the shunt resistance in the circuit in Fig. 4.5(a), we find that for very high light intensities the output current I approaches but cannot exceed $V_{oc}/(R_L + r_s)$. In order to keep the cell temperature as constant as possible during the measurement, it is best to flash the illumination. That, of course, makes the method available only to those who have a flash solar simulator. If we approximate the voltage by $V_{oc} \approx I (R_L + r_s)$, then by varying the load resistance at constant light intensity, we find

$$r_s \approx \frac{(I_2 R_{L2} - I_1 R_{L1})}{(I_1 - I_2)} \qquad (4.29)$$

where I_1 and I_2 are the load currents for load resistances R_{L1} and R_{L2}. Series resistances as low as 7 to 9 mΩ have been determined with this method for GaAs concentrator solar cells at light intensities approaching 9000 suns with 1-ms light pulses.[26] The value of the load resistances should be on the order of the series resistance.

4.4.3   Shunt Resistance

The shunt resistance r_{sh} can be determined by some of the curve-fitting approaches discussed in the previous section, or it can be determined independently. It is sometimes found from the slope of the reverse-biased current-voltage characteristic before breakdown. Most solar cells, however, exhibit large reverse currents at voltages well below breakdown because solar cells are not designed to be operated under reverse voltages. This makes it difficult, if not impossible, to obtain reasonable values for r_{sh} by this method. Furthermore a solar cell in the dark under reverse bias is a poor representation of a solar cell operating in the light under forward bias.

An alternate method is to rewrite Eq. (4.19) in terms of V_{oc} and I_{sc} as

$$I_{sc}\left(1 + \frac{r_s}{r_{sh}}\right) - \frac{V_{oc}}{r_{sh}} = I_0\left[\exp\left(\frac{qV_{oc}}{nkT}\right) - \exp\left(\frac{qI_{sc}r_s}{nkT}\right)\right] \qquad (4.30)$$

This equation is rather complex, but it can be simplified for $r_s \ll r_{sh}$. If the measurement is made under low light intensities where $I_{sc}r_s \ll nkT/q$, then Eq. (4.30) becomes

$$I_{sc} - I_0\left[\exp\left(\frac{qV_{oc}}{nkT}\right) - 1\right] = \frac{V_{oc}}{r_{sh}} \qquad (4.31)$$

This approximation is valid for $I_{sc} \leq 3$ mA for series resistances on the order of 0.1 Ω. When measurements of r_{sh} were made under these conditions, r_{sh} was found to be highly sensitive to I_0 and n, values that may not be known accurately.[27] This problem was alleviated by making the measurements at very low light intensities, allowing the second term on the left side of Eq. (4.31) to be neglected. Equation (4.31) then simply becomes

$$I_{sc} \approx \frac{V_{oc}}{r_{sh}} \qquad (4.32)$$

The $I_{sc} - V_{oc}$ plot has a linear region of slope $1/r_{sh}$. The curve becomes nonlinear at higher light intensities, and the method becomes invalid. Measurements on 2-inch and 3-inch diameter cells, showed that for I_{sc} in the 0 to 200 μA and V_{oc} in the 0 to 50 mV range, the shunt resistances were 65 to 1170 Ω.[27]

4.5 BIPOLAR JUNCTION TRANSISTORS

An *npn* junction-isolated, integrated-circuit bipolar junction transistor (BJT) with parasitic series resistances is shown in Fig. 4.8. The n^+ emitter and the *p*-base are formed in an *n*-epitaxial collector layer grown on a *p*-substrate. An n^+ buried layer is provided to reduce the collector resistance. The transistor is isolated from adjacent transistors by *p* isolation regions.

The parasitic resistances and their measurement are the most important aspect for our purpose. The emitter resistance r_E is primarily determined by the emitter contact resistance. The base resistance r_B is composed of the intrinsic base resistance r_{Bi}, under the emitter, and the extrinsic base resistance r_{Bx}, from the emitter to the base contact including the base contact resistance. The collector resistance r_C is comprised of three components: (1) the resistance of the epitaxial collector under the base r_{C1}, (2) the resistance of the buried layer r_{C2}, and (3) the resistance from the buried layer to the contact r_{C3}, including the collector contact resistance. Since the resistances are generally functions of the device operating point, it is important not only to be able to measure the resistances, but equally important to decide which resistance value to use.

A common method to display the base and collector current is a semilog plot of the logarithm of the current plotted against the emitter-base voltage. Such a plot, shown in Fig. 4.9, is known as a *Gummel plot*. The two currents are expressed as a function of the emitter-base voltage V_{EB} by

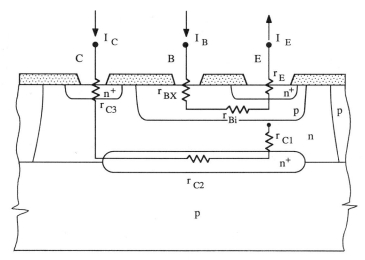

Fig. 4.8 An *npn* bipolar junction transistor and its parasitic series resistances.

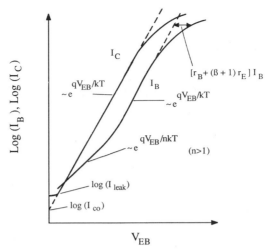

Fig. 4.9 Schematic Gummel Plots showing the effects of base space-charge region recombination and series resistances.

$$I_B = I_{B0} \exp\left[\frac{q(V_{EB} - I_B r_B - I_E r_E)}{nkT}\right] \qquad (4.33)$$

$$I_C = I_{C0} \exp\left[\frac{q(V_{EB} - I_B r_B - I_E r_E)}{kT}\right] \qquad (4.34)$$

I_{B0} depends on whether the dominant recombination mechanism is space-charge region or quasi-neutral region recombination. I_{C0} is a constant at low and intermediate voltages, but it decreases due to conductivity modulation at high injection levels.

The collector current Gummel plot is linear with slope of q/kT over most of its range. It saturates at the collector-base junction leakage current at low voltages and deviates from linearity at high voltages due to series resistances. Additional deviations from linearity at high voltages due to high level injection are not shown for simplicity.

The base current generally exhibits two linear regions. At low voltages the current is dominated by emitter-base space-charge region recombination with a slope of q/nkT, where $n \approx 1.5$ to 2. At intermediate voltages the slope is q/kT just as it is for the collector current, and at higher voltages the curve deviates from linearity due to series resistances. High level injection effects are again not shown for clarity.

The external voltage drop between the emitter and the base terminals V_{EB} is

$$V_{EB} = V'_{EB} + I_B r_B + I_E r_E = V'_{EB} + [r_B + (\beta + 1)r_E]I_B \qquad (4.35)$$

where β is the common emitter current gain, $I_C = \beta I_B$, $I_E = I_C + I_B = (\beta + 1)I_B$, and V'_{EB} is the potential drop across the emitter-base junction. Although r_E is generally small, the $(\beta + 1)$ multiplier can make it appreciable. The emitter and base resistances depress the currents below their ideal values, shown by the curves below the extrapolated dashed lines in Fig. 4.9.

BJT resistance measurement techniques fall into two main categories: dc methods and ac methods. The dc methods are generally fast and easy to implement. The ac techniques are time consuming and require measurement frequencies of typically 50 to 1000 MHz, necessitating a careful consideration of parasitics and of the distributed nature of BJT parameters.

4.5.1 Emitter Resistance

The emitter resistance in discrete BJT's is around $1\,\Omega$ and for small-area IC transistors it is around 5 to $10\,\Omega$. A popular method to determine r_E is based on a measurement of the collector-emitter voltage V_{CE} given by[28-30]

$$V_{CE} = \frac{kT}{q} \ln\left[\frac{I_B + I_C(1 - \alpha_R)}{\alpha_R[I_B - I_C(1 - \alpha_F)/\alpha_F]} \right] + r_E(I_B + I_C) + r_C I_C \qquad (4.36)$$

neglecting the small reverse saturation current. Here α_F and α_R are the large-signal forward and reverse common base current gains. When the collector is open circuited, $I_C = 0$ and Eq. (4.36) becomes

$$V_{CE} = \frac{kT}{q} \ln\left(\frac{1}{\alpha_R} \right) + r_E I_B \qquad (4.37)$$

A plot of I_B versus V_{CE}, measured with the setup of Fig. 4.10(a), is shown in Fig. 4.10(b). The curve is linear with a V_{CE}-axis intercept of $(kT/q) \ln(1/\alpha_R)$ and a slope of $1/r_E$. This behavior is indeed observed for discrete transistors.[29-31] The base current should not be too small for unambiguous measurements. For example, base currents around 10 mA are suitable for $r_E = 1\,\Omega$, and it is important to ensure that zero or very low collector currents are drawn during the measurement. Commercial curve tracers can be used for the measurement. A suitable connection is the following: BJT base connected to the *collector* terminal, BJT emitter connected to the *emitter* terminal, and BJT collector connected to the *base* terminal of the curve tracer.[32]

Departures of the I_B-V_{CE} curve from linearity occur when α_R is current dependent. This generally happens at low and high currents. Hence an r_E determination may not yield one unique value. The slope of the curve increases sometimes at high base currents.[32-33] Intermediate base currents usually give good linearity. Additional complications can arise for integrated

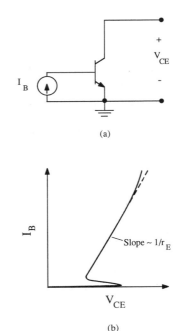

Fig. 4.10 Emitter resistance measurement: (a) measurement setup, (b) I_B-V_{CE} plot.

circuit transistors where part of the buried layer resistance can add to the emitter resistance due to internally circulating currents even for zero external collector current. The base resistance can also contribute to r_E.[34] The accuracy of this method is also dependent on the sensitivity of the base charge with respect to base current.[35]

4.5.2 Collector Resistance

A problem with collector resistance measurements is the strong dependence of collector resistance on the device operating point. The collector resistance can be determined by the same I_B-V_{CE} method of Section 4.5.1 by inter-changing the collector and emitter terminals. With $E \rightarrow C$ and $C \rightarrow E$, the I_B-V_{CE} curve has a V_{CE}-axis intercept of $(kT/q)\ln(1/\alpha_F)$ and a slope of $1/r_C$. Another method utilizes the parasitic substrate *pnp* transistor that exists in the structure of Fig. 4.8 and the reverse transistors associated with the *npn* transistor to determine the internal voltages of the *npn* BJT.[36] This allows r_C to be determined.

Another method uses the transistor output characteristics for r_C determination. Typical output I_C-V_{CE} curves are shown in Fig. 4.11. The two lines labelled $1/r_{Cnorm}$ and $1/r_{Csat}$ represent the two limiting values of r_C. The $1/r_{Cnorm}$ line is drawn through the knee of each curve, where the output curves tend to horizontal. The collector resistance obtains for the device in

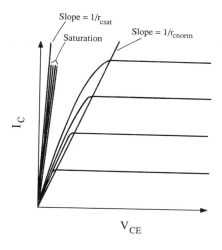

Fig. 4.11 Common emitter output characteristics. The two lines show the limiting values of r_C.

its normal, active mode of operation. The $1/r_{Csat}$ line gives the appropriate collector resistance for the transistor in saturation. A good discussion of this measurement technique using a curve tracer is given in Getreu.[32]

4.5.3 Base Resistance

The base resistance is difficult to determine accurately because it depends on the device operating point and because it is influenced by the emitter resistance through the term $(\beta + 1)r_E$. The base current flows laterally in BJT's giving lateral voltage drops in the base. This in turn causes V_{EB} to be a function of position. Small V_{EB} variations give rise to large current variations since I_C and I_B depend exponentially on V_{EB}. Most of the emitter current flows at the emitter edge nearest the base contact. This is referred to as *emitter crowding*. It reduces the distance for base current flow with increased emitter current, thereby decreasing r_{Bi} with current.

A simple method to determine the total series resistance between emitter and base is shown in Fig. 4.9. The experimental base current deviates from the extrapolated straight line by the voltage drop

$$\Delta V_{EB} = [r_B + (\beta + 1)r_E]I_B \tag{4.38}$$

A plot of $\Delta V_{EB}/I_B$ versus β has a slope of r_E and an intercept on the $\Delta V_{EB}/I_B$-axis of $r_B + r_E$. The current gain β must be varied in this measurement. There are two ways to do this. (1) Choose a device with varying β over some operating range, or (2) use different devices from the same lot. The first method ensures that only one device is measured, but conductivity modulation and other second-order effects may distort the measurement since the current must be varied to get different β's. In order to avoid conductivity modulation and other second-order effects, one should make the measurement at a constant emitter current. But of course a constant I_E implies constant β. In that case one must resort to the use of different devices from the same lot whose β's vary over some appropriate range, assuming the resistances to be the same for all devices from that lot. The latter method has been successfully used.[37]

A variation on this method is based on rewriting Eqs. (4.33) as[33]

$$\frac{nkT}{qI_C} \ln\left(\frac{I_{B1}}{I_B}\right) = \left(r_E + \frac{r_{Bi}}{\beta}\right) + \frac{r_E + r_{Bx}}{\beta} \qquad (4.39)$$

where $r_B = r_{Bi} + r_{Bx}$ and $I_{B1} = I_{B0} \exp(qV_{EB}/nkT)$. r_{Bi}/β is constant if r_{Bi} is proportional to β.[38] A plot of $(kT/qI_C)\ln(I_{B1}/I_B)$ versus $1/\beta$, for $n = 1$, has a slope of $r_E + r_{Bx}$ and an intercept on the $(kT/qI_C)\ln(I_{B1}/I_B)$-axis of $r_E + r_{Bi}/\beta$, as shown in Fig. 4.12. The intrinsic base resistance can be calculated. For a rectangular emitter of width Z and length L with a base contact on one side $r_{Bi} = Z\rho_{si}/3L$, where ρ_{si} is the intrinsic base sheet resistance. For a rectangular emitter with two base contacts, $r_{Bi} = Z\rho_{si}/12L$. For square emitters with contacts on all sides, $r_{Bi} = \rho_{si}/32$ and for circular emitters with a base contact all around $r_{Bi} = \rho_{si}/8\pi$.[33] The method based on Eq. (4.39) does not take into account lateral voltage drops along the intrinsic base current path. This condition is satisfied for sufficiently low collector currents of less than 10 to 20 mA for scaled digital BJT's.[33]

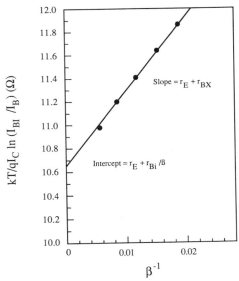

Fig. 4.12 Measured device characteristics according to Eq. (4.39) for a self-aligned, high-speed digital BJT. Reprinted after Ning and Tang[33] by permission of IEEE (© 1984, IEEE).

The method of Eq. (4.39) must be used with caution for polysilicon emitter contacts when a thin insulating barrier exists between the polysilicon and the single crystal emitter. This can cause the $(kT/qI_C) \ln(I_{B1}/I_B)$ versus $1/\beta$ curve to be nonlinear for low $1/\beta$ values. The slope of this plot can even become negative. This behavior cannot be explained by a resistive drop, but is attributed to an interfacial layer between the polysilicon contact and the single crystal emitter.[39]

Several techniques to measure r_B are based on frequency measurements. In the *input impedance circle method*, the emitter-base input impedance is measured as a function of frequency and plotted on the complex impedance plane for zero ac collector voltage.[40-41] The locus of this plot is a semicircle whose real axis intersections at low and high frequencies are

$$R_{in,lf} = r_\pi + r_B + (1 + \beta)r_E \tag{4.40a}$$

$$R_{in,hf} = r_\pi + r_B \tag{4.40b}$$

Resistance r_π can be calculated from the relationships $r_\pi = \beta/g_m$ with $g_m = qI_C/nkT$. This method allows both r_B and r_E to be determined. The effect of r_π on the measurement of r_B can be reduced by making the measurement at low temperatures, where r_π is reduced according to the relationships $r_\pi = nkT\beta/qI_C$.[42] The semicircle is sometimes distorted due to parasitic capacitances, making the interpretation more difficult. Furthermore the measurement is very time-consuming and loses accuracy at low collector current when the circle diameter is large. The method is more accurate for $r_B > 4\Omega$ and $I \geq 1$ mA.[43]

A variation of this technique is the *phase cancellation method* in which a common base transistor is connected to an impedance bridge, and the input impedance is measured as a function of collector current at a constant frequency of a few MHz. The collector current is varied until the input capacitance becomes zero, and the input impedance is purely resistive at collector current I_{C1}. The input impedance is $Z_i = r_B + r_E$, and the base resistance is given by[41]

$$r_B = \frac{nkT}{qI_{C1}} \tag{4.41}$$

The phase cancellation method does not lend itself to BJT's with low β's ($\beta < 10$) commonly found in lateral *pnp* transistors, and the base resistance in this method obtains for one value of collector current only. However, the method is fast and relatively unaffected by the emitter resistance since r_E appears in the input impedance as r_E directly, not as $(\beta + 1)r_E$.

In another method the frequency response of $\beta(f)$ and $y_{fb}(f)$, the forward transfer admittance of the BJT in the common base configuration, are measured. The base resistance is given by[44]

$$r_B = \frac{\beta(0)f_\beta}{y_{fb}(0)f_y} \tag{4.42}$$

where $\beta(0)$ is the low frequency β, $y_{fb}(0)$ is the low frequency y_{fb}, $f_\beta = 3$ db frequency of β, and $f_y = 3$ db frequency of y_{fb}. The 3 db frequency is the frequency at which the respective quantity has decreased to 0.7 of its low frequency value. The advantage of this technique is that Eq. (4.42) is relatively unaffected by collector and emitter resistances and that the measurement of y_{fb} is relatively insensitive to stray capacitance. However, it does require measurements of β and y_{fb} over a wide frequency range. In a variation on one of the ac methods, the input impedance of common emitter BJT's is measured at 10 to 50 MHz and r_{Bi}, r_{Bx}, and r_E are extracted from the measurement.[45] The method is suitable for low emitter-base voltages, where high current effects are negligible. A further variation using a single frequency but varying the emitter-base voltage allows not only the base and emitter resistances but also the emitter-base and the collector-base capacitances to be determined.[46]

The base resistance can also be determined from a pulse measurement similar to the method shown in Fig. 4.3. The base current of a common emitter BJT is pulsed to zero, and the resulting V_{EB} is determined.[47] The base resistance is determined from the sudden drop of the emitter-base voltage $\Delta V_{EB} = r_B I_B$. For this method to be successful, the base resistance must be sufficiently large for the emitter-base voltage drop to be measurable.

The base resistance can be determined through the use of an added external resistance R, as shown in Fig. 4.13(a). The principle of the method consists of measuring the voltage drop across r_B caused by the capacitive current flowing through C_C from the collector to the base terminal. First switch S is closed and the voltage drop across r_B, the small-signal emitter voltage $v_{e1} = j\omega C_C r_B v_C$, is measured across the forward-biased emitter-base terminal by a high impedance voltmeter. Then switch S is opened, and the voltage $v_{e2} = j\omega C_C r_B v_C + j\omega(C_C + C_L)R v_C$ is measured. From these two expressions one finds[48]

$$r_B = R\left(1 + \frac{C_L}{C_C}\right)\frac{v_{e1}}{(v_{e2} - v_{e1})} \tag{4.43}$$

where C_C represents that part of the base junction capacitance facing the active base region and C_L represents the remaining portion. One may replace $(1 + C_L/C_C)$ by A_E/A_C. This simplification allows r_B to be determined from v_{e1}, v_{e2}, R, and the geometry of the transistor.

The frequency of the measurement must be sufficiently low that $(r_B + R) \ll 1/[j\omega(C_C + C_L)]$, and it must be sufficiently high to overrule the Early effect. One way to check this is to change the frequency of the sinusoidal signal source v_c and to verify that v_e changes proportionally. A plot of r_B as

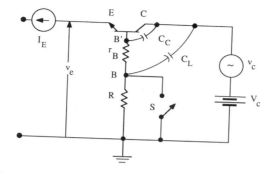

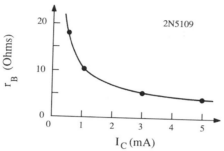

4.13 (a) Base resistance measurement by means of the voltage drop across the base resistance caused by the capacitive collector current flowing through C_c, (b) base resistance measured as a function of collector current. Reprinted with permission after Jespers.[48]

a function of collector current measured by this technique is shown in Fig. 4.13(b).

A somewhat more elaborate experimental technique to determine r_B is a measurement of the noise in a BJT.[49-50] The base resistance may be calculated from the noise figure in the "white" part of the noise spectrum at low source impedance. It can also be obtained from the low-frequency noise spectrum.[43] Noise methods become inaccurate when current crowding occurs.

4.6 MOSFET's

4.6.1 Series Resistance and Channel Length—*I-V*

The MOSFET series resistance and the *effective* channel length are frequently determined with one measurement technique. The resistance between source and drain consists of the source resistance, the channel resistance, and the drain resistance. The source resistance r_S and drain resistance r_D are shown in Fig. 4.14(a). They are the result of the source and drain contact resistance, the sheet resistance of the source and drain, the spreading resistance at the transition from the source diffusion to the channel, and any additional "wire" resistance. The channel resistance is contained in the MOSFET symbol and is not shown explicitly.

Current crowding in the source in the vicinity of the channel gives rise to the spreading resistance R_{sp}. A first-order expression for R_{sp} for a diffused source of constant resistivity is given by

$$R_{sp} = \frac{0.64\rho}{Z} \ln\left[\frac{\xi x_j}{x_{ch}}\right] \tag{4.44}$$

where Z is the channel width, ρ the source resistivity, x_j the junction depth, x_{ch} the channel thickness, and ξ a factor that has been given as 0.37,[51] 0.58,[52] 0.75,[53] and 0.9.[54] Its exact value is not that important since it appears in the "ln" term. More realistic expressions for R_{sp} have been

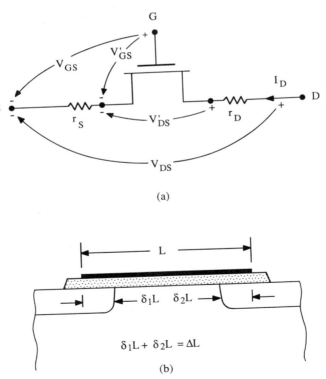

(a)

(b)

Fig. 4.14   (a) A MOSFET with source and drain resistances, (b) device cross section showing the physical gate length L and gate shortening ΔL with $L_{eff} = L - \Delta L$.

derived for diffused junctions where the non-uniform dopant profile of the source junction is considered.[51] The effective channel length L_{eff} differs from the mask-defined channel length L and even from the device channel length due to source and drain junction encroachment under the gate, as shown in Fig. 4.14(b). In that figure we show L as the physical channel length. Frequently L is taken to be the mask-defined channel length which may be different from that shown on the figure.

Neglecting the body effect of the ionized bulk charge in the MOSFET space-charge region for simplicity, the MOSFET current-voltage equation is

$$I_D = k(V'_{GS} - V_T - 0.5V'_{DS})V'_{DS} \tag{4.45}$$

where $k = k_0/[1 + \theta(V'_{GS} - V_T)]$, $k_0 = Z_{eff}\mu_0 C_{ox}/L_{eff}$, $Z_{eff} = Z - \Delta Z$, $L_{eff} = L - \Delta L$, V_T is the threshold voltage, V'_{GS} and V'_{DS} are defined in Fig. 4.14, θ is the mobility degradation coefficient, Z is the channel width, L is the channel length, C_{ox} is the oxide capacitance/unit area, and μ_0 is the low-field mobility. Z and L are sometimes used for the physical gate width and length, respectively. However, Z and L usually refer to the mask dimensions. Both θ and μ_0 are discussed in Section 5.6.1 with reference to Eq. (5.63). With $V_{GS} = V'_{GS} + I_D r_S$ and $V_{DS} = V'_{DS} + I_D(r_S + r_D)$, we can express Eq. (4.45) in terms of the terminal voltages V_{GS} and V_{DS} as

$$I_D = k[V_{GS} - V_T - I_D r_S - 0.5(V_{DS} - I_D(r_S + r_D))][V_{DS} - I_D(r_S + r_D)] \tag{4.46}$$

Equation (4.46) is quite complex and does not lend itself to easy interpretation since k is a function of V'_{GS}, which in turn is a function of I_D. For series resistance measurements the drain voltage is usually very low ($V_D \approx 50 - 100$ mV) ensuring device operation in the *linear* region, and the device is turned on strongly [$(V_{GS} - V_T) \geq 1$ V]. These conditions ensure that $r_S V_{DS} \ll r_T(V_{GS} - V_T)$, where $r_T = r_S + r_D$. If furthermore $2k_0 r_S V_{DS} \ll 1$, Eq. (4.46) becomes[55]

$$I_D \approx \frac{k_0(V_{GS} - V_T)V_{DS}}{1 + (\theta + k_0 r_T)(V_{GS} - V_T)} \tag{4.47}$$

Equation (4.47) has been used in several implementations to find r_T.

Hsu suggests the addition of an external resistance r_X in series with the source.[56] The total source resistance becomes effectively $r_S + r_X$, and Eq. (4.47) becomes

$$\frac{1}{I_D} = \frac{\theta + k_0(r_T + r_X)}{k_0 V_{DS}} + \frac{1}{k_0(V_{GS} - V_T)V_{DS}} \tag{4.48}$$

$1/I_D$ is plotted as a function of r_X for several different gate voltages. This plot has a slope of $1/V_{DS}$ and an intercept on the r_X axis of $r_X = -[r_T + \theta/k_0 + 1/(k_0(V_{GS} - V_T))]$; that is, $r_X \sim -(V_{GS} - V_T)^{-1}$. A second plot of $-r_X$ versus $(V_{GS} - V_T - V_{DS}/2)^{-1}$ has a slope given by $1/k_0$ and an intercept of $(r_T + \theta/k_0)$, as shown in Fig. 4.15.

In the original work the mobility degradation factor was neglected.[56] This can introduce an appreciable error since the intercept would be interpreted as $-r_T$ instead of $-(r_T + \theta/k_0)$. Typically $k_0 \approx 4 \times 10^{-5} Z_{eff}/L_{eff}$ A/V^2 and $\theta \approx 0.05 - 0.1$ V^{-1} for $W_{ox} = 500$ Å. For $Z_{eff}/L_{eff} = 5$, we find $\theta/k_0 \approx 250$ Ω using $\theta = 0.05$ V^{-1}. Since θ can be determined from a μ_{eff} versus $(V_{GS} - V_T)$ measurement, as discussed in Section 5.6.1, it is not necessary to neglect it in the analysis.

A different representation of Eq. (4.47) is[55]

$$\frac{(V_{GS} - V_T)}{I_D/V_{DS}} = \frac{1 + [\theta + k_0(r_T + r_X)](V_{GS} - V_T)}{k_0} \equiv E \qquad (4.49)$$

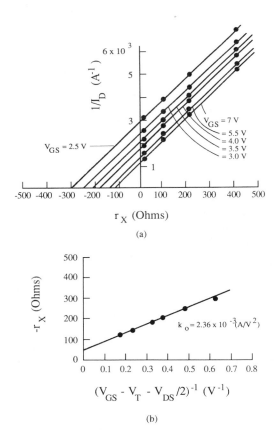

(a)

(b)

Fig. 4.15 (a) $1/I_D$ versus r_x as a function of gate voltage, (b) intercept $-r_x$ versus inverse gate voltage. Reprinted with permission after Hsu.[56]

This equation can be analyzed in one of two ways: (1) E is plotted against $(V_{GS} - V_T)$ as a function of r_X requiring only one device. The slope is $[(r_T + r_X) + \theta/k_0]$ and the intercept on the E-axis is $1/k_0$. From the intercept we find k_0 and from the mobility-gate voltage curve, we find θ, allowing r_T to be determined; (2) E is plotted against $(V_{GS} - V_T)$ as a function of channel length for $r_X = 0$ requiring several devices with different channel lengths. The slopes of these plots are $[r_T + \theta/k_0]$ and the intercepts on the E-axis are $1/k_0$. k_0 varies since several devices with varying channel lengths are used. As in (1) θ can be determined from a μ_{eff} versus $(V_{GS} - V_T)$ curve. It is also possible to plot the slope $\Delta E/\Delta(V_{GS} - V_T)$ versus $1/k_0$. The slope of this plot is θ and the intercept on the $\Delta E/\Delta(V_{GS} - V_T)$ axis is r_T.

Several other techniques are based on a simplification of Eq. (4.46). For strong inversion but low V_{DS}, the gate voltage satisfies the inequality $(V_{GS} - V_T) \gg I_D r_S + (V_{DS} - I_D r_T)/2$, allowing Eq. (4.46) to be written as

$$I_D \approx k(V_{GS} - V_T)(V_{DS} - I_D r_T) \tag{4.50}$$

The measured resistance, defined by $r_m = V_{DS}/I_D$, is[57]

$$r_m = r_T + \frac{1}{k(V_{GS} - V_T)} = r_T + \frac{[1 + \theta(V_{GS} - V_T)](L - \Delta L)}{Z_{eff}\mu_0 C_{ox}(V_{GS} - V_T)} \tag{4.51}$$

The channel length L in this equation is generally the mask-defined channel length. Plotting r_m versus L for several values of $(V_{GS} - V_T)$ gives a series of straight lines. In the ideal case these lines all intersect with the point of intersection giving r_T on the r_m axis and ΔL on the L axis, as shown in Fig. 4.16. These values of r_T and ΔL are obtained whether L is the mask-defined

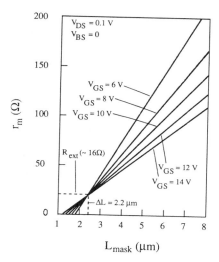

Fig. 4.16 Measured MOSFET resistance versus mask channel length as a function of gate voltage. Reprinted after Chern et al.[57] by permission of IEEE (© 1980, IEEE).

channel length or the physical gate length of the device. A method related to Eq. (4.51), in which the total resistance is measured and plotted as a function of mask channel length, was one of the first methods to determine the effective channel length.[58]

If the r_m versus L lines fail to intersect at a common point, one can carry this technique one step further by writing Eq. (4.51) as

$$r_m = r_T + AL_{eff} = (r_T - A\Delta L) + AL = B + AL \qquad (4.52)$$

where B is given by

$$B = r_T - A\Delta L \qquad (4.53)$$

A and B depend implicitly on $(V_{GS} - V_T)$, and they can be fitted for various gate voltages with a least squares technique. Such a linear regression can be used to extract both r_T and ΔL, with no requirement for a common intersection point.[59]

The difference between L and L_{eff} is particularly important for short-channel devices. But short-channel devices also have a channel length-dependent threshold voltage. It is therefore important that the threshold voltage of each device be determined independently. Measuring the threshold voltage on a device with one channel length and then using that same threshold voltage for all devices with different channel lengths is generally insufficient. Furthermore both the series resistance and the effective channel length may depend on the gate voltage.[60] The effective channel length increases and the series resistance decreases with increasing gate voltage. This is the result of channel broadening in which L_{eff} is modulated by the gate voltage. The effective channel is considered to lie between the transitional points where the current flows from the lateral spread of the source and the drain diffusion to the inversion layer. The end of the channel is where the conductivity of the diffusion resistance is approximately equal to the incremental inversion layer conductivity. Since the inversion layer conductivity increases with gate voltage, it follows that L_{eff} increases with higher gate voltage. Simultaneously the series resistance decreases with increasing gate voltage.

The dependence of L_{eff} and r_T on gate voltage is particularly acute for lightly doped drain–source (LDD) devices. LDD MOSFET's contain lightly doped regions between the source and the channel and between the drain and the channel.[61] The effect of gate voltage-dependent L_{eff} and r_T is a failure of the r_T versus L lines to intersect at a common point. As a result no unique value of these two parameters can be obtained. A suggested method to ensure that the lines intersect at one point is to vary V_T in Eq. (4.51) instead of varying V_{GS}.[62] This is most conveniently done by varying the substrate bias V_{BS}, maintaining the gate voltage constant at $V_{GS} \approx 1$ to 2 V.

Later studies showed the substrate bias technique to yield unreliable data because substrate bias changes the threshold voltage of MOSFET's of different channel lengths by different amounts. An improved method is a combination substrate/gate bias technique.[63] The gate voltage of the longest channel device is held constant while its threshold voltage is changed by substrate bias modulation. When measuring the resistance of shorter-channel devices, the gate voltage is reduced by the amount the threshold voltage has decreased from the long-channel value. This ensures constant gate drive for all devices. Yet another variation on the r_T versus L method utilizes a "paired gate voltage" approach.[64] Two r_T versus L lines are determined for two gate voltages, one being typically 0.5 V lower than the other. The intersection of these two lines gives a good approximation of r_T and L_{eff}. The gate voltage dependence of r_T and L_{eff} can be found using various V_{GS} pairs.

A further variation of Eq. (4.50) for devices with two different channel lengths is the drain current ratio[65]

$$\frac{I_{D1}}{I_{D2}} \approx \frac{k_1}{k_2} \left(1 - \frac{(I_{D1} - I_{D2})r_T}{V_{DS}} \right) \tag{4.54}$$

for $V_{DS1} \gg I_{D1}r_T$ and $V_{DS2} \gg I_{D2}r_T$ and equal mobilities and equal threshold voltages for the two devices. A plot of I_{D1}/I_{D2} versus $(I_{D1} - I_{D2})$ has a slope of $k_1 r_T / k_2 V_{DS}$ and an intercept on the I_{D1}/I_{D2} axis of k_1/k_2. This method does not work if the conditions $V_{DS1} \gg I_{D1}r_T$ and $V_{DS2} \gg I_{D2}r_T$ are not satisfied. For that case a modification has been proposed, consisting of a plot of $(V_{DS2}/I_{D2} - V_{DS1}/I_{D1})$ versus V_{DS1}/I_{D1}.[66] This plot gives a straight line with an intercept on the V_{DS1}/I_{D1} axis of r_T. It also yields the effective channel length.

A derivative technique to determine L_{eff} that overcomes some of the difficulties of the intersection point methods is based on transconductance measurements.[67] It utilizes low gate and drain voltages ensuring $r_{ch} \gg r_T$, where r_{ch} is the channel resistance. $V_{GS} \approx 0.8$ to 1.1 V and $V_{DS} \approx 0.1$ V are suitable for $V_T \approx 0.6$ V. This method requires devices of constant gate width and varying gate lengths and one large square device with $L = Z = 100$ μm, for example. The devices operate in the linear current–voltage region given by

$$I_D = \mu_{eff} C_{ox} \left(\frac{Z_{eff}}{L_{eff}} \right) (V_{GS} - V_T - 0.5 V_{DS}) V_{DS} \tag{4.55}$$

The transconductance, defined by $g_m = \partial I_D / \partial V_{GS}$ at constant V_{DS}, is given by

$$g_m = \frac{\mu_{eff} C_{ox} Z_{eff} V_{DS}}{L_{eff}} \tag{4.56}$$

A quantity K is defined by

$$K = \frac{g_m}{V_{DS}} \frac{L}{Z} \qquad (4.57)$$

where L and Z are the mask-defined channel length and width. With the help of Eq. (4.56), K can be written as

$$K = \mu_{eff} C_{ox} \frac{L}{Z} \frac{Z_{eff}}{L_{eff}} = \mu_{eff} C_{ox} \frac{(1 - \Delta Z/Z)}{(1 - \Delta L/L)} \qquad (4.58)$$

where $L_{eff} = L - \Delta L$ and $Z_{eff} = Z - \Delta Z$.

For devices with varying L but constant width $Z = Z_0$, K becomes

$$K_L = \mu_{eff} C_{ox} \frac{(1 - \Delta Z/Z_0)}{(1 - \Delta L/L)} \qquad (4.59)$$

For the square device there is a similar expression denoted by K_{100}. Taking the ratio K_{100}/K_L and neglecting second- and higher-order terms, we find

$$\left(1 - \frac{\Delta Z}{100}\right) - \left(1 - \frac{\Delta Z}{Z_0}\right)\frac{K_{100}}{K_L} = \Delta L \left(\frac{1}{L} - \frac{K_{100}}{100 K_L}\right) \qquad (4.60)$$

where ΔZ, ΔL, Z_0, and L are in units of microns. We assume μ_{eff} and C_{ox} to be identical for all devices of a given test chip.

Equation (4.60) is the key equation for this technique. K_L is determined for each of several devices with constant gate width Z_0 and different gate lengths, and the ratio K_{100}/K_L is determined for each. Equation (4.60) is solved for ΔL by initially assuming some value for ΔZ ($\Delta Z = 0$ is acceptable). Then, by iteration with Eq. (4.70), a more precise value of ΔL can be found. Usually two iterations are sufficient. By operating the MOSFET's in the linear region, it is permissible to neglect effects due to mobility reduction, short-channel and narrow-width threshold voltage variations as well as series resistances.

The transconductance is also used in the transresistance method.[68-69] The transconductance g_m and the drain conductance $g = I_D/V_D$ are measured in the linear MOSFET region at drain voltages of 25 to 50 mV. The transresistance r is defined by

$$r = \frac{g_m}{g^2} \qquad (4.61)$$

Two devices are required for the measurement. One is a long-channel device with channel length L_{ref} ($L_{ref} = 30$ to $100\ \mu m$) and the other is a short-channel device with known mask channel length. L. The transresistance is determined for each device and a parameter $\Delta\lambda$ is calculated from the two channel lengths and the two transresistances as

$$\Delta\lambda = \frac{(Lr_{\text{ref}} - L_{\text{ref}}r)}{(r_{\text{ref}} - r)} \tag{4.62}$$

$\Delta\lambda$ is plotted against $(V_{\text{GS}} - V_{\text{T}})$ and the extrapolated intercept on the $\Delta\lambda$ axis is ΔL. The series resistance depends on the channel lengths and the drain conductances as

$$r_{\text{T}} = \frac{L_{\text{ref}}/g - L/g_{\text{ref}} + \Delta L/g_{\text{ref}} - \Delta L/g}{(L_{\text{ref}} - L)} \tag{4.63}$$

The transresistance technique has been shown to be very accurate when compared to some of the other methods discussed in this chapter.[70]

Other methods of determining the series resistance are based on fitting the current-voltage characteristics. In one method, a two-dimensional device simulator is employed to determine L_{eff}.[71] The source and drain resistances are then predicted from comparison of experimental data with the simulated curves. In another approach, an analytic equation is fit to the experimental data and series resistance as well as mobility degradation factor are extracted.[72]

A combination of resistance measurements, plots, and curve fitting is employed in one method in which ΔL, r_{T}, $\mu_0 C_{\text{ox}}$, and θ are extracted.[73] It is based on Eq. (4.51), which can be written as

$$r_{\text{m}} = r_{\text{T}} + \frac{(L - \Delta L)}{Z_{\text{eff}}\mu_0 C_{\text{ox}}(V_{\text{GS}} - V_{\text{T}})} + \frac{\theta(L - \Delta L)}{Z_{\text{eff}}\mu_0 C_{\text{ox}}} \tag{4.64}$$

First $r_{\text{m}} = V_{\text{DS}}/I_{\text{D}}$ is plotted against $(V_{\text{GS}} - V_{\text{T}})^{-1}$. The slope of this plot is $S = (L - \Delta L)/Z_{\text{eff}}\mu_0 C_{\text{ox}}$, and the intercept on the r_{m} axis is $r_i = [r_{\text{T}} + \theta(L - \Delta L)/Z_{\text{eff}}\mu_0 C_{\text{ox}}] = r_{\text{T}} + \theta S$. Next S is plotted against L. This plot has a slope of $1/Z_{\text{eff}}\mu_0 C_{\text{ox}}$ and an intercept on the L axis of ΔL, allowing $\mu_0 C_{\text{ox}}$ and ΔL to be determined. Lastly, r_i is plotted against S. This plot gives θ from the slope and r_{T} from the intercept on the r_i axis.

Two devices suffice for these measurements. The channel lengths of the device pair should be selected to minimize the error in ΔL associated with the extrapolation of the S versus L plot because errors in the plotted S points are magnified by extrapolation. Errors in ΔL are minimized by choosing channel lengths that differ by about a factor of ten. Furthermore $(V_{\text{GS}} - V_{\text{T}})$ should be chosen to cover a wide range. One bias point should be for low $(V_{\text{GS}} - V_{\text{T}})$ (about 1 V) where $\mu_0 C_{\text{ox}}$ is dominant. A second bias point should be for high $(V_{\text{GS}} - V_{\text{T}})$ (about 3–5 V), where θ and r_{T} dominate.

An ac method utilizes Eq. (4.47) with the device biased at its dc operating point. An ac gate signal is applied and the first and second harmonics of the drain current are measured to determine k_0, θ, and r_{T}.[74] A comprehensive study of the various mechanisms limiting the accuracy of channel length extraction techniques especially for lightly doped drain MOSFET's has been carried out.[75] Low gate overdrives and consistent threshold voltage measurements are very important for reliable channel length extraction.

4.6.2 Channel Length—C-V

The current-voltage methods of Section 4.6.1 are by far the most popular methods to determine series resistance and effective channel length, largely because of their measurement simplicity. Nevertheless, capacitance techniques are occasionally used to determine L_{eff}. Series resistance cannot be determined by C-V techniques. In one method, the capacitance is measured between the gate and the source-drain connected together.[76] With the device biased into strong accumulation, the area directly under the gate, the intrinsic gate area, given by $L_{eff}Z_{eff} = (L - \Delta L)Z_{eff}$ is decoupled from the source and drain, and only the overlap capacitance is measured. With the device biased into strong inversion, the intrinsic gate area is coupled to the source and the drain by the inversion layer and the intrinsic plus overlap capacitances are measured. By subtracting the overlap capacitance, determined in strong accumulation, the intrinsic capacitance is determined. It is given by

$$C_{ox} = \frac{K_{ox}\varepsilon_0 Z_{eff}(L - \Delta L)}{W_{ox}}$$ (4.65)

Plotting C_{ox} versus L gives a straight line with slope $K_{ox}\varepsilon_0 Z_{eff}/W_{ox}$ and intercept on the L axis of ΔL.

For small-area MOSFET's this results in very small capacitances. For example, for $L_{eff} = 5$ μm, $Z_{eff} = 50$ μm, and $W_{ox} = 300$ Å, we find $C_{ox} \approx 0.3$ pF. The overlap capacitance is still lower making for difficult measurements. The measurement problems can be alleviated by connecting many devices in parallel, thereby making the effective area much larger. In one design 3200 transistors were connected in parallel.[77] Two measurements are made. In the first, the capacitance between the gate and the source-drain substrate, all connected together, is measured with the surface heavily inverted. This gives the intrinsic and the overlap capacitance. Next, the capacitance between the gate and the source-drain, connected together, is measured with the surface in strong accumulation giving the overlap capacitance. The difference is the intrinsic capacitance. The effective channel length is then given by

$$L_{eff} = \frac{C_{ox}W_{ox}}{nK_{ox}\varepsilon_0 Z_{eff}}$$ (4.66)

where n is the number of MOSFET's connected in parallel.

4.6.3 Channel Width

The methods to determine the channel width Z are similar to those for channel length. An early technique used a plot of the MOSFET drain conductance as a function of Z for devices with constant channel length.[78] If source and drain resistances are neglected, then from Eq. (4.45) the drain conductance $g_D = \partial I_D / \partial V_{DS}$ at constant gate voltage is given by

$$g_D \approx \frac{(Z - \Delta Z)\mu_0 C_{ox}(V_{GS} - V_T)}{L_{eff}[1 + \theta(V_{GS} - V_T)]} \tag{4.67}$$

A plot of g_D against Z has an intercept on the Z-axis of ΔZ at $g_D = 0$. This method neglects the source and drain resistances which is more problematic than it is for channel length measurements. Although it is a reasonably good assumption to take r_S and r_D as constants for devices with varying channel *lengths*, this is no longer true for devices with varying channel *widths*. Both source and drain resistances depend on channel width. Taking $r_T = r_S + r_D$ into account allows the measured drain resistance to be written as [see Eq. (4.51)]

$$r_m = r_T + \frac{1}{k(V_{GS} - V_T)} = r_T + \frac{[1 + \theta(V_{GS} - V_T)](L - \Delta L)}{Z_{eff}\mu_0 C_{ox}(V_{GS} - V_T)} \tag{4.68}$$

A simple interpretation of Eq. (4.68) is no longer possible for $r_T \neq$ constant.

A method that does not rely on current-voltage measurements, and therefore is not affected by series resistance, is the capacitance method. The oxide capacitance of a MOSFET is given by

$$C_{ox} = \frac{K_{ox}\varepsilon_0 Z_{eff} L_{eff}}{W_{ox}} \tag{4.69}$$

where $Z_{eff} = Z - \Delta Z$. A plot of C_{ox} as a function of Z for transistors with identical gate lengths gives a straight line with slope $K_{ox}\varepsilon_0 L_{eff}/W_{ox}$ and intercept on the Z-axis at $Z = \Delta Z$.[79] The difficulty is that very small capacitances, in the 0.01 to 0.1 pF range, must be measured and stray and bonding pad capacitances must be carefully considered. Capacitance measurements are discussed in Chapter 6.

The method summarized by Eq. (4.60) is also suitable for effective channel width determination. Using a set of MOSFET's with varying channel widths and constant channel length, L_0 gives the relationship[67]

$$\left(1 - \frac{\Delta L}{100}\right) - \left(1 - \frac{\Delta L}{L_0}\right)\frac{K_Z}{K_{100}} = \Delta Z\left(\frac{1}{Z} - \frac{K_Z}{100 K_{100}}\right) \tag{4.70}$$

The measurement procedure is similar to that described in Section 4.6.1. K_Z is determined for each of several devices with constant gate length and different gate widths and the ratio K_Z/K_{100} is determined for each. Equation (4.70) is solved for ΔZ by initially assuming some value for ΔL ($\Delta L = 0$ is acceptable). Then, by iteration, a more precise value of ΔZ can be found. Usually two iterations are sufficient. Effects due to mobility reduction, short-channel and narrow-width threshold voltage variations as well as series resistances can be neglected by operating the MOSFET's in the linear current-voltage region.

4.7   MESFET's AND MODFET's

A MESFET (metal-semiconductor field-effect transistor) consists of a source, channel, drain, and gate, as shown in Fig. 4.17(a). Majority carriers flow from source to drain in response to a drain voltage. The drain current is modulated by a reverse bias on the metal-semiconductor junction gate. With sufficient reverse bias, the space-charge region of the metal-semiconductor contact extends to the insulating substrate and the channel is pinched off. The gate voltage necessary to drive the current to zero is the pinch-off voltage V_p. The output current-voltage characteristics resemble those of a MOSFET. However, in contrast to MOSFET's, the MESFET metal-semiconductor input junction can be forward biased leading to high input currents. A MODFET (modulation-doped FET) is similar to a MESFET. The main difference between the two lies in the method of forming the active lightly doped layers. For our purposes we will not distinguish between these two structures.

The ability to forward bias the gate of a MESFET allows additional measurements that are not possible with a MOSFET. With the gate forward biased, the equivalent circuit in Fig. 4.17(b) shows the drain-source voltage to be

$$V_{DS} = (r_{ch} + r_S + r_D)I_D + (\alpha r_{ch} + r_S)I_G \qquad (4.71)$$

where α accounts for the fact that the gate current flows only through a portion of the channel resistance from the gate to the source. The gate-source voltage is

$$V_{GS} = \left(\frac{nkT}{q}\right)\ln\left(\frac{I_G}{I_s}\right) + r_S(I_D + I_G) \qquad (4.72)$$

where $I_G = I_s \exp(qV_{GS}/nkT)$ is the expression for the forward-biased gate with no resistance.

A plot of I_D versus V_{DS} as a function of I_G has a slope given by $1/(r_{ch} + r_S + r_D)$, and V_{DS}/I_G gives $(\alpha r_{ch} + r_S)$ for $I_D = 0$. Furthermore, from the forward-biased I_G-V_{GS} curves as a function of I_D, we find $\Delta V_{GS}/\Delta I_D = r_S$ for $I_G =$ constant. This allows r_S, r_D, and r_{ch} to be determined.

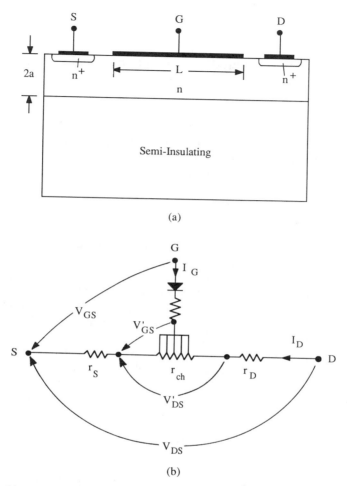

Fig. 4.17 (a) MESFET and (b) equivalent circuit showing the distributed nature of the channel resistance.

When the gate resistance is included in Fig. 4.17(b), it can be determined from a measurement of the gate current with a voltage between gate and source. However, $\log(I_G)$ is plotted against V_{GD}, not V_{GS}, with the drain open circuited. The deviation of this semilog plot from a straight line is caused by the gate resistance.[80]

Another method relies on a measure of the gate current as a function of the drain-source voltage. The source is grounded, and the gate current flows from the gate to the source. The gate current flowing through the source resistance and through a portion of the channel resistance r_{ch} creates a voltage drop. The drain acts as a voltage probe of this voltage drop. The "end" resistance is defined as

$$r_{end} = \frac{\partial V_{DS}}{\partial I_G} \tag{4.73}$$

From Eq. (4.71) the "end" resistance is approximately

$$r_{end} = \alpha r_{ch} + r_S \tag{4.74}$$

In one "end" resistance measurement method, the drain current is zero, and the drain contact floats electrically. This gives $\alpha = 0.5$. In another version drain current does flow, but it is constant during the measurement, and the drain does not float. For $I_G \ll I_D$,[81]

$$r_{end} = r_S + \frac{nkT}{qI_D} \tag{4.75}$$

A plot of r_{end} versus $1/I_D$ has a slope nkT/q and intercept on the r_{end} axis of r_S. It turns out that this plot has a rather limited straight-line portion. Deviation from a straight line at high I_D is the result of the drain current being not much lower than the saturation drain current. At low I_D there is a deviation due to a violation of the $I_G \ll I_D$ requirement, rendering the method of rather limited usefulness. A refinement of this method is given in Chaudhuri and Das.[82]

An entirely different method relies on a measurement of the drain-source resistance. The current-voltage expression of a MESFET, identical to that of a junction FET (JFET), is given by[83]

$$I_D = k\left\{V'_{DS} - \frac{2}{3}(V_{bi} - V_P)\left[\left(\frac{V'_{DS} - V_{bi} - V'_{GS}}{V_{bi} - V_P}\right)^{3/2} - \left(\frac{V_{bi} - V'_{GS}}{V_{bi} - V_P}\right)^{3/2}\right]\right\} \tag{4.76}$$

where $k = 2qZ\mu_n N_D a/L$, V_{bi} is the built-in voltage, V_P is the pinch-off voltage, $V'_{DS} = V_{DS} - (r_S + r_D)I_D$, and $V'_{GS} = V_{GS} - r_S(I_D + I_G)$. In Eq. (4.76), $0 \leq V_{DS} \leq V_{DS,sat}$ and $V_P \leq V_{GS} \leq 0$. The pinch-off voltage is determined by extrapolating the I_D versus V_{DS} curve, measured in the linear region of the device, to $I_D = 0$.

Under normal MESFET operation the gate is reverse biased, and the gate current can be neglected. Furthermore, if the device is operated in its linear region where $V_{DS} \ll (V_{bi} - V_{GS})$, Eq. (4.76) simplifies to

$$I_D \approx k(V_{DS} - r_T I_D)\left(1 - \sqrt{\frac{V_{bi} - V_{GS}}{V_{bi} - V_P}}\right) \tag{4.77}$$

From Eq. (4.77) the resistance $r_m = V_{DS}/I_D$ is

$$r_m = r_T + \frac{1}{k(1 - \sqrt{(V_{bi} - V_{GS})/(V_{bi} - V_P)})} \tag{4.78}$$

A plot of r_m as a function of $1/\{1-[(V_{bi}-V_{GS})/(V_{bi}-V_P)]^{1/2}\}$ gives a straight line with slope $1/k$ and intercept on the r_m axis of $r_T = r_S + r_D$.[84] Another method uses two drain currents at constant gate current with the gate under forward bias. The shift in the $I_G - V_{GS}$ curves corresponding to these two conditions is related to the source resistance.[85] A technique, related to the end contact resistance method, uses the gate electrode instead of the source and drain contacts to measure the source and drain resistances.[86]

4.8 THRESHOLD VOLTAGE

The threshold voltage V_T is an important MOSFET parameter required for circuit applications and for the channel length/width and series resistance measurements of this chapter. The threshold voltage for large-geometry devices on uniformly doped substrates with no short- or narrow-channel effects, when measured from gate to source, is given by

$$V_T = V_{FB} + 2\phi_F + \frac{\sqrt{2qK_s\varepsilon_0 N_A (2\phi_F - V_{BS})}}{C_{ox}} \qquad (4.79)$$

where V_{BS} is the substrate-source voltage, V_{FB} is the flatband voltage, and $\phi_F = (kT/q)\ln(N_A/n_i)$. The threshold voltage for non-uniformly doped, ion-implanted devices depends on the implant dose as well. Additional corrections obtain for short- and narrow-channel devices. We are not concerned with threshold voltage definitions here but with its measurement.

The most common threshold voltage measurement method is the *linear extrapolation* method in which the drain current is measured as a function of gate voltage at a low drain voltage of typically 50 mV to ensure operation in the linear MOSFET region.[75,87-89] According to Eq. (4.46) the drain current is zero for $(V_{GS} - V_T - 0.5V_D) = 0$. But Eq. (4.46) is valid only above threshold. The drain current is not zero below threshold and approaches zero only asymptotically. Hence the I_D versus V_{GS} curve is extrapolated to $I_D = 0$, and the threshold voltage is determined from the extrapolated or intercept gate voltage V_{GSi} by

$$V_T = V_{GSi} - \frac{V_{DS}}{2} \qquad (4.80)$$

Equation (4.80) is strictly only valid for negligible series resistance.[90] Fortunately series resistance is usually negligible at the low drain currents where threshold voltage measurements are made, but the resistance can be appreciable in LDD devices.

The $I_D - V_{GS}$ curve deviates from a straight line at gate voltages below V_T due to subthreshold currents and above V_T due to series resistance and mobility degradation effects. It is common practice to find the point of maximum slope on the $I_D - V_{GS}$ curve by a maximum in the transconductance, $g_m = \partial I_D / \partial V_{GS}$, fit a straight line to the curve at that point and extrapolate to $I_D = 0$, as illustrated in Fig. 4.18(a). The linear extrapolation method is sensitive to series resistance and mobility degradation.[75,90–91] Failure to correct for θ and r_T leads to an underestimate in V_T.

The threshold voltage can also be determined in the MOSFET saturation regime by plotting $I_D^{1/2}$ versus V_{GS} and extrapolating the curve to zero drain current, illustrated in Fig. 4.18(b).[92–93] Assumptions of negligible mobility degradation and negligible series resistance hold in this method as well but are more easily violated due to the higher drain and gate voltages. For the device of Fig. 4.18 there is good threshold voltage agreement between the two methods as shown in Table 4.1. The linear extrapolation technique can also be used for threshold voltage measurements of depletion-mode or buried channel MOSFET's.[94]

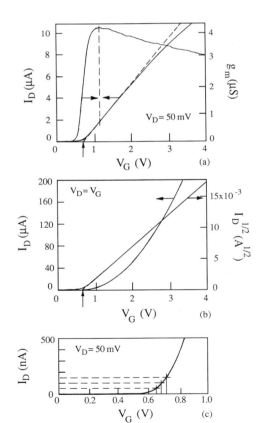

Fig. 4.18 (a) I_D and g_m versus V_G in the linear region, (b) I_D and $I_D^{1/2}$ versus V_G in the saturation region, (c) I_D versus V_G in the linear region. $L = Z = 25\ \mu m$, $W_{ox} = 390\ \text{Å}$. Courtesy of D. Feldbaumer, Motorola and B. Hussain, Arizona State University.

TABLE 4.1   Threshold Voltages for the Device of Fig. 4.18

Method	Threshold Voltage (V)
Linear I_D	0.69
$I_D^{1/2} - V_{GS}$	0.65
$g_D/g_m^{1/2} - V_{GS}$	0.72
$I_T = 50\,nA$	0.64
$I_T = 100\,nA$	0.68
$I_T = 150\,nA$	0.71
Subthreshold I_D	0.69

The derivative of the transconductance with gate voltage $\partial g_m/\partial V_{GS}$ is determined at low drain voltage ($V_{DS} \approx 5\,mV$) and plotted versus gate voltage in the *transconductance change* method. The peak of this curve occurs at $V_{GS} = V_T$. The method is not affected by series resistance and mobility degradation factors.[90]

A slightly more complex procedure, but one immune to series resistance and mobility degradation, is based on Eq. (4.47). The drain conductance $g_D = \partial I_D/\partial V_{DS}$ and transconductance $g_m = \partial I_D/V_{GS}$ are determined, and the ratio

$$\frac{g_D}{\sqrt{g_m}} = \frac{\sqrt{k_0}(V_{GS} - V_T)}{\sqrt{V_{DS}}} \tag{4.81}$$

is plotted against V_{GS}.[91] The intercept gives the threshold voltage. This method is valid provided the gate voltage is confined to small variations near V_T and the assumptions $V_{DS}/2 \ll (V_{GS} - V_T)$ and $\partial r_T/\partial V_{GS} = 0$, used in the derivation of Eq. (4.81), are satisfied.

It is obvious from Fig. 4.18(a) that the drain current at the threshold voltage is larger than zero. This fact is utilized in the *constant drain current* method where the gate voltage at a specified drain current is taken to be the threshold voltage. This technique is fast with only one voltage measurement necessary and it can be implemented with the circuit of Fig. 4.19 or by digital means.[92] It lends itself readily to threshold voltage mapping. The threshold current I_T is forced at the MOSFET source terminal, and the op-amp adjusts its output voltage to be equal to the threshold voltage consistent with that I_T. The problem with this method is the choice of the threshold current which should be independent of device geometry. Typically $I_T = I_D/(Z_{eff}/L_{eff})$ is specified at a current around 10 to 50 nA but other values have been used.[91–92] The effective channel length and width should be used in the expression. In Fig. 4.18(c) $I_T \approx 100\,nA$ gives a threshold voltage close to that obtained by the linear extrapolation method.

In the *subthreshold* method the drain current is measured as a function of gate voltage below threshold and plotted as $\log(I_D)$ versus V_{GS}. The

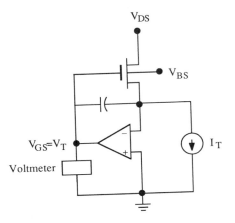

Fig. 4.19 Measurement circuit for the constant drain current threshold voltage method. Reprinted after Lee et al.[92] by permission of IEEE (© 1982, IEEE).

subthreshold current depends linearly on gate voltage in such a plot. The gate voltage at which the plot departs from linearity is defined as the threshold voltage.

In the *split C-V* method both gate and bulk currents are measured during a quasi-static *C-V* measurement with the transistor in a gate-controlled diode configuration.[95] The threshold voltage is the gate voltage where the two currents are equal. In all threshold voltage measurements it is important to state the sample temperature during measurement since V_T does depend on temperature. A typical V_T temperature coefficient is -2 mV/°C, but it can be higher.[96]

4.9   STRENGTHS AND WEAKNESSES

The techniques of this chapter cover such a variety of characterization techniques that it is difficult to summarize the strengths and weaknesses of each method here. Instead, we have chosen to mention the strengths and weaknesses throughout the chapter itself.

APPENDIX 4.1   SCHOTTKY DIODE CURRENT-VOLTAGE EQUATION

The current-voltage equation of a Schottky diode with series resistance is generally expressed as

$$I = AA^{**}T^2 e^{-q\phi_B/kT}(e^{q(V-Ir_s)/nkT} - 1) \tag{A4.1}$$

It has been proposed that Eq. (A4.1) is incorrect because it predicts the nonideality, included through the parameter n, to affect only the current flow from the semiconductor to the metal but not from the metal to the semiconductor.[97] This is obvious from Eq. (A4.1). For large forward bias the first term in the bracket is important only, and it contains the factor n. For reverse bias the second term is important, and it does not contain n.

To overcome this problem, we consider the voltage dependence of the barrier height. The barrier height ϕ_B depends on voltage due to image force barrier lowering, due to voltage drops across any interfacial layers that may exist between the metal and the semiconductor, and other possible effects. Assuming the barrier height depends linearly on voltage according to

$$\phi_B(V) = \phi_{BO} + \gamma(V - Ir_s) \tag{A4.2}$$

where γ is positive because the barrier height increases with increased forward bias. Equation (A4.1) then becomes

$$I = AA^{**}T^2 e^{-q\phi_{BO}/kT} e^{-q\gamma(V-Ir_s)/kT} (e^{q(V-Ir_s)/kT} - 1) \tag{A4.3}$$

Defining the ideality factor n by

$$\frac{1}{n} = 1 - \gamma = 1 - \frac{\partial \phi_B}{\partial V} \tag{A4.4}$$

allows Eq. (A4.3) to be written as

$$I = AA^{**}T^2 e^{-q\phi_{BO}/kT} e^{q(V-Ir_s)/nkT} (1 - e^{-q(V-Ir_s)/kT}) \tag{A4.5}$$

To determine n, it is common practice to limit the current-voltage characteristic to that range of the $\log(I) - V$ plot where the series resistance is negligible ($V \ll Ir_s$). Under those restrictions Eq. (A4.5) becomes

$$I = AA^{**}T^2 e^{-q\phi_{BO}/kT} e^{qV/nkT} (1 - e^{-qV/kT}) \tag{A4.6}$$

Instead of plotting $\log(I)$ versus V as is usually done, Eq. (A4.6) predicts that $\log[I/(1 - \exp(-qV/kT))]$ versus V should be plotted. Such a plot exhibits a straight line all the way to $V = 0$, which Eq. (A4.1) does not do due to the -1 term in the bracket. This gives a wider range of the curve from which n is determined.[98]

The ideality factor is near unity for well-behaved Schottky diodes. However, it can and does deviate from unity as a result of current flow due to mechanisms other than thermionic emission. For example, thermionic-field emission current, interface damage, and interfacial layers all tend to increase n above unity and a correlation has been found between n and ϕ_B.[99]

REFERENCES

[1] J. S. Escher, H. M. Berg, G. L. Lewis, C. D. Moyer, T. U. Robertson, and H. A. Wey, "Junction-Current-Confinement Planar Light-Emitting Diodes and Optical Coupling into Large-Core Diameter Fibers Using Lenses," *IEEE Trans. Electron Dev.* **ED-29**, 1463–1469, Sept. 1982.

[2] K. Schuster and E. Spenke, "The Voltage Step at the Switching of Alloyed *pin* Rectifiers," *Solid-State Electron.* **8**, 881–882, Nov. 1965.

[3] S. M. Sze, *Physics of Semiconductor Devices*, 2d ed., Wiley, New York, 1981, pp. 256–263.

[4] S. K. Cheung and N. W. Cheung, "Extraction of Schottky Diode Parameters from Forward Current-Voltage Characteristics," *Appl. Phys. Lett.* **49**, 85–87, July 1986.

[5] H. Norde, "A Modified Forward *I-V* Plot for Schottky Diodes with High Series Resistance," *J. Appl. Phys.* **50**, 5052–5053, July 1979.

[6] N. T. Tam and T. Chot, "Experimental Richardson Constant of Metal-Semiconductor Schottky Barrier Contacts," *Phys. Stat. Sol.* **93a**, K91–K95, Jan. 1986.

[7] N. Toyama, "Variation in the Effective Richardson Constant of a Metal-Silicon Contact due to Metal Film Thickness," *J. Appl. Phys.* **63**, 2720–2724, April 1988.

[8] C. D. Lien, F. C. T. So, and M. A. Nicolet, "An Improved Forward *I-V* Method for Non-ideal Schottky Diodes with High Series Resistance," *IEEE Trans. Electron Dev.* **ED-31**, 1502–1503, Oct. 1984.

[9] K. Sato and Y. Yasumura, "Study of the Forward *I-V* Plot for Schottky Diodes with High Series Resistance," *J. Appl. Phys.* **58**, 3655–3657, Nov. 1985.

[10] T. Chot, "A Modified Forward *I-U* Plot for Schottky Diodes with High Series Resistance," *Phys. Stat. Sol.* **66a**, K43–K45, July 1981.

[11] R. J. Bennett, "Interpretation of Forward Bias Behavior of Schottky Barriers," *IEEE Trans. Electron Dev.* **ED-34**, 935–937, April 1987.

[12] D. K. Schroder and D. L. Meier, "Solar Cell Contact Resistance—A Review," *IEEE Trans. Electron Dev.* **ED-31**, 637–647, May 1984.

[13] M. A. Green, "General Solar Cell Curve Factors Including the Effects of Ideality Factor, Temperature and Series Resistance," *Solid-State Electron.* **20**, 265–266, March 1977.

[14] M. Wolf and H. Rauschenbach, "Series Resistance Effects on Solar Cell Measurements," *Adv. Energy Conv.* **3**, 455–479, Apr./June 1963.

[15] R. J. Handy, "Theoretical Analysis of the Series Resistance of a Solar Cell," *Solid-State Electron.* **10**, 765–775, Aug. 1967.

[16] G. M. Smirnov and J. E. Mahan, "Distributed Series Resistance in Photovoltaic Devices; Intensity and Loading Effects," *Solid-State Electron.* **23**, 1055–1058, Oct. 1980.

[17] S. K. Agarwal, R. Muralidharan, A. Agarwala, V. K. Tewary, and S. C. Jain, "A New Method for the Measurement of Series Resistance of Solar Cells," *J. Phys. D.* **14**, 1643–1646, Sept. 1981.

[18] K. Rajkanan and J. Shewchun, "A Better Approach to the Evaluation of the Series Resistance of Solar Cells," *Solid-State Electron.* **22**, 193–197, Feb. 1979.

[19] G. L. Araujo and E. Sanchez, "A New Method for Experimental Determination of the Series Resistance of a Solar Cell," *IEEE Trans. Electron Dev.* **ED-29**, 1511–1513, Oct. 1982.

[20] J. C. H. Phang, D. S. H. Chan, and Y. K. Wong, "Comments on the Experimental Determination of Series Resistance in Solar Cells," *IEEE Trans. Electron Dev.* **ED-31**, 717–718, May 1984.

[21] K. L. Kennerud, "Analysis of Performance Degradation in CdS Solar Cells," *IEEE Trans. Aerosp. Electr. Syst.* **AES-5**, 912–917, Nov. 1969.

[22] D. S. H. Chan, J. R. Phillips, and J. C. H. Phang, "A Comparative Study of Extraction Methods for Solar Cell Model Parameters," *Solid-State Electron.* **29**, 329–337, March 1986.

[23] J. P. Charles, M. Abdelkrim, Y. H. Muoy, and P. Mialhe, "A Practical Method for Analysis of the *I-V* Characteristics of Solar Cells," *Solar Cells* **4**, 169–178, Sept. 1981.

[24] P. Mialhe, A. Khoury, and J. P. Charles, "A Review of Techniques to Determine the Series Resistance of Solar Cells," *Phys. Stat. Sol.* **83a**, 403–409, May 1984.

[25] D. Fuchs and H. Sigmund, "Analysis of the Current-Voltage Characteristic of Solar Cells," *Solid-State Electron.* **29**, 791–795, Aug. 1986.

[26] J. E. Cape, J. R. Oliver, and R. J. Chaffin, "A Simplified Flashlamp Technique for Solar Cell Series Resistance Measurements," *Solar Cells* **3**, 215–219, May 1981.

[27] D. S. Chan and J. C. H. Phang, "A Method for the Direct Measurement of Solar Cell Shunt Resistance," *IEEE Trans. Electron Dev.* **ED-31**, 381–383, March 1984.

[28] J. J. Ebers and J. L. Moll, "Large Signal Behavior of Junction Transistors," *Proc. IRE* **42**, 1761–1772, Dec. 1954.

[29] B. Kulke and S. L. Miller, "Accurate Measurement of Emitter and Collector Resistances in Transistors," *Proc. IRE* **45**, 90, Jan. 1957.

[30] W. Filensky and H. Beneking, "New Technique for Determination of Static Emitter and Collector Series Resistances of Bipolar Transistors," *Electron. Lett.* **17**, 503–504, July 1981.

[31] L. J. Giacoletto, "Measurement of Emitter and Collector Series Resistances," *IEEE Trans. Electron Dev.* **ED-19**, 692–693, May 1972.

[32] I. Getreu, *Modeling the Bipolar Transistor*, Tektronix, Beaverton, OR, 1976. This book provides a very nice discussion of BJT characterization methods, especially for the use of curve tracers.

[33] T. H. Ning and D. D. Tang, "Method for Determining the Emitter and Base Series Resistances of Bipolar Transistors," *IEEE Trans. Electron Dev.* **ED-31**, 409–412, April 1984.

[34] H. G. Rudenberg, "On the Effect of Base Resistance and Collector-to-Base Overlap on the Saturation Voltages of Power Transistors," *Proc. IRE* **46**, 1304–1305, June 1958.

[35]   J. Choma, Jr., "Error Minimization in the Measurement of Bipolar Collector and Emitter Resistances," *IEEE J. Solid-State Circ.* **SC-11**, 318–322, April 1976.

[36]   W. D. Mack and M. Horowitz, "Measurement of Series Collector Resistance in Bipolar Transistors," *IEEE J. Solid-State Circ.* **SC-17**, 767–773, Aug. 1982.

[37]   J. Logan, "Characterization and Modeling for Statistical Design," *Bell Syst. Tech. J.* **50**, 1105–1147, April 1971.

[38]   D. D. Tang, "Heavy Doping Effects in *pnp* Bipolar Transistors," *IEEE Trans. Electron Dev.* **ED-27**, 563–570, March 1980.

[39]   B. Ricco, J. M. C. Stork, and M. Arienzo, "Characterization of Non-Ohmic Behavior of Emitter Contacts of Bipolar Transistors," *IEEE Electron Dev. Lett.* **EDL-5**, 221–223, July 1984.

[40]   J. Lindmayer, "Power Gain of Transistors at High Frequencies," *Solid-State Electron.* **5**, 171–175, Jan. 1962.

[41]   W. M. C. Sansen and R. G. Meyer, "Characterization and Measurement of the Base and Emitter Resistances of Bipolar Transistors," *IEEE J. Solid-State Circ.* **SC-7**, 492–498, Dec. 1972.

[42]   T. E. Wade, A. van der Ziel, E. R. Chenette, and G. Roig, "Base Resistance Measurements on Bipolar Junction Transistors via Low Temperature Bridge Techniques," *Solid-State Electron.* **19**, 385–388, May 1976.

[43]   R. T. Unwin and K. F. Knott, "Comparison of Methods Used for Determining Base Spreading Resistance," *Proc. IEE Pt.I* **127**, 53–61, April 1980.

[44]   G. C. M. Meijer and H. J. A. de Ronde, "Measurement of the Base Resistance of Bipolar Transistors," *Electron. Lett.* **11**, 249–250, June 1975.

[45]   A. Neugroschel, "Measurement of the Low-Current Base and Emitter Resistances of Bipolar Transistors," *IEEE Trans. Electron Dev.* **ED-34**, 817–822, April 1987; "Corrections to 'Measurement of Low-Current Base and Emitter Resistances of Bipolar Transistors'," *IEEE Trans. Electron Dev.* **ED-34**, 2568–2569, Dec. 1987.

[46]   J. S. Park and A. Neugroschel, "Parameter Extraction for Bipolar Transistors," *IEEE Trans. Electron Dev.* **ED-36**, 88–95, Jan. 1989.

[47]   P. Spiegel, "Transistor Base Resistance and Its Effect on High Speed Switching," *Solid State Design*, 15–18, Dec. 1965; ref. 41.

[48]   P. G. A. Jespers, "Measurements for Bipolar Devices," in *Process and Device Modeling for Integrated Circuit Design* (F. van de Wiele, W. L. Engl, and P. G. Jespers, eds.) Noordhoff, Leyden, 1977, pp. 307–363.

[49]   R. C. Jaeger and A. J. Brodersen, "Low Frequency Noise Sources in Bipolar Junction Transistors," *IEEE Trans. Electron Dev.* **ED-17**, 128–134, Feb. 1970.

[50]   S. T. Hsu, "Noise in High-Gain Transistors and Its Application to the Measurement of Certain Transistor Parameters," *IEEE Trans. Electron Dev.* **ED-18**, 425–431, July 1971.

[51]   K. K. Ng and W. T. Lynch, "Analysis of the Gate-Voltage-Dependent Series Resistance of MOSFET's," *IEEE Trans. Electron Dev.* **ED-33**, 965–972, July 1986.

[52]   K. K. Ng, R. J. Bayruns, and S. C. Fang, "The Spreading Resistance of MOSFET's," *IEEE Electron Dev. Lett.* **EDL-6**, 195–198, April 1985.

[53] G. Baccarani and G. A. Sai-Halasz, "Spreading Resistance in Submicron MOSFET's," *IEEE Electron Dev. Lett.* **EDL-4**, 27–29, Feb. 1983.

[54] J. M. Pimbley, "Two-Dimensional Current Flow in the MOSFET Source-Drain," *IEEE Trans. Electron Dev.* **ED-33**, 986–996, July 1986.

[55] P. I. Suciu and R. L. Johnston, "Experimental Derivation of the Source and Drain Resistance of MOS Transistors," *IEEE Trans. Electron Dev.* **ED-27**, 1846–1848, Sept. 1980.

[56] S. T. Hsu, "A Simple Method to Determine Series Resistance and *k* Factor of an MOS Field Effect Transistor," *RCA Rev.* **44**, 424–429, Sept. 1983.

[57] J. G. J. Chern, P. Chang, R. F. Motta, and N. Godinho, "A New Method to Determine MOSFET Channel Length," *IEEE Electron Dev. Lett.* **EDL-1**, 170–173, Sept. 1980.

[58] K. Tereda and H. Muta, "A New Method to Determine Effective MOSFET Channel Length, *Japan. J. Appl. Phys.* **18**, 953–959, May 1979.

[59] S. E. Laux, "Accuracy of an Effective Channel Length/External Resistance Extraction Algorithm for MOSFET's," *IEEE Trans. Electron Dev.* **ED-31**, 1245-1251, Sept. 1984.

[60] K. L. Peng, S. Y. Oh, M. A. Afromowitz, and J. L. Moll, "Basic Parameter Measurement and Channel Broadening Effect in the Submicrometer MOS-FET," *IEEE Electron Dev. Lett.* **EDL-5**, 473–475, Nov. 1984.

[61] S. Ogura, P. J. Tsang, W. W. Walker, D. L. Critchlow, and J. F. Shepard, "Design and Characteristics of the Lightly Doped Drain-Source (LDD) Insulated Gate Field-Effect Transistor," *IEEE J. Solid. State Circ.* **SC-15**, 424–432, Aug. 1980.

[62] B. J. Sheu, C. Hu, P. Ko, and F. C. Hsu, "Source-and-Drain Series Resistance of LDD MOSFET's," *IEEE Electron Dev. Lett.* **EDL-5**, 365–367, Sept. 1984; C. Duvvury, D. A. G. Baglee, and M. P. Duane, "Comments on 'Source-and-Drain Series Resistance of LDD MOSFET's'," *IEEE Electron Dev. Lett.* **EDL-5**, 533–534, Dec. 1984; B. J. Sheu, C. Hu, P. Ko, and F. C. Hsu, "Reply to "Comments on 'Source-and-Drain Series Resistance of LDD MOSFET's'," *IEEE Electron Dev. Lett.* **EDL-5**, 535, Dec. 1984.

[63] M. R. Wordeman, J. Y. C. Sun, and S. E. Laux, "Geometry Effects in MOSFET Channel Length Extraction Algorithms," *IEEE Electron Dev. Lett.* **EDL-6**, 186–188, April 1985.

[64] G. J. Hu, C. Chang, and Y. T. Chia, "Gate-Voltage-Dependent Effective Channel Length and Series Resistance of LDD MOSFET's" *IEEE Trans. Electron Dev.* **ED-34**, 2469–2475, Dec. 1987.

[65] K. L. Peng and M. A. Afromowitz, "An Improved Method to Determine MOSFET Channel Length," *IEEE Electron Dev. Lett.* **EDL-3**, 360–362, Dec. 1982.

[66] J. D. Whitfield, "A Modification on 'An Improved Method to Determine MOSFET Channel Length'," *IEEE Electron Dev. Lett.* **EDL-6**, 109–110, March 1985.

[67] L. Chang and J. Berg, "A Derivative Method to Determine a MOSFET's Effective Channel Length and Width Electrically," *IEEE Electron Dev. Lett.* **EDL-7**, 229–231, April 1986.

[68] S. Jain, "A New Method for Measurement of MOSFET Channel Length," *Japan. J. Appl. Phys.* **27**, L1559–L1561, Aug. 1988.

[69] S. Jain, "Generalized Transconductance and Transresistance Methods for MOSFET Characterization," *Solid-State Electron.* **32**, 77–86, Jan. 1989.

[70] S. Jain, "Equivalence and Accuracy of MOSFET Channel Length Measurement Techniques," *Japan. J. Appl. Phys.* **28**, 160–166, Feb. 1989.

[71] M. H. Seavey, "Source and Drain Resistance Determination for MOSFET's," *IEEE Electron Dev. Lett.* **EDL-5**, 479–481, Nov. 1984.

[72] C. Duvvury, D. Baglee, M. Duane, A. Hyslop, M. Smayling, and M. Maekawa, "An Analytical Method for Determining Intrinsic Drain/Source Resistance of Lightly Doped Drain (LDD) Devices," *Solid-State Electron.* **27**, 89–96, Jan. 1984.

[73] F. H. De La Moneda, H. N. Kotecha, and M. Shatzkes, "Measurement of MOSFET Constants," *IEEE Electron Dev. Lett.* **EDL-3**, 10–12, Jan. 1982.

[74] M. J. Thoma and C. R. Westgate, "A New AC Measurement Technique to Accurately Determine MOSFET Constants," *IEEE Trans. Electron Dev.* **ED-31**, 1113–1116, Sept. 1984.

[75] J. Y.-C. Sun, M. R. Wordeman, and S. E. Laux, "On the Accuracy of Channel Length Characterization of LDD MOSFET's," *IEEE Trans. Electron Dev.* **ED-33**, 1556–1562, Oct. 1986.

[76] B. J. Sheu and P. K. Ko, "A Capacitance Method to Determine Channel Lengths for Conventional and LDD MOSFET's," *IEEE Electron Dev. Lett.* **EDL-5**, 491–493, Nov. 1984.

[77] P. Vitanov, U. Schwabe, and I. Eisele, "Electrical Characterization of Feature Sizes and Parasitic Capacitances Using a Single Test Structure," *IEEE Trans. Electron Dev.* **ED-31**, 96–100, Jan. 1984.

[78] Y. R. Ma and K. L. Wang, "A New Method to Electrically Determine Effective MOSFET Channel Width," *IEEE Trans. Electron Dev.* **ED-29**, 1825–1827, Dec. 1982.

[79] B. J. Sheu and P. K. Ko, "A Simple Method to Determine Channel Widths for Conventional and LDD MOSFET's," *IEEE Electron Dev. Lett.* **EDL-5**, 485–486, Nov. 1984.

[80] P. Urien and D. Delagebeaudeuf, "New Method for Determining the Series Resistances in a MESFET or TEGFET," *Electron. Lett.* **19**, 702–703, Aug. 1983.

[81] K. Lee, M. S. Shur, A. J. Valois, G. Y. Robinson, X. C. Zhu, and A. van der Ziel, "A New Technique for Characterization of the 'End' Resistance in Modulation-Doped FET's," *IEEE Trans. Electron Dev.* **ED-31**, 1394–1398, Oct. 1984.

[82] S. Chaudhuri and M. B. Das, "On the Determination of Source and Drain Series Resistances of MESFET's," *IEEE Electron Dev. Lett.* **EDL-5**, 244–246, July 1984.

[83] R. F. Pierret, *Field Effect Transistors*, Addison-Wesley, Reading, MA, 1983, p. 13.

[84] H. Fukui, "Determination of the Basic Device Parameters of a GaAs MESFET," *Bell Syst. Tech. J.* **58**, 771–797, March 1979.

[85] L. Yang and S. I. Long, "New Method to Measure the Source and Drain Resistance of the GaAs MESFET," *IEEE Electron Dev. Lett.* **EDL-7**, 75–77, Feb. 1986.

[86] R. P. Holmstrom, W. L. Bloss, and J. Y. Chi, "A Gate Probe Method of Determining Parasitic Resistance in MESFET's," *IEEE Electron Dev. Lett.* **EDL-7**, 410–412, July 1986.

[87] S. C. Sun and J. D. Plummer, "Electron Mobility in Inversion and Accumulation Layers on Thermally Oxidized Silicon Surfaces," *IEEE Trans. Electron Dev.* **ED-27**, 1497–1508, Aug. 1980.

[88] R. V. Booth, M. H. White, H. S. Wong, and T. J. Krutsick, "The Effect of Channel Implants on MOS Transistor Characterization," *IEEE Trans. Electron Dev.* **ED-34**, 2501–2509, Dec. 1987.

[89] ASTM Standard F617, "Standard Method for Measuring MOSFET Linear Threshold Voltage," *1988 Annual Book of ASTM Standards*, Am. Soc. Test. Mat., Philadelphia, 1988.

[90] H. S. Wong, M. H. White, T. J. Krutsick, and R. V. Booth, "Modeling of Transconductance Degradation and Extraction of Threshold Voltage in Thin Oxide MOSFET's," *Solid-State Electron.* **30**, 953–968, Sept. 1987.

[91] S. Jain, "Measurement of Threshold Voltage and Channel Length of Submicron MOSFET's," *Proc. IEE* Pt. I **135**, 162–164, Dec. 1988.

[92] H. G. Lee, S. Y. Oh, and G. Fuller, "A Simple and Accurate Method to Measure the Threshold Voltage of an Enhancement-Mode MOSFET," *IEEE Trans. Electron Dev.* **ED-29**, 346–348, Feb. 1982.

[93] ASTM Standard F1096, "Standard Method for Measuring MOSFET Saturated Threshold Voltage," *1988 Annual Book of ASTM Standards*, Am. Soc. Test. Mat., Philadelphia, 1988.

[94] S. W. Tarasewicz and C. A. T. Salama, "Threshold Voltage Characteristics of Ion-Implanted Depletion MOSFETs," *Solid-State Electron.* **31**, 1441–1446, Sept. 1988.

[95] C. G. Sodini, T. W. Ekstedt, and J. L. Moll, 'Charge Accumulation and Mobility in Thin Dielectric MOS Transistors," *Solid-State Electron.* **25**, 833–841, Sept. 1982.

[96] F. M. Klaassen and W. Hes, "On the Temperature Coefficient of the MOSFET Threshold Voltage," *Solid-State Electron.* **29**, 787–789, Aug. 1986.

[97] E. H. Rhoderick, "Metal-Semiconductor Contacts," *Proc. IEE* Pt. I **129**, 1–14, Feb. 1982; E. H. Rhoderick and R. H. Williams, *Metal-Semiconductor Contacts*, 2d ed., Clarendon, Oxford, 1988.

[98] J. D. Waldrop, "Schottky-Barrier Height of Ideal Metal Contacts to GaAs," *Appl. Phys. Lett.* **44**, 1002–1004, March 1984.

[99] L. F. Wagner, R. W. Young, and A. Sugerman, "A Note on the Correlation between the Schottky-Diode Barrier Height and the Ideality Factor as Determined from *I-V* Measurements," *IEEE Electron Dev. Lett.* **EDL-4**, 320–322, Sept. 1983; D. P. Verret, "The Problem of Correlating Schottky-Diode Barrier Height with an Ideality Factor Using *I-V* Measurements," *IEEE Electron Dev. Lett.* **EDL-5**, 142–144, May 1984.

CHAPTER 5

MOBILITY

5.1 INTRODUCTION

The carrier mobility is an important device parameter. It influences the device behavior through its frequency response or time response in two ways. The carrier velocity is proportional to the mobility for low electric fields. Hence, a higher mobility material is likely to have a higher frequency response. At high fields the velocity becomes a constant, and the concept of mobility loses its meaning. Second, the device current depends on the mobility and higher mobility materials have higher current. Higher currents charge capacitances more rapidly, resulting in a higher frequency response.

There are several mobilities in use. The fundamental mobility is the *microscopic mobility*, which is calculated from basic concepts and describes the mobility of the carriers in their respective band. The *conductivity mobility* is the mobility that determines the conductivity or the resistivity of a semiconducting material. The *Hall mobility* is determined from the Hall effect and differs from the conductivity mobility by a factor dependent on the scattering mechanisms. The *drift mobility* is the mobility measured when minority carriers drift in an electric field. It is a device-oriented mobility and therefore very useful. But it is not as easy to measure as the Hall mobility, for example, and is not used as extensively for that reason.

The geometry has a major influence on the mobility in some devices. For example, the effect of surface scattering has a major influence in reducing the mobility in MOS field effect transistors. The resulting mobility, determined from the device current-voltage characteristic, is termed the *effective mobility*. In addition there are considerations that cause further division between *majority carrier mobility* and *minority carrier mobility*.

Momentum considerations show that electron-electron and hole-hole scattering has no first-order effect on the mobility. However, electron-hole scattering does reduce the mobility, since electrons and holes have opposite average drift velocities. Hence the minority carrier mobility should be lower than the majority carrier mobility, if carrier–carrier scattering is significant compared to lattice and ionized impurity scattering. We address measurement techniques for the most commonly used mobilities in this chapter.

5.2 CONDUCTIVITY MOBILITY

The conductivity of a semiconductor σ is given by

$$\sigma = q(\mu_n n + \mu_p p) \tag{5.1}$$

For reasonably extrinsic p-type semiconductors $p \gg n$, and the hole mobility from Eq. (5.1) is

$$\mu_p = \frac{\sigma}{qp} \tag{5.2}$$

or in terms of the resistivity ρ the majority carrier or *conductivity mobility* is

$$\mu_p = \frac{1}{qp\rho} \tag{5.3}$$

Measurement of the conductivity and carrier concentration was one of the first means of determining the semiconductor mobility, namely, the conductivity mobility.[1,2] The main reasons for its use are the ease of the measurement and the fact that the Hall scattering coefficient, to be discussed in the following chapter, need not be known. To determine the conductivity mobility, it is necessary to measure the majority carrier concentration and either the conductivity or the resistivity of the sample independently. The method is not used much anymore.

5.3 HALL EFFECT AND MOBILITY

5.3.1 Basic Equations for Uniform Layers or Wafers

The Hall effect was discovered by Hall in 1879 when he investigated the nature of the force acting on a conductor carrying a current in a magnetic field.[3] In particular, he measured the transverse voltage on gold foils. Suspecting the magnet may tend to deflect the current, he wrote "that in this case there would exist a state of stress in the conductor, the electricity pressing, as it were, toward one side of the wire . . . I thought it necessary to test for a difference of potential between points on opposite sides of the conductor." A nice discussion of the discovery of the Hall effect including excerpts from Hall's unpublished notebook is given by Sopka.[4]

Discussions of the Hall effect can be found in many solid-state and semiconductor books. A comprehensive treatment is given by Putley.[5] The Hall effect measurement technique has found wide application in the characterization of semiconductor materials because it gives the *resistivity*, the *carrier concentration*, and the *mobility*. The use of the Hall effect for resistivity measurements is discussed in Chapter 1, and its use in carrier concentration characterization is discussed in Chapter 2. In this chapter we give a more detailed discussion of the Hall effect and its application to mobility measurements.

Hall found that a magnetic field applied to a conductor perpendicular to the current flow direction produces an electric field perpendicular to both the magnetic field and the current. Consider the *p*-type semiconductor sample shown in Fig. 5.1. A current I flows in the *x*-direction, indicated by the holes flowing to the right and a magnetic field B is applied in the *z*-direction. The current is given by

$$I = qwtp v_x \qquad (5.4)$$

The voltage measured along the *x*-direction and indicated by V_ρ is given by

$$V_\rho = \frac{\rho s I}{wt} \qquad (5.5)$$

from which the resistivity is derived as

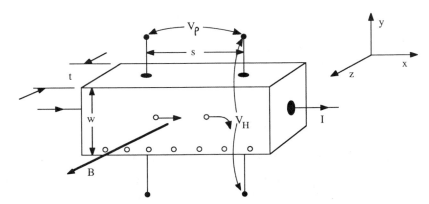

Fig. 5.1   A schematic demonstrating the Hall effect in a *p*-type sample. The current and the holes flow in the *x*-direction, the magnetic field is in the *z*-direction, and the Hall voltage is developed in the *y*-direction. Electrons in an *n*-type sample are also deflected in the *y*-direction to the bottom of the sample.

$$\rho = \frac{wt}{s} \frac{V_\rho}{I} \tag{5.6}$$

The force on the holes is given by

$$\mathbf{F} = q(\mathscr{E} + \mathbf{v} \times \mathbf{B}) \tag{5.7}$$

The magnetic field in conjunction with the current flow forces some holes to be deflected to the bottom of the sample, as indicated in Fig. 5.1 by the hole with the curved arrow. The holes accumulate at the bottom surface. In the y-direction there is no net force on the holes since no current can flow in that direction and therefore $F_y = 0$. Combining Eqs. (5.7) and (5.4) allows us to write the electric field as

$$\mathscr{E}_y = B\upsilon_x = \frac{BI}{qwtp} \tag{5.8}$$

The electric field in the y-direction produces the Hall voltage V_H

$$\int_0^{V_H} dV = V_H = \int_0^w \mathscr{E}_y \, dy = \int_0^w \frac{BI}{qwtp} \, dy = \frac{BI}{qtp} \tag{5.9}$$

The Hall coefficient R_H is defined as

$$R_H = \frac{tV_H}{BI} \quad \text{m}^3/\text{C} \tag{5.10}$$

giving

$$p = \frac{1}{qR_H} \tag{5.11}$$

A similar derivation for n-type samples gives

$$n = -\frac{1}{qR_H} \tag{5.12}$$

When both holes and electrons are present, then the Hall coefficient becomes[6]

$$R_H = \frac{[(p - b^2n) + (\mu_n B)^2(p - n)]}{q[(p + bn)^2 + (\mu_n B)^2(p - n)^2]} \tag{5.13}$$

This expression is relatively complex and depends sensitively on the mobility ratio $b = \mu_n/\mu_p$ and on the magnetic field strength B. For $B \to 0$

$$R_H = \frac{(p - b^2 n)}{q(p + bn)^2} \tag{5.14}$$

and for $B \to \infty$

$$R_H = \frac{1}{q(p - n)} \tag{5.15}$$

For Eq. (5.14) to hold in the low field limit, $B \ll 1/\mu_n$ if $p \gg n$ or $B \ll 1/\mu_p$ if $n \gg p$. For a mobility of $1000 \text{ cm}^2/\text{V} \cdot \text{s} = 0.1 \text{ m}^2/\text{V} \cdot \text{s}$ this requires $B \ll 10 \text{ T}$ ($1 \text{ T} = 10{,}000 \text{ G}$). For mobilities of $10^5 \text{ cm}^2/\text{V-s}$, this requirement becomes more severe, with $B \ll 0.1 \text{ T}$. The high-field limit of Eq. (5.15) requires $B \gg 1/\mu_n$ if $p \gg n$ or $B \gg 1/\mu_p$ if $n \ll p$. Hence magnetic fields much larger than 10 T or 0.1 T, respectively, are necessary in this example.

For semiconductors with modest mobilities in the 100 to $1000 \text{ cm}^2/\text{V} \cdot \text{s}$ range and with mobility ratios of $b \approx 3$ to 10, the Hall coefficient is generally found to vary little with magnetic field and Eq. (5.15) or Eqs. (5.11) to (5.12) are used. However, for those semiconductors with high mobilities and high b the Hall coefficient is found to vary with magnetic field. In addition it is found that the Hall coefficient changes sign as a function of temperature. Such behavior is found in semiconductors like HgCdTe. An example of such a Hall coefficient variation is shown in Fig. 5.2(a) for a p-type HgCdTe with $E_g = 0.15 \text{ eV}$.[7] Electron conduction dominates for temperatures of 220 to 300 K, with $n = n_i^2/p \gg p$, because n_i^2 is high for narrow band gap materials. $R_H = -1/qn$ in this temperature range, and it is independent of B. For $T \approx 100$ to 200 K holes begin to participate and mixed conduction causes R_H to decrease and be magnetic field dependent. Hole conduction dominates at lower temperatures. The Hall coefficient is positive and becomes magnetic field independent. This figure exhibits the temperature and magnetic field dependent behavior of mixed conduction very nicely. Fig. 5.2(b) shows the Hall coefficient for GaAs, and there is neither magnetic field dependence nor mixed conduction.[8] Sometimes it is necessary to consider the contribution of light and heavy holes.[9]

We will define the units since magnetic units are generally not as familiar as electric units. With t in meters, V_H in volts, B in Teslas (T) ($1 \text{ T} = 1 \text{ Weber}/\text{m}^2 = 1 \text{ V} \cdot \text{s}/\text{m}^2$), and I in amperes, the Hall coefficient has the units m^3/C. If, as is frequently done, the magnetic field is expressed in Gauss (G) ($1 \text{ T} = 10^4 \text{ G}$), t in centimeters, V_H in volts, and I in amperes, then

$$R_H = \frac{10^8 t V_H}{BI} \quad \text{cm}^3/\text{C} \tag{5.16}$$

For $B = 5000 \text{ G}$, $I = 0.1 \text{ mA}$, and $p = 10^{15} \text{ cm}^{-3}$, we find $V_H = 3.1/t$, with t in

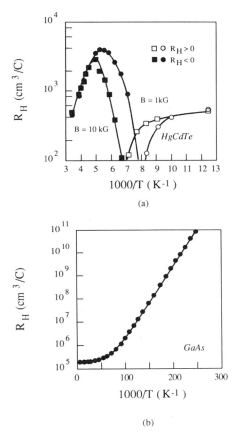

Fig. 5.2 (a) Temperature and magnetic field-dependent Hall coefficient for HgCdTe showing typical mixed conduction behavior. Reprinted with permission after Zemel et al.[7] (b) Hall coefficient for GaAs adapted from Stillman and Wolfe.[8]

μm. For a wafer of thickness $t = 500$ μm, this gives a Hall voltage $V_H \approx 6$ mV and a Hall coefficient $R_H \approx 60,000$ cm^3/C.

Equations (5.11) to (5.15) are derived under simplifying assumptions of energy-independent scattering mechanisms. With this assumption relaxed, the expressions for the hole and electron concentrations become[5–6]

$$p = \frac{r}{qR_H}$$

$$n = -\frac{r}{qR_H} \qquad (5.17)$$

where r is the Hall scattering factor, defined by $r = \langle \tau^2 \rangle / \langle \tau \rangle^2$, with τ being the mean time between collisions for the carriers. The scattering factor depends on the scattering mechanisms in the semiconductor and generally lies between 1 and 2. For lattice scattering, $r = 3\pi/8 = 1.18$, for ionized impurity scattering $r = 315\pi/512 = 1.93$, and for neutral impurity scattering

$r = 1$.[6,10] The scattering factor is also a function of the magnetic field and of the temperature. In the high magnetic field limit $r \to 1$, and r can be determined by measuring R_H in the high field limit; that is, $r = R_H(B)/R_H(B = \infty)$. The scattering factor has been measured in n-type GaAs as a function of magnetic field and was found to vary from 1.17 at $B = 0.1$ kG, as expected from lattice scattering, to 1.006 at $B = 83$ kG.[11] The high fields necessary for r to approach unity are not achievable in some laboratories, and $r > 1$ for most Hall measurements. Typical magnetic fields used for Hall measurements lie between 0.5 and 10 kG.

The *Hall mobility* μ_H is defined by

$$\mu_H = \frac{|R_H|}{\rho} = |R_H|\sigma \tag{5.18}$$

The Hall mobility is not identical to the conductivity mobility. Substituting Eq. (5.1) into Eq. (5.18) gives

$$\mu_H = r\mu_p , \quad \mu_H = r\mu_n \tag{5.19}$$

for extrinsic p- and n-type semiconductors. Hall mobilities can differ significantly from conductivity mobilities since r is generally larger than unity. For most Hall-determined mobilities, r is taken as unity, but this assumption should be carefully specified.

The schematic Hall sample of Fig. 5.1 has a variety of practical implementations. One of these is the geometry shown in Fig. 5.3(a). It is, in principle, identical to Fig. 5.1 but has four "legs" for making the contacts and is known as a bridge-type Hall bar. The current flows into 1 and out of 4, the Hall voltage is measured between 2 and 6 or between 3 and 5 in the presence of a magnetic field. The resistivity is determined in the absence of the magnetic field by measuring between 2 and 3 or between 6 and 5. The equations developed above apply for this geometry. For bulk samples Hall bars are cut out of a larger wafer with ultrasonic cutting tools.

A more general geometry is the irregularly shaped sample shown in Fig. 5.3(b). The theoretical foundation of Hall measurement evaluation for irregularly shaped samples is based on conformal mapping developed by van der Pauw.[12–13] He showed how the resistivity, carrier concentration, and mobility of a flat sample of arbitrary shape can be determined without knowing the current pattern if the following conditions are met: (1) the contacts are at the circumference of the sample, (2) the contacts are sufficiently small, (3) the sample is uniformly thick, and (4) the sample surface is singly connected (i.e., the sample does not contain isolated holes).

For the sample of Fig. 5.3(b) the resistivity is given by[12]

$$\rho = \frac{\pi t}{\ln(2)} \frac{(R_{12,34} + R_{23,41})}{2} F \tag{5.20}$$

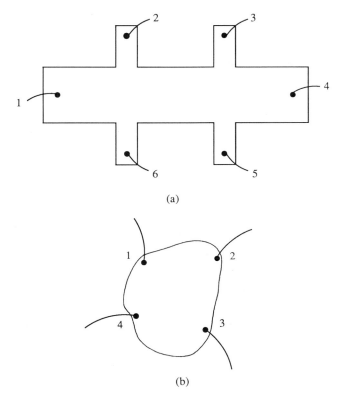

Fig. 5.3   (a) A bridge-type Hall sample and (b) a lamella-type van der Pauw Hall sample.

where $R_{12,34} = V_{34}/I$. The current I enters the sample though contact 1 and leaves through contact 2 and $V_{34} = V_4 - V_3$ is the voltage between contacts 4 and 3. $R_{23,41}$ is similarly defined. F is a function of the ratio $R_{12,34}/R_{23,41}$ only and satisfies the relation

$$\frac{R_r - 1}{R_r + 1} = \frac{F}{\ln(2)} \operatorname{arcosh}\left\{ \frac{\exp[\ln(2)/F]}{2} \right\} \tag{5.21}$$

The F function is plotted in Fig. 5.4. $F = 1$ for symmetrical samples like circles or squares. Most van der Pauw samples are symmetric.

The van der Pauw Hall mobility is given by the same expression as the Hall bar mobility

$$\mu_H = \frac{|R_H|}{\rho} = |R_H|\sigma \tag{5.22}$$

with the Hall coefficient given by

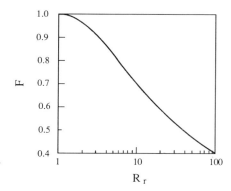

Fig. 5.4 The van der Pauw F factor plotted against R_r.

$$R_{\mathrm{H}} = \frac{t\,\Delta V_{34}}{2BI} \tag{5.23}$$

where $\Delta V_{34} = V_{34}(\text{for} + B) - V_{34}(\text{for} - B)$ with I flowing into terminal 1 and out of terminal 2.

These equations are for carrier concentrations per unit volume and for resistivity ρ (ohm-cm). Occasionally it is useful to determine carrier concentrations per unit area and sheet resistance ρ_s (ohms/square). Then the *sheet Hall coefficient* R_{Hs} is defined as

$$R_{\mathrm{Hs}} = \frac{R_{\mathrm{H}}}{t} \tag{5.24}$$

and $\mu_{\mathrm{H}} = |R_{\mathrm{Hs}}|/\rho_s$, where $\rho_s = \rho/t$.

The thickness is well defined for bulk samples. For epitaxial or implanted thin layers on substrates of opposite conductivity or on semi-insulating substrates, the active film thickness is not necessarily the total film thickness. If depletion effects caused by Fermi level pinned band bending or surface charges and by band bending at the layer–substrate interface are not considered, the Hall coefficient can be in error as will those semiconductor parameters derived from it.[14–15] For sufficiently lightly doped films it is possible for the surface space-charge region to deplete the entire film. Hall effect measurements then indicate a semi-insulating film. For semiconducting films on insulating substrates, the mobility is frequently observed to decrease toward the substrate. Surface depletion forces the current to flow in the low-mobility portion of the film giving apparent mobilities that are lower than actual mobilities.[15] Even the temperature dependence of the surface and interface space-charge regions should be considered for unambiguous temperature-dependent mobility and carrier concentration measurements.[16]

5.3.2 Nonuniform Layers

Hall effect measurements are simple to interpret for wafers and for uniformly doped films. Nonuniformly doped layer measurements are more difficult to interpret. If the doping concentration varies with film thickness, then its resistivity and mobility also vary with thickness. A Hall effect measurement gives the *average* resistivity, carrier concentration, and mobility. For spatially varying mobility $\mu_p(x)$ and carrier concentration $p(x)$, the Hall sheet coefficient R_{Hs}, the sheet resistance ρ_s, and the average Hall mobility $\langle \mu_H \rangle$ for a *p*-type film of thickness *t* are given by[17–18]

$$R_{Hs} = \frac{\int_0^t p(x)\mu_p^2(x)\, dx}{q[\int_0^t p(x)\mu_p(x)\, dx]^2} \tag{5.25}$$

$$\rho_s = \frac{1}{q \int_0^t p(x)\mu_p(x)\, dx} \tag{5.26}$$

$$\langle \mu_H \rangle = \frac{\int_0^t p(x)\mu_p^2(x)\, dx}{\int_0^t p(x)\mu_p(x)\, dx} \tag{5.27}$$

assuming that $r = 1$.

To determine resistivity and mobility *profiles*, Hall measurements must be made as a function of film thickness. The film thickness is varied by two major techniques: (1) remove thin portions of the film by etching and measure the Hall coefficient repeatedly, and (2) make portions of the film electrically inactive by a reverse-biased space-charge region.

In principle, one can use chemical etching to remove thin layers of the film to be profiled. In practice, it is difficult to remove thin layers reproducibly by chemical etching. The electrochemical profiler, discussed in more detail in Section 2.2.4, has been successfully used to remove thin layers by electrolytic etching of GaAs in Tiron (1,2 dihydroxybenzene-3,5 disulphonic acid, disodium salt in an aqueous solution).[19] Hall effect measurements are made after each etch. A more common method for reliable layer removal is anodic oxidation and subsequent oxide etch.[17–18,20–24] Anodic oxidation consumes a fraction of the semiconductor during oxidation. When the oxide is subsequently etched, that portion of the semiconductor consumed during oxidation is also removed. This method provides for very reproducible semiconductor removal and since the oxidation is performed at room temperature it does not alter the doping profile. A more detailed discussion of anodic oxidation is given in Section 1.3.1.

A second method utilizes a junction formed on the upper surface of the film to be profiled. The film must be sufficiently thin for the reverse-biased space-charge region to be able to deplete it completely, or at least most of its total thickness. This implies that the layer to be profiled must be bounded at its lower surface by an insulator or a junction. The upper junction may be a *pn* junction, a Schottky barrier junction, a MOSFET, or an MOS capacitor. An example, shown in Fig. 5.5, consists of a *p*-layer on an insulator. The layer is provided with a Schottky gate. The zero-biased metal-semiconductor junction induces a space-charge region of width W under the metal. The insulator could be replaced by a semi-insulating substrate or by an *n*-substrate. The square sample is laterally isolated by etching but could be isolated by surrounding it with an *n*-type film. Four contacts provide for current and voltage probes. When ion-implanted samples are to be profiled, they are usually formed by defining the appropriate Hall sample shape photolithographically. A *p*-type implant is then automatically isolated from the *n*-type substrate by the resulting *pn* junction.

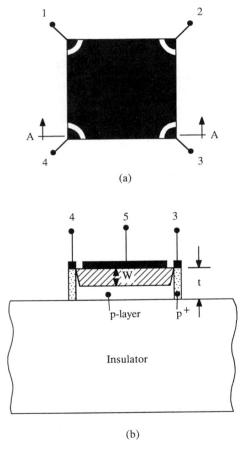

(a)

(b)

Fig. 5.5 A schematic Schottky-gated thin film van der Pauw sample: (a) top view, (b) cross section along line *A–A* showing the gate (5), two contacts (3 and 4), and the space-charge region of width W.

Van der Pauw measurements provide information on the undepleted film of thickness t-W, where t is the total film thickness. A single measurement gives the mobility, the resistivity, and the carrier concentration averaged over t-W. When the Schottky barrier junction is reverse biased, its space-charge region extends into the film reducing the width of the neutral portion of the film. By measuring the Hall effect as a function of reverse-bias voltage, one can determine mobility, resistivity, and carrier concentration profiles of the underlying layer. This method has been implemented with MOSFET's for thin Si films on sapphire,[25–27] for Si-on-insulator with the insulator formed by an implanted oxide layer,[28] and with Schottky diodes for GaAs on semi-insulating substrates.[29–30] A comparison of the destructive "anodize-etch-measure" with the "gated" technique has shown the "gated" method to give more reliable mobilities and to have higher spatial resolution.[31]

The spatially varying Hall mobility is determined from the spatially varying sheet Hall coefficient and sheet conductance $\sigma_s = 1/\rho_s$ by the relationship[17–18,32]

$$\mu_H(x) = \frac{d(R_{Hs}\sigma_s^2)/dx}{d\sigma_s/dx} \tag{5.28}$$

and the spatially varying carrier concentration is

$$p(x) = \frac{r}{q}\frac{(d\sigma_s/dx)^2}{d(R_{Hs}\sigma_s^2)/dx} \tag{5.29}$$

Equations (5.28) and (5.29) are useful when R_{Hs} versus x and σ_s versus x curves have been generated.

The mobility and carrier concentration profiles can also be determined after each layer removal step by making Hall measurements after each layer stripping and using the Hall measured values of adjacent layers in the calculations. The sheet conductivity measured on layer i after stripping the first (i-1) layers is given by

$$\sigma_{s_i} = q\int_{x_i}^{t} p(x)\mu(x)\,dx \tag{5.30}$$

The mobility of the ith layer is

$$\mu_H(x_i) = \frac{R_{Hs_i}\sigma_{s_i}^2 - R_{Hs_{i+1}}\sigma_{s_{i+1}}^2}{\sigma_{s_i} - \sigma_{s_{i+1}}} \tag{5.31}$$

and the carrier concentration is

$$p(x_i) = \frac{\sigma_{s_i} - \sigma_{s_{i+1}}}{q\mu(x_i)\,\Delta x_i} \tag{5.32}$$

where Δx_i is the thickness of the ith layer. The average values of mobility and carrier concentration may differ from the true values if there are large inhomogeneities in the sample. To reduce this effect, it is necessary to make Δx_i small to approximate the nonuniform film by a uniform film. For ion-implanted and fully annealed samples with no mobility anomalies, the error between the measured and real values in mobility and carrier concentration is less than 1% if $\Delta x_i < 0.5 \Delta R_p$, where ΔR_p is the standard deviation of the implanted profile.[33] A concentration profile of a boron layer implanted into Si is shown in Fig. 5.6 where the Hall measured profile is compared with the profile determined by secondary ion mass spectrometry.[34]

Difficulties can arise when there are large mobility variations through the film. Let us consider a film consisting of two layers of equal thickness. The upper layer has a carrier concentration of P_1 holes/cm^2 with mobility μ_1 and the lower one has P_2 holes/cm^2 and μ_2.[35] The total hole concentration is $P_1 + P_2$. The Hall effect measures weighted averages given by[18]

$$P = \frac{(P_1\mu_1 + P_2\mu_2)^2}{P_1\mu_1^2 + P_2\mu_2^2} \tag{5.33}$$

$$\mu_H = \frac{P_1\mu_1^2 + P_2\mu_2^2}{P_1\mu_1 + P_2\mu_2} \tag{5.34}$$

P will be significantly less than $(P_1 + P_2)$, and μ_H will lie between μ_1 and μ_2 for $P_1 > P_2$ and $P_1\mu_1^2 < P_2\mu_2^2$. For $P_1 = 10P_2$ and $\mu_2 = 10\mu_1$, we find $P \approx 4P_2$ and $\mu_H = 0.55\mu_2$. For inhomogeneous samples it is possible for the mobility to be higher than the expected bulk mobility. One cause of abnormally high

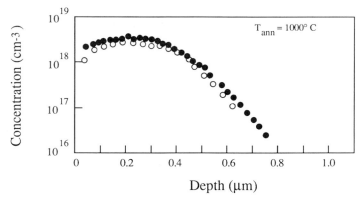

Fig. 5.6 Concentration profile of B implanted into Si. The solid circles were measured by SIMS and the open circles by Hall measurements. Implant conditions: 10^{14} cm^{-2}, 70 keV, annealed at 1000°C. Reprinted with permission after Maes et al.[18]

mobilities is the inclusion of metallic precipitates in the crystal. A thorough discussion of this effect has been given by Wolfe and Stillman.[36]

5.3.3 Multilayers

The discussion in the previous section dealt with the measurement of non-uniform films on an "inert" substrate. By "inert" we mean a substrate that does not contribute to the measurement. An insulating substrate fits such a description. A semi-insulating substrate also approximates this situation and for most practical purposes can be considered insulating. A p-film on an n-substrate or an n-film on a p-substrate might be thought to be in the same category, with the space-charge region (scr) between two semiconductors of opposite conductivity considered an insulating boundary. But this is a more precarious situation. For example, a leaky junction can no longer be considered an insulator. Even if the insulator properties of the scr are sufficiently good, there may be leakage paths along the surface. Or, even worse, the heavily doped contacts may be diffused into the substrate, providing a leakage path. Film characterization is then no longer unique to the film, and the substrate properties are reflected in the measurements.

This problem was originally addressed by Nedoluha and Koch[37] and by Petritz.[38] Petritz considered a substrate whose surface is inverted by surface charges. For example, an n-type inversion layer may be formed on a p-type substrate with a scr between the inversion layer and the substrate. The two-layer interacting configuration was later extended.[39–40] For a simple two-layer structure with an upper layer having thickness t_1 and conductivity σ_1 and a substrate of thickness t_2 and σ_2, the Hall constant is given by[37]

$$R_H = \frac{t[(R_{H1}\sigma_1^2 t_1 + R_{H2}\sigma_2^2 t_2) + R_{H1}R_{H2}\sigma_1^2\sigma_2^2(R_{H1}t_2 + R_{H2}t_1)B^2]}{(\sigma_1 t_1 + \sigma_2 t_2)^2 + \sigma_1^2\sigma_2^2(R_{H1}t_2 + R_{H2}t_1)^2 B^2} \tag{5.35}$$

which becomes[38,40]

$$R_H = \frac{t(R_{H1}\sigma_1^2 t_1 + R_{H2}\sigma_2^2 t_2)}{(\sigma_1 t_1 + \sigma_2 t_2)^2} = R_{H1}\frac{t_1}{t}\left(\frac{\sigma_1}{\sigma}\right)^2 + R_{H2}\frac{t_2}{t}\left(\frac{\sigma_2}{\sigma}\right)^2 \tag{5.36}$$

in the low magnetic field limit, and

$$R_H = \frac{tR_{H1}R_{H2}}{R_{H1}t_2 + R_{H2}t_1} \tag{5.37}$$

in the high magnetic field limit. In these equations R_{H1} is the Hall constant of layer 1, R_{H2} is the Hall constant of substrate 2, $t = t_1 + t_2$, and σ is given by

$$\sigma = \frac{t_1}{t}\sigma_1 + \frac{t_2}{t}\sigma_2 \qquad (5.38)$$

The magnetic field dependence of Eq. (5.35) can be used to advantage to gain additional information by measuring the Hall coefficient as a function of magnetic field. This is illustrated in Fig. 5.7 for a sample consisting of a p-substrate and an n-layer where $R_{H1} = -1/qn_1$ and $R_{H2} = 1/qp_2$. The Hall coefficients are of opposite sign, making it possible for the measured Hall coefficient to reverse its sign with magnetic field. The Hall coefficient is plotted against the $n_1 t_1$ product. For low $n_1 t_1$ the Hall coefficient is dominated by the p-substrate and is magnetic field independent. Both p_2 and μ_2 can be determined from R_H. For intermediate values of $n_1 t_1$ the Hall coefficient becomes field dependent. Conduction is initially dominated by holes, and then by electrons as the Hall coefficient changes its sign. The carrier concentration and mobility of both the n-layer and the p-substrate can be deduced from an analysis of the field-dependent R_H using the two-layer model of Eq. (5.35). For high $n_1 t_1$ values the Hall coefficient is negative, conduction is dominated by the n-layer, and R_H becomes again magnetic field independent. A good discussion of this effect can be found in Zemel et al.[7]

For $t_1 = 0$ we have $t = t_2$, $\sigma = \sigma_2$, and $R_H = R_{H2}$ with the substrate being characterized. If the upper layer is more heavily doped than the substrate or is formed by inversion through surface states, for example, and the carriers in the substrate freeze out at low temperatures making σ_2 very small, then

$$\sigma \approx \frac{t_1}{t}\sigma_1, \quad R \approx R_{H1}\frac{t}{t_1} \qquad (5.39)$$

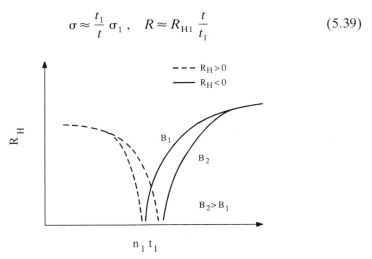

Fig. 5.7 Hall coefficient of a p-type substrate with an n-type layer as a function of electron concentration (cm^{-2}) for two magnetic fields.

and the Hall measurement characterizes the surface layer. This problem can be especially serious if the existence of the upper layer is not suspected and it is believed that the substrate is being characterized. Since both conductivity and the Hall coefficient can be erroneous, the resulting Hall mobility will be in error. Examples of an n-type skin on a p-type bulk, an n-type film on p-type bulk, and an n-type skin on n-type bulk are given for HgCdTe and InSb.[40–41]

5.3.4 Sample Shapes and Measurement Circuits

Hall samples come in two basic geometries: bridge-type and lamellar. The parallelepiped sample shape of Fig. 5.1 is not recommended because contacts have to be directly soldered to the sample. To ease the contact problem, the Hall bridge has extended arms as shown in Fig. 5.8(i).[42] Both six- and eight-arm geometries can be used, and the dimensions should be[42]

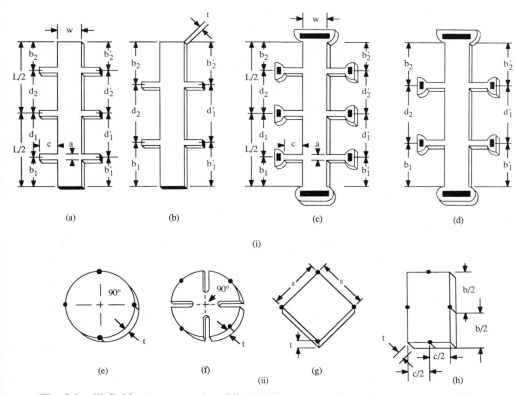

Fig. 5.8 (i) Bridge-type samples, (ii) lamella-type samples, where contact positions are indicated by the small dots. Reproduced from Standard Practice F76[42] with permission from the American Society for Testing and Materials, 1916 Race St., Philadelphia, PA 19103.

Six-Arm Specimen	Eight-Arm Specimen
$L \geq 5w$	$L \geq 4w$
$w \geq 3a$	$w \geq 3a$
$b_1, b_2 \geq 2w$	$b_1 \leq b_2 \geq w$
$t \leq 0.1$ cm	$t \leq 0.1$ cm
$c \geq 0.1$ cm	$c \geq 0.1$ cm
1 cm $\leq L \leq 1.5$ cm	1 cm $\leq L \leq 1.5$ cm

The lamellar specimen may be of arbitrary shape, but a symmetrical configuration is preferred. The sample must be free of geometrical holes, and typical shapes are shown in Fig. 5.8(ii). The peripheral length should be $L_p \geq 1.5$ cm and $t \leq 0.1$ cm. Such a large L_p is not always possible, and many other shapes have been used. Whereas the bridge-type specimen have extensions for contact placement, the lamella-type specimen of Fig. 5.8(b) have no such provision. It is important for the contacts to be small and to be placed as close to the periphery as possible.

A few of the common lamella or van der Pauw shapes are shown in Fig. 5.9. The shapes beyond simple circles or squares are usually fabricated by photolithographic methods where it becomes possible to provide the contact extensions shown in Fig. 5.9(a) to (c). In Fig. 5.9(a) a large area was implanted, and the Hall sample was isolated from the remainder of the wafer by trench etching. In Fig. 5.9(b), photolithography was used to provide the patterns for the contact diffusions 1 to 4. The implant can then be done over the entire wafer, provided the oxide is sufficiently thick to mask against the implant. Any implant into the p-diffused regions is of no consequence. A transfer length contact resistance test pattern has been used for Hall measurements. In addition to the contact resistance, specific contact resistance, and sheet resistance, the mobility in the implanted layer and under the contacts, as well as the sheet carrier concentration, were extracted by applying a magnetic field.[43]

The size and placement of the contacts is important. For van der Pauw samples the contacts should be point contacts located symmetrically on the periphery. This is not achievable in practice, and some error is introduced thereby. A few cases were treated by van der Pauw.[12] He considered circular samples with contacts spaced at 90° intervals. The contacts are equipotential areas, and three cases are shown in Fig. 5.10. In each case there are three ideal contacts, with the fourth being nonideal. The fourth contact is either of length ℓ and larger than a point contact or is displaced by a distance ℓ from the periphery. Also indicated for each geometry is the relative error in resistivity $\Delta\rho/\rho$ and in mobility $\Delta\mu_H/\mu_H$ introduced by the nonideal contact, valid for small $1/d$ and small $\mu_H B$. The errors are additive to first order if more than one contact is nonideal. Nonideal contact effects are reduced by removing the contacts from the "active" area. One implementation is the use of some form of cloverleaf geometry shown in Fig.

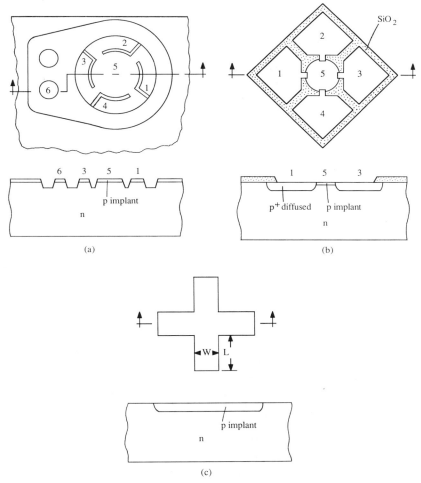

Fig. 5.9 van der Pauw Hall sample shapes.

5.8(ii-f) and Fig. 5.9. The errors due to displaced contacts on square specimen are discussed in Perloff,[44] and David and Buehler.[45] The placement of the contacts on square samples is better at the midpoint of the sides than at the corners.[44] The *Greek cross* in Fig. 5.9(c) makes use of this type of geometry, where for $L \geq 1.02W$ less than 0.1% error is introduced.[45] For square samples with sides of length L having square and triangular contacts of contact length δ in the four corners, less than 10% error was introduced for Hall measurements as long as $\delta/L < 0.1$.[46]

The contacts need not be exactly opposite one another, since the magnetic field reversal routinely made during Hall measurements tends to cancel any unbalanced voltage. But for an unbalanced voltage higher than

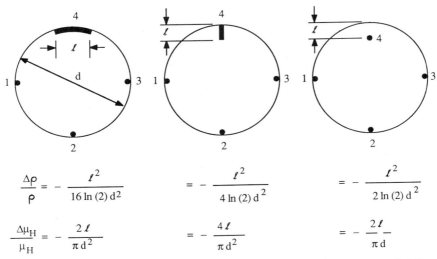

$$\frac{\Delta\rho}{\rho} = -\frac{\ell^2}{16\ln(2)d^2} \qquad = -\frac{\ell^2}{4\ln(2)d^2} \qquad = -\frac{\ell^2}{2\ln(2)d^2}$$

$$\frac{\Delta\mu_H}{\mu_H} = -\frac{2\ell}{\pi d^2} \qquad = -\frac{4\ell}{\pi d^2} \qquad = -\frac{2\ell}{\pi d}$$

Fig. 5.10 Effect of nonideal contact length or contact location on the resistivity and mobility for van der Pauw samples. Reprinted with permission from van der Pauw.[12]

the Hall voltage, the Hall voltage is the difference of two large numbers, and errors are likely to be introduced.

Some samples use a geometry close to that of a conventional semiconductor device. For example, a MOSFET fabricated in a thin film on an insulating substrate has the general shape of the Hall sample in Fig. 5.11, where the p^+ regions 1 and 2 are the source and the drain and 7 is the gate. The contact regions 3–6 are added for Hall measurements. The Hall voltage is developed between contacts 3–4 and 5–6. However, the sample is shorted at the ends by the source and the drain. This has a significant influence on the interpretation of the measured Hall voltage V_{Hm}. For $L \leq 3Z$, V_{Hm} is less than the Hall voltage for samples with $L > 3Z$. The Hall voltage V_H for sample dimensions of $L \gg Z$ used in the earlier equations in this chapter is related to the measured Hall voltage for short samples by $V_H = V_{Hm}/G$, where G is shown in Fig. 5.12(b).[47] The curves in Fig. 5.12(b) are calculated for the Hall voltage measured across the sample at $x = L/2$. Note that for sample lengths $L \geq 3Z$, the shorting effect is negligible, and the measured voltage is the usual Hall voltage.

ASTM-recommended measurement circuits are shown in Fig. 5.13 for six-contact bridge-shaped and van der Pauw samples. For a detailed discussion of the measurement procedure and for measurement precautions see ASTM standard F76.[42] The current and the magnetic field are reversed and the readings averaged for more accurate measurements. Many Hall systems are now computerized, but the principle of the measurement is no different.

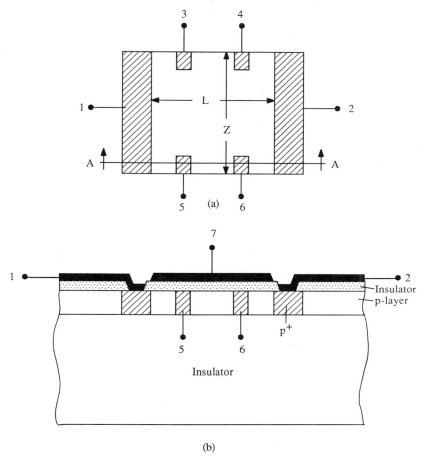

Fig. 5.11   A Hall sample with electrically shorted regions at the ends: (a) top view with the gate not shown, (b) cross section along line A–A.

Special precautions are necessary to eliminate current leakage paths and sample loading by the voltmeter for high-resistance samples. The *guarded* approach utilizes high input impedance unity gain amplifiers between each probe on the sample and the external circuitry.[48] The unity gain outputs drive the shields on the leads between the amplifier and the sample to reduce leakage currents and system time constant by effectively eliminating the stray capacitance in the leads. Measurements of resistances up to 10^{12} ohms have been made with such a system, and the *guarded* approach has been automated.[49] Measurements on semi-insulating GaAs have been made by illuminating a slit across a "dark" wafer and introducing a dark spot within the illuminated slit.[50–51] A resistance measurement along the slit determines essentially the resistance of the small dark spot since the dark

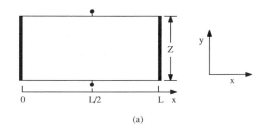

(a)

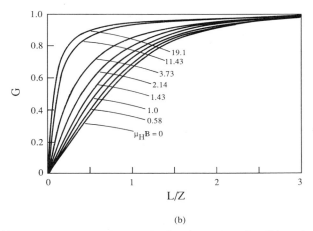

(b)

Fig. 5.12 (a) Hall sample with electrically shorted ends, (b) ratio of measured voltage V_m to Hall voltage V_H, $G = V_{Hm}/V_H$. Reprinted with permission after Lippmann and Kuhrt.[47]

spot resistance is much larger than the resistance of the illuminated strip. A resistance map can be obtained by moving the dark spot.

Hall effect profiling measurements have other possible errors. For example, the bottom *pn* junction may be leaky, causing smaller Hall voltages than would be measured for perfect isolation of the film from the substrate. The upper junction in a Schottky contact configuration may also be leaky. Junction leakage currents can be reduced by sample cooling[21] but that may not always be possible. If the upper junction is forward biased to reduce the space-charge region width in order to be able to profile closer to the surface, considerable error is introduced due to the high forward-biased junction current.[29] Although the effect of injected gate current can be corrected,[52] the correction is large and the accuracy of the corrected results is questionable. Instead of conventional dc measurement circuits, ac circuits can be employed.[29-30] The device is driven with an ac current at one frequency and a gate voltage containing both a dc bias to vary the scr width and an ac component of a frequency different from the current. Lock-in amplifier

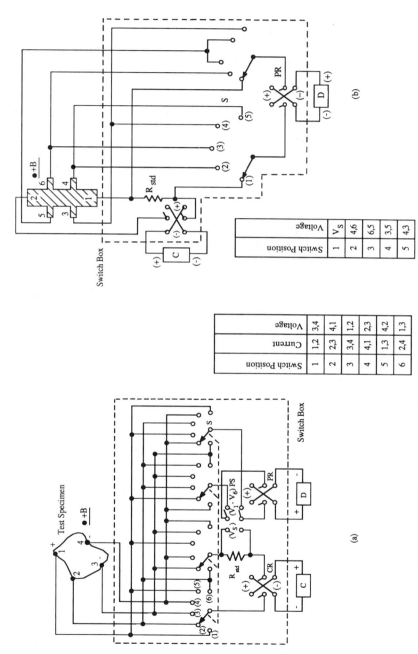

Switch Position	Voltage V_S
1	
2	4,6
3	6,5
4	3,5
5	4,3

Switch Position	Current	Voltage
1	1,2	3,4
2	2,3	4,1
3	3,4	1,2
4	4,1	2,3
5	1,3	4,2
6	2,4	1,3

Fig. 5.13 Measurement circuit for (a) bridge-type specimen and (b) lamellar specimen. Reproduced from Standard Practice F76[42] with permission from the American Society for Testing and Materials, 1916 Race St., Philadelphia, PA 19103.

methods are employed to measure the appropriate ac components without interference from the dc leakage current. In a recent implementation the magnetic field and the current had frequencies of 60 and 200 Hz, respectively.[53] The Hall voltage is detected by a lock-in amplifier at the sum frequency of 260 Hz. This eliminates most thermoelectric and thermomagnetic errors associated with dc measurements and Hall voltages as low as 10 µV have been measured.

5.4   MAGNETORESISTANCE MOBILITY

Typical Hall-effect structures are either long or of the van der Pauw variety. They require four or more contacts. A long Hall bar is shown schematically in Fig. 5.14(a) with $L \gg Z$. Certain semiconductor devices, such as field-effect transistors (FET's) are short with $L \ll Z$, shown in Fig. 5.14(b). The Hall electric field, resulting from an applied magnetic field, is nearly shorted by the long contacts, and FET structures do not lend themselves well to Hall

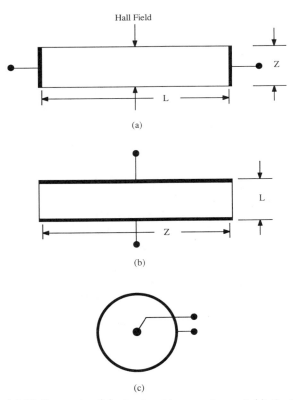

Fig. 5.14   (a) Hall sample, (b) short, wide sample, and (c) Corbino disk.

measurements. The extreme of this short geometry is when one contact is in the center of a circular sample and the other contact is at the periphery, shown in Fig. 5.14(c). The Hall electric field in this *Corbino disk*[54] is shorted, and no Hall voltage exists. The geometries of Fig. 5.14(b) and (c), however, lend themselves well to *magnetoresistance effect* measurements.

The *resistivity* of a semiconductor generally increases when the sample is placed in a magnetic field. This is known as the *physical magnetoresistance effect* (PMR). It occurs if the conduction is anisotropic, if conduction involves more than one type of carrier, and if carrier scattering is energy dependent. The *resistance* of a semiconductor is also influenced by the presence of a magnetic field.[55] The magnetic field causes the path of the charge carriers to deviate from a straight line. The deviation of the current paths from straight lines raises the sample resistance. This depends on the sample geometry and is known as the *geometrical magnetoresistance* (GMR). The resistance change as a result of the magnetic field is due to resistivity changes of the semiconductor as well as due to geometrical effects and is larger the higher the sample mobility is. Geometrical effects dominate. For example, in GaAs at room temperature and in a magnetic field of 10 kG, the PMR is about 2%, whereas the GMR is about 50%. The geometric magnetoresistance mobility μ_{GMR} is related to the Hall mobility μ_H by

$$\mu_{GMR} = \xi \mu_H \qquad (5.40)$$

where ξ is the magnetoresistance scattering factor given by $\xi = (\langle \tau^3 \rangle \langle \tau \rangle / \langle \tau^2 \rangle^2)^2$.[10] For τ independent of energy, the mean time between collisions becomes isotropic, $\xi = 1$ and $\mu_{GMR} = \mu_H$. The physical magneto-resistivity change ratio $\Delta \rho_{PMR} = (\rho_B - \rho_0)/\rho_0$ becomes zero under those conditions, where ρ_B is the resistivity in the presence and ρ_0 is the resistivity in the absence a magnetic field.

The dependence of the resistance ratio R_B/R_0 is shown in Fig. 5.15 as a function of $\mu_{GMR}B$ for rectangular samples of varying L/Z ratios.[56] Here R_B is the resistance with $B \neq 0$ and R_0 is the resistance with $B = 0$. For long rectangular samples with contacts at the ends of the long sample as in Fig. 5.14(a), the ratio is near unity, and the magnetoresistance effect is very small. The ratio is higher for short, wide samples. The highest ratio is obtained for the Corbino disk with $L/Z = 0$. Figure 5.15 shows the magneto-resistance and the Hall effect to be complementary. When one decreases, the other increases. For example, we showed in Fig. 5.12 a Hall voltage reduction for short, wide samples. But those same sample shapes produce maximum magnetoresistance. Magnetoresistance measurements are suitable for field-effect transistors that are short and wide.

The current flow in a Corbino disk is radial from the center to the periphery for $B = 0$. With a magnetic field perpendicular to the sample, the current streamlines become logarithmic spirals and the resistance ratio becomes

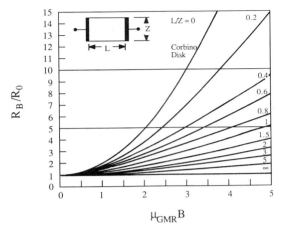

Fig. 5.15 Geometric magnetoresistance ratio of rectangular samples versus $\mu_{GMR}B$ as a function of the length-to-width ratio. $L/Z = 0$ corresponds to the Corbino disk geometry. Reprinted with permission after Lippmann and Kuhrt.[56]

$$\frac{R_B}{R_0} = \frac{\rho_B}{\rho_0}\left[1 + (\mu_{GMR}B)^2\right] \qquad (5.41)$$

Equation (5.41) represents the Corbino disk's curve in Fig. 5.15. Generally the magnetoresistance scattering factor ξ is taken as unity just as the Hall scattering factor is generally taken to be unity. This is done for simplicity and because the scattering mechanisms are not known precisely. Measurements of μ_{GMR} on a modified Corbino disk geometry and of μ_H on Hall samples from the Corbino disk showed ξ to be unity for GaAs within experimental error.[57–58] The measurements were performed for magnetic fields up to 0.7 T and temperatures from 77 to 400 K.[58] Under those conditions $\rho_B \approx \rho_0$ and $\mu_{GMR} \approx \mu_H$. Making the additional assumption of $\mu_H \approx \mu_p$, the mobility is given by

$$\mu_p \approx \frac{1}{B}\sqrt{\frac{R_B}{R_0} - 1} \qquad (5.42)$$

The mobility is obtained from the slope of a plot of $(R_B/R_0 - 1)^{1/2}$ versus B. The mobility can be profiled by using a Corbino disk with a Schottky gate and measuring the resistance as a function of the gate voltage.[59]

The use of Corbino disks is inconvenient because of its special geometrical configuration. However, as is evident from Fig. 5.15, rectangular sample shapes with low L/Z ratios are equally suitable for magnetoresistance measurements.[60] For rectangular samples with low L/Z ratios and $\mu_{GMR}B < 1$, Eq. (5.41) becomes[56–57]

$$\frac{R_B}{R_0} = \frac{\rho_B}{\rho_0}\left[1 + \mu_{GMR}^2 B^2\left(1 - 0.54\,\frac{L}{Z}\right)\right] \qquad (5.43)$$

If the error in the determination of μ_{GMR} is to be less than 10%, then the aspect ratio L/Z must be less than 0.4. For typical FET structures with $L/Z \ll 1$, Eq. (5.43) is a close approximation to Eq. (5.41), and it is for that reason that Eq. (5.41) is generally used in GMR measurements. Magnetoresistance measurements were first used for GaAs Gunn effect devices.[57,61] It is a rapid technique which can be used for functional devices, requiring no special test structures. Instead of measuring the resistance as a function of the FET gate voltage, it is also possible to determine the mobility from transconductance measurements with and without a magnetic field.[62]

The magnetoresistance mobility measurement method has been applied to metal-semiconductor FET's (MESFET's) as well as to modulation-doped FET's (MODFET's). By using the magnetic field dependence of the GMR effect, it is possible to extract the mobilities of the various conducting regions and subbands in MODFET's.[63] The method has been used to determine the mobility dependence on gate electric field.[64] Effects of gate currents for Schottky-gate devices and series resistance effects must be corrected.[52,62,65] Gate current corrections are particularly important when the gate becomes forward biased. Contact resistance, which is of only secondary importance for Hall measurements, is very important for GMR measurements because it adds to the measured resistance, and contact resistance is relatively independent of magnetic field. When the mobility is measured as a function of gate bias, the average mobility is measured for each value of gate voltage. The determination of both the *average* and the *differential mobilities* from transconductance measurements of MODFET's is discussed in Lui and Das.[66]

The GMR effect is not universally applicable the way the Hall effect is, shown by Eq. (5.41). Assuming that $\rho_B/\rho_0 \approx 1$, which is a reasonable assumption, we find that in order to observe a resistance change, $\Delta R/R_0 = (R_B - R_0)/R_0$ of, say, 10%, the condition $\mu_{GMR} \ge 0.3/B$ must be met. For typical magnetic fields of 0.1 to 1 T, this requires $\mu_{GMR} \ge 30,000$ to $3000\ cm^2/V\text{-s}$. These are the kinds of mobilities found in MESFET's and MODFET's made in III–V materials, especially at low temperatures. These are the very materials that have been successfully characterized by GMR. For higher magnetic fields, as obtained in superconducting magnets, lower mobilities can be determined. Silicon, whose mobility lies in the 500–1300 $cm^2/V\text{-s}$ range, is unsuitable for magnetoresistance measurements because its GMR is negligibly small for typical laboratory magnetic fields.

5.5 TIME-OF-FLIGHT DRIFT MOBILITY

The *time-of-flight* method to determine the mobility of minority carriers was first demonstrated in the well-known *Haynes-Shockley experiment*.[67–69] The first comprehensive mobility measurements for Ge and Si were made using this technique by Prince.[70–71] The principle of the method is demonstrated with the *p*-type semiconductor bar in Fig. 5.16(a). A drift voltage V_{dr} produces an electric field $\mathcal{E} = V_{dr}/L$ along the bar. Minority electrons are injected by negative polarity pulses at the *n*-emitter. The injected electron packets drift from the emitter to the collector in the applied electric field to be collected by the collector. A voltage proportional to the electron packet is displayed on the oscilloscope, shown in Fig. 5.16(b). The five Gaussian-shaped output waveforms correspond to the five input pulses shown at $t = 0$. The input pulses marked 1 to 5 at $t = 0$ are merely shown to demonstrate the changing input amplitudes. They are not meant to show the true amplitude of the input voltage, nor are they observed in the output voltage. In the actual output waveform there is only a small feedthrough pulse at $t = 0$ that is used for zero-time reference.

The electrons are injected as a narrow pulse at $t = 0$. For $t > 0$ they diffuse and recombine with majority holes as they drift along the bar. Consequently the minority carrier pulse broadens due to diffusion, and its area decreases by recombination. The pulse shape is given as a function of space and time by the expression[72]

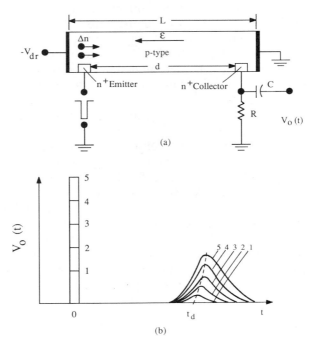

(a)

(b)

Fig. 5.16 (a) Drift mobility measurement arrangement, (b) the output voltage pulses at the right. The pulses at the left show the input pulse magnitudes.

$$\Delta n(x, t) = \frac{N}{(4\pi D_n t)^{1/2}} \exp\left[-\frac{(x - vt)^2}{4D_n t} - \frac{t}{\tau_n}\right] \tag{5.44}$$

where N is the electron density (electrons/cm^2) in the packet at $t = 0$ at the point of injection. The first term in the exponent describes diffusion and drift, and the second term describes recombination.

The time for the electron packet to drift from the point of injection to the collector is given by $t_d = d/v$, where d is the spacing between contacts shown in Fig. 5.16(a) and v the electron packet velocity. The delay time t_d is determined by measuring the output pulses for varying amplitude input pulses and extrapolating to zero injection, as illustrated in Fig. 5.16(b). Alternately, the injection pulse amplitude can be reduced until the peak position of the output pulse no longer shifts in time. This ensures low-level injection with the injected carrier concentration well below the majority carrier equilibrium concentration, eliminating any local disturbance of the electric field by the minority carrier pulse. With the velocity given by $v = \mu_n \mathscr{E}$, the drift mobility is determined from the relationship

$$\mu_n = \frac{d}{t_d \mathscr{E}} \tag{5.45}$$

The time-of-flight method actually measures the *minority carrier velocity* or the *minority carrier mobility*. It is therefore useful for the determination of the carrier velocity–electric field behavior. This relationship is difficult to determine with other mobility measurement techniques.

To determine the diffusion constant D_n, the collected pulse width is measured at half its maximum amplitude. It can be shown that D_n is given by[73]

$$D_n = \frac{(d\,\Delta t)^2}{16\ln(2)t_d^3} \tag{5.46}$$

where Δt is the pulse width.

The lifetime is determined by measuring the collected electron packet pulse at times t_{d1} and t_{d2}, corresponding to the two drift voltages, V_{dr1} and V_{dr2}. In the ideal case with no minority carrier trapping, the collected pulse has the predicted Gaussian shape and the lifetime is obtained by comparing the corresponding output pulse amplitudes V_{01} and V_{02}. The electron lifetime is then[72]

$$\tau_n = \frac{(t_{d2} - t_{d1})}{[\ln(V_{01}/V_{02}) - 0.5\ln(t_{d2}/t_{d1})]} \tag{5.47}$$

If carrier trapping is present, $\log(A_p)$ should be plotted against the delay time t_d, where A_p is the pulse area. The slope of such a plot is $-1/\tau_n$.

Electrical injection can be replaced by optical injection with the basic method unchanged. Optical injection is followed by electrical detection. A variation on this technique combines optical injection with *optical detection*. A laser pulse creates ehp's that drift and diffuse just as electrically injected ehp's do. Electron–hole pair recombination is accompanied by photon emission, especially in III–V materials where radiative recombination dominates. It is this radiative recombination that is detected in the "photon in–photon out" time-of-flight method. In one particular scheme quantum wells are used as time markers for both GaAs/AlGaAs[74] and InGaAs/InP.[75] The ehp's can also be created by a pulsed electron beam[76] or by placing the sample into a microwave circuit. In the latter case the electron beam is deflected at microwave frequencies across the sample, and the resulting microwave current is detected.[77-78] The drift velocity is determined from the amplitude and phase of the microwave current.[77,79]

Mobility or carrier velocity measurements as a function of electric field are not difficult with the circuit of Fig. 5.16(a) at *low electric fields*. Measurements at *high electric fields* are more difficult due to sample heating when large voltages are applied to the semiconductor sample. To overcome this limitation, a configuration is employed in which the electric field is developed in the scr of a reverse-biased junction. The drift of carriers through such a high electric field region is more suitable for determining the drift velocity rather than the mobility as a function of the electric field, however.

We demonstrate the principle of the method in Fig. 5.17(a). That arrangement consists of two parallel plates. Voltage $-V_1$ is applied to the cathode. Electrons, liberated at the cathode, drift with velocity v_n from the cathode to the anode in the electric field generated by V_1. The electron charge $Q_N = qN \text{ coul/cm}^2$ induces charges Q_C and Q_A in the cathode and anode, respectively, with $Q_N = Q_C + Q_A$. The arrows represent electric field lines from Q_C and Q_A terminating on Q_N. The electric field lines due to the applied voltage are not shown.

The charge on both plates redistributes itself continuously as the charge drifts from the cathode to the anode. The anode charge is $Q_A = 0$ at $t = 0$ and $Q_A = Q_N$ at $t = t_t$, where t_t is the transit time defined by

$$t_t = \frac{W}{v_n} \qquad (5.48)$$

When Q_A changes from zero to Q_N, the charge flows through the external circuit as current $I(t)$, given by[80-81]

$$I(t) = \frac{Q_N A}{t_t} = \frac{Q_N A v_n}{W}, \quad 0 \le t \le t_t \qquad (5.49a)$$

$$I(t) = 0, \quad t > t_t \qquad (5.49b)$$

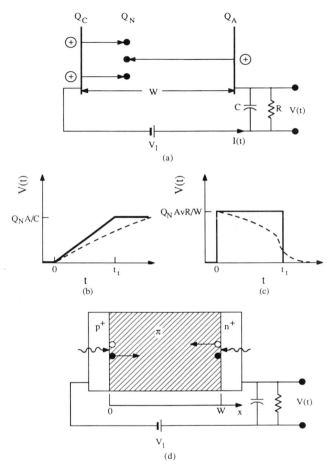

Fig. 5.17 (a) Time-of-flight measurement schematic, (b) output voltage for $t_t \ll RC$, (c) output voltage for $t_t \gg RC$, (d) implementation with a $p^+\pi n^+$ diode.

where A is the electrode area. The current flows only during the transit time.

The sample, connecting leads, and input to the voltage sensing circuit all contain capacitances. These are all lumped into C. R is the load resistance in Fig. 5.17(a). A current $I(t)$ flowing through impedance $Z = R/(1 + j\omega RC)$ produces the output voltage

$$V(t) = \frac{Q_N A v_n R}{W} (1 - e^{-t/RC}) \qquad (5.50)$$

Equation (5.50) has two limits that are of interest for transit time measurements: For $t_t \ll RC$, the voltage becomes

$$V(t) \approx \frac{Q_N A v_n t}{WC} , \quad 0 \leq t \leq t_t \qquad (5.51a)$$

$$V(t) = \frac{Q_N A}{C} , \quad t > t_t \qquad (5.51b)$$

provided that $t \ll RC$. In this approximation the RC circuit acts as an integrator, and the resulting voltage is shown in Fig. 5.17(b).

For $t_t \gg RC$, the voltage becomes

$$V(t) \approx \frac{Q_N A v_n R}{W} , \quad 0 \leq t \leq t_t \qquad (5.52a)$$

$$V(t) = 0 , \quad t > t_t \qquad (5.52b)$$

The RC time constant in this approximation is so small that the capacitor charges instantly and $V(t) \approx RI(t)$. The resulting voltage is shown in Fig. 5.17(c). The transit time can be determined for either case and the carrier velocity is extracted from t_t.

This time-of-flight method can be implemented with the $p^+ \pi n^+$ junction in Fig. 5.17(d). The π region is a lightly doped p-region. Bias voltage V_1 depletes the π region entirely. Shallow penetration excitation (high energy light or an electron beam) from the *left* creates ehp's near $x = 0$. The holes flow into the p^+ contact layer and the electrons drift to $x = W$, allowing the *electron velocity* to be determined. With excitation from the *right*, holes drift to the left, and the *hole velocity* is measured. This test structure can be used for both kinds of carriers.

Two slightly different implementations of time-of-flight measurement geometries are shown in Fig. 5.18. Both use *pn* diodes combined with MOS structures. Figure 5.18(a) shows a gate-controlled diode with both diode and gate biased to V_1.[82–83] This ensures deep depletion under the gate so that an inversion layer cannot form. The gate of the gate-controlled diode consist of a high-resistivity polysilicon film whose sheet resistance is around 10 kohms/square. The voltage pulse V_2 with 200 ns pulse length and 10 kHz repetition rate creates a periodic voltage along the gate as well as along the semiconductor. This voltage in turn generates a lateral electric field in the semiconductor. Optical pulses, from a mode-locked Nd:YAG laser, are directed to two openings defined in a metal gate. The absorbed photons create electron–hole pairs in the semiconductor. The holes drift into the substrate and the electrons drift along the surface to the collecting diode to produce a current pulse in the output circuit. By injecting minority carriers at two locations, defined by optical apertures, the difference in arrival times is used to determine the drift velocity. To study the field dependence of the mobility, the *lateral* or *tangential* electric field is varied by changing V_2. To study the gate voltage dependence of the mobility, the *normal* or *vertical* electric field is varied by adjusting V_1. A detailed description of the technique is given in Cooper and Nelson.[82]

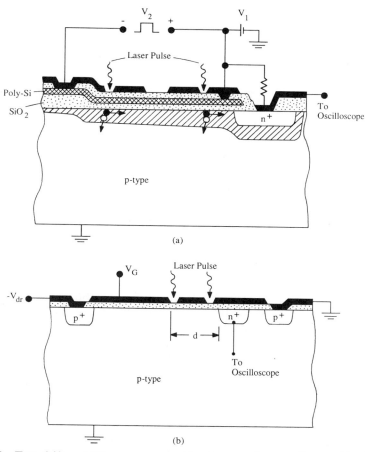

Fig. 5.18   Two drift mobility measurement implementations as discussed in the text.

The electric field in the semiconductor in Fig. 5.18(b) is obtained not from a voltage drop along a polysilicon gate, but from a voltage applied between two p^+ contacts in the semiconductor itself.[84] The Al gate is used to set the surface potential, but the lateral electric field is independent of the vertical electric field since the lateral field does not originate from a gate voltage. The continuous gate is also a light shield with two slits for the laser pulses to generate ehp's. Optical pulses with 70 ps pulse widths from a mode-locked Nd-YAG laser have been used. The minority carrier packets are collected by the n^+ collector and displayed on a sampling oscilloscope. The circuits of Figs. 5.16 and 5.18 are, in principle, very similar. The chief difference lies in the method of minority carrier injection. In Fig. 5.16 minority carriers are injected electrically, and in Fig. 5.18 they are injected optically.

In all of these techniques it is important to ascertain that carrier trapping is either eliminated or accounted for in the data analysis.[75,85] The dashed lines in Fig. 5.17(b) and (c) indicate the effects of trapping.[81]

5.6   MOSFET MOBILITY

The conductivity, Hall, and magnetoresistance mobilities are *bulk* mobilities. Surfaces play a relatively minor role in their determination. The carriers are contained in a bulk wafer or in an epitaxial layer, and a mobility, averaged over the sample thickness, is measured. The main scattering events determining the mobility are *lattice* or *phonon scattering* and *ionized impurity scattering*. *Neutral impurity scattering* becomes important at low temperatures. For some semiconductors there is additionally *piezoelectric scattering*. Each scattering mechanism is associated with a mobility. The net mobility is determined by $1/\mu = 1/\mu_1 + 1/\mu_2 + 1/\mu_3 + \cdots$ according to Mathiessen's rule and the lowest mobility dominates.[86]

In this section we are concerned with additional scattering mechanisms that occur when the current carriers are confined within a narrow region as in an inversion layer in a MOSFET. The location of the carriers at the oxide-semiconductor interface introduces additional scattering mechanisms like Coulomb scattering from oxide charges and interface states, as well as surface roughness scattering. These additional scattering sources reduce the mobility of MOSFET's below the bulk mobility.[87] The quantization of the carriers in inversion layers further reduces the mobility.[88–89]

For an *n*-channel MOSFET on a substrate doped with N_A dopant atoms/cm^3, the drain current is given as a function of the drain voltage V_{DS} and gate voltage V_{GS}, both measured with respect to the source as[90]

$$I_{DS} = \frac{Z\bar{\mu}_n C_{ox}}{L} \left\{ \left[V_{GS} - V_T - \frac{V_{DS}}{2} \right] V_{DS} \right. $$
$$\left. - \gamma \left[\frac{2}{3}(V_{DS} + 2\phi_F')^{3/2} - \frac{2}{3}(2\phi_F')^{3/2} - V_{DS}(2\phi_F')^{1/2} \right] \right\} \quad (5.53)$$

where $\bar{\mu}_n$ is the MOSFET mobility and V_T is the *threshold voltage*, which is defined as

$$V_T = V_{FB} + 2\phi_F + \gamma(2\phi_F')^{1/2} \quad (5.54)$$

The *flatband voltage* is given by

$$V_{FB} = \phi_{MS} - \frac{Q_f}{C_{ox}} - \frac{Q_{it}(\phi_s = 0)}{C_{ox}} - \frac{1}{C_{ox}} \int_0^{W_{ox}} \frac{x}{W_{ox}} Q_m(x)\, dx \quad (5.55)$$

The Fermi potential ϕ_F' and the *body factor* γ, which accounts for the substrate doping concentration charge in the space-charge region, are given by

$$\phi_F' = \phi_F - V_{BS} = \frac{kT}{q} \ln\left(\frac{N_A}{n_i}\right) - V_{BS}$$

$$\gamma = \frac{\sqrt{2qK_s\varepsilon_0 N_A}}{C_{ox}} \qquad (5.56)$$

where $V_{BS} = V_B - V_S$ is the substrate voltage with respect to the source. $V_{BS} < 0$ for *n*-channel and $V_{BS} > 0$ for *p*-channel MOSFET's.

5.6.1 Effective Mobility

The *effective* MOSFET mobility is determined from the drain conductance measured at low drain voltages. The drain conductance is defined as

$$g_D = \left.\frac{\partial I_D}{\partial V_{DS}}\right|_{V_{GS}=\text{constant}} \qquad (5.57)$$

From Eq. (5.53) we find the drain conductance to be given by

$$g_D = \frac{Z\bar{\mu}_n C_{ox}}{L} \left\{ [V_{GS} - V_T - V_{DS}] - \gamma[\sqrt{V_{DS} + 2\phi_F'} - \sqrt{2\phi_F'}] \right\} \qquad (5.58)$$

When Eq. (5.58) is solved for the mobility, the mobility (known as the effective mobility μ_{eff}) is given by

$$\boxed{\mu_{eff} = \frac{Lg_D}{ZC_{ox}\{V_{GS} - V_T - V_{DS} - \gamma[(V_{DS} + 2\phi_F')^{1/2} - (2\phi_F')^{1/2}]\}}} \qquad (5.59)$$

V_{DS} is kept low for μ_{eff} measurements ($V_{DS} \approx 50\,\text{mV}$). With $V_{DS} \ll 2\phi_F'$, Eq. (5.59) reduces to

$$\mu_{eff} = \frac{Lg_D}{ZC_{ox}\{V_{GS} - V_T - V_{DS}[1 + \gamma/(8\phi_F')^{1/2}]\}} \qquad (5.60)$$

Usually $V_{DS} \ll (V_{GS} - V_T)$, and if additionally $\gamma \ll (8\phi_F')^{1/2}$, then Eq. (5.60) becomes

$$\mu_{eff} \approx \frac{Lg_D}{ZC_{ox}(V_{GS} - V_T)} \qquad (5.61)$$

The threshold voltage is required for the effective mobility determination according to Eq. (5.61). Inaccuracies in the threshold voltage reflect directly in the mobility. Threshold voltage measurements are discussed in Chapter 4.

The effective mobility, determined from Eq. (5.61), is shown in Fig. 5.19. It decreases with gate voltage. This has been attributed to enhanced surface roughness scattering with increased gate voltage[91] and to quantization effects.[88-89] Since it has not been definitely determined what causes the mobility reduction with gate voltage, it is common practice to express the effective mobility through an empirical relationship. The effective mobility is frequently given by

$$\frac{1}{\mu_{eff}} = \frac{1}{\mu_0} + \frac{1}{\mu_s} \tag{5.62}$$

where μ_0 is the low-field mobility and μ_s the surface mobility. The low-field mobility is essentially the mobility at zero electric field. Equation (5.62) gives $\mu_{eff} = \mu_0/(1 + \mu_0/\mu_s)$ which has been expressed as[92]

$$\mu_{eff} = \frac{\mu_0}{1 + \alpha/L + \beta/Z + \theta(V_{GS} - V_T)} \tag{5.63}$$

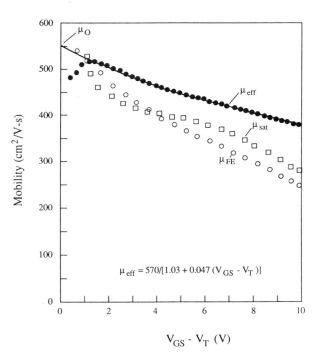

Fig. 5.19 Effective, field-effect and saturation mobilities, n-channel Si MOSFET, $L = Z = 50\ \mu m$, $W_{ox} = 330\ Å$, $V_T = 0.66\ V$ for $V_{BS} = 0$.

where $\alpha \approx 0.34 \, \mu\text{m}$, $\beta \approx 0.06 \, \mu\text{m}$, and $\theta \approx 0.05 - 0.06 \, \text{V}^{-1}$ for one particular device.[92] The mobility degradation factor θ varies with gate oxide thickness and with doping concentration.[93] The low-field mobility μ_0 is obtained by extrapolating the experimental $\mu_{\text{eff}} - V_{\text{GS}}$ plot to $V_{\text{GS}} = V_{\text{T}}$, shown in Fig. 5.19. The constants α/L, β/Z, and θ are obtained from a plot of μ_0/μ_{eff} versus $(V_{\text{GS}} - V_{\text{T}})$, as shown in Fig. 5.20. θ is the slope of such a plot, and $(1 + \alpha/L + \beta/Z)$ is the intercept at $(V_{\text{GS}} - V_{\text{T}}) = 0$ on the μ_0/μ_{eff} axis.

A number of variations of the expression in Eq. (5.63) have been proposed to agree with experimental data. Frequently the α/L and the β/Z terms can be neglected to first order, and only the gate voltage-dependent θ term remains in the denominator. Some expressions include series resistance,[94] others include mobility reduction due to lateral electric fields.[95-96] This latter effect is important only for short-channel devices in which the drain voltage or the lateral electric field affects the mobility.

The effective mobility of *depletion-mode* devices can be measured by the same drain conductance method. Depletion-mode devices have the additional advantage that mobility profiles can be obtained by varying the gate voltage. In order to extract depth-dependent mobilities, it is necessary to determine the carrier concentration independently. That is usually done by capacitance-voltage measurements. A combination capacitance-voltage and current-voltage measurements gives the mobility profile.[97-98]

The dependence of the effective mobility on gate voltage is sometimes expressed as the dependence on the normal or vertical electric field, according to

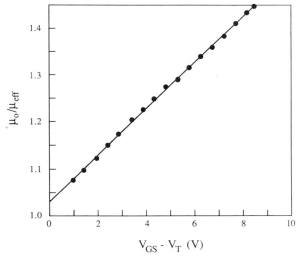

Fig. 5.20 μ_0/μ_{eff} as a function of $V_{\text{GS}} - V_{\text{T}}$ for the device of Fig. 5.19. The plot gives $\theta = 0.05 \, \text{V}^{-1}$ and $(1 + \alpha/L + \beta/Z) = 1.03$.

$$\mu_{eff} = \frac{\mu_0}{1 + (\alpha \mathscr{E}_{eff})^{\kappa}} \qquad (5.64)$$

Equation (5.64) has the advantage of producing "universal" mobility–electric field curves if the electric field produced by the gate voltage is expressed as the electric field due to the space-charge region and the inversion layer charges as[99–100]

$$\mathscr{E}_{eff} = \frac{(Q_B + Q_N/2)}{K_s \varepsilon_0} \qquad (5.65)$$

where Q_B and Q_N are the change densities (C/cm^2) in the space-charge region and the inversion layer, respectively. The 2 in the inversion layer charge accounts for averaging of the electric field over the electron distribution in the inversion layer.[99] A large body of experimental room temperature Si data agrees closely with the empirical relations[83]

$$\mu_{eff}(n\text{-Si}) = \frac{1105}{1 + (\mathscr{E}_{eff}/30.5)^{0.657}} \qquad (5.66a)$$

$$\mu_{eff}(p\text{-Si}) = \frac{342}{1 + (\mathscr{E}_{eff}/15.4)^{0.617}} \qquad (5.66b)$$

This mode of presentation is appealing from its universality point of view, but it is more difficult to arrive at from an operational point of view. It is, after all, the gate voltage that is measured experimentally, not the electric field. Conversion of measured voltages to electric field requires a knowledge of the doping concentration under the gate and of the inversion charge.

5.6.2 Field-Effect Mobility

While the effective mobility is derived from the drain conductance, the *field-effect mobility* is determined from the transconductance. The MOSFET transconductance is defined by

$$g_m = \frac{\partial I_D}{\partial V_{GS}}\bigg|_{V_{DS}=constant} \qquad (5.67)$$

From Eq. (5.53) we find the transconductance to be

$$g_m = \frac{Z \bar{\mu}_n C_{ox}}{L} V_{DS} \qquad (5.68)$$

When this expression is solved for the mobility, it is known as the *field-effect mobility* μ_{FE} and given by

$$\mu_{FE} = \frac{Lg_m}{ZC_{ox}V_{DS}}$$

(5.69)

The threshold voltage need not be known for the determination of μ_{FE}. This would appear to make the field-effect mobility easier to determine since only a knowledge of V_{DS} is required. However, with the field-effect mobility defined by Eq. (5.69), it is found experimentally that $\mu_{FE} < \mu_{eff}$, as shown in Fig. 5.19. This is rather disturbing since it is the same device that is measured under similar bias conditions. This discrepancy between μ_{eff} and μ_{FE} is due to the neglect of the electric field dependence of the mobility in the derivation of Eq. (5.69).[101–102]

5.6.3 Saturation Mobility

Occasionally the MOSFET mobility is derived from the drain current–drain voltage curves with the device in saturation. The saturation drain current can be expressed as

$$I_{DS,sat} = \frac{BZ\bar{\mu}_n C_{ox}}{2L}(V_{GS} - V_T)^2$$

(5.70)

where B represents the body effect which is weakly dependent on the gate voltage. When Eq. (5.70) is solved for the mobility, this mobility is sometimes called the *saturation mobility*. It is defined by

$$\mu_{sat} = \frac{2Lm^2}{BZC_{ox}}$$

(5.71)

where m is the slope of a plot of $(I_{DS,sat})^{1/2}$ against $(V_{GS} - V_T)$. The saturation mobility defined by Eq. (5.71) is also shown on Fig. 5.19. Similar to μ_{FE}, it is lower than μ_{eff} because the gate voltage dependence of the mobility is neglected in Eq. (5.71). Furthermore additional error is introduced because the factor B is not well known. All of these considerations render this method questionable, and it is rarely used.

5.7 STRENGTHS AND WEAKNESSES

- *Conductivity Mobility* The weakness of the conductivity mobility method is the requirement of both sample resistivity and carrier concentration. Both of these require independent measurements. Its strength lies in that it is directly defined from the sample resistivity or conductivity and no correction factors are required in its analysis.

- *Hall Effect Mobility*   The weakness of the Hall method lies in the special sample and test apparatus requirements and the inability to predict a precise value for the Hall scattering factor. The usual assumption of $r = 1$ introduces an error into the measured mobility. Although appropriate sample geometries exist for profiling, the method is awkward for mobility profiling. The strength of the Hall technique lies in its common use and the availability of mobilities determined by this method for all of the common semiconductors. It is a common mobility measurement method.

- *Magnetoresistance Mobility*   The weakness of the magnetoresistance technique lies in its limited use and the inability to characterize low-mobility semiconductors. For example, it does not work well for Si. In common with the Hall effect, it is difficult to determine the magnetoresistance scattering factor, and the assumption $\xi = 1$ introduces an error. Its strength is the ability to measure devices requiring no special test structures. MESFET's and MESFET-like devices can be easily characterized.

- *Time-of-Flight or Drift Mobility*   The weakness of this method is the requirement of special test structures and high speed electronics and/or optics. This puts the method into the hands of specialists in a few laboratories. Its strengths lies in the ability to measure the mobility and the carrier velocity at high electric fields. Most of the experimental data of velocity–electric field curves were generated by this method.

- *MOSFET Mobility*   This method is only suitable for MOSFET's, MESFET's, and MODFET's, and an operational mobility is extracted. Depending on how the mobility is measured, different experimental values are obtained. The effective mobility is the most common one of these and also the least ambiguous. Both the field-effect and saturation mobility, as usually defined, yield low mobilities and should not be used to characterize a device, unless they are modified.

APPENDIX 5.1   SEMICONDUCTOR MOBILITIES

Silicon   The dependence of the mobility on carrier concentration that gives good agreement with experiment at room temperature is given by the empirical expressions[103]

$$\mu_n = \mu_0 + \frac{(\mu_{max} - \mu_0)}{[1 + (n/C_r)^a]} - \frac{\mu_1}{[1 + (C_s/n)^b]} \tag{A5.1}$$

$$\mu_p = \mu_0 e^{-p_c/p} + \frac{\mu_{max}}{[1 + (p/C_r)^a]} - \frac{\mu_1}{[1 + (C_s/p)^b]} \tag{A5.2}$$

TABLE A5.1   Mobility Fit Parameters for Silicon

Parameter	Arsenic	Phosphorus	Boron
μ_0 (cm^2/V-s)	52.2	68.5	44.9
μ_{max} (cm^2/V-s)	1417	1414	470.5
μ_1 (cm^2/V-s)	43.4	56.1	29.0
C_r (cm^{-3})	9.68×10^{16}	9.20×10^{16}	2.23×10^{17}
C_s (cm^{-3})	3.43×10^{20}	3.41×10^{20}	6.10×10^{20}
a	0.680	0.711	0.719
b	2.00	1.98	2.00
p_c (cm^{-3})	—	—	9.23×10^{16}

Source: Masetti et al.[103]

The parameters that give the best fit to experimental data are given in Table A5.1.[103]

These two expressions are plotted in Fig. A5.1. Equation (A5.1) agrees with experimental data over the carrier concentration range $10^{14} \leq n \leq 5 \times 10^{21}$ cm^{-3}, and (A5.2) agrees with experimental data over the carrier concentration range $10^{14} \leq p \leq 10^{21}$ cm^{-3}. The experimental points are not shown in Fig. A5.1 for clarity but are found in Masetti.[103] The doping concentration dependence of the mobility is often expressed as[104]

$$\mu = \mu_{min} + \frac{\mu_0}{[1 + (N/N_{ref})^\alpha]} \tag{A5.3}$$

where μ is either the electron or hole mobility and N is the donor or acceptor concentration. The temperature dependence of the various parameters in Eq. (A5.3) has the form

$$A = A_0(T/300)^n \tag{A5.4}$$

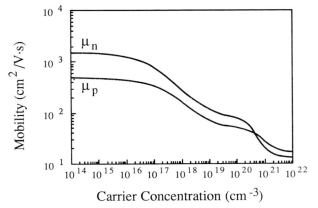

Fig. A5.1   Electron and hole mobilities in Si calculated from Eqs. (A5.1) and (A5.2).

TABLE A5.2   Mobility Fit Parameters for Silicon

Parameter	Temperature-Independent Prefactor		Temperature Exponent
	Electrons	Holes	
μ_0 (cm^2/V-s)	1268	406.9	-2.33 electrons
			-2.23 holes
μ_{min} (cm^2/V-s)	92	54.3	-0.57
N_{ref} (cm^{-3})	1.3×10^{17}	2.35×10^{17}	2.4
α	0.91	0.88	-0.146

Source: Baccarani and Ostoja[105] and Arora et al.[106]

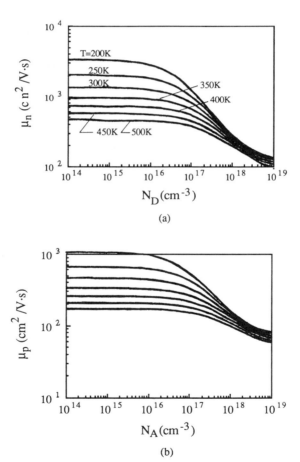

(a)

(b)

Fig. A5.2   (a) Electron and (b) hole mobilities in Si as a function of temperature calculated from Eqs. (A5.3) and (A5.4).

The parameters that give the best fit to experimental data are given in Table A5.2.[105–106]

Equation (A5.3) is plotted in Fig. A5.2 for *n*-Si and *p*-Si as a function of temperature. Experimental data from Li and Thurber[107] and Li[108] agree reasonably well with the mobilities in Fig. A5.2. Other mobility expressions have also been proposed.[109–110]

Gallium Arsenide   The mobilities of *n*- and *p*-type GaAs are shown in Fig. A5.3 at $T = 300$ K.

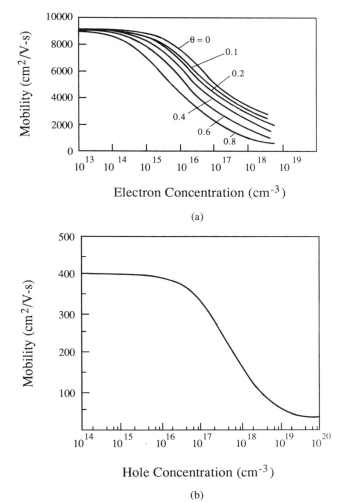

(a)

(b)

Fig. A5.3   (a) Electron Hall mobilities, (b) hole Hall mobility of GaAs at $T = 300$ K. θ is the compensation ratio. Reprinted with permission after Lancefield et al.[111] and Wiley.[112]

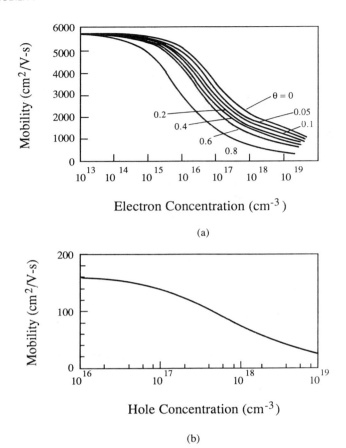

Fig. A5.4 (a) Electron Hall mobilities, (b) hole Hall mobility of InP at $T = 300$ K. θ is the compensation ratio. Reprinted with permission after Lancefield et al.[111] and Wiley.[112]

Indium Phosphide   The mobilities of n- and p-type InP are shown in Fig. A5.4 at $T = 300$ K.

REFERENCES

[1]   F. J. Morin, "Lattice-Scattering Mobility in Germanium," *Phys. Rev.* **93**, 62–63, Jan. 1954.

[2]   F. J. Morin and J. P. Maita, "Electrical Properties of Silicon Containing Arsenic and Boron," *Phys. Rev.* **96**, 28–35, Oct. 1954.

[3]   E. H. Hall, "On a New Action of the Magnet on Electric Currents," *Amer. J. Math.* **2**, 287–292, 1879.

[4] K. R. Sopka, "The Discovery of the Hall Effect: Edwin Hall's Hitherto Unpublished Account," in *The Hall Effect and Its Applications* (C. L. Chien and C. R. Westgate, eds.), Plenum, New York, 1980, pp. 523–545.

[5] E. H. Putley, *The Hall Effect and Related Phenomena*, Butterworths, London, 1960; "The Hall Effect and Its Application," *Contemp. Phys.* **16**, 101–126, *March* 1975.

[6] R. A. Smith, *Semiconductors*, Cambridge University Press, Cambridge, 1959, Ch. 5.

[7] A. Zemel, A. Sher, and D. Eger, "Anomalous Hall Effect in p-Type $Hg_{1-x}Cd_xTe$ Liquid-Phase-Epitaxial Layers," *J. Appl. Phys.* **62**, 1861–1868, Sept. 1987.

[8] G. E. Stillman and C. M. Wolfe, "Electrical Characterization of Epitaxial Layers," *Thin Solid Films* **31**, 69–88, Jan. 1976.

[9] M. C. Gold and D. A. Nelson, "Variable Magnetic Field Hall Effect Measurements and Analyses of High Purity, Hg Vacancy (p-type) HgCdTe," *J. Vac. Sci. Technol.* **A4**, 2040–2046, July/Aug. 1986.

[10] A. C. Beer, *Galvanomagnetic Effects in Semiconductors*, Academic Press, New York, 1963, p. 308.

[11] D. L. Rode, C. M. Wolfe, and G. E. Stillman, "Magnetic-Field Dependence of the Hall Factor of Gallium Arsenide," in *GaAs and Related Compounds* (G. E. Stillman, ed.) Conf. Ser. No. 65, Inst. Phys., Bristol, 1983, pp. 569–572.

[12] L. J. van der Pauw, "A Method of Measuring Specific Resistivity and Hall Effect of Discs of Arbitrary Shape," *Phil. Res. Rep.* **13**, 1–9, Feb. 1958.

[13] L. J. van der Pauw, "A Method of Measuring the Resistivity and Hall Coefficient on Lamellae of Arbitrary Shape," *Phil. Tech. Rev.* **20**, 220–224, Aug. 1958.

[14] A. Chandra, C. E. C. Wood, D. W. Woodward, and L. F. Eastman, "Surface and Interface Depletion Corrections to Free Carrier-Density Determinations by Hall Measurements," *Solid-State Electron.* **22**, 645–650, July 1979.

[15] W. E. Ham, "Surface Charge Effects on the Resistivity and Hall Coefficient of Thin Silicon-on-Sapphire Films," *Appl. Phys. Lett.* **21**, 440–443, Nov. 1972.

[16] T. R. Lepkowski, R. Y. DeJule, N. C. Tien, M. H. Kim, and G. E. Stillman, "Depletion Corrections in Variable Temperature Hall Measurements," *J. Appl. Phys.* **61**, 4808–4811, May 1987.

[17] R. Baron, G. A. Shifrin, O. J. Marsh, and J. W. Mayer, "Electrical Behavior of Group III and V Implanted Dopants in Silicon," *J. Appl. Phys.* **40**, 3702–3719, Aug. 1969.

[18] H. Maes, W. Vandervorst, and R. Van Overstraeten, "Impurity Profile of Implanted Ions in Silicon," in *Impurity Doping Processes in Silicon* (F. F. Y. Wang, ed.) North-Holland, Amsterdam, 1981, pp. 443–638.

[19] T. Ambridge and C. J. Allen, "Automatic Electrochemical Profiling of Hall Mobility in Semiconductors," *Electron. Lett.* **15**, 648–650, Sept. 1979.

[20] J. W. Mayer, O. J. Marsh, G. A. Shifrin, and R. Baron, "Ion Implantation of Silicon; II Electrical Evaluation Using Hall-Effect Measurements," *Can. J. Phys.* **45**, 4073–4089, Dec. 1967.

[21] N. G. E. Johannson, J. W. Mayer, and O. J. Marsh, "Technique Used in Hall Effect Analysis of Ion Implanted Si and Ge," *Solid-State Electron.* **13**, 317–335, March 1970.

[22] N. D. Young and M. J. Hight, "Automated Hall Effect Profiler for Electrical Characterisation of Semiconductors," *Electron. Lett.* **21**, 1044–1046, Oct. 1985.

[23] H. Müller, F. H. Eisen, and J. W. Mayer, "Anodic Oxidation of GaAs as a Technique to Evaluate Electrical Carrier Concentration Profiles," *J. Electrochem. Soc.* **122**, 651–655, May 1975.

[24] L. Buoro and D. Tsoukalas, "Determination of Doping and Mobility Profiles by Automated Electrical Measurements and Anodic Stripping," *Phys. E: Sci. Instrum.* **20**, 541–544, May 1987.

[25] A. C. Ipri, "Variation in Electrical Properties of Silicon Films on Sapphire Using the MOS Hall Technique," *Appl. Phys. Lett.* **20**, 1–2, Jan. 1972.

[26] A. B. M. Elliot and J. C. Anderson, "An Investigation of Carrier Transport in Thin Silicon-on-Sapphire Films Using MIS Deep Depletion Hall Effect Structures," *Solid-State Electron.* **15**, 531–545, May 1972.

[27] P. A. Crossley and W. E. Ham, "Use of Test Structures and Results of Electrical Tests for Silicon-on-Sapphire Integrated Circuit Processes," *J. Electron. Mat.* **2**, 465–483, Aug. 1973.

[28] S. Cristoloveanu, J. H. Lee, J. Pumfrey, J. R. Davies, R. P. Arrowsmith, and P. L. F. Hemment, "Profiling of Inhomogeneous Carrier Transport Properties with the Influence of Temperature in Silicon-on-Insulator Films Formed by Oxygen Implantation," *J. Appl. Phys.* **60**, 3199–3203, Nov. 1986.

[29] T. L. Tansley, "AC Profiling by Schottky-Gated Cloverleaf," *J. Phys. E: Sci. Instrum.* **8**, 52–54, Jan. 1975.

[30] C. W. Farley and B. G. Streetman, "The Schottky-Gated Hall-Effect Transistor and Its Application to Carrier Concentration and Mobility Profiling in GaAs MESFET's," *IEEE Trans. Electron. Dev.* **ED-34**, 1781–1787, Aug. 1987.

[31] P. R. Jay, I. Crossley, and M. J. Caldwell, "Mobility Profiling of FET Structures," *Electron. Lett.* **14**, 190–191, March 1978.

[32] H. H. Wieder, *Laboratory Notes on Electrical and Galvanomagnetic Measurements*, Elsevier, Amsterdam, 1979, chs. 5–6.

[33] H. Ryssel, K. Schmid, and H. Müller, "A Sample Holder for Measurement and Anodic Oxidation of Ion Implanted Silicon," *J. Phys. E: Sci. Instrum.* **6**, 492–494, May 1973.

[34] W. K. Hofker, H. W. Werner, D. P. Oosthoek, and N. J. Koeman, "Boron Implantation in Si: A Comparison of Charge Carrier and Boron Concentration Profiles," *Appl. Phys.* **4**, 125–131, Feb. 1974.

[35] J. W. Mayer, L. Eriksson, and J. A. Davies, *Ion Implantation in Semiconductors; Silicon and Germanium*, Academic Press, New York, 1970.

[36] C. M. Wolfe and G. E. Stillman, "Apparent Mobility Enhancement in Inhomogeneous Crystals," in *Semiconductors and Semimetals* (R. K. Willardson and A. C. Beer, eds.) Academic Press, New York, **10**, 175–220, 1975.

[37] A. Neduloha and K. M. Koch, "On the Mechanism of the Resistance Change in a Magnetic Field (in German)," *Z. Phys.* **132**, 608–620, 1952.

[38]   R. L. Petritz, "Theory of an Experiment for Measuring the Mobility and Density of Carriers in the Space-Charge Region of a Semiconductor Surface," *Phys. Rev.* **110**, 1254–1262, June 1958.

[39]   R. D. Larrabee and W. R. Thurber, "Theory and Application of a Two-Layer Hall Technique," *IEEE Trans. Electron Dev.* **ED-27**, 32–36, Jan. 1980.

[40]   L. F. Lou and W. H. Frye, "Hall Effect and Resistivity in Liquid-Phase-Epitaxial Layers of HgCdTe," *J. Appl. Phys.* **56**, 2253–2267, Oct. 1984.

[41]   A. Zemel and J. R. Sites, "Electronic Transport near the Surface of Indium Antimonide Films," *Thin Solid Films* **41**, 297–305, March 1977.

[42]   ASTM Standard F76, "Standard Method for Measuring Hall Mobility and Hall Coefficient in Extrinsic Semiconductor Single Crystals," *1988 Annual Book of ASTM Standards*, Am. Soc. Test. Mat., Philadelphia, 1988.

[43]   D. C. Look, "Bulk and Contact Electrical Properties by the Magneto-Transmission-Line Method: Application to GaAs," *Solid-State Electron.* **30**, 615–618, June 1987.

[44]   D. S. Perloff, "Four-Point Sheet Resistance Correction Factors for Thin Rectangular Samples," *Solid-State Electron.* **20**, 681–687, Aug. 1977.

[45]   J. M. David and M. G. Buehler, "A Numerical Analysis of Various Cross Sheet Resistor Test Structures," *Solid-State Electron.* **20**, 539–543, June 1977.

[46]   R. Chwang, B. J. Smith, and C. R. Crowell, "Contact Size Effects on the van der Pauw Method for Resistivity and Hall Coefficient Measurement," *Solid-State Electron.* **17**, 1217–1227, Dec. 1974.

[47]   H. J. Lippmann and F. Kuhrt, "The Geometrical Influence of Rectangular Semiconductor Plates on the Hall Effect (in German)," *Z. Naturforsch.* **13a**, 474–483, 1958; I. Isenberg, B. R. Russell, and R. F. Greene, "Improved Method for Measuring Hall Coefficients," *Rev. Sci. Instrum.* **19**, 685–688, Oct. 1948.

[48]   P. M. Hemenger, "Measurement of High Resistivity Semiconductors Using the van der Pauw Method," *Rev. Sci. Instrum.* **44**, 698–700, June 1973.

[49]   L. Forbes, J. Tillinghast, B. Hughes, and C. Li, "Automated System for the Characterization of High Resistivity Semiconductors by the van der Pauw Method," *Rev. Sci. Instrum.* **52**, 1047–1050, July 1981.

[50]   R. T. Blunt, S. Clark, and D. J. Stirland, "Dislocation Density and Sheet Resistance Variations across Semi-insulating GaAs Wafers," *IEEE Trans. Electron Dev.* **ED-29**, 1038–1045, July 1982.

[51]   K. Kitahara and M. Ozeki, "Nondestructive Resistivity Measurement of Semi-Insulating GaAs Using Illuminated n^+-GaAs Contacts," *Japan. J. Appl. Phys.* **23**, 1655–1656, Dec. 1984.

[52]   D. C. Look, "Schottky-Barrier Profiling Techniques in Semiconductors: Gate Current and Parasitic Resistance Effects," *J. Appl. Phys.* **57**, 377–383, Jan. 1985.

[53]   P. Chu, S. Niki, J. W. Roach, and H. H. Wieder, "Simple, Inexpensive Double ac Hall Measurement System for Routine Semiconductor Characterization," *Rev. Sci. Instrum.* **58**, 1764–1766, Sept. 1987.

[54]   O. M. Corbino, "Electromagnetic Effects Resulting from the Distortion of the Path of Ions in Metals Produced by a Field (in German)," *Physik. Zeitschr.* **12**, 561–568, July 1911.

[55] H. Weiss, "Magnetoresistance," in *Semiconductors and Semimetals* (R. K. Willardson and A. C. Beer, eds.) Academic Press, New York, **1**, 315–376, 1966.

[56] H. J. Lippman and F. Kuhrt, "The Geometrical Influence on the Transverse Magnetoresistance Effect for Rectangular Semiconductor Plates (in German)," *Z. Naturforsch.* **13a**, 462–474, 1958.

[57] T. R. Jervis and E. F. Johnson, "Geometrical Magnetoresistance and Hall Mobility in Gunn Effect Devices," *Solid-State Electron.* **13**, 181–189, Feb. 1970.

[58] P. Blood and R. J. Tree, "The Scattering Factor for Geometrical Magneto-resistance in GaAs," *J. Phys. D: Appl. Phys.* **4**, L29–L31, Sept. 1971.

[59] H. Poth, "Measurement of Mobility Profiles in GaAs at Room Temperature by the Corbino Effect," *Solid-State Electron.* **21**, 801–805, June 1978.

[60] J. R. Sites and H. H. Wieder, "Magnetoresistance Mobility Profiling of MESFET Channels," *IEEE Trans. Electron Dev.* **ED-27**, 2277–2281, Dec. 1980.

[61] R. D. Larrabee, W. A. Hicinbothem, Jr., and M. C. Steele, "A Rapid Evaluation Technique for Functional Gunn Diodes," *IEEE Trans. Electron Dev.* **ED-17**, 271–274, April 1970.

[62] P. R. Jay and R. H. Wallis, "Magnetotransconductance Mobility Measurements of GaAs MESFET's," *IEEE Electron. Dev. Lett.* **EDL-2**, 265–267, Oct. 1981.

[63] D. C. Look and G. B. Norris, "Classical Magnetoresistance Measurements in $Al_xGa_{1-x}As/GaAs$ MODFET Structures: Determination of Mobilities," *Solid-State Electron.* **29**, 159–165, Feb. 1986.

[64] W. T. Masselink, T. S. Henderson, J. Klem, W. F. Kopp, and H. Morkoç, "The Dependence of 77 K Electron Velocity-Field Characteristics on Low-Field Mobility in AlGaAs–GaAs Modulation-Doped Structures," *IEEE Trans. Electron Dev.* **ED-33**, 639–645, May 1986.

[65] D. C. Look and T. A. Cooper, "Schottky-Barrier Mobility Profiling Measurements with Gate-Current Corrections," *Solid-State Electron.* **28**, 521–527, May 1985.

[66] S. M. J. Liu and M. B. Das, "Determination of Mobility in Modulation-Doped FET's Using Magnetoresistance Effect," *IEEE Electron. Dev. Lett.* **EDL-8**, 355–357, Aug. 1987.

[67] J. R. Haynes and W. Shockley, "Investigation of Hole Injection in Transistor Action," *Phys. Rev.* **75**, 691, Feb. 1949.

[68] J. R. Haynes and W. Shockley, "The Mobility and Life of Injected Holes and Electons in Germanium," *Phys. Rev.* **81**, 835–843, March 1951.

[69] J. R. Haynes and W. C. Westphal, "The Drift Mobility of Electrons in Silicon," *Phys. Rev.* **85**, 680–681, Feb. 1952.

[70] M. B. Prince, "Drift Mobilities in Semiconductors. I Germanium," *Phys. Rev.* **92**, 681–687, Nov. 1953.

[71] M. B. Prince, "Drift Mobilities in Semiconductors. II Silicon," *Phys. Rev.* **93**, 1204–1206, March 1954.

[72] J. P. McKelvey, *Solid State and Semiconductor Physics*, Harper & Row, New York, 1966, p. 342.

[73] A. Bar-Lev, *Semiconductors and Electronic Devices*, 2d ed., Prentice-Hall, Englewood Cliffs, NJ, 1984, pp. 49–55.

[74] H. Hillmer, G. Mayer, A. Forchel, K. S. Löchner, and E. Bauser, "Optical Time-of-Flight Investigation of Ambipolar Carrier Transport in GaAlAs Using GaAs/GaAlAs Double Quantum Well Structures," *Appl. Phys. Lett.* **49**, 948–950, Oct. 1986.

[75] D. J. Westland, D. Mihailovic, J. F. Ryan, and M. D. Scott, "Optical Time-of-Flight Measurement of Carrier Diffusion and Trapping in an InGaAs/InP Heterostructure," *Appl. Phys. Lett.* **51**, 590–592, Aug. 1987.

[76] C. B. Norris, Jr., and J. F. Gibbons, "Measurement of High-Field Carrier Drift Velocities in Silicon by Time-of-Flight Technique," *IEEE Trans. Electron Dev.* **ED-14**, 38–43, Jan. 1967.

[77] A. G. R. Evans and P. N. Robson, "Drift Mobility Measurements in Thin Epitaxial Semiconductor Layers Using Time-of-Flight Techniques," *Solid-State Electron.* **17**, 805–812, Aug. 1974.

[78] P. M. Smith, M. Inoue, and J. Frey, "Electron Velocity in Si and GaAs at Very High Electric Fields," *Appl. Phys. Lett.* **37**, 797–798, Nov. 1980.

[79] T. H. Windhorn, L. W. Cook, and G. E. Stillman, "High-Field Electron Transport in InGaAsP ($\lambda_g = 1.2$ μm)," *Appl. Phys. Lett.* **41**, 1065–1067, Dec. 1982.

[80] W. Shockley, "Currents to Conductors Induced by a Moving Point Charge," *J. Appl. Phys.* **9**, 635–636, Oct. 1938.

[81] W. E. Spear, "Drift Mobility Techniques for the Study of Electrical Transport Properties in Insulating Solids," *J. Non-Cryst. Sol.* **1**, 197–214, April 1969.

[82] J. A. Cooper, Jr., and D. F. Nelson, "High-Field Drift Velocity of Electrons at the Si–SiO$_2$ Interface as Determined by a Time-of-Flight Technique," *J. Appl. Phys.* **54**, 1445–1456, March 1983.

[83] J. A. Cooper, Jr., D. F. Nelson, S. A. Schwarz, and K. K. Thornber, "Carrier Transport at the Si-SiO$_2$ Interface," in *VLSI Electronics Microstructure Science* (N. G. Einspruch and R. S. Bauer, eds.), Academic Press, Orlando, FL, **10**, 1985, pp. 323–361.

[84] D. D. Tang, F. F. Fang, M. Scheuermann, and T. C. Chen, "Time-of-Flight Measurements of Minority-Carrier Transport in *p*-Silicon," *Appl. Phys. Lett.* **49**, 1540–1541, Dec. 1986.

[85] C. Canali, M. Martini, G. Ottaviani, and K. R. Zanio, "Transport Properties of CdTe," *Phys. Rev.* **B4**, 422–431, July 1971.

[86] C. Kittel, *Introduction to Solid State Physics*, 4th ed., Wiley, New York, 1975, p. 261.

[87] J. R. Schrieffer, "Effective Carrier Mobility in Surface-Space Charge Layers," *Phys. Rev.* **97**, 641–646, Feb. 1955.

[88] M. S. Lin, "The Classical versus the Quantum Mechanical Model of Mobility Degradation due to the Gate Field in MOSFET Inversion Layers," *IEEE Trans. Electron Dev.* **ED-32**, 700–710, March 1985.

[89] A. Rothwarf, "A New Quantum Mechanical Channel Mobility Model for Si MOSFET's," *IEEE Electron Dev. Lett.* **EDL-8**, 499–502, Oct. 1987.

[90] S. M. Sze, *Physics of Semiconductor Devices*, 2d ed., Wiley, New York, 1981, pp. 440.

[91] S. M. Goodnick, D. K. Ferry, C. W. Wilmsen, Z. Liliental, D. Fathy, and O. L. Krivanek, "Surface Roughness at the Si(100)-SiO$_2$ Interface," *Phys. Rev.* **B32**, 8171–8186, Dec. 1985.

[92] P. P. Wang, "Device Characteristics of Short-Channel and Narrow-Width MOSFET's," *IEEE Trans. Electron Dev.* **ED-25**, 779–786, July 1978.

[93] K. Y. Fu, "Mobility Degradation due to the Gate Field in the Inversion Layer of MOSFET's," *IEEE Electron Dev. Lett.* **EDL-3**, 292–293, Oct. 1982.

[94] L. Risch, "Electron Mobility in Short-Channel MOSFET's with Series Resistances," *IEEE Trans. Electron Dev.* **ED-30**, 959–961, Aug. 1983.

[95] N. Herr and J. J. Barnes, "Statistical Circuit Simulation Modeling of CMOS VLSI," *IEEE Trans. Comp.-Aided Des.* **CAD-5**, 15–22 Jan. 1986.

[96] M. H. White, F. van de Wiele, and J. P. Lambot, "High-Accuracy MOS Models for Computer-Aided Design," *IEEE Trans. Electron Dev.* **ED-27**, 899–906, May 1980.

[97] S. T. Hsu and J. H. Scott, Jr., "Mobility of Current Carriers in Silicon-on-Sapphire (SOS) Films," *RCA Rev.* **36**, 240–253, June 1975.

[98] R. A. Pucel and C. A. Krumm, "Simple Method of Measuring Drift-Mobility Profiles in Thin Semiconductor Films," *Electron. Lett.* **12**, 240–242, May 1976.

[99] A. G. Sabnis and J. T. Clemens, "Characterization of the Electron Mobility in the Inverted (100) Si Surface," *IEEE Int. Electron Dev. Meet.*, Washington, DC, 1979, 18–21.

[100] S. C. Sun and J. D. Plummer, "Electron Mobility in Inversion and Accumulation Layers on Thermally Oxidized Silicon Surfaces," *IEEE Trans. Electron Dev.* **ED-27**, 1497–1508, Aug. 1980.

[101] F. F. Fang and A. B. Fowler, "Transport Properties of Electrons in Inverted Silicon Surfaces," *Phys. Rev.* **169**, 619–631, May 1968.

[102] J. S. Kang, D. K. Schroder, and A. R. Alvarez, "Effective and Field-Effect Mobilities in Si MOSFET's," *Solid-State Electron.* **32**, 679–681, Aug. 1989.

[103] G. Masetti, M. Severi, and S. Solmi, "Modeling of Carrier Mobility against Carrier Concentration in Arsenic, Phosphorus-, and Boron-Doped Silicon," *IEEE Trans. Electron Dev.* **ED-30**, 764–769, July 1983, and references therein.

[104] D. M. Caughey and R. E. Thomas, "Carrier Mobilities in Silicon Empirically Related to Doping and Field," *Proc. IEEE* **55**, 2192–2193, Dec. 1967.

[105] G. Baccarani and P. Ostoja, "Electron Mobility Empirically Related to the Phosphorus Concentration in Silicon," *Solid-State Electron.* **18**, 579–580, June 1975.

[106] N. D. Arora, J. R. Hauser, and D. J. Roulston, "Electron and Hole Mobilities in Silicon as a Function of Concentration and Temperature," *IEEE Trans. Electron Dev.* **ED-29**, 292–295, Feb. 1982.

[107] S. S. Li and W. R. Thurber, "The Dopant Density and Temperature Dependence of Electron Mobility and Resistivity in n-Type Silicon," *Solid-State Electron.* **20**, 609–616, July 1977.

[108] S. S. Li, "The Dopant Density and Temperature Dependence of Hole Mobility and Resistivity in Boron-Doped Silicon," *Solid-State Electron.* **21**, 1109–1117, Sept. 1978.

[109] J. M. Dorkel and Ph. Leturcq, "Carrier Mobilities in Silicon Semi-Empirically Related to Temperature, Doping and Injection Level," *Solid-State Electron.* **24**, 821–825, Sept. 1981.

[110] Y. Sasaki, K. Itoh, E. Inoue, S. Kishi, and T. Mitsuishi, "A New Experimental Determination of the Relationship between the Hall Mobility and the Hole Concentration in Heavily Doped *p*-Type Silicon," *Solid-State Electron.* **31**, 5–12, Jan. 1988.

[111] D. Lanefield, A. R. Adams, and M. A. Fisher, "Reassessment of Ionized Impurity Scattering and Compensation in GaAs and InP Including Correlation Scattering," *J. Appl. Phys.* **62**, 2342–2359, Sept. 1987, and references therein.

[112] J. D. Wiley, "Mobility of Holes in III-V Compounds," in *Semiconductors and Semimetals* (R. K. Willardson and A. C. Beer, eds.) Academic Press, New York, **10**, 91–174, 1975, and references therein.

CHAPTER 6

OXIDE AND INTERFACE
TRAPPED CHARGE

6.1 INTRODUCTION

The discussions in this chapter are applicable to all oxide–semiconductor systems. But the examples are generally directed at the SiO_2–Si system since that is the most important one. There are four general types of charges associated with the SiO_2–Si system shown on Fig. 6.1. They are *fixed oxide charge, mobile oxide charge, oxide trapped charge*, and *interface trapped charge*. This nomenclature was standardized in 1978. The abbreviations of the various charges are given in brackets below. In each case Q is the net effective charge per unit area at the SiO_2–Si interface, C/cm^2, N is the net number of charges per unit area at the SiO_2–Si interface, $number/cm^2$, and D_{it} is given in units of $number/cm^2 \cdot eV$. $N = |Q/q|$, where Q can be positive or negative, but N is always positive. The charges are described as:[1]

Fixed Oxide Charge (Q_f, N_f): Positive charge, due primarily to structural defects (ionized silicon) in the oxide layer less than 25 Å from the Si–SiO_2 interface. The density of this charge, whose origin is related to the oxidation process, depends on the oxidation ambient and temperature, on cooling conditions, and on silicon orientation. Since the fixed oxide charge cannot be determined unambiguously in the presence of moderate densities of interface trapped charge, it is only measured after a low-temperature ($\sim450°C$) hydrogen anneal which minimizes interface trapped charge. The fixed oxide charge is not in electrical communication with the underlying silicon.

It has been found that Q_f depends on the final oxidation temperature. The higher the oxidation temperature, the lower is Q_f. However, if it is

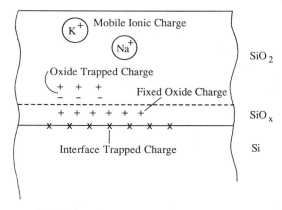

Fig. 6.1 Charges and their location for thermally oxidized silicon. Reprinted after Deal[1] by permission of IEEE (© 1980, IEEE).

not permissible to oxidize at high temperatures, it is possible to lower Q_f by annealing the oxidized wafer in a nitrogen or argon ambient after oxidation. This has resulted in the well-known "Deal triangle" shown in Fig. 6.2, which clearly shows the relationship between Q_f and oxidation and anneal treatments.[2] It shows that the processes are reversible in that "an oxidized sample may be prepared at any temperature and then subjected to dry oxygen at any other temperature, with the resulting value of Q_f being associated with the final temperature" and also that "any Q_f value resulting from a previous oxidation . . . can be reduced to a constant value . . ."[2]

This latter statement was later challenged by pointing out that Q_f is dependent on the anneal temperature and that there is a difference in behavior between nitrogen and an inert gas like argon.[3–4] During nitrogen anneal, Q_f reaches a minimum rapidly but then increases again.

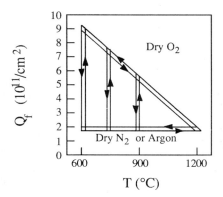

Fig. 6.2 "Deal triangle" showing the reversibility of heat treatment effects on Q_f. Reprinted after Deal et al.[2] with permission of the publisher, the Electrochemical Society, Inc.

This is believed to be due to a reaction of nitrogen with silicon at the surface. Fixed charge has often been designated by Q_{ss} in the past.

Mobile Oxide Charge (Q_m, N_m): Primarily due to ionic impurities such as Na^+, Li^+, K^+, and possibly H^+. Negative ions and heavy metals may contribute to this charge even though they are not mobile below 500°C.

Oxide Trapped Charge (Q_{ot}, N_{ot}): May be positive or negative due to holes or electrons trapped in the bulk of the oxide. Trapping may result from ionizing radiation, avalanche injection, or other similar processes. Unlike fixed charge, oxide trapped charge is generally annealed by low-temperature (<500°C) treatments, although neutral traps may remain.

Interface Trapped Charge (Q_{it}, N_{it}, D_{it}): Positive or negative charges, due to (1) structural, oxidation-induced defects, (2) metal impurities, or (3) other defects caused by radiation or similar bond breaking processes (e.g., hot electrons). The interface trapped charge is located at the Si–SiO$_2$ interface. Unlike fixed, mobile, or trapped charge, interface trapped charge is in electrical communication with the underlying silicon and can thus be charged or discharged, depending on the surface potential. Most of the interface trapped charge can be neutralized by low-temperature (450°C) hydrogen annealing. This charge type in the past has been called *surface states*, *fast states*, *interface states*, etc. It has been designated by N_{ss}, N_{st}, and other symbols in the past.

6.2 FIXED, OXIDE TRAPPED, AND MOBILE CHARGE

6.2.1 Capacitance-Voltage Curves

The various charges can be measured by many methods, the most popular being the capacitance-voltage $(C\text{-}V)$ measurement of a metal-oxide-semiconductor capacitor (MOS-C). Before discussing measurement methods, it is necessary to derive the appropriate capacitance relationships and to describe the C-V curves.

The energy band diagram of an MOS capacitor on a p-type substrate is shown in Fig. 6.3. The intrinsic energy level E_i is taken as the zero reference energy and zero reference potential. The surface potential is measured from this reference level. The capacitance is defined as

$$C = \frac{dQ}{dV} \tag{6.1}$$

It is the change of charge due to a change of voltage. During a capacitance measurement a small-signal ac voltage is applied to the device. The resulting charge variation gives rise to the capacitance. Looking at an MOS-C from the gate, we find $C = dQ_G/dV_G$, where Q_G and V_G are the gate charge and

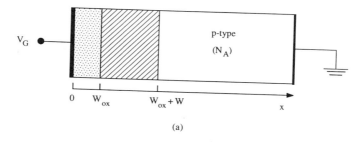

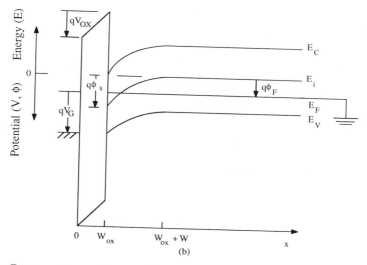

Fig. 6.3 Cross section and energy (potential) band diagram of an MOS capacitor.

the gate voltage, respectively. Since the total charge in the device must be zero, $Q_G = -(Q_S + Q_{it})$ assuming no oxide charge. The gate voltage is partially dropped across the oxide and partially across the semiconductor. This gives $V_G = V_{ox} + \phi_s$, where V_{ox} is the oxide voltage and ϕ_s the semiconductor voltage or surface potential, allowing Eq. (6.1) to be rewritten as

$$C = -\frac{dQ_S + dQ_{it}}{dV_{ox} + d\phi_s} \qquad (6.2)$$

The semiconductor charge Q_S, in general, consists of hole accumulation charge Q_P, space-charge region bulk charge Q_B, and electron inversion charge Q_N. With $Q_S = Q_P + Q_B + Q_N$, Eq. (6.2) becomes

$$C = -\frac{1}{[dV_{ox}/(dQ_S + dQ_{it})] + [d\phi_s/(dQ_P + dQ_B + dQ_N + dQ_{it})]} \qquad (6.3)$$

Utilizing the general capacitance definition of Eq. (6.1), Eq. (6.3) becomes

$$C = \frac{1}{(1/C_{ox}) + [1/(C_P + C_B + C_N + C_{it})]} = \frac{C_{ox}(C_P + C_B + C_N + C_{it})}{C_{ox} + C_P + C_B + C_N + C_{it}}$$

(6.4)

The positive accumulation charge Q_P dominates for negative oxide voltage and negative surface potentials. When V_{ox} and ϕ_s are positive, the semiconductor charges are negative. The minus sign cancels in either case.

Equation (6.4) is represented by the equivalent circuit in Fig. 6.4(a). For negative gate voltages, the surface is heavily accumulated, and Q_P dominates. C_P is very high approaching a short circuit. Hence the four capacitances are shorted as shown in Fig. 6.4(b). For small positive gate voltages, the surface is depleted, and the space-charge region charge, $Q_B = -qN_AW$, dominates. Trapped interface charge capacitance also contributes. The total capacitance is the combination of C_{ox} in series with C_B, which in turn is in parallel with C_{it}. In weak inversion C_N begins to appear. Figure 6.4(c) shows the equivalent circuit for weak inversion. For strong inversion, C_N dominates because Q_N is very high. If Q_N is able to follow the applied ac voltage, the low-frequency equivalent circuit [Fig 6.4(d)] becomes the oxide capacitance again. This gives the *low-frequency C-V* curve. When the inversion charge is unable to follow the ac voltage, the circuit in Fig. 6.4(e) applies in inversion, with $C_B = K_s\varepsilon_0/W_f$ giving the *high-frequency C-V* curve. W_f is the final space-charge region width discussed in Chapter 2.

The inversion capacitance dominates only if the inversion charge is able to follow the frequency of the applied ac voltage also called the ac probe frequency. With the MOS-C biased at some dc bias point, the ac voltage drives the device periodically above and below the dc bias point. During the phase when the device is driven to a slightly higher bias voltage, electron-hole pairs (ehp) are thermally generated in the space-charge region (scr) bulk, at the scr surface, and in the quasi-neutral bulk to try to restore equilibrium.[5] The scr generation current density, given by $J_{scr} = qn_iW/\tau_g$ and discussed in more detail in Chapter 8, dominates at room temperature in silicon. While J_{scr} flows in the semiconductor, the current flowing through the oxide is the displacement current density $J_d = C\, dV/dt$. In order for the inversion charge to be able to respond, the scr current must be able to supply the required displacement current or $J_d \leq J_{scr}$. This leads to

$$\frac{dV}{dt} \leq \frac{qn_iW}{\tau_g C_{ox}}$$

(6.5)

with C approximated by C_{ox}. For Si at $T = 300\,\text{K}$ with $W = 1\,\mu\text{m}$ and $C_{ox} = 3.45 \times 10^{-8}\,\text{F/cm}^2$, corresponding to $W_{ox} = 1000\,\text{Å}$, we find

$$\frac{dV}{dt} \leq \frac{6.5 \times 10^{-6}}{\tau_g}$$

(6.6)

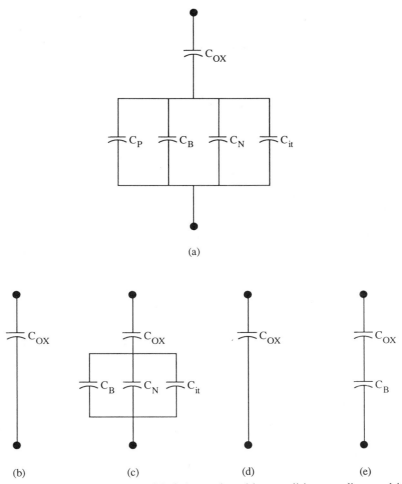

Fig. 6.4 Capacitances of an MOS-C for various bias conditions as discussed in the text.

Typical values for the generation lifetime lie in the $10 \, \mu s$ to $10 \, ms$ range. For $\tau_g = 10 \, \mu s$, $dV/dt \leq 0.65 \, V/s$, not a severe constraint. However, for $\tau_g = 1 \, ms$, $dV/dt \leq 6.5 \, mV/s$, a very severe constraint. Defining an effective frequency as $f_{\text{eff}} = (dV/dt)/v$, where v is the ac voltage, we find $f_{\text{eff}} = 45 \, Hz$ for the former and $0.4 \, Hz$ for the latter using $v = 15 \, mV$. These first-order numbers show that extremely low frequencies are required to obtain low-frequency C-V curves at room temperature. Increased generation rates at higher temperatures allow higher frequencies to be used. Experimentally determined frequencies lie in this range with $f_{\text{eff}} \approx 10 \, Hz$ at $300 \, K$ and $f_{\text{eff}} \approx 10 \, kHz$ at $450 \, K$.[6] Since typical C-V measurement frequencies lie in the 0.1–1 MHz range, it is obvious that high-frequency curves are almost always measured.

The *low-frequency* semiconductor capacitance $C_{S,LF}$ is given by

$$C_{S,LF} = \hat{U}_S \frac{K_s \varepsilon_0}{2 L_{Di}} \frac{e^{U_F}(1 - e^{-U_S}) + e^{-U_F}(e^{U_S} - 1)}{F(U_S, U_F)} \tag{6.7}$$

where the dimensionless electric field at the semiconductor surface F is defined by

$$F(U_S, U_F) = \sqrt{\exp(U_F)[\exp(-U_S) + U_S - 1] + \exp(-U_F)[\exp(U_S) - U_S - 1]} \tag{6.8}$$

The U's are normalized potentials, defined by $U_S = q\phi_s/kT$ and $U_F = q\phi_F/kT$, where the surface potential ϕ_s and the Fermi potential $\phi_F = (kT/q)\ln(N_A/n_i)$ are defined in Fig. 6.3. The symbol $\hat{U}_S$ stands for the sign of the surface potential and is given by

$$\hat{U}_S = \frac{|U_S|}{U_S} \tag{6.9}$$

$\hat{U}_S = 1$ if $U_S > 0$ and $\hat{U}_S = -1$ if $U_S < 0$. The intrinsic Debye length L_{Di} is

$$L_{Di} = \sqrt{\frac{K_s \varepsilon_0 kT}{2q^2 n_i}} \tag{6.10}$$

The *high-frequency* C-V curve is observed when the inversion charge is unable to follow the ac voltage. The high-frequency semiconductor capacitance in inversion is more difficult to calculate exactly, although exact expression do exist.[7] An expression, which is accurate to 0.02% in strong inversion, is[7]

$$C_{S,HF} = \sqrt{\frac{q^2 K_s \varepsilon_0 N_A}{2kT\{2|U_F| - 1 + \ln[1.15(|U_F| - 1)]\}}} \tag{6.11}$$

When the dc bias voltage is changed rapidly with insufficient time for inversion charge generation, then the *deep-depletion* curve results. Its capacitance is[8]

$$C_{S,DD} = \frac{C_{ox}}{[\sqrt{1 + 2(V_G - V_{FB})/V_0} - 1]} \tag{6.12}$$

where $V_0 = qK_s\varepsilon_0 N_A/C_{ox}^2$.

The total capacitance is given by

$$C = \frac{C_{ox}C_S}{C_{ox} + C_S} \qquad (6.13)$$

To calculate a capacitance-voltage curve, it is furthermore necessary to relate the gate voltage to the oxide voltage, the surface potential, and the flatband voltage V_{FB}. That relationship is

$$V_G = V_{FB} + \phi_s + V_{ox} = V_{FB} + \phi_s + \hat{U}_S \cdot \frac{kTK_sW_{ox}F(U_S,U_F)}{qK_{ox}L_{Di}} \qquad (6.14)$$

An ideal low-frequency *C-V* curve (LF), consisting of $C_{S,LF}$ in series with C_{ox}, is shown in Fig. 6.5. It is ideal in the sense that Q_{it} and V_{FB} are both zero. The normalized flatband capacitance is 0.7 for this example, as shown by the cross. Ideal high-frequency (HF) and deep-depletion (DD) curves are also shown in Fig. 6.5. They coincide with the low-frequency curve at negative gate voltages but deviate for positive gate voltages because the inversion charge is unable to follow the applied ac voltage for the HF case and does not exist for the DD case. These curves are for *p*-type substrates. For *n*-type substrates the curves are a mirror image about a vertical line through $V_G = 0$.

Which of these three curves is obtained during a *C-V* measurement? That depends on the measurement conditions. Let us consider an MOS-C with the dc gate voltage swept from negative to positive voltages. Superimposed on the dc voltage is a small-amplitude ac voltage of typically 10–15 mV amplitude. The ac voltage is necessary to measure the capacitance, while the dc voltage determines the bias condition. All three capacitance-voltage curves are identical in accumulation and depletion. The curves deviate from one another when the device enters inversion. If the dc voltage is swept

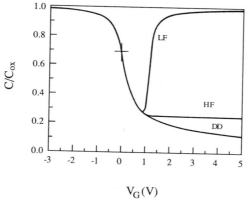

Fig. 6.5 Ideal low-frequency (LF), high-frequency (HF), and deep-depletion ((DD) capacitances of an MOS-C. $N_A = 10^{15}$ cm^{-3}, $W_{ox} = 1000$ Å, $T = 300$ K.

sufficiently slowly to allow the inversion charge to form and if the ac voltage is of a sufficiently low frequency for the inversion charge to be able to respond to the ac probe frequency, then the low-frequency curve is obtained. If the dc voltage is swept sufficiently slowly to allow the inversion charge to form but the ac probe frequency is not sufficiently low for the inversion charge to be able to respond, then the high-frequency curve is obtained. The deep-depletion curve is obtained if the dc sweep rate is too high and no inversion charge can form.

The most commonly measured curve is the high-frequency curve. However, the true HF curve is not always easy to obtain. Consider the C-V curve in Fig. 6.6. The true curve is shown by the solid line. If the bias is swept from the left $(-V_G \rightarrow +V_G)$, there is a tendency for the C-V curve to go into partial deep depletion, and the resulting curve will be *below* the true curve. This is especially true for high lifetime material where very low dc bias sweep rates are required. We showed the limitation on the ac frequency to be $dV/dt \leq qn_iW/\tau_g C_{ox}$. This limitation also holds for the dc bias sweep rate; the sweep rate for high lifetime material must be extremely low.

When the bias is swept from right to left $(+V_G \rightarrow -V_G)$, inversion charge is injected into the substrate. The inversion layer/space-charge region/substrate junction becomes forward biased, and the resulting capacitance will be *above* the true curve. The true curve is, in general, only obtained by setting the bias voltage to some value and wait for the device to come to equilibrium. Then repeat this procedure to generate the C-V curve point by point. This is rarely done but is necessary if the true curve is desired. If the point-by-point procedure is inconvenient, then the $+V_G \rightarrow -V_G$ sweep is preferred since the deviation of the capacitance from its true value is generally less than it is for the $-V_G \rightarrow +V_G$ sweep.

The flatband voltage is determined by the metal-semiconductor work function difference ϕ_{MS} and the various oxide charges through the relation

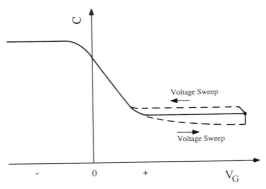

Fig. 6.6   Effect of sweep direction on the HF MOS-C capacitance.

$$V_{FB} = \phi_{MS} - \frac{Q_f}{C_{ox}} - \frac{\gamma Q_m}{C_{ox}} - \frac{\gamma Q_{ot}}{C_{ox}} - \frac{Q_{it}(\phi_s)}{C_{ox}} \qquad (6.15)$$

The fixed charge Q_f is located very near the Si–SiO$_2$ interface and can be considered to be at the interface. Mobile and oxide charges, however, may be distributed throughout the oxide. The effect on flatband voltage is greatest when the charge is located at the oxide-semiconductor interface because then it images all of its charge in the semiconductor. When the charge is located at the gate-insulator interface, it images all of its charge in the gate and has no effect on the flatband voltage. To account for possible charge distribution throughout the oxide, the factor γ is introduced. It is defined by

$$\gamma = \frac{\int_0^{W_{ox}} (x/W_{ox})\rho(x)\, dx}{\int_0^{W_{ox}} \rho(x)\, dx} \qquad (6.16)$$

where $\rho(x)$ is the oxide trapped or mobile charge per unit volume. The x-axis is defined in Fig. 6.3. For $x = 0$ the charge is located at the oxide–gate interface and $\gamma = 0$. For $x = W_{ox}$ the charge is located at the oxide-semiconductor interface and $\gamma = 1$. Q_m and Q_{ot} in Eq. (6.15) are considered to be constant during the measurement. However, Q_{it} is designated as $Q_{it}(\phi_s)$ because the occupancy of the interface trapped charge depends on the surface potential.

The flatband voltage of Eq. (6.15) is for a uniformly doped substrate, where the gate voltage is referenced to the grounded back contact. For a wafer consisting of an epitaxial layer of doping concentration N_{epi} on a substrate of doping concentration N_{sub}, there is a built-in potential at the epi-substrate junction. The flatband voltage is modified by this built-in voltage to[9]

$$V_{FB}(\text{epi wafer}) = V_{FB}(\text{uniform wafer}) \pm \frac{kT}{2q} \ln \frac{N_{sub}}{N_{epi}} \qquad (6.17)$$

The plus sign in Eq. (6.17) is for p-type and the minus sign for n-type material. It is assumed in Eq. (6.17) that the substrate and the epitaxial layer doping concentrations are of the same type, either both acceptors or both donors.

To determine the various charges, one compares theoretical and experimental capacitance-voltage curves. The experimental curves are usually shifted with respect to the theoretical curves as a result of the charges and the work function difference given in Eq. (6.15). The voltage shift can be measured at any capacitance; however, it is frequently measured at the *flatband capacitance* C_{FB} and is designated the *flatband voltage* V_{FB}. V_{FB} is

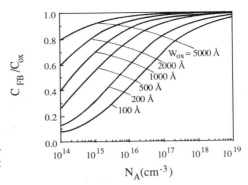

Fig. 6.7 C_{FB}/C_{ox} versus N_A as a function of W_{ox} for the SiO$_2$–Si system at $T = 300$ K.

zero for theoretical curves. The flatband capacitance is given by Eq. (6.13) with $C_s = K_s \varepsilon_0 / L_D$. $L_D = [kT K_s \varepsilon_0 / q^2 (p + n)]^{1/2}$ is the Debye length, as defined in Eq. (2.11). It is common to replace $(p + n)$ by N_A in extrinsic p-type and by N_D in extrinsic n-type substrates. For Si with SiO$_2$ as the insulator, C_{FB} normalized by C_{ox} is given as

$$\frac{C_{FB}}{C_{ox}} = \frac{1}{1 + (136\sqrt{T/300}/W_{ox}\sqrt{N_A \text{ or } N_D})} \tag{6.18}$$

with W_{ox} in cm and N_A (N_D) in cm^{-3}. C_{FB}/C_{ox} is plotted in Fig. 6.7 versus N_A as a function of the oxide thickness.

The flatband capacitance can be easily calculated when the doping concentration is uniform and when the wafer is sufficiently thick. The calculation becomes more difficult when the doping is non-uniform, and numerical techniques may have to be employed.[10] For thin silicon layers, as found in silicon-on-insulators, the active semiconductor layer may be so thin that it cannot accommodate the space-charge region of the MOS-C. Then special precautions must be used to determine C_{FB}. Graphical[11–12] and analytical methods have been used. The analytical methods rely on a measure of the capacitance which is 90 or 95% of the oxide capacitance. The voltage for this capacitance is then related to the flatband voltage.[13–14]

6.2.2 Fixed Charge

The fixed charge is determined by comparing the flatband voltage shift of an experimental capacitance-voltage (C-V) curve with a theoretical curve, as shown in Fig. 6.8. To measure the flatband voltage, it is necessary to determine the flatband capacitance. C_{FB} is calculated from Eq. (6.18) or taken from Fig. 6.7, provided the oxide thickness and the doping concentration are known. The oxide thickness may be known from process

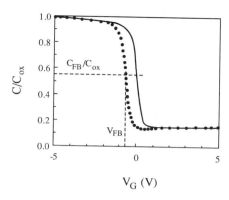

Fig. 6.8 Ideal (solid line) and experimental (points) Al–SiO$_2$–Si MOS-C C-V curves. $N_A = 2.6 \times 10^{14}$ cm^{-3}, $W_{ox} = 1000$ Å, $T = 296$ K, $C_{FB}/C_{ox} = 0.54$, $V_{FB} = -0.7$ V.

specifications, it can be measured by ellipsometry, or it can be calculated from the oxide capacitance. The doping concentration can be determined by one of the methods described in Chapter 2.

Q_f is related to the measured flatband voltage by the equation

$$Q_f = (\phi_{MS} - V_{FB})C_{ox} \qquad (6.19)$$

where ϕ_{MS} must be known in order to determine Q_f. Methods to determine ϕ_{MS} and values of ϕ_{MS} are given in Section 6.2.3. The normalized flatband capacitance is 0.54, and $V_{FB} = -0.7$ V for the example in Fig. 6.8. Since ϕ_{MS} is required to determine Q_f from C-V flatband voltage shifts, there is as much uncertainty in the fixed charge as there is in ϕ_{MS}. For example, the uncertainty in $N_f = Q_f/q$, according to Eq. (6.19), is related to the uncertainty in ϕ_{MS} by the relation

$$\Delta N_f = \frac{K_{ox}\varepsilon_0}{qW_{ox}}\Delta\phi_{MS} = \frac{2.2 \times 10^6}{W_{ox}}\Delta\phi_{MS} \qquad (6.20)$$

For an uncertainty in the metal-semiconductor work function difference of $\Delta\phi_{MS} = 0.1$ V, $\Delta N_f = 2.2 \times 10^{10}$ cm^{-2} for $W_{ox} = 1000$ Å and $\Delta N_f = 4.4 \times 10^{10}$ cm^{-2} for $W_{ox} = 500$ Å. This kind of uncertainty is, in some cases, on the order of the actual fixed charge showing the importance of knowing ϕ_{MS} accurately.

A second method to determine Q_f dispenses with a knowledge of ϕ_{MS}. Rewriting Eq. (6.19) as

$$V_{FB} = \phi_{MS} - \frac{Q_f}{C_{ox}} = \phi_{MS} - \frac{Q_f W_{ox}}{K_{ox}\varepsilon_0} \qquad (6.21)$$

gives another method to determine Q_f. A plot of V_{FB} versus W_{ox} has a slope of $Q_f/K_{ox}\varepsilon_0$. This method, to be described in more detail in the next section, is more laborious because a number of MOS capacitors with differing W_{ox} must be fabricated. However, it is inherently more accurate because it is independent of ϕ_{MS}. Since the published literature shows a variation of ϕ_{MS} by as much as 0.5 V, it is obviously important to determine ϕ_{MS} for a given process and not to rely on published values.

To determine Q_f, it is necessary to eliminate or at least to reduce the effects of all other oxide charges and to reduce the interface trapped charge to as low a value as possible. Q_{it} is reduced by an anneal in a hydrogen ambient at temperatures around 450°C. Pure hydrogen can be, but is rarely used, due to its explosive nature. Forming gas, a hydrogen-nitrogen mixture is more commonly utilized. Typical mixtures are 5 to 10% hydrogen in nitrogen. Devices with Al gates are frequently annealed in nitrogen. It is believed that any moisture in the nitrogen gas reacts with the aluminum and releases hydrogen species at the Al-SiO$_2$ interface.[15] These hydrogen species diffuse through the oxide to the SiO$_2$-Si interface to passivate Q_{it}. When Al is absent, hundreds of hours are necessary for interface trapped charge annealing in N_2, while in the presence of Al only a few minutes are required.[16] Magnesium gates also exhibit Q_{it} reductions in nitrogen anneals, but gold and platinum do not. When the SiO$_2$ is covered by Si$_3$N$_4$, for example, interface trapped charge annealing is more difficult due to the imperviousness of the nitride.[16]

Although we mentioned in the Introduction that Q_f was fixed and depended only on the oxidizing and subsequent annealing ambient, the fixed charge can in fact be increased by negative gate voltages at temperatures of 100 to 400°C. The increase is proportional to the initial Q_f and to the negative electric field.[18] Q_f also increases for long annealing times (>60 min) at high temperatures (~1200°C) in a nitrogen ambient.[18] Nitrogen anneals for about 30 min gives the lowest fixed charge.[19]

6.2.3 Metal-Semiconductor Work Function Difference

The metal-semiconductor work function difference ϕ_{MS} is shown in Fig. 6.9. The figure shows a metal-oxide-semiconductor potential band diagram with no oxide charges under flatband conditions; that is, a gate voltage equal to the flatband voltage ($V_G = V_{FB}$) is applied such that the bands in the semiconductor and in the oxide are flat. For no oxide or interface charge, $V_{FB} = \phi_{MS}$ from Eq. (6.15). Note that all quantities are given in potentials in Fig. 6.9, not in energies as is usually the case. ϕ_M and ϕ_M' are the metal work function and effective work function, ϕ_S is the semiconductor work function, and χ and χ' are the electron and effective electron affinity. All other symbols have their usual meanings. From Fig. 6.9 we find

$$\phi_{MS} = \phi_M - \phi_S = \phi_M' - \left(\chi' + \frac{E_C - E_F}{q}\right) \qquad (6.22)$$

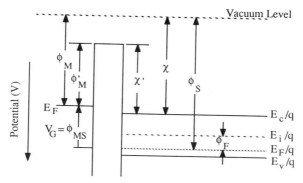

Fig. 6.9 Potential band diagram of a metal-oxide-semiconductor system.

ϕ'_M, χ', and $(E_C - E_F)/q$ are constants for a given gate material, semiconductor, and temperature. For p-substrates, Eq. (6.22) becomes

$$\phi_{MS} = K - \phi_F = K - \frac{kT}{q} \ln \frac{N_A}{n_i} \qquad (6.23a)$$

where $K = \phi'_M - \chi' - (E_C - E_i)/q$ and $(E_C - E_F)/q = (E_C - E_i)/q + \phi_F = (E_C - E_i)/q + (kT/q) \ln(N_A/n_i)$. For n-substrates

$$\phi_{MS} = K + \phi_F = K + \frac{kT}{q} \ln \frac{N_D}{n_i} \qquad (6.23b)$$

ϕ_{MS} depends not only on the semiconductor and the gate material but also on the substrate doping type and concentration.

Figure 6.10 shows the band diagrams for an n^+ polysilicon–p substrate and a p^+ polysilicon–n substrate MOS-C. Since both gate and substrate are silicon with the same electron affinity, we find

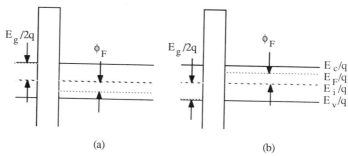

(a)   (b)

Fig. 6.10 Potential band diagram of (a) n^+ polysilicon/p silicon and (b) p^+ polysilicon/n silicon systems.

$$\phi_{MS} = \phi_F(\text{poly}) - \phi_F(\text{substrate}) \tag{6.24}$$

The Fermi level for heavily doped gates coincides approximately with the conduction band for n^+ polysilicon gates and with the valence band for p^+ polysilicon gates. This gives ϕ_{MS} (n^+ gates) $\approx -E_g/2q - (kT/q)\ln(N_A/n_i)$ and ϕ_{MS} (p^+ gates) $\approx E_g/2q + (kT/q)\ln(N_D/n_i)$, where N_A and N_D are the substrate doping concentrations.

Early ϕ_{MS} determinations used photoemission measurements.[20–24] With a voltage applied between a semitransparent gate and the substrate, no current flows in the absence of light because of the insulating nature of the oxide. When photons of sufficient energy strike the gate, electrons are excited from the gate into the oxide. Some of these electrons drift through the oxide to be collected as photocurrent. Electrons are excited from the semiconductor into the oxide and flow to the gate for positive gate voltages, and the barrier height of the *semiconductor* is determined. For negative gate voltages, electrons are excited from the gate metal into the oxide and flow to the semiconductor, and the barrier height of the *gate material* is determined. In a related method the gate voltage is varied, and that gate voltage at which the photocurrent changes sign corresponds to a "zero oxide voltage" condition allowing ϕ_{MS} to be determined.[25–26]

Although photoemission measurements give reasonably good values for ϕ_{MS}, they rely on a technique that measures ϕ_{MS} only indirectly. A more direct measure utilizes Eq. (6.21), repeated here

$$V_{FB} = \phi_{MS} - \frac{Q_f}{C_{ox}} = \phi_{MS} - \frac{Q_f W_{ox}}{K_{ox}\varepsilon_0} \tag{6.25}$$

A measurement of V_{FB} as a function of oxide thickness has a slope of $Q_f/K_{ox}\varepsilon_0$ and an intercept on the V_{FB} axis of ϕ_{MS}.[2] This method is more direct because it measures the capacitance of MOS capacitors. Furthermore, since the flatband voltage is measured, it ensures zero electric field at the semiconductor surface, eliminating Schottky barrier lowering corrections. The oxide thickness can be varied by oxidizing the wafer to a given thickness, measuring V_{FB}, etching a portion of the oxide, remeasuring V_{FB}, etc. After each etch, gates can be formed by metal evaporation or by using a mercury probe. This method ensures that the same spot on the oxide is measured each time. Oxide etching does not affect the fixed charge since Q_f is located very near the SiO_2-Si interface. Sometimes the oxide is etched in strips to different thicknesses. Although each MOS-C is now on a different part of the wafer, the devices are at least all on the same wafer. Alternately, oxides can be grown to different thicknesses on different wafers and MOS capacitors formed. This method assumes that Q_f is the same for all samples.

An example of a $V_{FB} - W_{ox}$ plot is shown in Fig. 6.11. The three curves for n-Si have differing slopes but a constant intercept indicating varying Q_f.

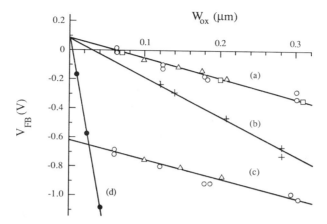

Fig. 6.11 Flatband voltage of Al–SiO$_2$–Si MOS capacitors as a function of oxide thickness. Curves (a), (b), and (d) are for n-Si and (c) is for p-Si. Reprinted with permission after Werner.[27]

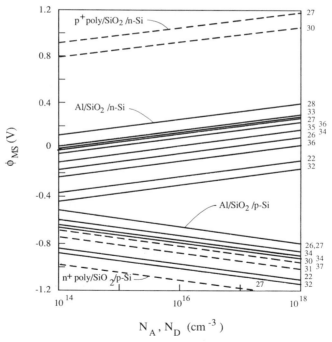

Fig. 6.12 ϕ_{MS} as a function of doping concentration for Al–SiO$_2$–Si and poly-Si–SiO$_2$–Si MOS capacitors. The numbers at right refer to references.

Curves (a) and (c) have identical slopes but different intercepts because (a) is *n*-Si and (c) is *p*-Si. The dependence of ϕ_{MS} on doping concentration is shown in Fig. 6.12 for the SiO_2-Si system with Al and polysilicon gates. Note that there is a fair amount of scatter in the data because some of the reported results were on CVD oxides, others on sputtered oxides; some were annealed in hydrogen or forming gas, but others were not.

There is a dependence of ϕ_{MS} on D_{it} anneals. D_{it} variations of 10^{10} to 3×10^{10} cm^{-2}eV^{-1} changed ϕ_{MS} by 0.25 to 0.35 V on Al–SiO_2–Si devices.[28] ϕ_{MS} also depends on oxidation temperature, wafer orientation, and on the low temperature Q_{it} anneal.[29] The work function of devices with polysilicon gates should depend on the doping concentration of the gate since the Fermi level of a polysilicon gate depends on doping. One report shows essentially no such dependence.[30] Another shows a ϕ_{MS} maximum at phosphorus and arsenic concentrations of 5×10^{19} cm^{-3}, with the work function difference decreasing above and below this concentration.[31]

6.2.4   Oxide Trapped Charge

Charge can become trapped in the oxide during device operation even if it is not introduced during device fabrication. Electrons or holes can be injected from the semiconductor or from the gate. It is also possible that energetic radiation produces electron-hole pairs in the oxide, and some of these electrons and/or holes are subsequently trapped in the oxide. The *flatband voltage shift* ΔV_{FB} due to oxide trapped charge Q_{ot} is obtained from Eq. (6.15) as

$$\Delta V_{FB} = V_{FB}(Q_{ot}) - V_{FB}(Q_{ot} = 0) = -\frac{\gamma Q_{ot}}{C_{ox}} \qquad (6.26)$$

assuming that all other charges remain unchanged during the oxide trapped charge introduction. Contrary to Q_f, the oxide trapped charge is usually not located at the oxide-semiconductor interface but is distributed through the oxide. The distribution of Q_{ot} must be known for proper interpretation of *C-V* curves. Trapped charge distributions are measured by several methods. The two most common are the etch-off method and the photo *I-V* method.

In the etch-off method, thin layers of the oxide are etched. The *C-V* curve is measured after each etch, and the oxide charge profile is determined from these *C-V* curves. It is generally difficult to profile closer than about 200 Å from the oxide-semiconductor interface due to shorting problems when portions of the oxide etch completely.[38] The photo *I-V* method is nondestructive and more accurate than the etch-off method. It is based on the optical injection of electrons from the gate or from the semiconductor into the oxide. Electron injection depends on the distance of the energy barrier from the injecting surface and on the barrier height. Both barrier distance and barrier height are affected by oxide charge and gate bias. Photo *I-V* curves allow information to be obtained on both the barrier distance and the barrier height. A good discussion of the method can be found in Nicollian and Brews,[32] and references therein. Occasionally the technique is useful to monitor the flatband voltage continuously. A circuit implementation is given in Li et al.[39]

A determination of the charge distribution in the oxide is tedious and therefore not routinely done. In the absence of such information, the flatband voltage shift is generally interpreted by assuming the charge to be located at the oxide-semiconductor interface, $\gamma = 1$, using the expression

$$Q_{ot} = -\Delta V_{FB} C_{ox}$$

(6.27)

There is sometimes a question of whether a measured flatband voltage shift is due to charge injected into the oxide or due to mobile charge within the oxide. A simple check to discriminate between the two is the following: Consider an MOS-C on a *p*-type substrate whose *C-V* curve is initially determined with moderate gate voltage excursions giving the *C-V* curve (a) in Fig. 6.13(a). We assume that as a result of the modest gate voltage excursion during the measurement of this curve, neither is charge injected into the oxide, nor does mobile charge move. Next a large positive gate voltage bias is applied. Two effects can happen: (1) electrons can be injected into the oxide, and (2) positively charged mobile ions distributed throughout the oxide can drift to the oxide-semiconductor interface. Negative charge injected into the oxide gives rise to a *positive* ΔV_{FB} for process 1, shown by curve (b). The opposite effect is observed for process 2 because

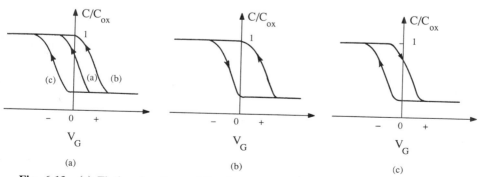

Fig. 6.13 (a) Flatband voltage shifts due to no charge [curve (a)], injected charge [curve(b)], and mobile charge [curve (c)], (b) *C-V* curves due to injected charge, (c) *C-V* curves due to mobile charge.

more positive charge at the oxide-semiconductor interface causes a more *negative* ΔV_{FB}, shown by curve (c).

This type of behavior is further shown by the C-V curves in Fig. 6.13(b) and (c). In (b) positive V_G causes electron injection into the oxide, and a sweep of $+V_G \rightarrow -V_G$ is shown by the arrow pointing to the left. For a large negative gate voltage, the trapped electrons can be ejected from the oxide, and it is even possible to inject holes into the oxide. A return sweep of $-V_G \rightarrow +V_G$ traces the curve indicated by the arrow pointing to the right. If mobile charge were the C-V curve-shifting culprit, the curves of Fig. 6.13(c) result. For large positive gate voltages, mobile ions drift to the oxide-semiconductor interface, giving a negative voltage shift. For large negative gate voltages, the mobile charge is attracted to the gate-oxide interface where it does not affect the C-V curve. The hysteresis loop direction in Fig. 6.13(c) is opposite that in (b).

6.2.5 Mobile Charge

Mobile charge in SiO_2 is due primarily to the ionic impurities Na^+, Li^+, K^+, and perhaps H^+. Sodium is the dominant contaminant. The practical application of MOSFET's was delayed due to mobile oxide charges in the early 1960s. MOSFET's were found to be very unstable for positive gate bias but relatively stable for negative gate voltages. Sodium was the first impurity to be related to this gate bias instability.[40] By intentionally contaminating MOS-C's and measuring the C-V shift after bias-temperature stress, it was shown that alkali cations could easily drift through thermal SiO_2 films. Chemical analysis of etched-back oxides by neutron activation analysis and flame photometry was used to determine the Na profile through the oxide film.[41]

The drift of sodium through SiO_2 is asymmetrical. The activation energy for drift from the metal-SiO_2 interface is larger than that from the Si-SiO_2 interface. This asymmetry is not present for polysilicon gate devices. This behavior led to the hypothesis that the kinetics of Na motion in thermal SiO_2 are governed by the emission of ions from traps at the interface and subsequent drift through the oxide. The traps at the metal-SiO_2 interface are deeper than those at the Si-SiO_2 interface making emission more difficult. The emission and subsequent drift have been measured with the isothermal transient ionic current method, the thermally stimulated ionic current method, and the triangular voltage sweep method.[42–52]

The mobility of the alkali metals Na, Li, and K is given by the expression[49]

$$\mu(T) = \mu_0 \exp\left(\frac{-E_a}{kT}\right) \quad (6.28)$$

where for Na $\mu_0 = 3.5 \times 10^{-4}$ cm^2/V·s (within a factor of 10) and $E_a =$

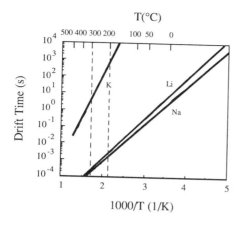

Fig. 6.14 Drift time for Na, Li, and K for an oxide electric field of 10^6 V/cm and $W_{ox} = 1000$ Å.

0.44 ± 0.09 eV, for Li $\mu_0 = 4.5 \times 10^{-4}$ cm^2/V·s (within a factor of 10) and $E_a = 0.47 \pm 0.08$ eV, and for K $\mu_0 = 2.5 \times 10^{-3}$ cm^2/V·s (within a factor of 8) and $E_a = 1.04 \pm 0.1$ eV. The oxide electric field is given by V_G/W_{ox}, neglecting the small voltage drop across the semiconductor. The drift velocity of mobile ions through the oxide is $v_d = \mu V_G/W_{ox}$, and the transit time $t_t = W_{ox}/v_d$ is

$$t_t = \frac{W_{ox}^2}{\mu V_G} = \frac{W_{ox}^2}{\mu_0 V_G} \exp\left(\frac{E_a}{kT}\right) \tag{6.29}$$

Equation (6.29) is plotted in Fig. 6.14 for the three alkali ions. For this plot the oxide electric field is 10^6 V/cm, which is a common oxide field during mobile charge measurements, and the oxide thickness is 1000 Å. For thinner or thicker oxides, the transmit time changes accordingly as given by Eq. (6.29). Note that both Na and Li drift very rapidly through the oxide. Typical measurement temperatures lie in the 200 to 300°C range and only a few milliseconds are sufficient for the charge to transit the oxide. Mobile charge densities in the low 10^{10} cm^{-2} range are generally acceptable in today's integrated circuits.

Capacitance-Voltage   The *C-V* method is the most popular technique to determine the mobile charge. However, in contrast to the room-temperature *C-V* method for Q_f determination, for mobile charge measurements the measurement temperature must be sufficiently high for the charge to be mobile. Typically the device is heated to 200 to 300°C, a gate bias to produce an oxide field of around 10^6 V/cm is applied for a time sufficiently long for the charge to drift to one oxide interface. Typical times are several minutes, say, 10 min or so. The device is then cooled to room temperature under bias, and a *C-V* curve is measured. The procedure is then repeated with the opposite bias polarity.[40] The mobile charge is determined from the flatband voltage shift, according to the equation

$$Q_m = -\Delta V_{FB} C_{ox}$$

(6.30)

In the simplest case there is merely a parallel shift of the two curves, and ΔV_{FB} is easily read from the plot. If there is significant change in the interface trapped charge as a result of the mobile charge drift, one of the two $C\text{-}V$ curves may become distorted, and ΔV_{FB} is more difficult to determine. The triangular voltage sweep method does not have this limitation.

Triangular Voltage Sweep   In the triangular voltage sweep (TVS) method the current is measured instead of the capacitance.[42,51–52] The MOS-C is held at an elevated, constant temperature of 200 to 300°C, and a triangular voltage ramp is applied to the gate as shown in Fig. 6.15. The op-amp circuit connected to the MOS-C substrate is an ammeter. The ramp frequency must be sufficiently low that the mobile charge can drift through the oxide in one direction as the ramp voltage increases, and then in the other direction as the ramp voltage decreases.

When a changing voltage is applied to a capacitor, the resulting current is given by $I = dQ/dt = (dQ/dV)(dV/dt) = C\,dV/dt$. For a linear voltage ramp $dV/dt = $ constant and $I \sim C$. The "mobile" charge is immobile with the MOS-C at room temperature, and the current is a measure of the capaci-

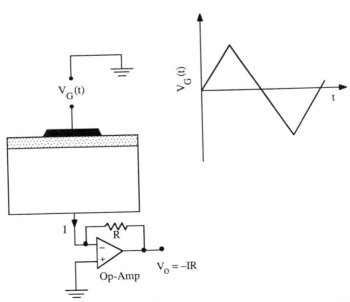

Fig. 6.15 Block diagram of a circuit to measure the current of an MOS capacitor.

tance as discussed in more detail in Section 6.3.1. With dV/dt sufficiently low, the low-frequency "C-V" curve A is obtained, as shown in Fig. 6.16(a), given by Eqs. (6.7) and (6.13). For an MOS-C *without* mobile charge, the curve changes to curve B in Fig. 6.16(a) at elevated measurement temperatures. The capacitance increase from curve A to B is merely due to the various temperature-dependent factors in Eq. (6.7). At elevated temperatures *with* mobile charge curve C is measured. Let us see how this curve comes about.

The current I is defined by

$$I = \frac{dQ_G}{dt} \tag{6.31}$$

With $Q_G = -(Q_s + Q_{it} + Q_f + Q_{ot} + Q_m)$, the current can be written as[38]

$$I = C_{LF}\left(\alpha - \frac{dV_{FB}}{dt}\right) \tag{6.32}$$

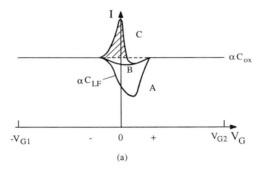

(a)

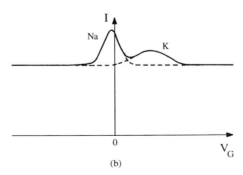

(b)

Fig. 6.16 Triangular sweep current-voltage curves: (a) room temperature A, elevated temperature with no mobile oxide charges B, elevated temperature with mobile oxide charges and C: (b) I-V curve showing two different mobile impurities.

where $\alpha = dV_G/dt$ is the gate voltage ramp rate. Integrating both sides from $-V_{G1}$ to $+V_{G2}$ gives

$$\int_{-V_{G1}}^{V_{G2}} (I/C_{LF} - \alpha)\, dV_G = -\alpha\{V_{FB}[t(V_{G2})] - V_{FB}[t(-V_{G1})]\} \qquad (6.33)$$

Let us assume that at $-V_{G1}$ all mobile charges are located at the metal–oxide interface ($x = 0$ and $\gamma = 0$) and that at V_{G2} all mobile charges are located at the semiconductor-oxide interface ($x = W_{ox}$ and $\gamma = 1$). Then considering mobile charge only, we find from Eq. (6.15)

$$-\alpha\{V_{FB}[t(V_{G2})] - V_{FB}[t(-V_{G1})]\} = \alpha\,\frac{Q_m}{C_{ox}} \qquad (6.34)$$

With the *LF* curve lying close to the oxide capacitance line, the ratio C_{LF}/C_{ox} is usually taken as unity, and Eqs. (6.33) and (6.34) lead to

$$\int_{-V_{G1}}^{V_{G2}} (C_{ox}I/C_{LF} - \alpha C_{ox})\, dV_G = \alpha Q_m \qquad (6.35)$$

The integral of Eq. (6.35) represents approximately the shaded area in Fig. 6.16(a), between the I-V_G curve and the straight line $I = \alpha C_{ox}$, where we have replaced the actual *LF* curve B by the straight line. C_{LF} is replaced by C_{ox} because frequently C_{LF} cannot be determined when it is obscured by the current peak from the mobile charge.

For the return gate voltage sweep, the current reverses direction and should be a mirror image of curve C in Fig. 6.16(a). This is not always observed. Sometimes one of the curves is more "smeared" out than the other, but the areas under the curves, representing the total charge transported through the oxide, are equal.[52] As mentioned earlier in this section, mobile charge motion through the oxide is governed by emission of the charge over an energy barrier from the interface followed by drift through the oxide. The energy barrier appears to be higher at the metal–oxide interface than at the oxide-semiconductor interface. An asymmetry in the I-V_G curves is not unexpected if emission dominates over drift. When there is leakage current flowing through the oxide, in addition to the currents discussed here, the I-V_G curves have a linear slope to them and are not horizontal as in Figs. 6.16(a). Instead of measuring the current, one can also measure the charge with an electrometer and plot it against gate voltage.

Sometimes double peaks are observed in I-V_G curves, as shown in Fig. 6.16(b). These have been attributed to mobile ions with different mobilities.[44,48] For an appropriate temperature and sweep rate, high-mobility ions (e.g., Na^+) drift at lower electric fields than low-mobility ions (e.g., K^+). Hence we would expect the Na peak to occur at lower gate voltages than the K peak. This is in fact observed. Such discrimination between different types of mobile impurities is not possible with the *C-V* method. It also explains why sometimes the total number of impurities determined by the *C-V* and the TVS methods are not identical. In the *C-V* method one usually waits long enough for all the mobile charge to drift through the oxide. If in the TVS method the temperature is too low or the gate ramp rate is too high, it is possible that only one type of charge is detected. For example, it is conceivable that high-mobility Na drifts but low-mobility K does not.

Other Methods   Electrical characterization methods are dominant because they are easily implemented and are very sensitive. The *C-V* method has a sensitivity of about $10^{10}\,cm^{-2}$ and the TVS method can detect as low as about $10^{9}\,cm^{-2}$. However, electrical methods cannot detect neutral impurities nor the sodium content in chemicals, furnace tubes, etc. Analytical methods that have been employed for sodium detection include radiotracer,[53] neutron activation analysis,[53-54] flame photometry,[55] and secondary ion mass spectrometry (SIMS). For SIMS it is important to take surface charging by the positive or negative ion beam into account because it can alter the ionic distribution and give erroneous distribution curves.[56]

6.3   INTERFACE TRAPPED CHARGE

The nature of interface trapped charge is not completely understood, but Q_{it} can be controlled to the extent that very low values are achieved in today's devices. A number of measurement techniques have been developed over the years, and we will describe the main ones in this chapter. A good overview of the nature of interface trapped charge and methods for its characterization can be found in Nicollian and Brews,[38] Goetzberger et al.,[57] and DeClerck.[58]

6.3.1   Low-Frequency (Quasi-static) Method

Current-Voltage   The low-frequency or *quasi-static method* is relatively simple and has become one of the most common interface trapped charge measurement methods. It uses standard semiconductor laboratory components, and the evaluation theory is simple. It provides information only on the interface trapped charge density, but not on the capture cross section of the traps. In this chapter we use the terms "interface trapped charge" and "interface traps" interchangeably. The effect of interface traps on both HF and LF *C-V* curves is shown in Fig. 6.17. The distortion in the HF curve labeled Ⓐ is the result of D_{it} near the valence band, Ⓑ is for D_{it} near midgap, and Ⓓ is for D_{it} near the conduction band. The broadening of the LF curve at its midpoint, labeled Ⓒ, is a result of the interface traps nearest the conduction band.

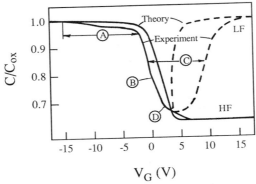

V_G (V)

Fig. 6.17   Comparison of theoretical and experimental LF and HF *C-V* curves. The symbols are explained in the text. $N_A = 10^{16} \, cm^{-3}$, $W_{ox} = 2000$ Å. Reprinted after Deal et al.[16] by permission of the publisher, the Electrochemical Society, Inc.

If interface traps cannot follow the ac probe frequency, they do not contribute additional capacitance and the equivalent circuits are those of Fig. 6.4 with $C_{it} = 0$. Nevertheless, the HF curve is not an ideal *C-V* curve because interface traps can follow the slowly varying dc bias. As the gate voltage is swept from accumulation to inversion, the gate charge is given $Q_G = -(Q_S + Q_{it})$ assuming, for simplicity, zero oxide charges. In contrast to the ideal case, where $Q_{it} = 0$, now both semiconductor and interface traps must be charged. The relationship of surface potential to gate voltage differs from Eq. (6.14) and the HF *C-V* curve stretches out as shown in Fig. 6.17. Note that this stretch-out is not the result of interface traps contributing excess capacitance, but rather it is the result of the *C-V* curve stretch-out along the gate voltage axis. Interface traps do respond to the probe frequency at low measurement frequencies and the curve distorts because the interface traps contribute interface trap capacitance C_{it}, and the curve stretches along the voltage axis, also shown in Fig. 6.17.

The basic theory of the quasistatic method was developed by Berglund.[59] The method utilizes a comparison of a low-frequency *C-V* curve with one free of interface trap effects. The latter can be a theoretical curve but is usually an HF *C-V* curve determined at a frequency where interface traps are assumed not to respond. "Low frequency" means that interface traps *and* minority carrier inversion charges must be able to respond to the measurement ac probe frequency. The constraints for minority carrier response are discussed in Section 6.2.1. The interface trap response has similar limitations. Fortunately the limitations are usually less severe, and frequencies low enough for inversion layer response are generally low enough for interface trap response.

The LF capacitance is given by Eq. (6.4) in depletion/inversion as

$$C_{LF} = \frac{1}{(1/C_{ox}) + 1/(C_S + C_{it})} \tag{6.36}$$

where we have replaced $C_B + C_N$ by C_S, the LF semiconductor capacitance. C_{it} is related to the interface trap density D_{it} by $D_{it} = C_{it}/q$, giving

$$D_{it} = \frac{1}{q}\left(\frac{C_{ox}C_{LF}}{C_{ox} - C_{LF}} - C_S\right) \tag{6.37}$$

$D_{it} = C_{it}/q$ derives from the definition $C_{it} = -dQ_{it}/d\phi_s$.[38] Equation (6.37) is suitable for interface trap density determination over the entire band gap.

C_{LF} and C_S must be known to determine D_{it}. C_{LF} is measured as a function of gate voltage, and C_S is calculated from Eq. (6.7). Note that in Eq. (6.7) the capacitance is calculated as a function of surface potential ϕ_s, but in Eq. (6.37) C_{LF} is measured as a function of gate voltage. Hence, we need a relationship between ϕ_s and V_G. Berglund proposed[59]

$$\phi_s = \int_{V_{G1}}^{V_{G2}}\left(1 - \frac{C_{LF}}{C_{ox}}\right)dV_G + \Delta \tag{6.38}$$

where Δ is an integration constant given by the surface potential at $V_G = V_{G1}$. The integrand is obtained by integrating the measured C_{LF}/C_{ox} versus V_G curve, with V_{G1} and V_{G2} arbitrarily chosen since the integration constant Δ is unknown. If the integration is carried out from strong accumulation to strong inversion, the integral should give a value $\phi_s(V_{G2}) - \phi_s(V_{G1})$ equal to or slightly less than the band gap. A value greater than E_g/q indicates gross nonuniformities in the oxide or at the oxide–semiconductor interface, making the analysis invalid.[59] Various approaches to determine the surface potential based on LF and HF *C-V* curves have been proposed.[57,59–61] Kuhn proposed fitting the experimental and theoretical C_{LF} versus ϕ_s curves in accumulation and strong inversion.[60] Alternatively, C_s can be plotted against ϕ_s. The slope of this curve gives N_A, and the intercept gives Δ. Direct electronic integration has also been suggested.[62–65] These methods are generally based on measuring charge using an operational amplifier with a capacitor in the feedback loop, which is similar to using an electrometer in the "charge" mode. In one circuit containing a divider, integrator, differential amplifiers, and logarithmic amplifiers, D_{it} is determined and plotted directly as a function of ϕ_s.[65]

The LF frequency *C-V* curve is very difficult to measure with a capacitance meter or capacitance bridge because the signal is very noisy at the low frequencies of around 1 Hz or so. The quasi-static or slow ramp voltage method was suggested by Castagné,[62] Kerr,[62] Kuhn,[60] and Kappallo and

Walsh[67] to overcome this problem. A slowly varying voltage ramp is applied to the MOS-C and the displacement current is measured, as given in Eq. (6.32). $dV_{FB}/dt = 0$ with no mobile charge or mobile charge unable to move, and

$$I = C_{LF} \frac{dV_G}{dt} \tag{6.39}$$

The determination of D_{it} from Eqs. (6.37) and (6.38) is quite time-consuming, and a simplified approach was proposed by Castagné and Vapaille.[66,68] It eliminates the uncertainty associated with the calculation of C_S in Eq. (6.37) and replaces it with a measured C_S. From the HF C-V curve we find from Eq. (6.13)

$$C_S = \frac{C_{ox} C_{HF}}{C_{ox} - C_{HF}} \tag{6.40}$$

Substituting Eq. (6.40) into (6.37) gives D_{it} in terms of the *measured* LF and HF C-V curves as

$$D_{it} = \frac{C_{ox}}{q} \left(\frac{C_{LF}/C_{ox}}{1 - C_{LF}/C_{ox}} - \frac{C_{HF}/C_{ox}}{1 - C_{HF}/C_{ox}} \right) \tag{6.41}$$

Equation (6.41) gives D_{it} over only a limited range of the band gap, typically from the onset of inversion but not strong inversion, to a surface potential where the ac measurement frequency equals the inverse of the interface trap emission time constant. Typically this corresponds to an energy about 0.2 eV from the majority carrier band edge. HF and LF curves are shown in Fig. 6.18(a). The interface trap charge densities determined from the C-V curves are shown in Fig. 6.18(b), where the solid line was determined by the approach of Eq. (6.37) covering essentially the entire band gap. The dashed curve is determined with Eq. (6.41) and covers a substantially smaller portion of the band gap.

Note the shape of the $D_{it} - \phi_s$ curves with a minimum near midgap and sharp increases toward either band edge. This is typical for interface traps in the SiO$_2$-Si system. It is very important, when using the technique based on Eq. (6.37), that the integration constant Δ be well known. Small errors in Δ have a large effect on D_{it} near the band edges.[69] Errors can also be introduced by surface potential fluctuations due to inhomogeneities in oxide charge and/or substrate doping concentration.[68–70]

It is not always necessary to determine D_{it} as a function of surface potential. For example, for process monitoring, it is frequently sufficient to determine D_{it} at one point on the C-V curve and then compare device to device or run to run. A convenient choice is the minimum capacitance of the LF C-V curve where the technique is most sensitive. This point corresponds to a surface potential in the light inversion region near midgap, that is

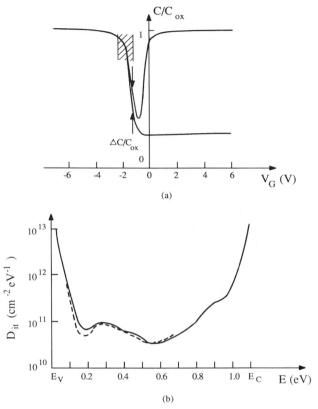

(a)

(b)

Fig. 6.18   (a) Experimental LF and HF *C-V* curves showing the approximate range over which Eq. (6.41) holds (shaded region); (b) D_{it} determined by Eq. (6.37) (solid curve) and by Eq. (6.41) (dashed curve). $N_A = 6 \times 10^{14}$ cm^{-3}, $W_{ox} = 1400$Å. Reprinted with permission after Castagné and Vapaille.[68]

$\phi_F < \phi_s < 2\phi_F$. To extract D_{it}, Eq. (6.41) is plotted in Fig. 6.19 for SiO$_2$ with $W_{ox} = 1000$ Å. To use the figure, measure C_{LF}/C_{ox} and C_{HF}/C_{ox}; then determine $\Delta C/C_{ox} = C_{LF}/C_{ox} - C_{HF}/C_{ox}$ and find D_{it} from the graph.[71] For oxide thicknesses other than 1000 Å, multiply the result by 1000/W_{ox}. Other graphical techniques have also been proposed.[72]

HF curves are generally obtained by conventional capacitance meter or bridge techniques. The measurement frequency must be sufficiently high that interface traps do not respond. The usual 1 MHz frequency may suffice, but for devices with high D_{it}, there will be some response due to interface traps. If possible, one should use higher frequencies. However, care must be used to ascertain that series resistance effects do not become important. The LF curve is determined by measuring the current with the *I-V* curve adjusted to coincide with the HF *C-V* curve in strong accumulation. With gate voltage

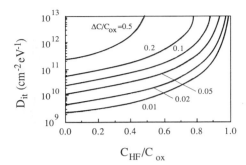

Fig. 6.19 Interface trapped charge density from the HF *C-V* curve and the offset $\Delta C/C_{ox}$.

sweep rates typically in the 10 to 100 mV/s range, it is important that the capacitance not be too small or the current will be exceedingly low and difficult to measure. The currents, typically in the 10^{-12} to 10^{-10} A range for $C = 10$ to $100\,pF$ and a $100\,mV$/sweep rate, are usually measured with a feedback ammeter having a virtual ground input to minimize the effects of cable and fixture impedances.

The current can of course be increased by increasing the capacitance or the sweep rate. Increasing the sweep rate is usually not tolerable since this can introduce nonequilibrium effects that distort the LF curve to give erroneous information, especially for high-lifetime devices.[73] Increasing the capacitance is acceptable. The LF curve is easier to measure when sweeping from inversion to accumulation because minority carriers need not be generated thermally since they already exist in the inversion layer. Series resistance and stray light can also influence the curve.[74] A detailed accounting of the errors in extracting D_{it} is given by Nicollian and Brews.[38] The lower limit of D_{it} that can be determined with the quasi-static technique lies around $10^{10}\,cm^{-2}eV^{-1}$, when care is taken with the measurement of the *I-V* and *C-V* curves.

Charge-Voltage In the quasi-static *I-V* method, shown in Fig. 6.20(a), the current should only be the displacement current. However, any leakage currents are also included in the *I-V* plot. Moreover the ammeter in conjunction with the capacitor is a differentiator and tends to exaggerate noise spikes or nonlinearities in the voltage ramp. To alleviate some of the limitations of the *I-V* quasi-static method, the *Q-V* quasi-static method was proposed. Initially the MOS-C was placed in the feedback loop of an op-amp, and it was charged with a constant current.[75–76] The method was later modified and extended.[77] Analog[78] and digital[79] implementations have been proposed, and a recent commercial version is shown schematically in Fig. 6.20(b).[80] This circuit is an integrator reducing the effects of spurious signals. The MOS-C, shown as capacitor C, is connected with its gate to the op-amp and its substrate to the voltage source to minimize effects of stray capacitance and noise.

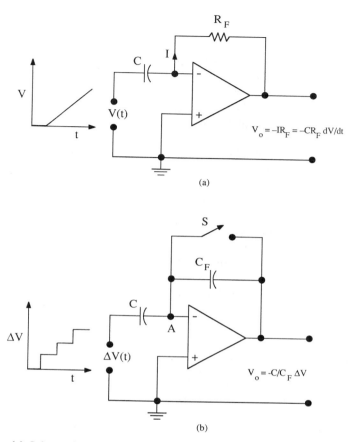

Fig. 6.20 (a) Schematic of the current-voltage quasi-static technique, (b) schematic of the charge-voltage quasi-static technique.

This technique, also called the *feedback charge method*, utilizes a voltage step input ΔV to the virtual ground op-amp. The capacitance is determined by measuring the transfer of charge in response to this voltage increment. The voltage ΔV appears essentially across the MOS-C since point A is at virtual ground. The feedback capacitor C_F is initially discharged by closing the low-leakage current switch S. When the measurement starts, S is opened and ΔV causes charge ΔQ to flow onto capacitor C_F, giving the output voltage

$$\Delta V_0 = -\frac{\Delta Q}{C_F} \tag{6.42}$$

Since $\Delta Q = C\,\Delta V$, we find

$$\Delta V_0 = -\frac{C}{C_F} \Delta V \qquad (6.43)$$

and the output voltage is proportional to the MOS-C capacitance. Gain is introduced into the measurement for $C > C_F$ by choosing the capacitance ratio C/C_F appropriately. By incrementing ΔV, a C_{LF} versus V curve is generated. Additionally, when Q changes, a current Q/t flows. This current should only flow during the transient time period until the device reaches equilibrium. Thereafter it should stop. Hence Q/t is a measure of whether equilibrium has been established and can be used to determine the time increments at which ΔV should be changed to measure the equilibrium low-frequency C-V curve.[80]

The method is well suited for MOS measurements since it has high noise immunity, because sizable voltages rather than low currents are measured, and since voltage steps rather than precisely linear voltage ramps are used. As detailed in Brews and Nicollian[11] the method is also suitable to determine the additive constant Δ of Eq. (6.38) by comparing experimental and theoretical ϕ_s versus W curves, where ϕ_s is the surface potential and W is the space-charge region width obtained from the experimental HF C-V curve.

6.3.2 Conductance Method

The conductance method, proposed by Nicollian and Goetzberger in 1967, is generally considered to be the most sensitive method to determine D_{it}.[81] Interface trap densities of 10^9 cm^{-2}eV^{-1} and lower can be measured. It is also the most complete method because it yields D_{it} in the depletion and weak inversion portion of the band gap, the capture cross sections for majority carriers, and information about surface potential fluctuations. But the measurement is quite tedious and time-consuming. It is based on the measurement of the equivalent parallel conductance G_p of an MOS-C as a function of bias and frequency. The conductance, representing the loss mechanism due to interface trap capture and emission of carriers, is a measure of the interface trap density.

The simplified equivalent circuit of an MOS-C appropriate for the conductance method is shown in Fig. 6.21(a). It consists of the oxide capacitance C_{ox}, the semiconductor capacitance C_S, and the interface trap capacitance C_{it}. The capture of carriers by D_{it} and emission of carriers from D_{it} is a lossy process, represented by the resistance R_{it}. For interface trap analysis it is convenient to replace the circuit of Fig. 6.21(a) by that in Fig. 6.21(b), where C_p and G_p are from a simple circuit conversion

$$C_p = C_S + \frac{C_{it}}{1 + (\omega\tau_{it})^2} \qquad (6.44)$$

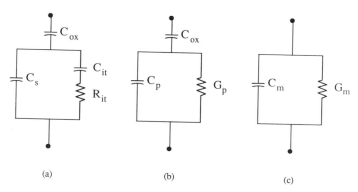

Fig. 6.21 Equivalent circuits for conductance measurements: (a) MOS-C with interface state time constant, (b) simplified circuit of (a), (c) measured circuit.

$$\frac{G_p}{\omega} = \frac{q\omega\tau_{it}D_{it}}{1 + (\omega\tau_{it})^2} \tag{6.45}$$

where $C_{it} = qD_{it}$, $\omega = 2\pi f$ (f = measurement frequency), and $\tau_{it} = R_{it}C_{it}$, the interface trap time constant, is given by $\tau_{it} = [v_{th}\sigma_p N_A \exp(-q\phi_s/kT]^{-1}$. G_p is divided by ω to make Eq. (6.45) symmetrical in $\omega\tau_{it}$. Time constants of this nature are discussed in more detail in Chapter 7. Equations (6.44) and (6.45) are for interface traps with a single energy level in the band gap. Interface traps at the SiO_2-Si interface, however, are continuously distributed in energy throughout the semiconductor band gap. Capture and emission occurs primarily by traps located within a few kT/q above and below the Fermi level for such a continuum of interface traps. This results in a time constant dispersion and gives the normalized conductance as[81]

$$\frac{G_p}{\omega} = \frac{qD_{it}}{2\omega\tau_{it}} \ln(1 + \omega^2\tau_{it}^2) \tag{6.46}$$

The units of conductance are S/cm^2 in these equations.

Equations (6.44) and (6.45) show that the conductance is easier to interpret than the capacitance because C_S is not required in Eq. (6.45). The conductance is measured as a function of frequency and plotted as G_p/ω versus ω or f. The function G_p/ω has a maximum at $\omega = 1/\tau_{it}$, and at that maximum $D_{it} = 2G_p/q\omega$. For Eq. (6.46) we find $\omega = 2/\tau_{it}$ and $D_{it} = 2.5G_p/q\omega$ at the maximum. Hence we determine D_{it} from the maximum G_p/ω and determine τ_{it} from ω at the peak conductance location on the ω-axis. G_p/ω versus f plots, calculated according to Eqs. (6.45) and (6.46), are shown in Fig. 6.22. The calculated curves are based on D_{it} values from a detailed interface extraction routine from the experimental data also shown on the figure. Note the much broader experimental peak.

Experimental G_p/ω versus ω curves are generally broader than predicted by Eq. (6.46). This is attributed to interface trap time constant dispersion caused by surface potential fluctuations due to nonuniformities in oxide charge, interface traps, and doping concentration. Surface potential fluctuations are more pronounced in p-Si than in n-Si.[82] Surface potential fluctuations complicate the analysis of the experimental data. When such fluctuations are taken into account, Eq. (6.46) becomes

$$\frac{G_p}{\omega} = \frac{q}{2} \int_{-\infty}^{\infty} \frac{D_{it}}{\omega \tau_{it}} \ln(1 + \omega^2 \tau_{it}^2) P(U_s) \, dU_s \qquad (6.47)$$

where $P(U_s)$ is a probability distribution of the surface potential fluctuation. The solid curve through the data points in Fig. 6.22 is calculated from Eq. (6.47). Note the good agreement between theory and experiment when ϕ_s fluctuations are considered. An approximate expression giving the interface trap density in terms of the measured maximum conductance is[81–82]

$$D_{it} \approx \frac{2.5}{q} \left[\frac{G_p}{\omega} \right]_{max} \qquad (6.48)$$

Capacitance meters and bridges generally assume the device to consist of the parallel $C_m - G_m$ combination in Fig. 6.21(c). A simple circuit comparison of Fig. 6.21(b) to 6.21(c) gives G_p/ω in terms of the measured capacitance C_m, the oxide capacitance, and the measured conductance G_m as

$$\frac{G_p}{\omega} = \frac{\omega G_m C_{ox}^2}{G_m^2 + \omega^2 (C_{ox} - C_m)^2} \qquad (6.49)$$

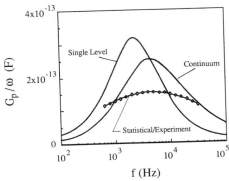

Fig. 6.22 G_p/ω vs. f for the single level [Eq. (6.45)], the continuum [Eq. (6.46)], and experimental data from DeClerck.[58] For the calculated curves: $D_{it} = 1.9 \times 10^9 \text{ cm}^{-2}\text{eV}^{-1}$, $\tau_{it} = 7 \times 10^{-5}$ s; for the experimental data $D_{it} = 1.9 \times 10^9 \text{ cm}^{-2}\text{eV}^{-1}$, $N_D = 5.25 \times 10^{15} \text{ cm}^{-3}$, $W_{ox} = 1295$ Å.

assuming negligible series resistance. In contrast to the quasi-static technique, the conductance measurement must be carried out over a wider frequency range. A comparison of interface traps determined by the quasi-static and the conductance techniques is shown in Fig. 6.23. Note the broad energy range over which the quasi-static method yields D_{it} and the good agreement over the narrower range where the conductance method is valid. The most accurate instrument to measure the conductance is a capacitance/conductance bridge. The frequency should be accurately determined with a frequency counter, and the signal amplitude should be kept at around 50 mV or less to prevent harmonics of the signal frequency giving rise to spurious conductances.

A number of models have been assumed to explain the experimental results.[81–91] In general, it is necessary to use one of these models to extract D_{it} and σ_p with confidence. Schemes have been proposed for analyzing data by taking pairs of values of G_p/ω having a predetermined relationship of either frequency[85,92] or magnitude.[93–95] For example in Nicollian et al.[85] and Simonne[92] G_p/ω curves are determined at two frequencies and the appropriate parameters are found from universal curves. Brews uses a single G_p/ω curve and determines the points where the curve has fallen to a fraction of its peak value and then utilizes universal curves to determine D_{it} and σ_p.[92] Noras presents an algorithm to extract the relevant parameters.[94–95] In yet another simplification, a single HF C-V and G-V curve suffice to determine D_{it}.[96]

It is possible to use a MOSFET instead of an MOS-C and measure the transconductance instead of the conductance but still use the concepts of the conductance method.[97] This permits interface trap density determination on devices with the small gate areas associated with MOSFET's without the need for special MOS-C test structures.

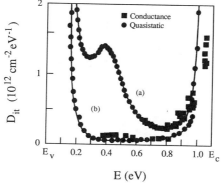

Fig. 6.23 Interface trapped charge density as a function of energy determined by the quasi-static and the conductance methods: (a) (111) n-Si oxidized in dry oxygen at 1140°C and annealed at 350°C in H_2,[82] (b) (100) n-Si, 960 Å SiO_2 plus 545 Å Si_3N_4 and annealed at 500°C for 15 min in forming gas.[83]

6.3.3 High-Frequency Methods

Terman Method The room-temperature, high-frequency capacitance method developed by Terman was one of the first methods for determining the interface trap density.[98] The method relies on a HF *C-V* measurement at a frequency sufficiently high that interface traps are assumed not to respond. They should therefore not contribute any capacitance.

How can one measure interface traps if they do not respond to the applied ac signal? Although interface traps do not respond to the ac probe frequency, they *do* respond to the slowly varying dc gate voltage and cause the HF *C-V* curve to stretch out along the gate voltage axis as interface trap occupancy changes with gate bias. In other words, for an MOS-C in depletion or inversion additional charge placed on the gate by a gate voltage induces additional semiconductor charge $Q_G = -(Q_B + Q_N + Q_{it})$. With

$$V_G = V_{FB} + \phi_s + V_{ox} = V_{FB} + \phi_s + \frac{Q_G}{C_{ox}} \tag{6.50}$$

it is obvious that for given surface potential ϕ_s, V_G increases when interface traps are present. This is the cause of the *C-V* "stretch-out" shown in Fig. 6.17. The stretch-out produces a *non-parallel* shift of the *C-V* curve. Interface traps distributed uniformly through the semiconductor band gap produce a fairly smoothly varying but distorted *C-V* curve. Interface traps with distinct structure, for example, peaked distributions, produce more abrupt distortions in the *C-V* curve.

The relevant equivalent circuit of the HF MOS-C is that in Fig. 6.4(c) with $C_{it} = 0$; that is, $C_{HF} = C_{ox}C_S/(C_{ox} + C_S)$, where $C_S = C_B + C_N$. C_{HF} is the same as that of a device without interface traps provided C_S is the same. The variation of C_S with surface potential is known for an ideal device. Knowing ϕ_s for a given C_{HF} in a device without Q_{it} allows us to construct a ϕ_s versus V_G curve of the actual capacitor. This is done as follows: From the ideal MOS-C *C-V* curve, find ϕ_s for a given C_{HF}. Then find V_G on the experimental curve for the same C_{HF}, giving one point of a ϕ_s versus V_G curve. Repeat for other points until a satisfactory $\phi_s - V_G$ curve is constructed. This $\phi_s - V_G$ curve contains the relevant interface trap information. Details of such construction are given in Nicollian and Brews.[38]

The experimental ϕ_s versus V_G curve is a stretched-out version of the theoretical curve, and the interface trap density is determined from this curve by[38,58]

$$D_{it} = \frac{C_{ox}}{q} \left[\frac{dV_G}{d\phi_s} - 1 \right] - \frac{C_S}{q} = \frac{C_{ox}}{q} \frac{d(\Delta V_G)}{d\phi_s} \tag{6.51}$$

where $\Delta V_G = V_G - V_G$ (ideal) is the voltage shift of the experimental from the ideal curve, with V_G being the experimental gate voltage. An approximate, simplified method that uses only two capacitance values and the associated gate voltages for D_{it} extraction is given in Jakubowski and Iniewski.[99]

The method is not widely used but is generally considered to be useful for measuring interface trap densities of 10^{10} cm^{-2}eV^{-1} and above. With careful measurements and analysis it has been used for interface trap densities in the low 10^{10} cm^{-2}eV^{-1} range.[100] The Terman method has been widely critiqued. Its limitations were originally pointed out to be due to inaccurate capacitance measurements and insufficiently high frequencies.[101] A later theoretical study concluded that D_{it} in the 10^9 cm^{-2}eV^{-1} range can be determined, provided the capacitance is measured to a precision of 0.001 to 0.002 pF.[102]

To compare experimental curves with theoretical curves, one needs to know the doping concentration exactly. Any dopant pileup or out-diffusion introduces errors. Surface potential fluctuations can cause fictitious interface trap peaks near the band edges.[58] The assumption that interface traps do not follow the ac probe frequency may not be satisfied for surface potentials near flatband and toward accumulation unless exceptionally high frequencies are used. Lastly, graphical differentiation of the ϕ_s versus V_G curve can cause errors. Large discrepancies were found for D_{it} determined by the Terman technique compared with deep-level transient spectroscopy (DLTS) and the Terman method is thought to be of questionable accuracy.[103]

Gray-Brown Method In the Gray-Brown method the high-frequency capacitance is measured as a function of temperature.[104–106] The Fermi level shifts with temperature. Lowering the temperature causes the Fermi level to shift toward the majority carrier band edge. Furthermore, the interface trap time constant τ_{it} increases at lower temperatures. Hence interface traps near the band edges should not respond to typical ac probe frequencies at low temperatures whereas at room temperature they do respond. This method should extend the range of interface traps measurements to D_{it} near the majority carrier band edge.

The HF *C-V* curves are measured as a function of temperature. The interface trap density is obtained from the flatband voltages at those temperatures. Just as the interface trap occupancy changes with gate voltage in the Terman method, so it changes with temperature in this method. It is this change that is analyzed, and D_{it} is extracted from the experimental data. The original measurements were made at 150 kHz and gave characteristic peaks of interface traps near the band edges. Theoretical calculations later indicated that these peaks were not real but were an artifact of the measurement technique produced by using too low ac probe frequencies.[107] Frequencies near 200 MHz should be used to maintain high-frequency conditions near the band edges. These requirements render quantitative results of the Gray-Brown technique inaccurate, so the method is rarely used today. Nevertheless, it is useful as a fast, qualitative indicator of interface traps. In particular, a HF C-V measurement at a temperature of 77 K shows a "ledge" in the curve.[106,108] This ledge voltage is related to the interface trap density over part of the band gap.

A method complementing the Gray-Brown method is one proposed by Arnold.[109] He uses a MOSFET and measures the transconductance as a function of gate voltage and temperature. Instead of sweeping through the majority part of the band gap, the minority part of the band gap near the band edge is sampled in this method, but interface traps near the middle of the band gap are not sensed.

6.3.4 Charge-Pumping Method

In the charge-pumping method, originally proposed in 1969,[110] a MOSFET is used as the test structure. The method is therefore suitable for interface trap measurements on small-geometry MOSFET's instead of large-diameter MOS capacitors. We explain the technique with reference to Fig. 6.24. Source and drain of the MOSFET are tied together and slightly reverse biased with voltage V_R. The square-wave gate voltage is of sufficient amplitude for the surface under the gate to be driven into inversion or accumulation. The pulse train can also be triangular or trapezoidal.

Let us begin by considering the MOSFET in heavy inversion shown in Fig. 6.24(a)—the result of a positive gate voltage. The corresponding semiconductor band diagram—from the Si surface into the substrate—is shown in Fig. 6.24(c). For clarity we show only the semiconductor on this energy band diagram. The interface traps, continuously distributed through the band gap, are represented by the eight small horizontal lines at the semiconductor surface with the filled circles representing electrons occupying interface traps. When the gate voltage changes from positive to negative potential, the surface changes from inversion to accumulation and finally ends up as in Fig. 6.24(b) and 6.24(f). However, the important processes take place during the transition from inversion to accumulation and from accumulation to inversion.

When the gate pulse falls from its positive to its negative value during its finite transition time, electrons in the inversion layer drift to both source and drain. In addition electrons captured by those interface traps near the conduction band are thermally emitted into the conduction band [Fig. 6.24(d)] and also drift to source and drain. Those electrons on interface traps deeper within the band gap do not have sufficient time to be emitted and will remain captured on interface traps. Once the hole barrier is reduced [Fig. 6.24(e)], holes flow to the surface where some are captured by those interface traps still filled with electrons. Holes are indicated by the open circles on the band diagrams. Finally, most traps are filled with holes as shown in Fig. 6.24(f). Then, when the gate returns to its positive voltage, the inverse process begins, and electrons flow into the interface to be captured.

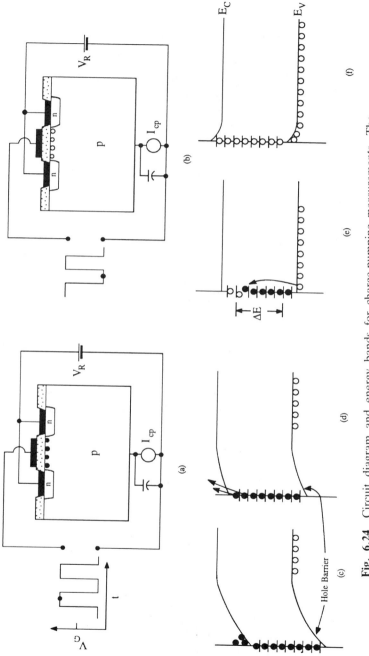

Fig. 6.24 Circuit diagram and energy bands for charge-pumping measurements. The figures are explained in the text.

281

The time constant for electron emission from interface traps is given by[8]

$$\tau_e = \frac{e^{E/kT}}{\sigma_n v_{th} N_c}$$ (6.52)

where E is the interface trap energy measured from the bottom of the conduction band with E_c being the reference energy ($E_c = 0$). The concepts of electron and hole capture, emission, time constants, etc., are discussed in detail in Chapter 7. For a square wave of frequency f, the time available for electron emission is half the period $\tau_e = 1/2f$. The energy interval over which electrons are emitted is from Eq. (6.52)

$$E = kT \ln\left(\frac{\sigma_n v_{th} N_c}{2f}\right)$$ (6.53)

For example, $E = 0.28\,\mathrm{eV}$ for $\sigma_n = 10^{-16}\,\mathrm{cm^2}$, $v_{th} = 10^7\,\mathrm{cm/s}$, $N_c = 10^{19}\,\mathrm{cm^{-3}}$, $T = 300\,\mathrm{K}$, and $f = 100\,\mathrm{kHz}$. Hence all electrons from E_c to $E_c - 0.28\,\mathrm{eV}$ are emitted, but those below $E_c - 0.28\,\mathrm{eV}$ are not emitted and therefore recombine with holes when holes come rushing in. The hole capture time constant is

$$\tau_c = \frac{1}{\sigma_p v_{th} p_s}$$ (6.54)

where p_s is the hole concentration $\mathrm{cm^{-3}}$ at the surface. τ_c is very small for any appreciable hole concentration. In other words, emission, not capture, is the rate limiting process.

During the reverse cycle when the surface changes from accumulation to inversion, the opposite process occurs. Holes within an energy interval

$$E - E_v = kT \ln\left(\frac{\sigma_p v_{th} N_v}{2f}\right)$$ (6.55)

are emitted into the valence band, and the remainder recombine with electrons flowing in from source and drain. Those electrons on interface traps within the energy interval ΔE,

$$\Delta E = E_g - kT\left[\ln\left(\frac{\sigma_n v_{th} N_c}{2f}\right) + \ln\left(\frac{\sigma_p v_{th} N_v}{2f}\right)\right]$$ (6.56)

recombine with holes.

We have then Q_N/q electrons/cm^2 flowing into the inversion layer from source/drain but only ($Q_N/q - D_{it}\,\Delta E$) electrons/cm^2 flowing back into the source/drain. $D_{it}\,\Delta E$ electrons/cm^2, the difference, recombine with holes. For each electron-hole pair recombination event, an electron and a hole must be supplied. Hence $D_{it}\,\Delta E$ holes/cm^2 also recombine. In other words, more holes flow into the semiconductor than leave, giving rise to the charge pumping current I_{cp} shown in Fig. 6.24. The capacitor in parallel with the ammeter averages the ac current to a dc current. $D_{it}\Delta E$ holes being supplied at rate of f Hz to a MOSFET with gate area A_G give rise to a current $I_{cp} = qA_G fD_{it}\Delta E$. In our example $\Delta E \approx 1.12 - 0.56 = 0.56$ eV, which is about half the band gap. Substituting numerical values, we find for a 10 μm × 10 μm gate area, a 100 kHz pump frequency, an interface trap density $D_{it} = 10^{10}$ cm^{-2}eV^{-1}, and $\Delta E = 0.56$ eV a current of $I_{cp} \approx 10^{-10}$ A. This current is sufficiently high to be easily measurable. As predicted, I_{cp} has been found to be proportional to both gate area and pump frequency. Charge-pumping current measurements have also been used to determine channel length variations across wafers by using its gate area dependence.[111]

A plot of charge pump current versus gate voltage as a function of source/drain voltage V_R is shown in Fig. 6.25. Note the sharp current increase beyond the threshold voltage. After the sharp rise the current saturates when the threshold voltage is exceeded. The I_{cp} decrease with increasing V_R is attributed to increased junction leakage currents that flow opposite to the charge pump current. The nonsaturating characteristic frequently observed for $V_R = 0$ has been attributed to the inability of all the mobile electrons to drift back to source and drain. This current is sometimes called the *geometrical component* of I_{cp}. The total charge pump current is given by

$$I_{cp} = A_G f[qD_{it}\Delta E + \alpha C_{ox}(V_G - V_T)] \tag{6.57}$$

where α is the fraction of the inversion charge that does not drift back to the source/drain. The geometrical component can be eliminated by using MOSFET's with short gate lengths ($L < 10$ μm) or by using gate pulse trains

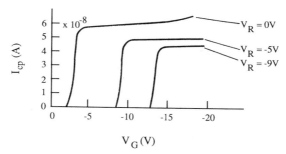

Fig. 6.25 Charge-pumping currents versus gate pulse amplitude as a function of source/drain voltage V_R. Reprinted with permission after Elliot.[112]

with moderate rise and fall times, giving the electrons sufficient time to drift back to source and drain.

The basic charge-pumping technique gives an average value of D_{it} over the energy interval ΔE. It does not give an energy distribution of the interface traps. Various refinements have been proposed to obtain energy-dependent interface trap distributions. Elliot varied the pulse base level from inversion to accumulation keeping the amplitude of the gate pulse constant.[112] Groeseneken[113] varied the rise and fall times of the gate pulses, whereas Wachnik[114–115] used small pulses with small rise and fall times to determine the energy distribution of D_{it}. A composite charge pumping method, utilizing pulse base level, pulse height, rise and fall times, and temperature variations, also gives interface trap distributions through the band gap.[116] The interface trap distribution through the band gap can be determined with a pulse shape that is not a simple square wave with finite rise and fall times but incorporates an intermediate voltage level. This switches the device from inversion to an intermediate state near midgap, and then to accumulation instead of from inversion to accumulation directly.[117–118]

6.3.5 MOSFET Subthreshold Current Method

The drain current of a MOSFET operated at gate voltages below threshold can be written as[119]

$$I_{D} = I_{1}e^{qV_{G}/NkT}(1 - e^{-qMV_{D}/NkT}) \qquad (6.58)$$

where I_1 is a constant that depends on temperature, device dimensions, and substrate doping concentration; $M = 1 + C_B/C_{ox}$, and $N = 1 + (C_B + C_{it})/C_{ox}$. M and N being larger than unity, account for the charge placed on the gate that does not result in inversion layer charge. Some gate charge is imaged as space-charge region charge and some as interface trap charge. The usual threshold plot is one of $\log(I_D)$ versus V_G for $V_D \gg NkT/qM$ with a slope $S = q/2.3NkT$, where the 2.3 accounts for the conversion from "ln" (logarithm to base e) to "log" (logarithm to base 10). At room temperature, with N close to unity, one expects a slope around $17V^{-1}$. This is frequently expressed as that gate voltage ΔV_G necessary to change the drain current by one decade and given as $\Delta V_G \approx 60$ mV for this example. Values of ΔV_G significantly higher than 60 mV are indicative of non-ideal behavior. The interface trap density obtained from a plot of $\log(I_D)$ versus V_G is

$$D_{it} = \frac{1}{q}\left[\left(\frac{q}{2.3SkT} - 1\right)C_{ox} - C_{B}\right] \qquad (6.59)$$

One can also measure the drain current as a function of drain voltage at a gate voltage that biases the MOSFET in the subthreshold regime. Equation (6.58) can be written as

$$I_D = I_{Dmax}(1 - e^{-qMV_D/NkT})$$ (6.60)

The best accuracy is obtained for the "$1/e$ point" at a drain voltage V_{De} where $I_D/I_{Dmax} = 0.632$, giving

$$D_{it} = \frac{C_{ox} + C_B}{q}\left(\frac{qV_{De}}{kT} - 1\right)$$ (6.61)

Both Eqs. (6.59) and (6.61) require an accurate knowledge of C_{ox} and C_B. An additional complication for the slope method is its strong dependence on surface potential fluctuations. The method is sometimes used to follow interface trap generation as a result of radiation-induced or voltage stress-induced interface trap generation.[120] Then it is only necessary to measure the original slope and determine subsequent changes. Alternatively, one can measure V_{De} as a function of D_{it} generation. It is even possible to separate interface traps from oxide trapped charge in the slope method.[120]

6.3.6 Other Methods

A sensitive method to determine D_{it} is deep-level transient spectroscopy. It is covered in Chapter 7. The charge transfer loss in charge-coupled devices (CCD) is also a sensitive indicator of interface trap densities.[121] However, it requires a CCD and is not practical if a CCD has to be specially fabricated as the test structure.

Crystallographic structural information on interface traps can be obtained from electron spin resonance (ESR) measurements.[122] But the method is relatively insensitive and densities $D_{it} \geq 10^{11}$ cm^{-2}eV^{-1} are required. From ESR measurements it appears that interface traps are trivalent silicon centers.

6.4 STRENGTHS AND WEAKNESSES

- *Mobile Charge* The strength of the *C-V method* is its simplicity requiring merely the measurement of a *C-V* curve, albeit at elevated temperatures. *C-V* methods are well established, and equipment exists in most semiconductor laboratories. Its weakness is that the total mobile charge is measured. Separation of various alkali species is not possible. Furthermore occasionally the *C-V* curve becomes distorted due to interface trapped charge, and the flatband voltage becomes difficult to determine.

The main strengths of the *triangular voltage sweep* method are the ability to differentiate between different mobile charge species, its higher sensitivity and the fact that the method is faster because the sample does not need to be heated and cooled, it just needs to be heated. Its weakness is the instrumentation. A low current measurement is not as readily available as a capacitance measurement setup.

- *Interface Trapped Charge* For MOS capacitors the choice for the most practical methods lies between the *conductance* and the *quasi-static methods*. These are the two most widely used techniques. The strength of the conductance method lies in its sensitivity and its ability to give the majority carrier capture cross sections. Its major weakness is the limited surface potential range over which D_{it} is obtained and the amount of work required to extract D_{it}, although simplified methods have been proposed.

The main strengths of the quasi-static method (both the *I-V* and the *Q-V*) are the relative ease of measurement and the large surface potential range over which D_{it} is obtained. A weakness for the *I-V* version is the current measurement requirement. The currents are usually low because the sweep rates must be low to ensure quasi-equilibrium. The recent *Q-V* version alleviates some of these problems, and commercial instruments are now available.

For MOSFET's the choice is the *charge-pumping* method. Its chief strength is the direct measurement of the current that is proportional to D_{it} and the fact that measurements can be made on regular MOSFET's with no need for special test structures. Its main weakness is that unless special measurement variations and interpretations are used, one gets a single value for an average interface trap density, not the energy distribution of D_{it}.

APPENDIX 6.1 CAPACITANCE MEASUREMENT TECHNIQUES

Most capacitance measurements are made with capacitance bridges or capacitance meters. In the vector voltage-current method of Fig. A6.1, an ac signal v_i is applied to the device under test (DUT), and the device impedance Z is calculated from the ratio of v_i to the sample current i_i. A high-gain operational amplifier with feedback resistor R_F operates as a current-to-voltage converter. With the input to the op-amp at virtual ground, the negative terminal is essentially at ground potential—called *virtual ground*—because the high input impedance allows no input current to the op-amp, $i_i = i_0$. With $i_i = v_i/Z$ and $i_0 = -v_0/R_F$, the device impedance can be derived from v_0 and v_i as

$$Z = -\frac{R_F v_i}{v_o} \qquad (A6.1)$$

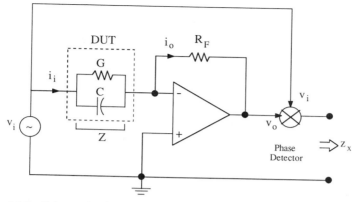

Fig. A6.1 Schematic circuit diagram of a capacitance-conductance meter.

where the device impedance of the parallel $G - C$ circuit in Fig. A6.1

$$Z = \frac{G}{[G^2 + (\omega C)^2]} - \frac{j\omega C}{[G^2 + (\omega C)^2]} \qquad (A6.2)$$

consists of a conductance—the first term—and a susceptance—the second term. The voltages v_0 and v_i are fed to a phase detector, and the conductance and susceptance of the sample are obtained by using the 0° and 90° phase angles of v_0 referenced to v_i. The 0° phase angle gives the conductance G, whereas the 90° phase angle gives the susceptance or the capacitance C.

Although this method uses a simple circuit configuration and has relatively high accuracy, it is difficult to design a feedback resistor amplifier with i_0 in exact proportion to i_i at high frequencies. An auto-balance circuit incorporating a null detector and a modulator overcomes this problem.[123] More detailed discussions of capacitance measurement circuits, probe stations, and other capacitance measurements hints can be found in the book by Nicollian and Brews.[38]

The capacitance meters, like the Boonton model 71 or 72 L-C meters and the Princeton Applied Research model 410 C-V plotter are three-terminal meters, whereas the more recent Hewlett Packard model 4275 LCR meter is a five-terminal instrument. One of the terminals in either instrument is grounded while the others connect to the device under test. The five-terminal operates much like a four-point probe, with the outer two terminals supplying the current and the inner two terminals measuring the potential.

The ground terminal on these instruments gives additional flexibility by eliminating stray capacitances. Two examples of making use of the ground terminal in a capacitance meter are shown in Fig. A6.2. Consider a three-terminal device with conductance G and capacitance C, which also has

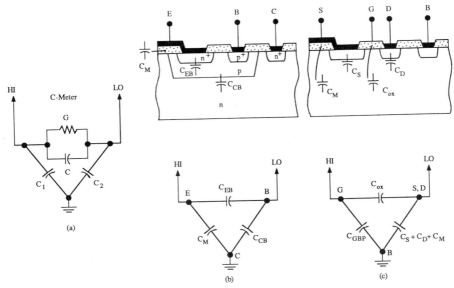

Fig. A6.2   Three-terminal capacitance measurement connections: (a) used to explain the principle, (b) a bipolar junction transistor, (c) a MOSFET.

stray capacitances C_1 and C_2 shown in Fig. A6.2(a). By connecting the DUT to the capacitance meter (C-Meter) and the two stray capacitances to ground, C_1 and C_2 are eliminated from the measurement by shunting them to ground. A bipolar junction transistor example is shown in Fig. A6.2(b). The emitter-base capacitance is measured while the collector-base and metal overlap capacitances are shunted to ground. In the MOSFET of Fig. A6.2(c), the gate voltage creates a channel connecting the source to the drain. With the connections shown, it is possible to measure the oxide capacitance, while shunting the source (C_S), drain (C_D), and source metal overlap (C_M) capacitances and the gate bonding pad capacitance C_{GBP} to ground.[124]

REFERENCES

[1]   B. E. Deal, "Standardized Terminology for Oxide Charges Associated with Thermally Oxidized Silicon," *IEEE Trans. Electron Dev.* **ED-27**, 606–608, March 1980.

[2]   B. E. Deal, M. Sklar, A. S. Grove, and E. H. Snow, "Characteristics of the Surface-State Charge (Q_{ss}) of Thermally Oxidized Silicon," *J. Electrochem. Soc.* **114**, 266–274, March 1967.

[3]   F. R. Badcock and D. R. Lamb, "Stability and Surface Charge in the MOS System," *Int. J. Electron.* **24**, 1–9, Jan. 1968; D. R. Lamb and F. R.

Badcock, "The Effect of Ambient, Temperature and Cooling Rate on the Surface Charge at the Silicon/Silicon Dioxide Interface," *Int. J. Electron.* **24**, 11–16, Jan. 1968.

[4] F. Montillo and P. Balk, "High-Temperature Annealing of Oxidized Silicon Surfaces," *J. Electrochem. Soc.* **118**, 1463–1468, Sept. 1971.

[5] S. R. Hofstein and G. Warfield, "Physical Limitations on the Frequency Response of a Semiconductor Surface Inversion Layer," *Solid-State Electron.* **8**, 321–341, Aug. 1965.

[6] C. H. Ling, C. Y. Kwok, E. G. Chan, and T. M. Tay, "Frequency Dependence of MOS Capacitance in Strong Inversion and at Elevated Temperatures," *Solid-State Electron.* **29**, 995–997, Sept. 1986.

[7] A. Berman and D. R. Kerr, "Inversion Charge Redistribution Model of the High-Frequency MOS Capacitance," *Solid-State Electron.* **17**, 735–742, July 1974.

[8] D. K. Schroder, *Advanced MOS Devices*, Addison-Wesley, Reading, MA, 1987.

[9] W. E. Beadle, J. C. C. Tsai, and R. D. Plummer, *Quick Reference Manual for Silicon Integrated Circuit Technology*, Wiley-Interscience, New York, 1985, pp. 14–28.

[10] H. El-Sissi and R. S. C. Cobbold, "Numerical Calculation of the Ideal *C/V* Characteristics of Nonuniformly Doped MOS Capacitors," *Electron. Lett.* **9**, 594–596, Dec. 1973.

[11] J. Hynecek, "Graphical Method for Determining the Flatband Voltage for Silicon on Sapphire," *Solid-State Electron.* **18**, 119–120, Feb. 1975.

[12] K. Lehovec and S. T. Lin, "Analysis of *C-V* Data in the Accumulation Regime of MIS Structures," *Solid-State Electron.* **19**, 993–996, Dec. 1976.

[13] F. P. Heiman, "Thin-Film Silicon-on-Sapphire Deep Depletion MOS Transistors," *IEEE Trans. Electron Dev.* **ED-13**, 855-862, Dec. 1966.

[14] K. Iniewski and A. Jakubowski, "New Method of Determination of the Flat-band Voltage in SOI MOS Structures, *Solid-State Electron.* **29**, 947–950, Sept. 1986.

[15] P. Balk, "Low-Temperature Annealing in the Al-SiO$_2$-Si System," Paper presented at the Buffalo Meeting of the Electrochem. Soc., Oct. 10–14, 1965, Abstract 111.

[16] B. E. Deal, E. L. MacKenna, and P. L. Castro, "Characteristics of Fast Surface States Associated with SiO$_2$-Si and Si$_3$N$_4$-SiO$_2$-Si Structures," *J. Electrochem. Soc.* **116**, 997–1005, July 1969.

[17] P. L. Castro and B. E. Deal, "Low-Temperature Reduction of Fast Surface States Associated with Thermally Oxidized Silicon," *J. Electrochem. Soc.* **118**, 280–286, Feb. 1971.

[18] B. E. Deal, "The Current Understanding of Charges in the Thermally Oxidized Silicon Structure," *J. Electrochem. Soc.* **121**, 198C–205C, June 1974.

[19] D. W. Hess and B. E. Deal, "Effect of Nitrogen and Oxygen/Nitrogen Mixtures on Oxide Charges in MOS Structures," *J. Electrochem. Soc.* **122**, 1123–1127, Aug. 1975.

[20] R. Williams, "Photoemission of Electrons from Silicon into Silicon Dioxide," *Phys. Rev.* **140**, A569–A575, Oct. 1965.

[21] A. M. Goodman, "Photoemission of Electrons from Silicon and Gold into Silicon Dioxide," *Phys. Rev.* **144**, 588–593, April 1966; "Photoemission of Electrons from *n*-Type Degenerate Silicon into Silicon Dioxide," *Phys. Rev.* **152**, 785–787, Dec. 1966.

[22] B. E. Deal, E. H. Snow, and C. A. Mead, "Barrier Energies in Metal-Silicon Dioxide-Silicon Structures," *J. Phys. Chem. Sol.* **27**, 1873–1879, Nov./Dec. 1966.

[23] R. Williams, "Properties of the Silicon-SiO$_2$ Interface," *J. Vac. Sci. Technol.* **14**, 1106–1111, Sept./Oct. 1977.

[24] R. J. Powell, "Interface Barrier Energy Determination from Voltage Dependence of Photoinjected Currents," *J. Appl. Phys.* **41**, 2424–2432, May 1970.

[25] H. M. Przewlocki, S. K. Krawczyk, and A. Jakubowski, "A Simple Technique of Work Function Difference Determination in MOS Structures," *Phys. Stat. Sol.* **65a**, 253–257, May 1981.

[26] S. K. Krawczyk, H. M. Przewlocki, and A. Jakubowski, "New Ways to Measure the Work Function Difference in MOS Structures," *Rev. Phys. Appl.* **17**, 473–480, Aug. 1982.

[27] W. M. Werner, "The Work Function Difference of the MOS-System with Aluminium Field Plates and Polycrystalline Silicon Field Plates," *Solid-State Electron.* **17**, 769–775, Aug. 1974.

[28] R. R. Razouk and B. E. Deal, "Hydrogen Anneal Effects on Metal-Semiconductor Work Function Difference," *J. Electrochem. Soc.* **129**, 806–810, April 1982.

[29] A. I. Akinwande and J. D. Plummer, "Process Dependence of the Metal Semiconductor Work Function Difference," *J. Electrochem. Soc.* **134**, 2297–2303, Sept. 1987.

[30] T. W. Hickmott and R. D. Isaac, "Barrier Heights and Polycrystalline Silicon-SiO$_2$ Interface," *J. Appl. Phys.* **52**, 3464–3475, May 1981.

[31] N. Lifshitz, "Dependence of the Work-Function Difference between the Polysilicon Gate and Silicon Substrate on the Doping Level in Polysilicon," *IEEE Trans. Electron Dev.* **ED-32,** 617–621, March 1985.

[32] S. Kar, "Determination of Si-Metal Work Function Differences by MOS Capacitance Technique," *Solid-State Electron.* **18**, 169–181, Feb. 1975; "Interface Charge Characteristics of MOS Structures with Different Metals on Steam Grown Oxides," *Solid-State Electron.* **18**, 723–732, Sept. 1975.

[33] K. Haberle and E. Fröschle, "On the Work Function Difference in the Al-SiO$_2$-Si System with Reactively Sputtered SiO$_2$," *J. Electrochem. Soc.* **126**, 878–880, May 1979.

[34] A. K. Gaind and L. A. Kasprzak, "Determination of Distributed Fixed Charge in CVD-Oxide and Its Virtual Elimination by the Use of HCL," *Solid-State Electron.* **22**, 303–309, March 1979.

[35] W. H. Krautschneider, J. Laschinski, W. Seifert, and H. G. Wagemann, "An Accurate MOS Measurement Procedure for Work Function Difference in the Al/SiO$_2$/Si System," *Solid-State Electron.* **29**, 571–578, May 1986.

[36] T. W. Hickmott, "Dipole Layers at the Metal-SiO$_2$ Interface," *J. Appl. Phys.* **51**, 4269–4281, Aug. 1980.

[37] D. B. Kao, K. C. Saraswat, and J. P. McVittie, "Annealing of Oxide Fixed Charges in Scaled Polysilicon Gate MOS Structures," *IEEE Trans. Electron Dev.* **ED-32**, 918–925, May 1985.

[38] E. H. Nicollian and J. R. Brews, *MOS Physics and Technology*, Wiley, New York, 1982.

[39] S. P. Li, M. Ryan, and E. T. Bates, "Rapid and Precise Measurement of Flatband Voltage," *Rev. Sci. Instrum.* **47**, 632–634, May 1976.

[40] E. H. Snow, A. S. Grove, B. E. Deal, and C. T. Sah, "Ion Transport Phenomena in Insulating Films," *J. Appl. Phys.* **36**, 1664–1673, May 1965.

[41] W. A. Pliskin and R. A. Gdula, "Passivation and Insulation," in *Handbook on Semiconductors*, Vol. 3 (S. P. Keller, ed.), North-Holland, Amsterdam, 1980, and references therein.

[42] N. J. Chou, "Application of Triangular Voltage Sweep Method to Mobile Charge Studies in MOS Structures," *J. Electrochem. Soc.* **118**, 601–609, April 1971.

[43] R. J. Kriegler and T. F. Devenyi, "Direct Measurement of Na$^+$ Ion Mobility in SiO$_2$ Films," *Thin Solid Films* **36**, 435–439, Aug. 1976.

[44] G. Derbenwick, "Mobile Ions in SiO$_2$: Potassium," *J. Appl. Phys.* **48**, 1127–1130, March 1977.

[45] J. P. Stagg, "Drift Mobilities of Na$^+$ and K$^+$ Ions in SiO$_2$ Films," *Appl. Phys. Lett.* **31**, 532–533, Oct. 1977.

[46] A. G. Tangena, N. F. de Rooij, and J. Middlehoek, "Sensitivity of MOS Structures for Contamination with H$^+$, Na$^+$, and K$^+$ Ions," *J. Appl. Phys.* **49**, 5576–5583, Nov. 1978.

[47] M. R. Boudry and J. P. Stagg, "The Kinetic Behavior of Mobile Ions in the Al-SiO$_2$-Si System," *J. Appl. Phys.* **50**, 942–950, Feb. 1979.

[48] M. W. Hillen, G. Greeuw, and J. F. Verweij, "On the Mobility of Potassium Ions in SiO$_2$," *J. Appl. Phys.* **50**, 4834–4837, July 1979.

[49] G. Greeuw and J. F. Verwey, "The Mobility of Na$^+$, Li$^+$ and K$^+$ Ions in Thermally Grown SiO$_2$ Films," *J. Appl. Phys.* **56**, 2218–2224, Oct. 1984.

[50] G. Greeuw and B. J. Hoenders, "Theoretical Solution of the Transient Current Equation for Mobile Ions in a Dielectric Film under the Influence of a Constant Electric Field," *J. Appl. Phys.* **55**, 3371–3375, May 1984.

[51] M. Yamin, "Charge Storage Effects in Silicon Dioxide Films," *IEEE Trans. Electron Dev.* **ED-12**, 88–96, March 1965.

[52] M. Kuhn and D. J. Silversmith, "Ionic Contamination and Transport of Mobile Ions in MOS Structures," *J. Electrochem. Soc.* **118**, 966–970, June 1971.

[53] T. M. Buck, F. G. Allen, J. V. Dalton, and J. D. Struthers, "Studies of Sodium in SiO$_2$ Films by Neutron Activation and Radiotracer Techniques," *J. Electrochem. Soc.* **114**, 862–866, Aug. 1967.

[54] E. Yon, W. H. Ko, and A. B. Kuper, "Sodium Distribution in Thermal Oxide on Silicon by Radiochemical and MOS Analysis," *IEEE Trans. Electron Dev.* **ED-13**, 276–280, Feb. 1966.

[55] B. Yurash and B. E. Deal, "A Method for Determining Sodium Content of Semiconductor Processing Materials," *J. Electrochem. Soc.* **115**, 1191–1196, Nov. 1968.

[56] H. L. Hughes, R. D. Baxter, and B. Phillips, "Dependence of MOS Device Radiation-Sensitivity on Oxide Impurities," *IEEE Trans. Nucl. Sci.* **NS-19**, 256–263, Dec. 1972.

[57] A. Goetzberger, E. Klausmann, and M. J. Schulz, "Interface States on Semiconductor/Insulator Interfaces," *CRC Crit. Rev. Solid State Sci.* **6**, 1–43, Jan. 1976.

[58] G. DeClerck, "Characterization of Surface States at the Si-SiO$_2$ Interface," in *Nondestructive Evaluation of Semiconductor Materials and Devices* (J. N. Zemel, ed.), Plenum, New York, 1979, pp. 105–148.

[59] C. N. Berglund, "Surface States at Steam-Grown Silicon-Silicon Dioxide Interfaces," *IEEE Trans. Electron Dev.* **ED-13**, 701–705, Oct. 1966.

[60] M. Kuhn, "A Quasi-static Technique for MOS *C-V* and Surface State Measurements," *Solid-State Electron.* **13**, 873–885, June 1970.

[61] E. A. Fogels and C. A. T. Salama, "Characterization of Surface States at the Si-SiO$_2$ Interface Using the Quasi-static Technique," *J. Electrochem. Soc.* **118**, 2002–2006, Dec. 1971.

[62] D. R. Kerr, "MIS Measurement Techniques Utilizing Slow Voltage Ramps," Int. Conf. on *Properties and Use of MIS Structures* (J. Bovel, ed.), Grenoble, 1969, pp. 303–307.

[63] D. Bräunig and H. G. Wagemann, "Experiments and Model of Nonequilibrium Behavior of MIS Varactors Using the Linear Ramp Technique," *IEEE Trans. Electron. Dev.* **ED-21**, 241–247, Apr. 1974.

[64] P. D. Tonner and J. G. Simmons, "Experimental Technique for Determining Surface Potential as a Function of Gate Voltage of a MOS Capacitor," *Rev. Sci. Instrum.* **51**, 1378–1380, Oct. 1980; "Non-equilibrium ψ_s vs. V_g Characteristics of MOS Capacitors and Related Effects," *Solid-State Electron.* **25**, 733–739, Aug. 1982.

[65] S. Nishimatsu and M. Ashikawa, "A Simple Method for Measuring the Interface State Density," *Rev. Sci. Instrum.* **45**, 1109–1112, Sept. 1984.

[66] R. Castagné, "Determination of Interface State Density of a Metal-Insulator-Semiconductor Capacitor with a Linearly Varying Voltage (in French)," *C.R. Acad. Sc. Paris* **267**, 866–869, Oct. 1968.

[67] W. K. Kappallo and J. P. Walsh, "A Current Voltage Technique for Obtaining Low-Frequency *C-V* Characteristics of MOS Capacitors," *Appl. Phys. Lett.* **17**, 384–386, Nov. 1970.

[68] R. Castagné and A. Vapaille, "Description of the SiO$_2$–Si Interface Properties by Means of Very Low Frequency MOS Capacitance Measurements," *Surf. Sci.* **28**, 157–193, Nov. 1971.

[69] G. DeClerck, R. Van Overstraeten, and G. Broux, "Measurement of Low Densities of Surface States at the Si-SiO$_2$ Interface," *Solid-State Electron.* **16**, 1451–1460, Dec. 1973.

[70] R. Castagné and A. Vapaille, "Apparent Interface State Density Introduced by the Spatial Fluctuations of Surface Potential in an MOS Structure," *Electron. Lett.* **6**, 691–694, Oct. 1970.

[71] S. Wagner and C. N. Berglund, "A Simplified Graphical Evaluation of High-Frequency and Quasistatic Capacitance-Voltage Curves," *Rev. Sci. Instrum.* **43**, 1775–1777, Dec. 1972.

[72] R. Van Overstraeten, G. DeClerck, and G. Broux, "Graphical Technique to Determine the Density of Surface States at the Si-SiO$_2$ Interface of MOS Devices Using the Quasi-static *C-V* Method," *J. Electrochem. Soc.* **120**, 1785–1787, Dec. 1973.

[73] M. Kuhn and E. H. Nicollian, "Nonequilibrium Effects in Quasi-static MOS Measurements," *J. Electrochem. Soc.* **118**, 370–373, Feb. 1971.

[74] A. D. Lopez, "Using the Quasi-static Method for MOS Measurements," *Rev. Sci. Instrum.* **44**, 200–204, Feb. 1972.

[75] J. Koomen, "The Measurement of Interface State Charge in the MOS System," *Solid-State Electron.* **14**, 571–580, July 1971.

[76] K. Ziegler and E. Klausmann, "Static Technique for Precise Measurements of Surface Potential and Interface State Density in MOS Structures," *Appl. Phys. Lett.* **26**, 400–402, Apr. 1975.

[77] J. R. Brews and E. H. Nicollian, "Improved MOS Capacitor Measurements Using the *Q-C* Method," *Solid-State Electron.* **27**, 963–975, Nov. 1984.

[78] E. H. Nicollian and J. R. Brews, "Instrumentation and Analog Implementation of the *Q-C* Method for MOS Measurements," *Solid-State Electron.* **27**, 953–962, Nov. 1984.

[79] D. M. Boulin, J. R. Brews, and E. H. Nicollian, "Digital Implementation of the *Q-C* Method for MOS Measurements," *Solid-State Electron.* **27**, 977–988, Nov. 1984.

[80] T. J. Mego, "Improved Feedback Charge Method for Quasi-static *CV* Measurements in Semiconductors," *Rev. Sci. Instrum.* **57**, 2798–2805, Nov. 1986.

[81] E. H. Nicollian and A. Goetzberger, "The Si-SiO$_2$ Interface—Electrical Properties as Determined by the Metal–Insulator–Silicon Conductance Technique," *Bell. Syst. Tech. J.* **46**, 1055–1133, July/Aug. 1967.

[82] M. Schulz, "Interface States at the SiO$_2$-Si Interface," *Surf. Sci.* **132**, 422–455, Sept. 1983.

[83] A. K. Aggarwal and M. H. White, "On the Nonequilibrium Statistics and Small Signal Admittance of Si-SiO$_2$ Interface traps in the Deep-Depleted Gated-Diode Structure," *J. Appl. Phys.* **55**, 3682–3694, May 1984.

[84] K. K. Hung and Y. C. Cheng, "Characterization of Si-SiO$_2$ Interface Traps in *p*-Metal-Oxide-Semiconductor Structures with Thin Oxides by Conductance Technique," *J. Appl. Phys.* **62**, 4204–4211, Nov. 1987.

[85] E. H. Nicollian, A. Goetzberger, and A. D. Lopez, "Expedient Method of Obtaining Interface State Properties from MIS Conductance Measurements," *Solid-State Electron.* **12**, 937–944, Dec. 1969.

[86] W. Fahrner and A. Goetzberger, "Energy Dependence of Electrical Properties of Interface States in Si-SiO$_2$ Interfaces," *Appl. Phys. Lett.* **17**, 16–18, July 1970.

[87] H. Deuling, E. Klausmann, and A. Goetzberger, "Interface States in Si-SiO$_2$ Interfaces," *Solid-State Electron.* **15**, 559–571, May 1972.

[88] J. R. Brews, "Admittance of an MOS Device with Interface Charge Inhomogeneities," *J. Appl. Phys.* **43**, 3451–3455, Aug. 1972.

[89] M. J. McNutt and C. T. Sah, "Effects of Spatially Inhomogeneous Oxide Charge Distribution on the MOS Capacitance-Voltage Characteristics," *J. Appl. Phys.* **45**, 3916–3921, Sept. 1974.

[90] J. A. Cooper, Jr., and R. J. Schwartz, "Electrical Characterization of the SiO_2-Si Interface near Midgap and in Weak Inversion," *Solid-State Electron.* **17**, 641–654, July 1974.

[91] P. A. Muls, G. J. DeClerck, and R. J. Van Overstraeten, "Influence of Interface Charge Inhomogeneities on the Measurement of Surface State Densities in Si-SiO_2 Interfaces by Means of the MOS ac Conductance Technique," *Solid-State Electron.* **20**, 911–922, Nov. 1977, and references therein.

[92] J. J. Simonne, "A Method to Extract Interface State Parameters from the MIS Parallel Conductance Technique," *Solid-State Electron.* **16**, 121–124, Jan. 1973.

[93] J. R. Brews, "Rapid Interface Parametrization Using a Single MOS Conductance Curve," *Solid-State Electron.* **26**, 711–716, Aug. 1983.

[94] J. M. Noras, "Extraction of Interface State Attributes from MOS Conductance Measurements, "*Solid-State Electron.* **30**, 433–437, April 1987.

[95] J. M. Noras, "Parameter Estimation in MOS Conductance Studies," *Solid-State Electron.* **31**, 981–987, May 1988.

[96] W. A. Hill and C. C. Coleman, "A Single-Frequency Approximation for Interface-State Density Determination," *Solid-State Electron.* **23**, 987–993, Sept. 1980.

[97] H. Haddara and G. Ghibaudo, "Analytical Modeling of Transfer Admittance in Small MOSFETs and Application to Interface State Characterisation," *Solid-State Electron* **31**, 1077–1082, June 1988.

[98] L. M. Terman, "An Investigation of Surface States at a Silicon/Silicon Oxide Interface Employing Metal-Oxide-Silicon Diodes," *Solid-State Electron.* **5**, 285–299, Sept./Oct. 1962.

[99] A. Jakubowski and K. Iniewski, "Technical Method of Determination of the Interface Trap Density," *Phys. Stat. Sol.* **89a**, 383–388, May 1985.

[100] C. C. H. Hsu and C. T. Sah, "Generation-Annealing of Oxide and Interface Traps at 150 and 298 K in Oxidized Silicon Stressed by Fowler-Nordheim Electron Tunneling," *Solid-State Electron.* **31**, 1003–1007, June 1988.

[101] K. H. Zaininger and G. Warfield, "Limitations of the MOS Capacitance Method for the Determination of Semiconductor Surface Properties," *IEEE Trans. Electron Dev.* **ED-12**, 179–193, April 1965.

[102] C. T. Sah, A. B. Tole, and R. F. Pierret, "Error Analysis of Surface State Density Determination Using the MOS Capacitance Method," *Solid-State Electron.* **12**, 689–709, Sept. 1969.

[103] E. Rosenecher and D. Bois, "Comparison of Interface State Density in MIS Structure Deduced from DLTS and Terman Measurements," *Electron. Lett.* **18**, 545–546, June 1982.

[104] P. V. Gray and D. M. Brown, "Density of SiO_2-Si Interface States," *Appl. Phys. Lett.* **8**, 31–33, Jan. 1966.

[105] D. M. Brown and P. V. Gray, "Si-SiO$_2$ Fast Interface State Measurements," *J. Electrochem. Soc.* **115**, 760–767, July 1968.

[106] P. V. Gray, "The Silicon-Silicon Dioxide System," *Proc. IEEE* **57**, 1543–1551, Sept. 1969.

[107] M. R. Boudry, "Theoretical Origins of N_{ss} Peaks Observed in Gray-Brown MOS Studies," *Appl. Phys. Lett.* **22**, 530–531, May 1973.

[108] D. K. Schroder and J. Guldberg, "Interpretation of Surface and Bulk Effects Using the Pulsed MIS Capacitor," *Solid-State Electron.* **14**, 1285–1297, Dec. 1971.

[109] E. Arnold, "Surface Charges and Surface Potential in Silicon Surface Inversion Layers," *IEEE Trans. Electron Dev.* **ED-15**, 1003–1008, Dec. 1968.

[110] J. S. Brugler and P. G. A. Jespers, "Charge Pumping in MOS Devices," *IEEE Trans. Electron Dev.* **ED-16**, 297–302, March 1969.

[111] T. J. Russell, C. L. Wilson, and M. Gaitan, "Determination of the Spatial Variation of Interface Trapped Charge Using Short-Channel MOSFET's," *IEEE Trans. Electron Dev.* **ED-30**, 1662–1671, Dec. 1983.

[112] A. B. M. Elliot, "The Use of Charge Pumping Currents to Measure Surface State Densities in MOS Transistors," *Solid-State Electron.* **19**, 241–247, March 1976.

[113] G. Groeseneken, H. E. Maes, N. Beltrán, and R. F. De Keersmaecker, "A Reliable Approach to Charge-Pumping Measurements in MOS Transistors," *IEEE Trans. Electron Dev.* **ED-31**, 42–53, Jan. 1984; P. Heremans, J. Witters, G. Groeseneken, and H. E. Maes, "Analysis of the Charge Pumping Technique and Its Application for the Evaluation of MOSFET Degradation," *IEEE Trans. Electron Dev.* **36**, 1318–1335, July 1989.

[114] R. A. Wachnik and J. R. Lowney, "A Model for the Charge-Pumping Current Based on Small Rectangular Voltage Pulses," *Solid-State Electron.* **29**, 447–460, April 1986.

[115] R. A. Wachnik, "The Use of Charge Pumping to Characterize Generation by Interface Traps," *IEEE Trans. Electron Dev.* **ED-33,** 1054–1061, July 1986.

[116] H. Haddara and S. Cristoloveanu, "Profiling of Stress Induced Interface States in Short Channel MOSFET's Using a Composite Charge Pumping Technique," *Solid-State Electron.* **29**, 767–772, Aug. 1986.

[117] W. L. Tseng, "A New Charge Pumping Method of Measuring Si-SiO$_2$ Interface States," *J. Appl. Phys.* **62**, 591–599, July 1987.

[118] F. Hofmann and W. H. Krautschneider, "A Simple Technique for Determining the Interface-Trap Distribution of Submicron Metal-Oxide-Semiconductor Transistors by the Charge Pumping Method," *J. Appl. Phys.* **65**, 1358–1360, Feb. 1989.

[119] P. A. Muls, G. J. DeClerck, and R. J. Van Overstraeten, "Characterization of the MOSFET Operating in Weak Inversion," *Adv. in Electron. and Electron Phys.* **47**, 197–266, 1978.

[120] P. J. McWhorter and P. S. Winokur, "Simple Technique for Separating the Effects of Interface Traps and Trapped-Oxide Charge in Metal-Oxide-Semiconductor Transistors," *Appl. Phys. Lett.* **48**, 133–135, Jan. 1986.

[121] R. J. Kriegler, T. F. Devenyi, K. D. Chik, and J. Shappir, "Determination of

Surface-State Parameters from Transfer-Loss Measurements in CCDs," *J. Appl. Phys.* **50**, 398–401, Jan. 1979.

[122] E. H. Poindexter and P. J. Caplan, "Characterization of Si/SiO_2 Interface Defects by Electron Spin Resonance," *Progr. Surf. Sci.* **14**, 201–294, 1983.

[123] Service manual for HP 4275-A Multi Frequency LCR Meter, Hewlett-Packard, 1983, p. 8-4.

[124] M. G. Buehler, *Lecture Notes.*

CHAPTER 7

DEEP-LEVEL IMPURITIES

7.1 INTRODUCTION

All semiconductors contain impurities. Some impurities are intentionally introduced as dopant atoms (shallow-level impurities), recombination centers (deep-level impurities) to reduce the device lifetime, or deep-level impurities to increase the substrate resistivity. Many impurities are unintentionally incorporated during crystal growth and device processing. The impurities may be foreign impurities (e.g., metals), crystallographic point defects (e.g., vacancies and interstitials), or structural defects (e.g., stacking faults and dislocations). Many impurities can be removed during processing through gettering.

The characterization of shallow-level or dopant impurities is discussed in Chapters 2 and 10. Shallow impurity concentrations are best measured electrically, but their energy levels are best determined optically. In this chapter we discuss predominantly the measurement of deep-level impurities whose concentrations and energy levels are best measured electrically. Milnes gives a good review of impurities in semiconductors[1] which he has updated with two recent review articles.[2-3] Jaros treats the theoretical aspects of deep-level impurities.[4]

7.2 GENERATION-RECOMBINATION STATISTICS

7.2.1 A Pictorial View

The band diagram of a perfect single crystal semiconductor consists of a valence band and a conduction band separated by the band gap. When the periodicity of the single crystal is perturbed by foreign atoms or crystal defects, discrete energy levels are introduced into the band gap, shown by the E_T lines in Fig. 7.1. Each line represents one such defect with energy E_T. Such defects are commonly called generation-recombination (G-R) centers or traps. G-R centers lie deep in the band gap and are known as deep energy level impurities, or simply *deep-level impurities*. They act as recombination centers when there are excess carriers in the semiconductor and as generation centers when the carrier density is below its equilibrium value as in the reverse-biased space-charge region (scr) of *pn* junctions or MOS-capacitors, for example.

For the common semiconductors silicon, germanium, and gallium arsenide, deep-level impurities are commonly metallic impurities like iron, gold, and copper. Deep-level impurities are also the result of crystal imperfections, such as dislocations, stacking faults, precipitates, vacancies, or interstitials. For the most part they are undesirable. Occasionally, however, they are deliberately introduced to alter a device characteristic. Most frequently it is the switching time of a device that is reduced by the introduction of controlled amounts of deep-level impurities. In some semiconductors like GaAs and InP, deep-level impurities raise the substrate resistivity—creating semi-insulating substrates.

Let us consider the deep level impurity shown in Fig. 7.1. It has an energy E_T and consists of N_T impurities/cm^3 uniformly distributed throughout the semiconductor. The energy E_T is an effective energy discussed in more detail in Appendix 7.1. The semiconductor has n electrons/cm^3 and p holes/cm^3 introduced by shallow-level dopants, not shown on the figure. To follow the various capture and emission processes, let the G-R center first capture an electron from the conduction band, shown in Fig. 7.1(a) and characterized by the capture coefficient c_n. After electron capture one of two events takes place. The center can either emit the electron back to the conduction band from where it came, called electron emission e_n and shown by Fig. 7.1(b), or it can capture a hole from the valence band, shown by Fig. 7.1(c) as c_p. After either of these events, the G-R center is occupied by a

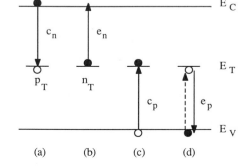

Fig. 7.1 Electron energy band diagram for a semiconductor with deep-level impurities. The capture and emission processes are described in the text.

hole and again has two choices. Either it emits the hole back to the valence band e_p [Fig. 7.1(d)] or captures an electron [Fig. 7.1(a)]. These are the only four possible events between the conduction band, the impurity, and the valence band. The process in Fig. 7.1(d) is sometimes viewed as electron emission from the valence band to the impurity shown by the dashed arrow. We will, however, use the hole emission process in (d) because it lends itself more readily to mathematical analysis.

A *recombination* event is Fig. 7.1(a) followed by (c) and a *generation* event is (b) followed by (d). The G-R center and *both* the conduction and valence bands participate in recombination and generation. These mechanisms are the topic of Chapter 8. A third event, which is neither recombination nor generation, is the *trapping* event (a) followed by (b) or (c) followed by (d). In either case a carrier is captured and subsequently emitted back to the band from which it came. Only one of the two bands and the center participate.

Whether an impurity acts as a trap or a G-R center depends on the location of the Fermi level in the band gap, the temperature, and the capture cross sections of the impurity. Generally those impurities whose energies lie near the middle of the band gap behave as G-R centers, whereas those near the band edges act as traps. However, for most of the measurements in this chapter, a center acts as a trap even if its energy level lies near the middle of the band gap, and we will refer to such impurities as G-R centers or traps. Generally the electron emission rate for centers in the upper half of the band gap is much higher than the hole emission rate. Similarly the hole emission rate is generally much higher than the electron emission rate for centers in the lower half of the band gap. For most centers one emission rate dominates, and the other can frequently be neglected.

7.2.2 A Mathematical Description

A G-R center can be in one of two charge states. When occupied by an electron, it is in the n_T state, and when occupied by a hole, it is in the p_T state (both are shown in Fig. 7.1). The concentration of G-R centers occupied by electrons n_T, and holes p_T must equal the total concentration N_T or $N_T = n_T + p_T$. In other words, a center is either occupied by an electron or a hole. When electrons and holes recombine or are generated, the electron concentration in the conduction band n, the hole concentration in the valence band p, and the charge state of the center n_T or p_T are all functions of time. For that reason we will first address the question, "What is the time rate of change of n, p, and n_T?" We develop the appropriate equations for electrons in detail and then merely state the equations for holes. The equations for holes are analogous, and their derivation follows similar paths. A thorough discussion of the equations and their derivations is given by Sah et al.[5]

The electron concentration in the conduction band is diminished by electron capture (process (a) in Fig. 7.1) and increased by electron emission (process (b) in Fig. 7.1). The time rate of change of n due to G-R mechanisms is given by[6-7]

$$\left.\frac{dn}{dt}\right|_{\text{G-R}} = (b) - (a) = e_n n_{\text{T}} - c_n n p_{\text{T}} \tag{7.1}$$

The subscript "G-R" signifies that we are only considering emission and capture processes through G-R centers. We are not considering radiative or Auger processes. However, later in the chapter we address briefly optical emission as a mechanism to excite carriers into or out of G-R centers. Electron emission depends on the concentration of G-R centers occupied by electrons and the emission rate through the relation $(b) = e_n n_{\text{T}}$. This relationship does not contain n because it is not necessary for there to be electrons in the conduction band during the emission process. But there must be G-R centers occupied by electrons. After all, if there are no electrons on the centers, none can be emitted.

The capture process is slightly more complicated because it depends on n, p_{T}, and the capture coefficient c_n through the relation $(a) = c_n n p_{\text{T}}$. The electron concentration n is important because, to capture electrons, there must be electrons in the conduction band. For holes we find the parallel expression

$$\left.\frac{dp}{dt}\right|_{\text{G-R}} = (d) - (c) = e_p p_{\text{T}} - c_p p n_{\text{T}} \tag{7.2}$$

The emission rate e_n represents the number of electrons emitted per second from electron-occupied G-R centers. The capture rate $c_n n$ represents the number of electrons captured per second from the conduction band. You may wonder how there can be more than one electron emitted from a G-R center. After an electron has been emitted, the center finds itself in the p_{T} state and subsequently emits a hole, returning it to the n_{T} state. Then the cycle repeats.

Where do the electrons and holes come from for this cycle to continue? Surely they cannot come from the center itself. It may be helpful to view the hole emission process as one of electron emission from the valence band to the G-R center, indicated by the dashed line in Fig. 7.1(d). In this picture the electron-hole emission process is nothing more than an electron jumping from the valence band to the conduction band with an intermediate stop at the E_{T} level. However, it is easier to deal with the equations if we consider hole and electron emission as shown by the solid lines in Fig. 7.1.

The capture coefficient c_n is a little more complicated than the emission coefficient. It is defined by

$$c_n = \sigma_n v_{\text{th}} \tag{7.3}$$

were v_{th} is the thermal velocity of the electrons and σ_n is the electron capture cross section of the G-R center. A physical explanation of c_n can be gleaned from Eq. (7.3). We know that electrons move randomly at their thermal velocity and that G-R centers remain immobile in the lattice. Nevertheless, it is helpful for this discussion to change the frame of reference by letting the electrons be immobile and the G-R centers move at velocity v_{th}. The centers then sweep out a volume per unit time of $\sigma_n v_{th}$, and those electrons that find themselves in that volume have a very high probability of being captured. Capture cross sections vary widely, depending on whether the center is neutral or negatively or positively charged. In general, a center with a negative or repulsive charge has a smaller electron capture cross section that one that is neutral or attractively charged. Neutral capture cross sections are on the order of 10^{-15} cm^2—roughly the physical size of the atom.

Whenever an electron or hole is captured or emitted, the center occupancy changes, and that rate of change is given by

$$\left.\frac{dn_T}{dt}\right|_{\text{G-R}} = \frac{dp}{dt} - \frac{dn}{dt} = (c_n n + e_p)(N_T - n_T) - (c_p p + e_n)n_T \qquad (7.4)$$

This equation is in general nonlinear, with n and p being time-dependent variables. If the equation can be linearized, it can be solved easily. Two cases allow this simplification: (1) In a reverse-biased space-charge region both n and p are small and can, to first order, be neglected. (2) In the quasi-neutral regions n and p are reasonably constant. Solving Eq. (7.4) for condition (2) gives $n_T(t)$ as

$$\boxed{n_T(t) = n_T(0)e^{-t/\tau} + \frac{e_p + c_n n}{e_n + c_n n + e_p + c_p p} N_T(1 - e^{-t/\tau})} \qquad (7.5)$$

where $n_T(0)$ is the concentration of G-R centers occupied by electrons at $t = 0$ and $\tau = 1/(e_n + c_n n + e_p + c_p p)$. The *steady-state concentration* as $t \to \infty$ is

$$n_T = \frac{e_p + c_n n}{e_n + c_n n + e_p + c_p p} N_T \qquad (7.6)$$

This equation shows the steady-state occupancy of n_T to be determined by the electron and hole concentrations as well as by the emission and capture rates. Equations (7.5) and (7.6) are the basis for most deep-level impurity measurements.

Equation (7.5) is difficult to solve in the general case because neither capture nor emission rates may be known and n and p vary with time, and generally also with distance in a device. Certain experimental simplifications are usually made to allow data interpretation. We will show the results of those simplifications here and the experimental implementations later.

Consider an n-type substrate where, to first order, p can be neglected. Equation (7.5) becomes

$$n_T(t) = n_T(0)e^{-t/\tau} + \frac{e_p + c_n n}{e_n + c_n n + e_p} N_T(1 - e^{-t/\tau}) \qquad (7.7)$$

where now $\tau = 1/(e_n + c_n n + e_p)$. There are two cases of particular interest for the Schottky diode on an n-substrate shown in Fig. 7.2. The diode is at zero bias in Fig. 7.2(a). With n mobile electrons, capture dominates emission, and the steady-state G-R center concentration from Eq. (7.7) is $n_T \approx N_T$. When the diode is pulsed from zero to reverse bias as shown in Fig. 7.2(b), with most G-R centers initially occupied by electrons for $t \leq 0$, electrons are emitted from the G-R centers for $t > 0$. Emission dominates during this reverse-bias phase because the emitted electrons are swept out of the reverse-biased space-charge region very quickly, thereby reducing the chance of being recaptured. Typically the electron sweep-out or transit time is $t_t \approx W/v_n$. For $v_n \approx 10^7$ cm/s and W being a few microns, t_t is a few tens of picoseconds. This time is significantly less than typical capture times. However, near the edge of the scr the mobile electron concentration tails off into the scr from the quasi-neutral region even under reverse bias. This implies that the $c_n n$ term in Eq. (7.7) is not negligible in that part of the scr and electron emission competes with electron capture. With n being spatially inhomogeneous, τ is not constant, and the time dependence of $n(t)$ can be non-exponential.

During the initial emission period, when hole emission can be neglected because few G-R centers are occupied by holes, the time dependence of n_T becomes

$$n_T(t) = n_T(0) \exp\left(-\frac{t}{\tau_e}\right) \approx N_T \exp\left(-\frac{t}{\tau_e}\right) \qquad (7.8)$$

where $\tau_e = 1/e_n$. Following electron emission from G-R centers, holes remain and are subsequently emitted followed by electron emission, etc. The steady-state G-R center concentration n_T in the reverse-biased scr is

$$n_T = \frac{e_p}{e_n + e_p} N_T \qquad (7.9)$$

Some G-R centers will be in the n_T, and some will be in the p_T state. When the diode is pulsed from reverse bias to zero bias, electrons rush in to be captured by G-R centers in the p_T state. The time dependence of n_T during the capture period is

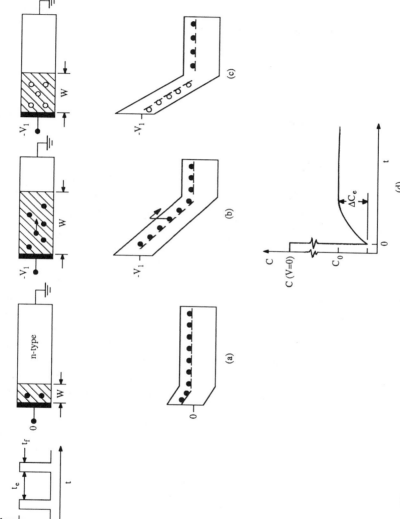

Fig. 7.2 A Schottky diode for (a) zero bias, (b) reverse bias at $t = 0$, (c) reverse bias as $t \to \infty$. The applied voltage waveform is shown in (a); (d) is the capacitance transient described in the text.

$$n_T(t) = N_T - [N_T - n_T(0)] \exp\left(-\frac{t}{\tau_c}\right) \tag{7.10}$$

where $\tau_c = 1/c_n n$ and $n_T(0)$ is the initial steady-state concentration given by Eq. (7.9).

Similar equations to those in this section also hold for interface trapped charge. The relevant electron and hole concentrations are those at the surface, the G-R centers are the interface traps, and the capture and emission coefficients are those of the interface traps. The concepts, however, are unchanged.

7.3 CAPACITANCE MEASUREMENTS

The equations in Section 7.2.2 describe the G-R centers in terms of their concentrations and their emission and capture coefficients. Of course the impurities are negatively charged, positively charged, or neutral. Capacitance measurements are well suited to determine charged impurities, and such measurements have found wide application. The capacitance of the Schottky diode of Fig. 7.2 is

$$C = A\sqrt{\frac{qK_s\varepsilon_0}{2}} \sqrt{\frac{N_{scr}}{(V_{bi} - V)}} = K\sqrt{\frac{N_{scr}}{(V_{bi} - V)}} \tag{7.11}$$

where N_{scr} is the ionized impurity concentration in the space-charge region. The ionized shallow-level donors in the scr are positively charged and $N_{scr} = N_D^+ - n_T^-$ for deep-level acceptor impurities that are negatively charged when occupied by electrons. When occupied by holes, the deep levels are neutral and $N_{scr} = N_D^+$. For shallow-level donors and deep-level donor impurities occupied by electrons, $N_{scr} = N_D^+$. $N_{scr} = N_D^+ + p_T^+$ for deep-level donors occupied by holes. N_{scr} *increases* in either case with time.

The time-dependent capacitance reflects the time dependence of $n_T(t)$ or $p_T(t)$. Two chief methods are utilized to determine deep-level impurities. In the first, the steady-state capacitance is measured at $t = 0$ and at $t = \infty$. In the second, the time-varying capacitance is monitored.

7.3.1 Steady-State Measurements

We saw in Chapter 2 that plots of $1/C^2$ versus V yield the doping concentration. It is possible to determine N_T from such plots. For shallow-level donors and deep-level acceptors, $1/C^2$ is given as

$$\frac{1}{C^2} = \frac{1}{K^2}\frac{V_{bi} - V}{N_D - n_T(t)} \tag{7.12}$$

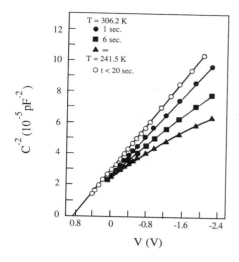

Fig. 7.3 C^{-2} versus V plots for a Au–GaAs Schottky barrier diode. The $T = 241.5$ K curve can be considered a $t = 0$ curve, and the others show the capacitance time dependence. Reprinted with permission after Senechal and Basinski.[12]

For the reverse-biased diode of Fig. 7.2, $n_T(t)$ is negatively charged when occupied by electrons. With time, as electrons are emitted and the G-R centers become neutral, $N_D - n_T(t)$ increases and $1/C^2$ decreases. In steady-state measurements the reverse-biased capacitance at $t = 0$ is compared with the reverse-biased capacitance as $t \to \infty$. Such a plot is shown in Fig. 7.3. If we define a slope $S(t) = -dV/d(1/C^2)$, then

$$S(\infty) - S(0) = K^2[n_T(0) - n_T(\infty)] \qquad (7.13)$$

For $n_T(0) \approx N_T$ and $n_T(\infty) \approx 0$, applicable for $e_n \gg e_p$, the difference of the two slopes gives the deep-level impurity concentration. This method was used during early C-V measurements.[8–12]

A slightly more detailed analysis takes account of those G-R centers whose energy levels lie below the Fermi level.[13] They do not emit and capture electrons in the same manner as those levels above the Fermi level, perturbing the charge distributions somewhat. But it is not a major effect.

7.3.2 Transient Measurements

It is obvious from Fig. 7.2 that the space-charge region width W changes as electrons are emitted from G-R centers. In transient measurements it is this time-varying W that is detected as a time-varying capacitance. From Eq. (7.11) we find

$$\boxed{C = A\sqrt{\frac{qK_s\varepsilon_0 N_D}{2(V_{bi} - V)}}\sqrt{1 - \frac{n_T(t)}{N_D}} = C_0\sqrt{1 - \frac{n_T(t)}{N_D}}} \qquad (7.14)$$

where C_0 is the capacitance without any deep-level impurities at reverse bias $-V$. It is of course possible to measure C and analyze the data as C^2 to avoid taking the square root. We address that method at the end of this section. However, for many cases it turns out that deep-level impurities form only a small fraction of the scr impurity concentration or $N_T \ll N_D$. In other words, one is looking for trace amounts of impurities. Using a first-order expansion of Eq. (7.14) gives

$$C = C_0\left(1 - \frac{n_T(t)}{2N_D}\right) \qquad (7.15)$$

It should be remembered that Eq. (7.15) is only valid for low impurity concentrations. But in most cases where N_T is to be determined, this equation is sufficiently accurate.

Emission—Majority Carriers   The most common measurements are emission measurements. The junction device is initially zero biased, allowing impurities to capture majority carriers [Fig. 7.2(a)]. The capacitance is the zero-biased value $C(V=0)$. Following a reverse bias pulse, majority carriers are emitted as a function of time [Fig. 7.2(b)]. Equation (7.8) is the appropriate equation. When substituted into Eq. (7.15), we find

$$C = C_0\left[1 - \left(\frac{n_T(0)}{2N_D}\right)\exp\left(-\frac{t}{\tau_e}\right)\right] \qquad (7.16)$$

Equation (7.16) is shown in Fig. 7.2(d) for $t>0$. The scr is widest and the capacitance is lowest immediately after the device is reverse biased. As majority carriers are emitted from the G-R centers [Fig. 7.2(b)], W decreases and C increases until steady state is attained [Fig. 7.2(c)]. The same time dependence of the capacitance is observed for deep-level donor impurities in n-type substrates. In that case the impurities are neutral, when initially occupied by electrons, and the scr impurity concentration at $t=0^+$ is N_D. As electrons are emitted, the centers become positively charged, and the final charge is $q[N_D + p_T(\infty)]$. Both charge and capacitance increase with time. The capacitance increases with time regardless whether the deep-level impurities are donors or acceptors. Using the same arguments, it is straightforward to convince yourself that this is also true for p-type substrates with either donor or acceptor G-R centers. *The capacitance increases with time for majority carrier emission.*

From the decay time constant of the C-t curve, one derives τ_e, and from the reverse-biased capacitance change, one obtains $n_T(0)$. Defining $\Delta C_e = C(t = \infty) - C(t = 0)$, we have

$$\Delta C_e = \frac{n_T(0)}{2N_D}C_0 \qquad (7.17)$$

Plotting the capacitance

$$C(\infty) - C(t) = \frac{n_T(0)}{2N_D} C_0 \exp\left(-\frac{t}{\tau_e}\right) \tag{7.18}$$

as $\ln[C(\infty) - C(t)]$ versus t, we find a slope of $-1/\tau_e$ and an intercept on the ln-axis of $\ln[n_T(0)C_0/2N_D]$. The emission time constant contains impurity parameters. To bring these out, we have to return to the capture and emission coefficients.

The capture and emission coefficients are related to each other through Eqs. (7.1) and (7.2). In equilibrium we invoke the *principle of detailed balance*, which states that under equilibrium conditions each fundamental process and its inverse must balance independent of any other process that may be occuring inside the material.[14-15] This requires fundamental process (a) in Fig. 7.1 to self-balance with its inverse process (b). Consequently $dn/dt = 0$ under *equilibrium conditions* and

$$e_{n0}n_{T0} = c_{n0}n_0 p_{T0} = c_{n0}n_0(N_T - n_{T0}) \tag{7.19}$$

where the subscript 0 stands for equilibrium. n_0 and n_{T0} are defined as[14]

$$n_0 = n_i \exp\left(\frac{E_F - E_i}{kT}\right) \tag{7.20a}$$

$$n_{T0} = \frac{N_T}{1 + \exp[(E_T - E_F)/kT]} \tag{7.20b}$$

Combining Eqs. (7.19) and (7.20) gives

$$e_{n0} = c_{n0}n_i \exp\left(\frac{E_T - E_i}{kT}\right) = c_{n0}n_1 \tag{7.21}$$

The derivation for holes is similar to Eq. (7.21). Then a crucial assumption is made: The emission and capture coefficients remain equal to their equilibrium values under non-equilibrium conditions. This gives

$$e_n = c_n n_1 \tag{7.22a}$$

$$e_p = c_p p_1 \tag{7.22b}$$

where

$$n_1 = n_i \exp\left(\frac{E_T - E_i}{kT}\right) \tag{7.23a}$$

$$p_1 = n_i \exp\left(-\frac{E_T - E_i}{kT}\right) \tag{7.23b}$$

The validity of the equilibrium assumption under non-equilibrium conditions is open to question. For small deviations from equilibrium, it may be assumed that the emission and capture coefficients do not deviate significantly from their equilibrium values.[16] Most certainly it is a poor approximation in the reverse-biased junction scr where high electric fields exist. But that is precisely where most capacitance transient measurements are made. Capture cross sections determined from emission measurements generally do not give true values, as discussed in Appendix 7.1. The equilibrium assumption is nevertheless an assumption that is commonly made, and any reported results are subject to its uncertainty.

We show the electric field effect in Fig. 7.4. An electron energy diagram at zero electric field is shown by the dashed lines. An energy $E_C - E_T$ is required for electron emission. The bands are slanted with an applied electric field, as shown by the solid lines, and the emission energy is reduced by ΔE. Poole-Frenkel emission over the lowered barrier is shown as (a).[17] Even less energy is required for phonon-assisted tunneling, shown as (b), in which the electron is excited by phonons for only part of the energy barrier and then tunnels through the remaining barrier. The electric field dependence of the emission coefficient for the gold acceptor level in silicon was found to be negligible for electric fields up to 10^4 V/cm, but for fields around 10^5 V/cm the emission coefficient increases by about a factor of two and continues to increase with higher fields.[18]

With $e_n = 1/\tau_e$ and $c_n = \sigma_n v_{th}$, we find the emission time constant to be

$$\tau_e = \frac{\exp[(E_i - E_T)/kT]}{\sigma_n v_{th} n_i} = \frac{\exp[(E_C - E_T)/kT]}{\sigma_n v_{th} N_C} \tag{7.24}$$

A similar expression for holes is

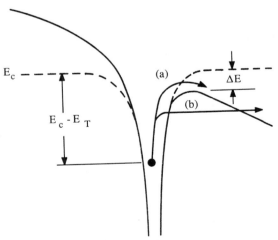

Fig. 7.4 Electron energy diagram in equilibrium (dashed lines) and in the presence of an electric field (solid lines) showing field-enhanced electron emission: (a) Poole-Frenkel emission, (b) phonon-assisted tunneling.

$$\tau_e = \frac{\exp[(E_T - E_i)/kT]}{\sigma_p v_{th} n_i} = \frac{\exp[(E_T - E_V)/kT]}{\sigma_p v_{th} N_V} \tag{7.25}$$

where the thermal velocities v_{th} are not identical for electrons and holes. Equation (7.24) shows the time constant τ_e to be dependent on the energy E_T and the capture cross section σ_n. It should be possible to extract both from a knowledge of τ_e. The electron thermal velocity is

$$v_{th} = \sqrt{3kT/m_n} \tag{7.26}$$

and the effective density of states in the conduction band is

$$N_C = 2\left(\frac{2\pi m_n kT}{h^2}\right)^{3/2} \tag{7.27}$$

allowing the emission time constant to be written as

$$\tau_e T^2 = \frac{\exp[(E_C - E_T)/kT]}{\gamma_n \sigma_n} \tag{7.28}$$

where $\gamma_n = (v_{th}/T^{1/2})(N_C/T^{3/2}) = 3.25 \times 10^{21}(m_n/m_0)$ cm^{-2} s^{-1} K^{-2}. m_n is the electron density-of-states effective mass.[19-20] The γ values for Si and GaAs[21-22] are given in Table 7.1.

Modified values $\gamma_n = 1.9 \times 10^{20}$ cm^{-2} s^{-1} K^{-2} and $\gamma_p = 1.8 \times 10^{21}$ cm^{-2} s^{-1} K^{-2} have been proposed, based on a critical evaluation of GaAs parameters.[23]

A plot of $\ln(\tau_e T^2)$ versus $1000/T$, shown in Fig. 7.5 for Au and Rh in Si, has a slope of $(E_C - E_T)/k$ and an intercept of $\ln[1/(\gamma_n \sigma_n)]$ at $1000/T = 0$. The energy levels are shown in Table 7.2. From the intercept we find σ_n. Although this method to determine the capture cross section is fairly common, the values so obtained should be viewed with caution. The cross sections are affected by the electric fields in the scr as well as by other effects discussed in Appendix 7.1.

TABLE 7.1 Coefficient γ for Si and GaAs

Semiconductor	$\gamma_{n,p}$ (cm^{-2} s^{-1} K^{-2})
n-Si	1.07×10^{21}
p-Si	1.78×10^{21}
n-GaAs	2.3×10^{20}
p-GaAs	1.7×10^{21}

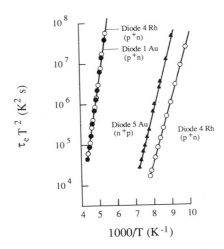

Fig. 7.5 $\text{Ln}(\tau_e T^2)$ versus $1000/T$ plots for Si diodes containing Au and Rh. Reprinted with permission after Pals.[34]

For the diodes in Fig. 7.5, we show in Table 7.2 the energy levels and the capture cross sections determined from the intercept of the $\ln(\tau_e T^2)$ versus $1000/T$ lines and by another method—the filling pulse method which is described later in the subsection "Capture—Majority Carriers." Note the large discrepancy between the two methods, with the intercept method giving values at least ten times larger. There are various reasons for this large discrepancy. Electric field enhanced emission tends to give larger cross sections. As discussed in more detail in Appendix 7.1, the term $(\gamma_n \sigma_n)$ contains possible degeneracy factors and entropy terms, rendering the extrapolated cross sections questionable.

The time constant τ_e can also be obtained by combining Eqs. (7.12), (7.13), and (7.8) as

$$S(\infty) - S(t) = K^2 n_T(t) = K^2 n_T(0) \exp\left(-\frac{t}{\tau_e}\right) \qquad (7.29)$$

and plotting $\ln[S(\infty) - S(t)]$ versus t. This was one of the earliest approaches.[11–12] However, the slope $-dV/d(1/C^2)$ is more complex to

TABLE 7.2   Energy Levels and Capture Cross Sections for the Diodes of Fig. 7.5

Diode	$E_C - E_T$ (eV)	$E_T - E_V$ (eV)	$\sigma_{n,p}$ (intercept) (cm^2)	$\sigma_{n,p}$ (filling pulse) (cm^2)
1	0.56		2.8×10^{-14}	1.3×10^{-16}
4	0.315		1.6×10^{-13}	3.6×10^{-15}
4	0.534		7.5×10^{-15}	4×10^{-15}
5		0.346	1.5×10^{-13}	1.6×10^{-15}

measure with automatic equipment than just C, and the method of Eq. (7.29) is rarely used today. However, Eq. (7.29) does not entail a small-signal expansion and is *not* subject to the limitation $N_T \ll N_D$.

Transient C-t data no longer follow a simple exponential time dependence when the emission rate is electric field dependent, when there are multiple exponentials due to several trapping levels with similar emission rates, and when the G-R center concentration is no longer negligibly small compared to the shallow-level dopant concentration. The analysis becomes more complicated for the last case, and we do not derive the relevant equations. This problem has been treated elsewhere.[24–28]

Emission—Minority Carriers   The considerations in the preceding section dealt with the capacitance response to majority carrier capture and emission when a Schottky diode is pulsed between zero and reverse bias. Similar results obtain when a *pn* junction is pulsed between zero and reverse bias. With the *pn* junction there is an additional option. Under forward bias minority carriers are injected over a distance of the minority carrier diffusion length from the space-charge region edges on both sides of the junction. Let us consider a p^+n junction and neglect the p^+ region in this discussion. During the forward-bias phase, holes are injected into the *n*-substrate and capture dominates over emission. The steady-state G-R center occupancy is from Eq, (7.6)

$$n_T = \frac{c_n n}{c_n n + c_p p} N_T \qquad (7.30)$$

which depends on both capture coefficients and both carrier concentrations. The occupancy is difficult to predict in general, but certainly the G-R centers are no longer solely occupied by electrons; a certain fraction is occupied by holes. Schottky diodes do not inject minority carriers efficiently, and *pn* junctions should be used for electrical minority carrier injection. It is, however, possible to inject minority carriers from high-barrier-height Schottky diodes with minority carrier storage at the inverted surface.[29] Iridium/*n*-Si devices with barrier heights of 0.9 V have been used for minority carrier injection.[30]

For the sake of our discussion here, let us assume that $c_p \gg c_n$ and $p \approx n$. Then most centers are occupied by holes at forward bias, and for the deep-level acceptor impurities we have considered so far, the centers are neutral, with $n_T \approx 0$ and $N_{scr} \approx N_D$ at $t = 0$ when the junction is pulsed from forward to reverse bias. As holes are emitted from the *G-R* centers, their charge changes from neutral to negative, and $N_{scr} \approx (N_D - n_T)$ for $t \rightarrow \infty$. The total ionized scr concentration *decreases*, the scr width increases, and the capacitance *decreases* with time. This is shown in Fig. 7.6 and is exactly opposite to majority carrier behavior. For simplicity, we assume in Fig. 7.6 all deep-level impurities to be filled with electrons (majority carrier emission) or

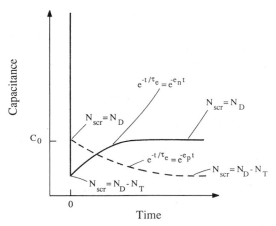

Fig. 7.6 The capacitance-time transient following majority carrier emission (solid curve) and minority carrier emission (dashed curve).

holes (minority carrier emission) at $t = 0$. The capacitance transient is still described by an expression of the type in Eq. (7.16). But the emission time constant is now $\tau_e = 1/e_p$.

G-R centers in the upper half of the band gap are generally detected with majority carrier pulses; those in the lower half of the band gap are observed with minority carrier pulses for n-type substrates. Centers with energies around the middle of the band gap can respond to either majority or minority carrier excitation. Minority carriers can also be created optically as discussed later on.

Capture—Majority Carriers Consider the Schottky diode of Fig. 7.2(c) which has been reverse biased sufficiently long that all majority carriers have been emitted and the G-R centers are in the p_T state. When the diode is pulsed from reverse bias [7.2(c)] to zero bias [7.2(a)], electrons rush into the scr to be captured by unoccupied G-R centers. The concentration of G-R centers able to capture majority carriers, for negligible emission, is given by

$$n_T(t_f) = N_T - [N_T - n_T(0)] \exp\left(-\frac{t_f}{\tau_c}\right) \qquad (7.31)$$

where t_f is the capture or "filling" time. If there is sufficient time (i.e., $t_f \gg \tau_c$), essentially all G-R centers capture electrons, and $n_T(t_f \rightarrow \infty) \approx N_T$. If the time available for electron capture is short, only a small fraction of the G-R centers will be occupied by electrons when the diode returns to reverse bias. In the limit of very short times (i.e., $t_f \ll \tau_c$), very few electrons are captured and $n_T(t_f \rightarrow 0) \approx 0$.

When the device is pulsed to reverse bias, $n_T(0)$ in Eq. (7.16) is given by

Eq. (7.31); that is, the initial concentration during the emission phase is equal to the final concentration of the capture phase. The reverse bias capacitance at $t = 0$ then depends on the filling pulse width, shown by substituting Eq. (7.31) into (7.16) to give

$$C = C_0 \left\{ 1 - \frac{\{N_T - [N_T - n_T(0)] \exp(-t_f/\tau_c)\}}{2N_D} \exp\left(-\frac{t - t_f}{\tau_e}\right) \right\} \tag{7.32}$$

Equation (7.32) is shown in Fig. 7.7(a).

The capture time τ_c can be determined by varying t_f, shown by the voltage waveform in Fig. 7.2. The capture time is much shorter than the emission time. We show the C-t curves during emission as a function of t_f in Fig. 7.7(b). The capitance at $t = t_f^+$ on these curves is dependent on the capture time and is given by

$$C = C_0 \left[1 - \frac{\{N_T - [N_T - n_T(0)] \exp(-t_f/\tau_c)\}}{2N_D} \right] \tag{7.33}$$

Equation (7.33) can be written as

$$\Delta C_c = C(t_f) - C(t_f = \infty) = \frac{[N_T - n_T(0)]}{2N_D} C_0 \exp\left(-\frac{t_f}{\tau_c}\right) \tag{7.34}$$

with ΔC_c shown on Fig. 7.7(b). t_f can be extracted from Eq. (7.34) by writing

$$\ln(\Delta C_c) = \ln\left\{ \frac{[N_T - n_T(0)]C_0}{2N_D} \right\} - \frac{t_f}{\tau_c} \tag{7.35}$$

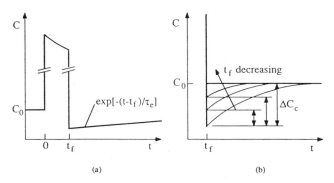

(a)   (b)

Fig. 7.7 (a) C-t response showing the capture and the initial part of the emission process; (b) the emission C-t response as function of capture pulse width.

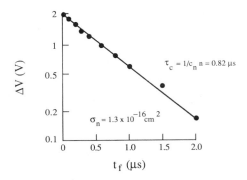

Fig. 7.8 Ln(ΔV) versus t_f plot for diode 1 of Fig. 7.5. Reprinted with permission after Pals.[34]

A plot of $\ln(\Delta C_c)$ versus t_f has a slope of $-1/\tau_c = -\sigma_n v_{th} n$ and an intercept on the $\ln(\Delta C_c)$ axis of $\ln\{[N_T - n_T(0)]C_0/2N_D\}$. Such a plot is obtained by varying the capture pulse width during the capacitance transient measurement. In this manner the capture cross section is determined from a capture, not an emission, process. The measurement is more difficult to implement because capture times are much shorter than emission times and the instrumentation is more demanding. Modifications to capacitance meters to accommodate the narrow pulses necessary for these measurements are given in Henry et al.[31] and Wang and Sah.[32] Sometimes one obtains nonlinear $\ln(\Delta C_c)$ versus t_f plots due to slow capture from carrier tails extending into the scr, and models to derive σ_n from these curves are frequently too imprecise or involve complicated curve fitting routines. Nevertheless, those kinds of analyses are required for nonlinear experimental data.[33]

A slight variation on this method is not to measure the capacitance as a function of time, but instead to keep the capacitance constant during the measurement through a feedback circuit and to measure the voltage required to keep the capacitance constant.[34-35] The data analysis is similar and a plot of the voltage change ΔV required for constant capacitance in Fig. 7.8 shows the expected semilogarithmic behavior.

Equation (7.31) predicts the capture time as $\tau_c = (\sigma_n v_{th} n)^{-1}$. The actual trap filling process is more complicated because not all traps are emptied during the emission process. Those G-R centers whose energy levels lie below the Fermi level will tend to remain occupied by electrons during the emission transient.[36] They therefore do not capture electrons during the filling pulse. This effect should be taken into account during the data analysis.

Capture—Minority Carriers   There are several methods to determine the capture properties of minority carriers. One method is very similar to that of the previous section, except that during the filling pulse the diode voltage is not zero but the diode is forward biased. Various pulse widths are used to determine the capture properties.[31,37-38] If we neglect carrier emission, we find the capture time constant during the filling pulse to be given by Eq. (7.5) as

$$\tau_c = \frac{1}{c_n n + c_p p} \qquad (7.36)$$

and the G-R center occupancy will be that of Eq. (7.30); it depends not only on n and p but also on c_n and c_p. The injected minority carrier concentration is varied by changing the injection level, and both c_n and c_p can be determined.[31] The narrow pulse widths (nanoseconds or lower) necessary to fill the centers partially are a decided disadvantage. A more fundamental limit is the turn-on time of junction diodes. pn junctions do not turn on instantly following a sharp pulse. The minority carrier concentration builds up in a time related to the minority carrier lifetime. For the narrow pulses required for the capture measurements, it is very likely that the minority carrier concentration does not reach its steady-state value.

In an alternate method the G-R centers are populated with minority carriers not with constant-amplitude, varying-width bias pulses, but with constant-width, varying-amplitude pulses.[32] The diode is forward biased with a long pulse, around 1 ms, and then reverse biased. The reverse-bias capacitance transient is observed. The minority carrier concentration is related to the injection current with the details given in Wang and Sah.[32] One must pay attention that recombination of minority carriers does not play an important role during the measurement.

It is also possible to inject minority carriers optically in pn junctions or Schottky diodes. We mention the method only briefly here and discuss it in more detail in Section 7.6.3. Consider a reverse-biased pn junction or Schottky barrier diode. A light pulse with photon energy $h\nu > E_g$ is flashed on the device, creating electron–hole pairs in the scr and in the quasi-neutral region. The minority carriers from the quasi-neutral region diffuse to the reverse-biased space-charge region to be captured by G-R centers. When the light is turned off, those captured minority carriers are emitted and detected as C-t or I-t transients. From the transient one determines E_T, σ_p, and N_T.[39–40]

7.4 CURRENT MEASUREMENTS

The carriers emitted from deep-level impurities can be detected as a capacitance, a charge, or a current.[5,41–44] We saw earlier that the capacitance, consisting of an initial capacitance followed by a transient, is given by Eq. (7.16). As the temperature changes, only the time constant changes; the initial capacitance step remains constant. That is not true for current measurements. The integral of the I-t curve represents the total charge emitted by the G-R centers. For high temperatures, the time constant is short, but the initial current is high. For low temperatures, the time constant increases, and the current decreases, but the area under the *I-t* curve remains constant. This makes current measurements difficult at low temperatures. By combining *C-t* measurements at the lower temperatures with *I-t* measurements at the higher temperatures, it is possible to obtain time constant data over ten orders of magnitude.[41]

Current measurements are more complicated because the current consists of emission current I_e, displacement current as the scr width changes I_d, and junction leakage current I_l. The electron emission current is given by

$$I_e = qA \int_0^W \left(\frac{dn}{dt} \right) dx \qquad (7.37)$$

The displacement current is[5]

$$I_d = qA \int_0^W \left(\frac{dn_T}{dt} \right) \left(\frac{x}{W} \right) dx \qquad (7.38)$$

The lower limit of the integral in Eqs. (7.37) and (7.38) should have been the zero-biased scr width. However, for simplicity we have set the lower limit to zero. With $dn/dt \approx e_n n_T$ [eq. (7.1)], $dn_T/dt \approx -e_n n_T$ [Eq. (7.4)], and electron emission dominating for the reverse-biased diode of Fig. 7.2, we find

$$\boxed{I = \frac{qAW(t)e_n n_T(t)}{2} + I_l = \frac{qA}{2\tau_e} \frac{W_0}{\sqrt{1 - n_T(t)/N_D}} n_T(t) + I_l} \qquad (7.39)$$

using

$$W(t) = \sqrt{\frac{2K_s \varepsilon_0 (V_{bi} - V)}{q[N_D - n_T(t)]}} = \sqrt{\frac{2K_s \varepsilon_0 (V_{bi} - V)}{qN_D[1 - n_T(t)/N_D]}} = \frac{W_0}{\sqrt{1 - n_T(t)/N_D}}$$

$$(7.40)$$

For $n_T \ll N_D$, and using Eq. (7.8), the current becomes

$$I = \frac{qAW_0}{2\tau_e} \frac{n_T(0) \exp(-t/\tau_e)}{[1 - (n_T(0)/2N_D) \exp(-t/\tau_e)]} + I_l \qquad (7.41)$$

The interpretation of the current measurements is more complex than capacitance measurements because the *I-t* curve does not possess a simple dependence on τ_e. τ_e appears in the numerator and the denominator. If the second term in the denominator is small compared to unity for $n_T(0) \ll N_D$ and may be neglected, the current exhibits the conventional exponential time dependence. The addition of the leakage current generally presents no problems since it is constant unless it is sufficiently high to mask the current transient. The instrumentation must be carefully designed to be able to handle the large current transients during the pulsing. The amplifier should be nonsaturable, or the large circuit transients must be eliminated from the current transient of interest. A circuit with these properties is described in Wang and Sah.[32]

Current transients do not allow a distinction between *majority* and *minority* carrier emission. Another feature of current DLTS is a shift of the peak to higher temperatures relative to capacitance DLTS for the same rate window because the current is inversely proportional to the emission time constant [see Eq. (7.41)], while the capacitance is not. This property causes the current to increase very rapidly with temperature, effectively skewing the line shape toward higher temperatures.

Current measurements are preferred when it is difficult to make capacitance measurements. For example, the capacitance of small-geometry MOSFET's or MESFET's is very small and difficult to measure. MOSFET's and MESFET's with $1\,\mu m \times 10\,\mu m$ gates have capacitances in the fF range, and the capacitance change is even smaller. In that case it is possible to detect the presence of deep-level impurities by pulsing the gate voltage and monitoring the drain–source current as a function of time. Consider a MOSFET biased to some drain voltage and pulsed from accumulation to inversion, that is, from "off" to "on." Traps capture majority carriers during the "off" state. A space-charge region is created when the device is turned "on," and drain current flows through the device. As carriers are emitted from traps, the scr width and the threshold voltage change. The drain current with a time-varying threshold voltage due to emission of carriers from traps is given by[45–46]

$$
I_D = \frac{Z\mu_{eff}C_{ox}}{L}\left(V_{GS} - V_T(t) - \frac{V_{DS}}{2}\right)V_{DS}
$$
$$
= \frac{Z\mu_{eff}C_{ox}}{L}\left\{V_{GS} - V_{FB} - 2\phi_F - \frac{Q_B}{C_{ox}}\left[1 - \frac{N_T}{2N_A}\exp\left(-\frac{t}{\tau_e}\right)\right] - \frac{V_{DS}}{2}\right\}V_{DS}
$$
$$(7.42)$$

The emission time constant is determined from the time-varying drain current, which in turn depends on carrier emission from traps.

Current measurements work best in devices in which the channel can be totally depleted. In a MESFET, for example, the gate is pulsed from zero to reverse bias, creating a deep space-charge region. Electron or hole emission from traps changes the scr width and is detected as a drain current change that can be detected with the gate voltage held constant, or the gate voltage change can be detected by holding the current constant with a feedback circuit.[47] Examples of MESFET drain current and capacitance measurements are shown in Fig. 7.9.[48] For these measurements it was necessary to

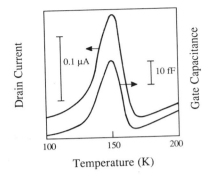

Fig. 7.9 Comparison of current and capacitance transient scans of a 100 μm × 150 μm gate MESFET. Reprinted with permission after Hawkins and Peaker.[48]

use gate areas of 100 μm × 150 μm to obtain sufficiently large capacitances to be measureable. However, current measurements for 1-μm-long gates present no problem, but capacitance measurements do.

Drain current measurements are relatively simple to implement, but they are more difficult to interpret than capacitance when absolute values of trap concentrations must be extracted because the measured current is a change in drain current brought about by a changing scr width. Interpretation of the data requires a knowledge of the mobility.[49] This difficulty is circumvented by holding the drain current constant, changing the gate voltage, and converting gate voltage changes to current changes through the device transconductance.[48] An interesting experiment on an Si MOSFET with a very small gate area (0.5 μm × 0.75 μm) showed drain current fluctuations that were interpreted by emission and capture kinetics from *individual* interface traps.[50]

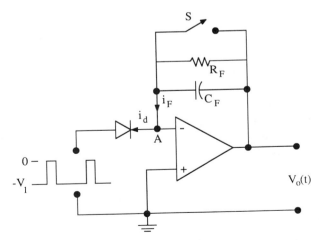

Fig. 7.10 Circuit for charge transient measurements.

7.5 CHARGE MEASUREMENTS

Carriers emitted from deep-level impurities can be detected directly as a charge with the measurement circuit shown in Fig. 7.10. Switch S is closed to discharge the feedback capacitor C_F. At $t = 0$ the diode is reverse biased, S is opened, and from Eq. (7.41), with the second term in the denominator neglected, the current through the diode for $t \geq 0$ is

$$I = \frac{qAW_0}{2\tau_e} n_T(0) \exp\left(-\frac{t}{\tau_e}\right) + I_1 \qquad (7.43)$$

With the input current into the op-amp essentially zero, the diode current must flow through the $R_F C_F$ feedback circuit. The output voltage is given by

$$V_0 = \frac{qAW_0 R_F n_T(0)}{2(t_F - \tau_e)} \left[\exp\left(-\frac{t}{t_F}\right) - \exp\left(-\frac{t}{\tau_e}\right) \right] + I_1 R_F \left[1 - \exp\left(-\frac{t}{t_F}\right) \right]$$

$$(7.44)$$

where $t_F = R_F C_F$. Choosing the feedback network such that $t_F \gg \tau_e$ reduces Eq. (7.44) to

$$V_0 \approx \frac{qAW_0 n_T(0)}{2C_F} \left[1 - \exp\left(-\frac{t}{\tau_e}\right) \right] + \frac{I_1 t}{C_F} \qquad (7.45)$$

Charge transient measurements have been implemented with the relatively simple circuit shown in Fig. 7.10.[51] The integrator replaces the high speed capacitance meter in C-t measurements or the high gain current amplifier in I-t measurements. The output voltage depends only on the total charge released during the measurement and is independent of τ_e. Charge measurements have also been used for MOS capacitor characterization.[52]

7.6 DEEP-LEVEL TRANSIENT SPECTROSCOPY

7.6.1 Conventional DLTS

Many of the C-t and I-t measurements and methods were developed by Sah and his students. They studied numerous impurities in Si and a summary of that work is given in Sah et al.[5] and Sah.[41] Although the basic techniques and theories were developed during that time, the implementation was rather time-consuming and tedious because most of the measurements were single-shot measurements. The power of emission and capture transient analysis was only fully realized when automated data acquisition techniques were adopted. The first of these was Lang's dual-gated integrator or double boxcar approach named *deep-level transient spectroscopy* (DLTS).[53–54]

Lang introduced the *rate window concept* to deep level characterization. If the *C-t* curve from a transient capacitance experiment is processed so that a selected decay rate produces a maximum output, then a signal whose decay time changes monotonically with time reaches a peak when the rate passes through the rate window of a boxcar averager or the frequency of a lock-in amplifier. When observing a repetitive *C-t* transient through such a rate window while varying the decay time constant by varying the sample temperature, a peak appears in the temperature versus output plot. Such a plot is named a *DLTS spectrum*.[55–56] The technique, being merely a method to extract a maximum in a decaying waveform, applies to capacitance, current, and charge transients.

We explain DLTS using capacitance transients. Assume that the *C-t* transient follows the exponential time dependence

$$C = C_0\left[1 - \frac{n_T(0)}{2N_D} \exp\left(-\frac{t}{\tau_e}\right)\right] \qquad (7.46)$$

with τ_e depending on temperature as

$$\tau_e = \frac{\exp[(E_C - E_T)/kT]}{\gamma_n \sigma_n T^2} \qquad (7.47)$$

The time constant τ_e increases with decreasing temperature, and a series of *C-t* curves are shown in Fig. 7.11 as a function of temperature.

The capacitance decay waveform is corrupted with noise, and the heart of DLTS is the extraction of the signal from the noise in some automated manner. The technique is a correlation technique, which is a signal-processing method where the input signal is multiplied by a reference signal, the weighting function $w(t)$, and the product is filtered (averaged) by a linear filter. The properties of the correlator depend strongly on the weighting function and on the filtering method. The filter can be an integrator or a low-pass filter. The correlator output is

$$\boxed{\delta C = \frac{1}{T} \int_0^T f(t)w(t)\, dt = \frac{C_0}{T} \int_0^T \left[1 - \frac{n_T(0)}{2N_D} \exp\left(-\frac{t}{\tau_e}\right)\right] w(t)\, dt}$$

$$(7.48)$$

where we use Eq. (7.46) for $f(t)$.

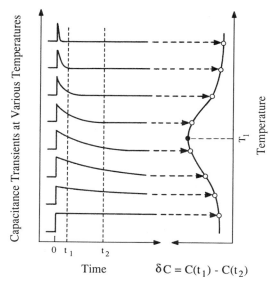

Fig. 7.11 Implementation of a rate window by a double boxcar integrator. The output is the average difference of the capacitance amplitudes at the sampling times t_1 and t_2. Required with permission after Miller et al.[55]

Boxcar DLTS   Suppose that the C-t waveforms in Fig. 7.11 are sampled or gated at times $t = t_1$ and $t = t_2$ and that the capacitance at t_2, $C(t_2)$, is subtracted from the capacitance at t_1, $C(t_1)$. Such a difference signal is a standard output feature of a double boxcar instrument. There is no difference between the capacitance at the two sampling times for very slow or for very fast transients, corresponding to low and high temperatures. A difference signal is generated when the time constant is on the order of the gate separation $t_2 - t_1$, and the capacitance difference passes through a maximum as a function of temperature. This is the DLTS peak. The capacitance difference, or DLTS signal, is obtained from Eq. (7.48), using the weighting function $w(t) = \delta(t - t_1) - \delta(t - t_2)$, as

$$\delta C = C(t_1) - C(t_2) = \frac{C_0 n_T(0)}{2N_D}\left[\exp\left(-\frac{t_2}{\tau_e}\right) - \exp\left(-\frac{t_1}{\tau_e}\right)\right] \quad (7.49)$$

where in Eq. (7.48), $T = t_1 - t_2$.

In Fig. 7.11 δC exhibits a maximum δC_{max} at temperature T_1. Differentiating Eq. (7.49) with respect to τ_e and setting the result equal to zero gives $\tau_{e,max}$ at δC_{max} as

$$\tau_{e,max} = \frac{t_2 - t_1}{\ln(t_2/t_1)} \quad (7.50)$$

Equation (7.50) is independent of the magnitude of the capacitance, and the baseline of the signal need not be known. By generating a series of C-t curves at different temperatures for a given gate setting t_1 and t_2, one value of τ_e corresponding to a particular temperature is generated, giving one datum point on a $\ln(\tau_e T^2)$ versus $1/T$ plot. The measurement sequence is then repeated for another t_1 and t_2 gate setting to generate another point. In this manner a series of points are obtained to generate an entire Arrhenius plot. Typically 5 to 10 points are required necessitating 5 to 10 temperature sweeps. The sampling times can be varied by (1) t_1 fixed, vary t_2; (2) t_2 fixed, vary t_1; and (3) t_2/t_1 fixed, vary t_1 and t_2. Method (3) is preferred because the peaks shift with temperature without much change of curve shape. This makes peak location easier. Additionally $\ln(t_2/t_1)$ remains constant. For methods (1) and (2) the peaks change both in size and in shape. We show on Fig. 7.12 a series of typical δC versus T DLTS spectra for varying t_1 and t_2 but constant t_2/t_1 ratio. Alternatively, one can vary $t_2 - t_1$ at a constant temperature with t_2/t_1 constant. Then one would change the temperature and repeat to generate an Arrhenius plot from a single temperature scan. This technique is usually implemented with computer-controlled equipment.

It has been found experimentally that the sampling or gate width should be relatively wide because the signal/noise ratio is proportional to the square root of the gate width.[56] Equation (7.50) needs to be slightly modified by changing t_1 to $(t_1 + \Delta t)$ and t_2 to $(t_2 + \Delta t)$, where Δt is the gate width.[57]

The DLTS signal does not give the capacitance step ΔC_e of Fig. 7.2, and the impurity concentration cannot be determined from the DLTS signal using Eq. (7.17). The impurity concentration, derived from the maximum capacitance δC_{max} of the $\delta C - T$ curves, is given by

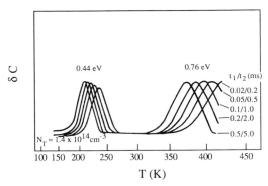

Fig. 7.12 Experimental DLTS spectra for hole traps in n-GaAs. The five spectra correspond to the five rate windows shown by the values of t_1 and t_2. The activation energies and trap concentration are shown on the figure. Reprinted with permission after Lang.[53]

$$N_T = \frac{\delta C_{max}}{C_0} \frac{2N_D \exp\{[r/(r-1)] \ln r\}}{(1-r)} = \frac{\delta C_{max}}{C_0} \frac{2N_D r^{r/(r-1)}}{(1-r)}$$

(7.51)

where $r = t_2/t_1$. Equation (7.51) is derived from Eqs. (7.49) and (7.50) with $\delta C_{max} = \delta C$, under the assumption that $n_T(0) = N_T$. For $r = 2$, a common ratio, we find $N_T = -8N_D\delta C_{max}/C_0$, and for $r = 10$, $N_T = -2.87N_D\delta C_{max}/C_0$. The minus sign accounts for the fact that $\delta C < 0$ for majority carrier traps.

Capacitance measurements are usually made with a capacitance bridge or a capacitance meter. Well-maintained DLTS systems can detect $\delta C_{max}/C_0 \approx 10^{-5}$ to 10^{-4}, allowing trap concentrations on the order of $(10^{-5}$ to $10^{-4})N_D$ to be determined. For substrate doping concentrations of 10^{15} cm^{-3}, one can determine trap concentrations to around 10^{11} cm^{-3}. High-sensitivity bridges allow measurements as low as $\delta C_{max}/C_0 \approx 10^{-6}$.[58] Capacitance meters typically have response times of 1 to 10 ms and should be modified to allow faster transients to be measured. The manufacturers of both the Boonton 71/72 and the Princeton Applied Research 410 capacitance meters provide these modifications. In addition to the inherent time response of the meter, difficulties arise from overloads during device pulsing. Overload recovery delays are avoided by installing a fast relay which grounds the input of the amplifier during the pulse, deactivating the internal overload detection circuitry.[59]

Several refinements of the basic boxcar DLTS technique have been implemented. In the *Double-Correlation DLTS* (D-DLTS) method, pulses of two different amplitudes are used instead of the one-amplitude pulse of the basic technique. However, D-DLTS retains the conventional DLTS rate window concepts as shown in Fig. 7.13.[60] The weighting function gives the signal

$$[C'(t_1) - C(t_1)] - [C'(t_2) - C(t_2)] = \Delta C(t_1) - \Delta C(t_2)$$

(7.52)

In the first correlation the transient capacitances after the two pulses are related to form the differences $\Delta C(t_1)$ and $\Delta C(t_2)$ at corresponding delay times after each pulse shown in Fig. 7.13. In a second step the correlation $[\Delta C(t_1) - \Delta C(t_2)]$ is performed as in conventional DLTS to resolve the time constant spectrum during the temperature scan. The measurement requires either a four-channel boxcar integrator or an external modification to a two-channel boxcar integrator.[61]

This added complexity sets an observation window within the space-charge region, allowing the impurities within this window to be detected. By setting the window well within the scr, away from the quasi-neutral region scr edge, all traps are well above the Fermi level, and the capacitance transient is determined by emission only. Traps near the Fermi level are excluded from the measurement. Furthermore all traps within the window experience approximately the same electric field. Trap concentration profiles

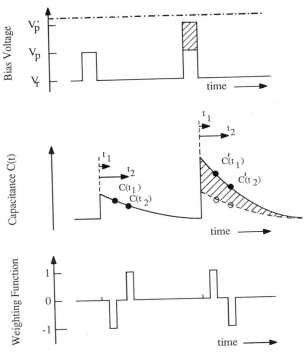

Fig. 7.13 Bias pulses, capacitance transients, and weighting functions for double correlation DLTS. Reprinted with permission after Lefèvre and Schulz.[60]

are obtained by varying the observation window or by changing the pulse amplitudes or the dc reverse bias.

Constant Capacitance DLTS A variation on the DLTS theme is *Constant Capacitance DLTS* (CC-DLTS) in which the capacitance is held constant during the carrier emission measurement by dynamically varying the applied voltage during the transient response through a feedback path.[34,62–63] The feedback method was pioneered by Miller who applied it originally to carrier concentration characterization as discussed in Chapter 2.[64] Just as the transient capacitance curve contains the trap information in the constant voltage method, so the time-varying voltage contains the trap information in the constant capacitance method. The capacitance transient expression in Eq. (7.15) is an approximation valid for $N_T \ll N_D$. For $N_T > 0.1 N_D$ large changes occur in W, and the C-t signal becomes non-exponential. From Eq. (7.14), which does not have this limitation, we find with no approximations

$$V = -\frac{q K_s \varepsilon_0 A}{2 C^2} \left[N_D - n_T(0) \exp\left(-\frac{t}{\tau_e}\right) \right] + V_{bi} \qquad (7.53)$$

This constant capacitance DLTS expression is valid for arbitrary N_T because the scr width is held constant and the resulting voltage change is directly proportional to the change in scr charge.

Equation (7.53) shows the V-t response to be exponential in time. Sometimes a non-exponential portion to the V-t curve is observed especially near $t = 0$. This can be caused by carrier capture even during the emission phase of the measurement. The majority carrier concentration does not drop abruptly to zero at the scr edge but tails into the scr, and electron emission competes with electron capture in that tail region. At the edge of the scr electron capture dominates, and most of the traps remain filled with electrons. This can give rise to a non-exponential V-t curve.[65]

One of the limitations of CC-DLTS is the slower circuit response due to the feedback circuits. An early implementation using a Boonton 72B capacitance meter was limited to transients with time constants on the order of a second.[66] The response time was reduced to about 10 ms for the same meter by using double feedback amplifiers.[67] The response time was further reduced to about 1 ms with feedback and sample-and-hold circuits.[68] Feedback circuitry generally degrades the sensitivity of CC-DLTS compared to constant voltage DLTS (CV-DLTS).

CC-DLTS is well suited for G-R center concentration depth profiling.[69] It has also been used for interface trapped charge measurements due to its high energy resolution, and it permits more accurate DLTS measurements of defect profiles for high G-R center concentrations. Further refinements are possible by combining D-DLTS with CC-DLTS.[70]

Lock-in Amplifier DLTS   The lock-in amplifier DLTS approach is attractive because lock-in amplifiers are more standard lab instruments than boxcar integrators and they are also cheaper.[71] Lock-in amplifier DLTS has also been shown to have a better signal/noise ratio than the boxcar DLTS.[72] Lock-in amplifiers use the square wave weighting function shown in Fig. 7.14(b) whose period is set by the frequency of the lock-in amplifier. A DLTS peak is observed when this frequency bears the proper relationship to the emission time constant. A lock-in amplifier can be thought of as a one-component Fourier analyzer to analyze a repetitive signal. The weighting function resembles that of a boxcar integrator but is wider. That increases the signal/noise ratio but also poses an overload problem.

Most DLTS capacitance signals are measured with commercial capacitance meters. The junction capacitance is very large during the forward bias phase and tends to overload the relatively slow (response time ~1 ms unless modified) capacitance meter. A lock-in amplifier is very sensitive to the meter transient and overloads easily since its square wave weighting function has unit magnitude at all times. The boxcar does not have this problem since the first sampling window is delayed past the transient as shown in Fig. 7.14. The lock-in's sensitivity to overloads can be reduced by preceding the weighting function by a narrow-band filter. This leads to an approximate

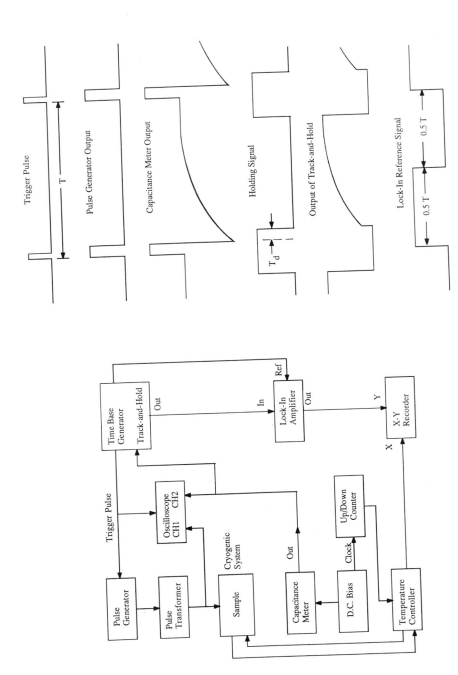

Fig. 7.14 (a) Block diagram of a lock-in amplifier DLTS system, (b) the various signals of the block diagram. Reprinted with permission after Rohatgi et al.[75]

sinusoidal weighting function. A better solution is to gate off the first 1 to 2 ms of the capacitance meter output, eliminating the overloading problems.[57,73–75] The analysis of the lock-in amplifier signal must include this gate-off time. The gate-off time also affects the base line which may become nonzero after the signal is suppressed part of the time.[76] In addition the phase setting affects the signal.[77–78] Details of three basic modes of lock-in DLTS operation and the relevant precautions to observe are discussed in Day et al.[57]. Choosing a gate-off time that is always the same fraction of the repetition rate avoids problems of erroneous DLTS peaks.[79]

A block diagram of a lock-in amplifier-based DLTS system is shown in Fig. 7.14(a). The track-and-hold circuit serves as the gate-off function to eliminate the large, but uninformative, portion of the capacitance signal during the capturing period. The DLTS system is synchronized by the three timing signals from the time-base generator: the reference signal for the lock-in amplifier, the hold signal for the track-and-hold circuit, and the trigger signal for the pulse generator. Figure 7.14(b) shows the signals at the various points. The holding interval given by $2T_d$ can be shifted in both directions and can also be expanded by adjusting the time base generator. T is the pulse period. A minimum delay interval, $T_d = 0.1T$, is sufficient to eliminate the capacitance transient.

For the weighting function $w(t) = 0$ for $0 \le t < T_d$, $w(t) = 1$ for $T_d < T/2$, $w(t) = -1$ for $T/2 < t < (T - T_d)$, and $w(t) = 0$ for $(T - T_d) < t < T$, the output from the lock-in amplifier is given by[72–73]

$$\delta C = -GC_0 \frac{n_T(0)}{N_D} \frac{\tau_e}{T} \exp\left(-\frac{T_d}{\tau_e}\right)\left[1 - \exp\left(-\frac{T - 2T_d}{2\tau_e}\right)\right]^2 \quad (7.54)$$

where G is the lock-in amplifier and capacitance meter gain, T is the pulse period, and the delay time T_d is the interval between the end of the bias pulse and the end of the holding interval, shown in Fig. 7.14(b). Equation (7.54) exhibits a maximum, similar to that of Eq. (7.49). Differentiating Eq. (7.54) with respect to τ_e and setting the result equal to zero allows $\tau_{e,max}$ to be determined from the transcendental equation

$$\left(1 + \frac{T_d}{\tau_{e,max}}\right) = \left(1 + \frac{T - T_d}{\tau_{e,max}}\right)\exp\left(-\frac{T - 2T_d}{2\tau_{e,max}}\right) \quad (7.55)$$

$\tau_{e,max} = 0.44T$ for a typical delay time of $T_d = 0.1T$. A $\ln(\tau_e T^2)$ versus $1/T$ plot is generated as described in the previous section once pairs of τ_e and T are known (here T is temperature). The trap concentration, derived from Eqs. (7.54) and (7.55) for $\delta C = \delta C_{max}$ under the assumption that $n_T(0) = N_T$ and $T_d = 0.1T$, is given by

$$N_T = -8 \frac{\delta C_{max}}{C_0} \frac{N_D}{G} \quad (7.56)$$

Instead of holding the lock-in frequency constant and varying the sample temperature, it is also possible to keep the temperature constant and vary the frequency to obtain the same trap information.[80]

Correlation DLTS   Correlation DLTS is based on optimum filter theory, which states that the optimum weighting function of an unknown signal corrupted by white noise has the form of the noise-free signal itself. This can be implemented in DLTS by mutiplying the exponential capacitance or current waveforms by a repetitive decaying exponential generated with an RC function generator and subsequently integrating the product.[72]

Correlation DLTS has a higher signal/noise ratio than either boxcar or lock-in DLTS.[81] Since the small capacitance transient rides on a dc

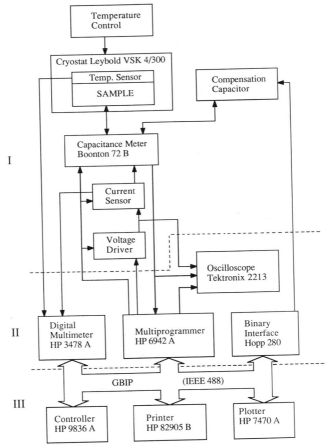

Fig. 7.15 Block diagram of an automated DLTS system. Reprinted with permission after Hölzlein et al.[90]

background, it is not sufficient to use a simple exponential because the weighting function[82] and baseline restoration are required.[83] The method has not found much application and the majority of DLTS measurements are made by either the boxcar or the lock-in approach, but correlation DLTS has been used to study impurities in high-purity germanium.[84]

Computer DLTS Computer DLTS refers to DLTS systems in which the capacitance waveform is digitized and stored electronically for further data management.[85–90] Only one temperature sweep of the sample is required since the entire *C-t* curve is obtained at each of a number of different temperatures. A block diagram is shown in Fig. 7.15. The entire *C-t* curve is acquired at typically several hundred data points per waveform. From a knowledge of the entire waveform, it is readily established whether the signal is exponential; this is not possible with the boxcar or lock-in methods since those methods only give maxima at selected temperatures but lose the waveform itself. Various signal processing functions can be performed on the *C-t* data: fast Fourier transforms, the method of moments to analyze simple and multiple exponential decays,[91–95] spectroscopic line fitting,[96] the covariance method of linear predictive modeling,[97] linear regression,[98] and an algorithm that allows the separation of closely spaced peaks.[99]

7.6.2 Interface Trapped Charge DLTS

The instrumentation for interface trapped charge DLTS is identical to that for bulk deep-level DLTS. However, the data interpretation is different because interface traps are continuously distributed in energy through the band gap, whereas bulk traps have discrete energy levels. We illustrate the interface trapped charge majority carrier DLTS concept for an MOS capacitor in Fig. 7.16(a). For a positive gate voltage most interface traps are occupied by majority electrons for *n*-substrates. A negative gate voltage drives the device into deep depletion, and electrons are emitted from interface traps. We assume negligible bulk states. The emitted electrons give rise to a charge, capacitance, or current transient. As evident from Fig. 7.16(a), electrons are emitted over a broad energy spectrum, but emission from interface traps in the upper half of the band gap dominates.

Interface trap characterization by DLTS was first implemented with MOSFET's.[100–103] MOSFET's, being three-terminal devices, have an advantage over MOS-C's. By reverse biasing the source/drain and pulsing the gate, majority electrons are captured and emitted without interference from minority holes which are collected by the source/drain. This allows interface trap majority–carrier characterization in the upper half of the band gap. With the source/drain forward biased, an inversion layer forms allowing interface traps to be filled with minority holes. Minority carrier characteriza-

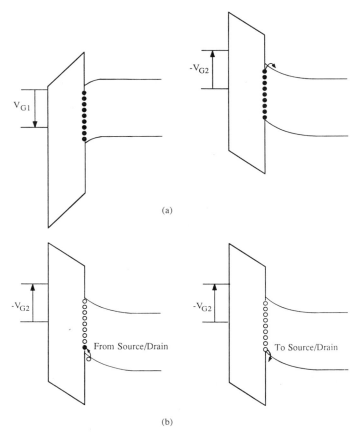

Fig. 7.16 (a) Majority carrier emission, (b) minority carrier emission from interface traps. In (a) the electrons flow into the n-substrate; in (b) the holes flow into the reverse-biased source and drain.

tion is then possible and the lower half of the band gap can be explored as shown in Fig. 7.16(b). This is not possible with MOS-C's because there is no ready source of minority carriers. When an inversion layer does form through thermal ehp generation, it can interfere with majority carrier trap DLTS measurements, especially at higher temperatures and at high ehp generation rates.

MOS capacitors are frequently used for interface trap characterization.[63,104–108] The derivation of the capacitance expression is more complex for MOS-C's than it is for diodes. We quote the main results whose derivations can be found in Johnson[63] and Yamasaki et al.[106] For $qD_{it} = C_{it} \ll C_{ox}$ and $\delta C = C_{HF}(t_1) - C_{HF}(t_2) \ll C_{HF}$

$$\delta C = -\frac{C_{HF}^3}{K_s \varepsilon_0 N_D C_{ox}} \int_{-\infty}^{\infty} D_{it} \left[\exp\left(-\frac{t_1}{\tau_e}\right) - \exp\left(-\frac{t_2}{\tau_e}\right) \right] dE_{it} \qquad (7.57)$$

where

$$\tau_e = \frac{\exp[(E_C - E_{it})/kT]}{\gamma_n \sigma_n T^2} \qquad (7.58)$$

E_{it} is the energy of the interface traps. The maximum emission time is $\tau_{e,max} = (t_2 - t_1)/\ln(t_2/t_1)$ from Eq. (7.50). In conjunction with Eq. (7.58) where $\tau_{e,max}$ corresponds to $E_{it,max}$, we find when the electron capture cross section is not a strong function of energy

$$E_{it,max} = E_C - kT \ln\left(\frac{\gamma_n \sigma_n T^2 (t_2 - t_1)}{\ln(t_2/t_1)}\right) \qquad (7.59)$$

$E_{it,max}$ is sharply peaked. If D_{it} varies slowly in the energy range of several kT around $E_{it,max}$, it can be considered reasonably constant and can be taken outside the integral of Eq. (7.57). The remaining integral becomes

$$\int_{-\infty}^{\infty} \left[\exp\left(-\frac{t_1}{\tau_{e,max}}\right) - \exp\left(-\frac{t_2}{\tau_{e,max}}\right) \right] dE_{it} = kT \ln\left(\frac{t_2}{t_1}\right) \qquad (7.60)$$

allowing Eq. (7.57) to be written as

$$\delta C \approx -\frac{C_{HF}^3}{K_s \varepsilon_0 N_D C_{ox}} \ln\left(\frac{t_2}{t_1}\right) kT D_{it} \qquad (7.61)$$

From Eq. (7.61) the interface trap density is

$$D_{it} = -\frac{K_s \varepsilon_0 N_D}{kT \ln(t_2/t_1)} \frac{C_{ox}}{C_{HF}^3} \delta C \qquad (7.62)$$

determined by electrons emitted from interface traps in time $(t_2 - t_1)$ in the energy interval $\Delta E = kT \ln(t_2/t_1)$ at energy $E_{it,max}$. A plot of D_{it} versus E_{it} is constructed by varying t_1 and t_2. For each t_1, t_2 combination an E_{it} is obtained from Eq. (7.59) and a D_{it} from Eq. (7.62). If the sample contains bulk as well as interface traps, it is possible to differentiate bulk traps from interface traps by the shape and the peak temperature of the DLTS plot.[106]

For the *constant capacitance* DLTS technique an equation analogous to Eq. (7.62) is[63]

$$D_{it} = \frac{C_{ox}}{qkTA \ln(t_2/t_1)} \Delta V_G \tag{7.63}$$

where A is the device area and ΔV_G is the gate voltage change required to keep the capacitance constant. Equation (7.63) is easier to use than (7.62) because neither the high-frequency capacitance nor the doping concentration needs to be known. Figure 7.17 shows the interface trap distribution for an n-type substrate, with D_{it} measured by the quasi-static and the CC-DLTS technique.[109] The discrepancy between the two curves may be due to the assumption of constant capture cross sections in the DLTS analysis. The DLTS techniques allows interface trap density determination as low as the mid 10^9 cm^{-2} eV^{-1} range.

MOS capacitors can also be measured by the current DLTS method. Using the small pulse method,[110–111] in which pulses of tens of millivolts are used, both interface trap density and capture cross sections can be measured.[112–113] Small filling pulses are applied as the quiescent bias is scanned at constant temperature and constant rate window. As the Fermi level scans the band gap, a DLTS peak is observed when τ_e in a small energy region around the Fermi level matches the rate window. Varying the rate window or the temperature gives the interface trap distribution. Although most MOS-C measurements are made on Si devices, MOS-C's made on GaAs and InP have also been characterized.[114]

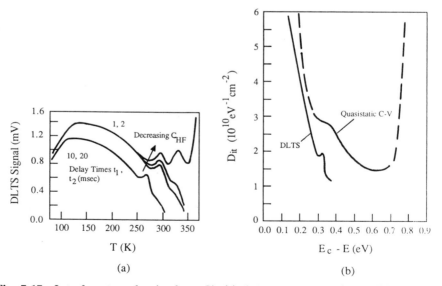

Fig. 7.17 Interface trap density for n-Si: (a) the constant capacitance DLTS signal for different bias conditions and delay times, (b) the interface trap distribution determined from (a) with constant capture cross section $\sigma_n = 10^{-18}$ cm^2 and from the quasi-static C-V technique. Reprinted with permission after Johnson et al.[109]

7.6.3 Optical and Scanning DLTS

Optical DLTS comes in various implementations. Light can be used (1) to determine optical properties of G-R centers, such as optical capture cross sections, (2) to create electron-hole pairs for minority carrier injection, (3) to create ehp's in semi-insulating materials, where electrical injection is difficult. Light does two basic things: (1) it imparts energy to a trapped carrier causing its emission from a G-R center to the conduction or to the valence band, and (2) it changes n and/or p by creating ehp's thereby changing the capture properties of the center. An electron beam in a scanning electron microscope also creates ehp's and can also be used for DLTS measurements. We will consider the boxcar data acquisition method here. Other methods follow similar arguments.

Optical Emission   Consider the conventional majority carrier emission process discussed earlier. A Schottky diode on an n-type substrate is zero biased, and traps are filled with electrons at low temperatures. Instead of raising the temperature and detecting the capacitance or current transient due to thermal emission, the sample is held at a sufficiently low temperature that thermal emission is negligibly low and optical emission takes place. Light is shone on the sample provided with a transparent or semitransparent contact. For $h\nu < E_C - E_T$ there is no optical absorption. For $h\nu > E_C - E_T$ photons excite electrons from the traps into the conduction band. Equation (7.8) holds, but the emission rate e_n becomes $e_n + e_n^0$, where e_n^0 is the *optical emission rate* $e_n^0 = \sigma_n^0 \Phi$, where σ_n^0 is the *optical* electron capture cross section and Φ the photon flux density. The G-R center concentration is obtained from the capacitance step just as it is during thermal emission measurements. Optical emission measurements were used very early during capacitance transient experiments.[5] The light was used in these experiments to determine optical properties of the *G-R* centers, such as the optical cross section, using either capacitance or current transients.[37,115-118]

It is possible to determine the multiplicity of charge states by varying the energy of the incident light. For a center with two donor levels, for example, one increases the light energy to excite electrons from the upper level into the conduction band, detected by a capacitance change. Increasing the energy further leaves the capacitance unchanged, provided all electrons have been excited out of that level, until the energy is sufficient to excite electrons from the second level into the conduction band. A second capacitance rise is seen. This method has been used to determine the double-donor nature of sulfur in silicon.[119]

In the two-wavelength method a steady-state background light creates a steady-state population of holes on traps below the Fermi level and of electrons on traps above the Fermi level. A variable-energy probe light excites carriers from the traps into either of the bands while the junction is pulsed electrically.[120] In another version ehp's are generated optically by above band-gap light.[121] Both electrons and holes can be captured by traps in the scr. When the light is turned off, the carriers are thermally emitted. In this method the light is merely used to generate ehp's; the transient is due to thermal emission. Other optical techniques were mentioned earlier when we discussed the use of light to generate ehp's for the measurement of the minority carrier capture cross sections.[31,39,122] These optical DLTS methods have not found wide application.

Photo-Induced Current Transient Spectroscopy   The optical techniques of the previous section supplement electrical measurements. The measurements can generally be done electrically; the optical input makes the measurement easier (minority carrier generation) or gives additional information (optical cross section). But purely electrical measurements are difficult to do in high resistivity or semi-insulating substrates. Optical inputs can then be a decided advantage and in some cases are the only way to obtain information on deep level impurities.

In the *photo-induced current transient spectroscopy* (PITS or PICTS) method the current is measured as a function of time, as shown in Fig. 7.18. The sample is provided with a top semitransparent ohmic contact. Capacitance cannot be measured because the substrate resistance is too high. During the PITS measurement light is pulsed on the sample, and the photocurrent rises to a steady-state value. The light pulse can have above band-gap[123–124] or below band-gap energy.[125–126] The photocurrent transient at the end of the light pulse consists of a rapid drop followed by a slower decay, shown in Fig. 7.18(b). The initial rapid drop is due to ehp recombination and the slow decay is due to carrier emission. The slow current transient can be analyzed by conventional DLTS rate window methods.[127–130] PITS spectra are shown in Fig. 7.18(c). It is sometimes possible to determine whether the level is an electron or a hole trap by measuring the peak height as the bias polarity is changed. However, this identification is not as simple as it is for capacitance transients.

If the trap is an electron trap and if the light intensity is sufficient to saturate the photocurrent, then the transient current is[131]

$$\delta I = \frac{K N_{\mathrm{T}}}{\tau_{\mathrm{e}}} \exp\left(-\frac{t}{\tau_{\mathrm{e}}}\right) \tag{7.64}$$

where K is a constant [see Eq. (7.43)]. δI exhibits a maximum for $t = \tau_{\mathrm{e}}$ when plotted against temperature, as determined by differentiating Eq. (7.64) with respect to temperature

$$\frac{d(\delta I)}{dT} = \frac{K N_{\mathrm{T}}}{\tau_{\mathrm{e}}^3} (t - \tau_{\mathrm{e}}) \exp\left(-\frac{t}{\tau_{\mathrm{e}}}\right) \frac{d\tau_{\mathrm{e}}}{dT} \tag{7.65}$$

and setting Eq. (7.65) equal to zero.

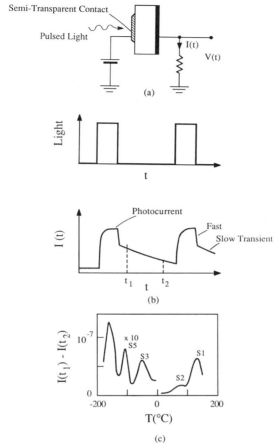

Fig. 7.18 Photo-induced transient spectroscopy: (a) circuit schematic, (b) light pulses and output voltage waveforms, (c) spectrum for semi-insulating GaAs. Reprinted with permission from Look.[130]

PITS is not well suited for trap concentration determination, and the reliability of information extracted from the data for trap identification falls off as the trap energy approaches the intrinsic Fermi level.[125] Additional complications occur when carriers emitted from traps recombine. The recombination lifetime for semi-insulating materials is usually quite low. In addition emitted carriers can be retrapped. All of these effects make the method difficult to analyze.[132–133] Unfortunately there are few techniques other than PITS to characterize such materials. PITS has been used to map deep-level impurities in GaAs at 1.2-mm resolution by scanning the light across the sample while making the measurement.[130] The method has been used for semi-insulating GaAs and InP to study deep-level impurities. The impurities Cr and EL2 are of main interest for GaAs and Fe and Cr for InP. Cu in GaAs has also been investigated.[134]

Scanning DLTS   Scanning DLTS (S-DLTS) is performed in a scanning electron microscope using the electron beam as the excitation source. The high spatial resolution—in the micron range—is its main advantage. This is also one of its disadvantages because such a small sampling area produces very small DLTS signals. For conventional DLTS the diode diameter is typically in the 0.5 to 1 mm range, and the entire area is active during the measurement. For S-DLTS the diode diameter is similar, giving rise to a large steady-state capacitance. But the emission-active area, defined by the electron beam diameter, can be much smaller and gives very small capacitance changes. The original S-DLTS work utilized current DLTS because it can be more sensitive than capacitance DLTS.[135] Equation (7.41) shows the current to be inversely proportional to the emission time constant. As T increases, τ_e decreases, and hence I increases. Later developments of an extremely sensitive capacitance meter with 10^{-6} pF sensitivity, consisting of a resonance-tuned LC bridge at 28 MHz with permanent slow automatic zero balance to ensure operation in a tuned state at all times, allowed capacitance DLTS measurements.[136] Quantitative measurements are difficult to implement in S-DLTS,[137] but one can map a distribution of a particular impurity by scanning the device area choosing an appropriate temperature and rate window. A few hundred impurity atoms per scanning point have been detected.[138]

7.6.4   Precautions

Leakage Current   A number of measurements precautions have already been mentioned throughout this chapter. Here we point out a few more. Devices sometimes exhibit high reverse bias leakage currents. During DLTS measurements of leaky MOS capacitors, it was observed that the DLTS peak amplitude decreased much more strongly with slower rate windows than expected. This was attributed to competition between carrier capture due to leakage current and thermal emission, leading to large errors in the trap energy extracted from an Arrhenius plot.[139] For leaky diodes an experimental system was developed with two diodes, having similar C-V and I-V characteristics, driven 180° out of phase.[140]

Series Resistance   Another device anomaly that can affect the DLTS response is the device series resistance. We discuss in Chapter 2 the effects of series resistance on capacitance and doping concentration characterization. For a diode consisting of a capacitance C in series with a resistance r_s, the measured capacitance is [see Eq. (2.30) with $G \approx 0$]

$$C_m = \frac{C}{1 + (2\pi f r_s C)^2} \tag{7.66}$$

A DLTS measurement records the change in capacitance given by

$$\Delta C_m = \frac{\Delta C}{[1 + (2\pi f r_s C)^2]} \left\{ 1 - \frac{2(2\pi f r_s C)^2}{[1 + (2\pi f r_s C)^2]} \right\} \tag{7.67}$$

$\Delta C_m = \Delta C$ for $r_s = 0$. However, as r_s increases, ΔC_m decreases. ΔC_m and the DLTS signal can become zero and even reverse sign. For example, the curly bracket is zero when $r_s \approx 1500\,\Omega$ for $C = 100\,pF$ and $f = 1\,MHz$. For $r_s > 1500\,\Omega$, the DLTS signal changes sign, and majority carrier traps can be mistaken for minority carrier traps.[95,141]

If series resistance is anticipated to be a problem, one can insert additional external resistance into the circuit and check for sign reversal.[142] If sign reversal is not observed, there is a good chance that it has already taken place without any additional external resistance, and the measured data must be carefully evaluated. Occasionally an additional capacitance is introduced by an oxide layer at the back of the sample. This can also lead to DLTS signal reversal.[143] Series resistance is not a particular problem for current DLTS because it is essentially a dc measurement, not requiring the high probe frequency of capacitance DLTS.

Incomplete Trap Filling We have assumed in most of the discussions and derivations that all traps fill with majority carriers during the capture time and emit majority carriers during the emission time. That is only an assumption as illustrated with the band diagram in Fig. 7.19.[144] Assume that the device has been reverse biased for some time and that all majority carriers have been emitted from traps in the scr. During the reverse bias to zero bias phase, Figs. 7.19(a) and (b), the scr collapses and electrons flow in. Those traps within W_1 do not fill because they are above the Fermi level; those traps to the right of W_1, but near W_1, fill more slowly than those further to the right because the electron concentration tails off. Consequently, if the filling pulse width is too narrow, not all traps to the right of W_1 become occupied by electrons. When the bias switches back to reverse bias, Fig. 7.19(c), those traps within λ do not emit electrons because they are below the Fermi level. λ is given by[56]

$$\lambda = \sqrt{\frac{2K_s\varepsilon_0(E_F - E_T)}{q^2 N_D}} \tag{7.68}$$

Only those traps within $(W - W_1 - \lambda)$ participate during the DLTS measurement.[36,145] W_1 is almost always neglected; frequently λ is neglected too. When λ is not neglected, the capacitance step ΔC_e of Eq. (7.17) becomes[56,146]

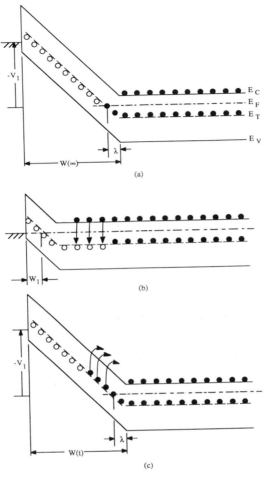

Fig. 7.19 Band diagram for a Schottky diode on an *n*-substrate: (a) diode under steady-state reverse bias, (b) diode at zero bias during the filling phase, (c) diode at reverse bias immediately after the filling phase. The scr width changes from $W(t)$ to $W(\infty)$ in (c).

$$\Delta C_e = \frac{n_T(0)}{2N_D}\, C_0 f(W) \tag{7.69}$$

where

$$f(W) = 1 - \frac{[2\lambda/W(V)][1 - C(V)/C(0)]}{1 - [C(V)/C(0)]^2} \tag{7.70}$$

where $C(0)$ and $C(V)$ are the capacitances at voltages zero and V, respectively. $f(W)$ becomes unity if the edge region can be neglected. However, with $f(W) < 1$, neglect of the edge region can introduce appreciable error.

7.7 THERMALLY STIMULATED CAPACITANCE AND CURRENT

Thermally stimulated capacitance (TSCAP) and current (TSC) measurements were used before DLTS became popular. The techniques were originally used for insulators and later adapted to lower resistivity semiconductors when it was recognized that the reverse-biased scr is a region of high resistance.[147] During the measurement the device is cooled, and the traps are filled with majority carriers at zero bias. Alternately, traps can be filled with minority carriers by optical injection or by forward biasing a *pn* junction. Then the device is reverse biased, heated at a constant rate, and the steady-state capacitance or current is measured as a function of temperature. Capacitance steps or current peaks are observed as traps emit their carriers, shown in Fig. 7.20.

Thermally stimulated current was first proposed by Driver and Wright,[148] and thermally stimulated capacitance originated with Carballes and Lebailly.[149] Much of the early work of Sah and his students utilized these same techniques.[5,41] The temperature of the TSC peak or the midpoint of the TSCAP step T_m is related to the activation energy $\Delta E = E_C - E_T$ or $\Delta E = E_T - E_V$ by[150]

$$\Delta E = kT_m \ln\left[\frac{\gamma_n \sigma_n k T_m^4}{\beta(\Delta E + 2kT_m)}\right] \tag{7.71}$$

For *p*-type samples the subscript *n* should be replaced by *p*. The trap

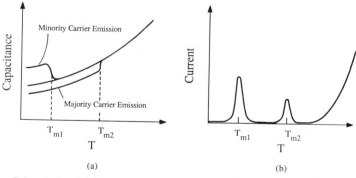

(a)

(b)

Fig. 7.20 Schematic of (a) TSCAP and (b) TSC for a sample containing a majority carrier trap of concentration N_T and a shallower minority carrier trap of concentration $2N_T$. The increase in the current at higher temperatures is due to thermally generated leakage current. Reprinted with permission after Lang.[56]

concentration is obtained from the area under the TSC curve or from the step height of the TSCAP curve.

The equipment is simpler than that for DLTS, but the information obtained from TSC and TSCAP is more limited and more difficult to interpret. Thermally stimulated techniques allow a quick sweep of the sample to survey the entire range of traps in a sample and work well for $N_T \geq 0.1 N_D$ and $\Delta E \geq 0.3$ eV. The TSC current peaks depend on the heating rate, but the TSCAP steps do not. TSC is influenced by leakage currents; TSCAP is much less influenced by diode quality. TSCAP allows discrimination between minority and majority carrier traps by the *sign* of the capacitance change as indicated in Fig. 7.20; TSC does not. Thermally stimulated measurements have been largely replaced with DLTS.

7.8 STRENGTHS AND WEAKNESSES

DLTS is the most common deep-level characterization technique today having replaced thermally stimulated current and capacitance. It lends itself to a number of different implementations and equipment is commercially available.

- *Capacitance Transient Spectroscopy* Its strength lies in the ease of measurement. Most systems use commercial capacitance meters or bridges and add signal-processing functions (lock-in amplifiers, boxcar integrators, or computers). One can distinguish between majority and minority carrier traps, and its sensitivity is independent of the emission time constant. Its major weakness is the inability to characterize high resistivity substrates. The fact that its sensitivity is independent of time constant can be a disadvantage because the sensitivity cannot be changed.

- *Current Transient Spectroscopy* Its strength lies in the ability to characterize conducting as well as semi-insulating substrates. The fact that the current depends inversely on the emission time constant allows the sensitivity of the method to be changed by changing the time constant. This led to its use in scanning DLTS. Its weakness is its dependence on diode quality, where leakage current can interfere with the measurement.

- *Optical DLTS* Its strength lies in the ability to create minority carriers without the need for *pn* junction. This allows materials to be characterized in which it is difficult to make *pn* junctions. O-DLTS is useful to determine impurity optical cross sections. Its major weakness lies in the requirement of light. The low temperature dewar must have transparent windows, and monochromators or pulsed light sources must be available.

APPENDIX 7.1 ACTIVATION ENERGY, CAPTURE CROSS SECTION

The relationship between the emission coefficient and the capture coefficient can be written, according to Eq. (7.24), as

$$e_n = \sigma_n v_{\mathrm{th}} N_C \exp\left(\frac{E_C - E_T}{kT}\right) \tag{A7.1}$$

The relationship is frequently used to determine E_T and σ_n. However, we mentioned earlier that when the capture cross section is determined from the intercept of a $\ln(\tau_e T^2)$ versus $1/T$ plot, considerable error can result. Let us look at that a little more closely.

From thermodynamics we find the following definitions[151]

$$G = H - TS \tag{A7.2a}$$

$$H = E + pV \tag{A7.2b}$$

where G is the Gibbs free energy, H enthalpy, E internal energy, T temperature, S entropy, p pressure, and V volume. The energy to excite an electron from a G-R center into the conduction band is given by ΔG_n.[152–153] Equation (A7.1) then becomes

$$e_n = \sigma_n v_{\mathrm{th}} N_C \exp\left(-\frac{\Delta G_n}{kT}\right) \tag{A7.3}$$

From Eq. (A7.2a), $\Delta G_n = \Delta H_n - T\,\Delta S_n$ for constant T. When substituted into Eq. (A7.3), the emission rate becomes

$$e_n = \sigma_n X_n v_{\mathrm{th}} N_C \exp\left(-\frac{\Delta H_n}{kT}\right) \tag{A7.4}$$

where $X_n = \exp(\Delta S_n/k)$ is an "entropy factor," which accounts for the entropy change accompanying electron emission from G-R centers to the conduction band. The entropy change can be expressed as $\Delta S_n = \Delta S_{ne} + \Delta S_{na}$, where ΔS_{ne} is the entropy change due to electronic degeneracy and ΔS_{na} is due to atomic vibrational changes. The electronic contribution may be expressed in terms of two degeneracy factors: g_0 is the degeneracy of the G-R center unoccupied by an electron, and g_1 is the degeneracy of the G-R center occupied by one electron, giving

$$X_n = \left(\frac{g_0}{g_1}\right)\exp\left(\frac{\Delta S_{na}}{k}\right) \tag{A7.5}$$

The degeneracy factors are not well known for deep-level impurities. Using values from shallow levels and with $\Delta S_{na} \approx$ a few k, X_n can easily be 10–100.

Equation (A7.4) states that the energy determined from a $\ln(\tau_e T^2)$ or $\ln(T^2/e_n)$ versus $1/T$ Arrhenius plot is an enthalpy, and the prefactor can be written as $\sigma_{n,\text{eff}} v_{\text{th}} N_C$, with $\sigma_{n,\text{eff}} = \sigma_n X_n$. In other words, the effective capture cross section differs from the true capture cross section by X_n. If that distinction is not made, then obviously the extracted cross section can be seriously in error. Effective cross sections larger by factors of 50 or more from true cross sections are not uncommon.[19] Examples are shown in Table 7.2.

Additional complications are introduced when σ_n is temperature dependent. Some cross sections follow the relationship

$$\sigma_n = \sigma_\infty \exp\left(-\frac{E_b}{kT}\right) \tag{A7.6}$$

where σ_∞ is the cross section at $T \to \infty$. Equation (A7.4) becomes

$$e_n = \sigma_n X_n v_{\text{th}} N_C \exp\left(-\frac{\Delta H_n + E_b}{kT}\right) \tag{A7.7}$$

Under these conditions the Arrhenius plot gives neither the G-R center energy level nor its extrapolated cross section correctly. If in addition the capture cross section is electric-field dependent, further inaccuracies are introduced. A good discussion of energy levels, enthalpies, entropies, capture cross sections, etc., can be found in the work of Lang et al.[19] Further thermodynamic derivations can be found in the work by Thurmond and Van Vechten.[154-155]

A non-thermodynamic approach defines the energy $\Delta E_T = E_C - E_T$ as being temperature dependent according to $\Delta E_T = \Delta E_{T0} - \alpha T$. The degeneracy ratio in Eq. (A7.5) is written as g_n.[156] Equation (A7.1) becomes

$$e_n = \sigma_n X_n v_{\text{th}} N_C \exp\left(-\frac{\Delta E_{T0}}{kT}\right) \tag{A7.8}$$

where now $X_n = g_n \exp(\alpha/k)$. We find the energy as that for $T \to 0\,\text{K}$ and the cross section is again $\sigma_n X_n$, although now X_n is defined somewhat differently.

APPENDIX 7.2 TIME CONSTANT EXTRACTION

The capacitance of a Schottky barrier or $p^+ n$ junction containing impurities is from Eq. (7.11)

$$C = K\sqrt{\frac{N_D - N_T \exp(-t/\tau_e)}{(V_{\text{bi}} - V)}} \tag{A7.9}$$

where $n_T(0) = N_T$, and we confine ourselves to emission transients for simplicity. How is τ_e determined?

One of the earliest method to extract τ_e is to take $dV/d(1/C^2)$ from Eq. (A7.9) as[11-12]

$$\left.\frac{dV}{d(1/C^2)}\right|_{t=\infty} - \left.\frac{dV}{d(1/C^2)}\right|_t = K^2 N_T \exp\left(-\frac{t}{\tau_e}\right) \qquad (A7.10)$$

and to plot the ln of the left side of Eq. (A7.10) versus t. The slope of this plot gives τ_e, and the intercept at $t = 0$ is $\ln(K^2 N_T)$. This method places no limitation on the magnitude of N_T with respect to N_D; that is, N_T need not be much smaller than N_D.

Another method also makes use of Eq. (A7.9) by defining $f(t) = C(t)^2 - C_0^2 = [-K^2 N_T/(V_{bi} - V)]\exp(-t/\tau_e)$, where C_0 is the capacitance in Eq. (A7.9) for $N_T = 0$. The measurement is performed at constant temperature. Differentiating $f(t)$ gives

$$t\frac{df}{dt} = \frac{K^2 N_T}{(V_{bi} - V)}\frac{t}{\tau_e}e^{-t/\tau_e} \qquad (A7.11)$$

which has a maximum of $K^2 N_T/e(V_{bi} - V)$ at $t = \tau_e$.[157] Hence determining the maximum in the curve gives the time constant.

If $N_T \ll N_D$, then we can write [see Eq. (7.16)]

$$C \approx C_0\left[1 - \frac{N_T}{2N_D}\exp\left(-\frac{t}{\tau_e}\right)\right] \qquad (A7.12)$$

Equation (A7.12) has been used in a number of implementations to extract τ_e. In the two-point method, the C-t exponential time-varying curve is sampled at two times, $t = t_1$ and $t = t_2$.[53] From Eq. (7.50) we have

$$\tau_e = \frac{t_2 - t_1}{\ln(t_2/t_1)} \qquad (A7.13)$$

In the three-point method, three points are measured on the C-t curve at a *constant* temperature, $C = C_1$ at $t = t_1$, $C = C_2$ at $t = t_2$, and $C = C_3$ at $t = t_3$.[158-159] Using Eq. (A7.12), we find

$$\frac{C_1 - C_2}{C_2 - C_3} = \frac{\exp(\Delta t/\tau_e) - 1}{1 - \exp(-\Delta t/\tau_e)} \qquad (A7.14)$$

where $\Delta t = t_2 - t_1 = t_3 - t_2$. A solution of Eq. (A7.14) for τ_e is

$$\tau_e = \frac{\Delta t}{\ln[(C_1 - C_2)/(C_2 - C_3)]} \qquad (A7.15)$$

A good choice for Δt is $\tau_e/2$, but of course τ_e is not known *a priori*, although a first-order value for it can be obtained from the "$1/e$ point" on the capacitance decay curve.

Another technique is based on a very different approach. Consider the function $y_1 = y(t) = A \exp(-t/\tau) + B$, that is, an exponentially decaying function superimposed on a dc background. We define a second function $y_2 = y(t + \Delta t) = A \exp[-(t + \Delta t)/\tau] + B$. The second function is obtained from the first by simply adding a constant increment Δt to the time t. A plot of y_2 versus y_1 is a straight line with a slope $m = \exp(-\Delta t/\tau)$ and an intercept on the y_2 axis of $B(1 - m)$.[160] τ is calculated from the slope, and Δt and B are found from the intercept and the slope. Δt should be smaller than τ, but not much smaller. $\Delta t \approx 0.1$ to 0.5τ are suitable values.

It is also possible to acquire the entire C-t curve by using data acquisition techniques. Then a number of signal processing functions can be performed on the decay curve. For example, a Fourier analysis allows the extraction of the time constant.[161] Other functions are discussed in Section 7.6.1 under "Computer DLTS."

APPENDIX 7.3 ARRHENIUS PLOTS FOR Si AND GaAs

Arrhenius plots for Si and GaAs are shown in Figs. A7.1 and A7.2. In Fig. A7.1 $(300/T)^2 e_n$ and $(300/T)^2 e_p$ are plotted instead of $\tau_n T^2$ and $\tau_p T^2$, giving negative slopes. The deep-level impurity elements are shown wherever possible, and the numbers listed below the elements are their energy levels calculated from the slopes. The superscripts are the references given in the review paper by Chen and Milnes.[2]

Table A7.1 lists typical trace contamination in Si most commonly produced during device processing or after 1-MeV electron beam irradiation.[162] The impurities were determined from transient capacitance spectroscopy. DLTS spectra have been correlated with metallic impurities, growth-related defects, oxidation, heat treatments, electron and proton irradiation, dislocation-related states, electronically stimulated defects, and laser anneal. Established temperature regimes of defect and impurity reactions are indicated.

An unknown DLTS peak can be compared with the data in the table by two methods.[162] First, an Arrhenius plot of $\tau_e T^2$ versus $1/T$ can be constructed using the point given by the temperature of the known peak (T) at a time constant of 1.8 ms (τ) and the slope given by the activation energy (E_T) in the table. Alternatively, the temperature at which a signal from a listed defect should occur using any time constant of the analyzing instrument can be determined by iteration. A simple computer program sets the ratio R,

$$R = \frac{\tau_1 T_1^2 \exp(-E_T/kT_1)}{\tau_2 T_2^2 \exp(-E_T/kT_2)} \tag{A7.16}$$

where subscript 1 refers to the value of Table A7.1 and subscript 2 refers to the value for the particular measurement. For $\tau_1 > \tau_2$ the temperature T_2 is increased; for $\tau_1 < \tau_2$ the temperature T_2 is decreased until $R = 1$.

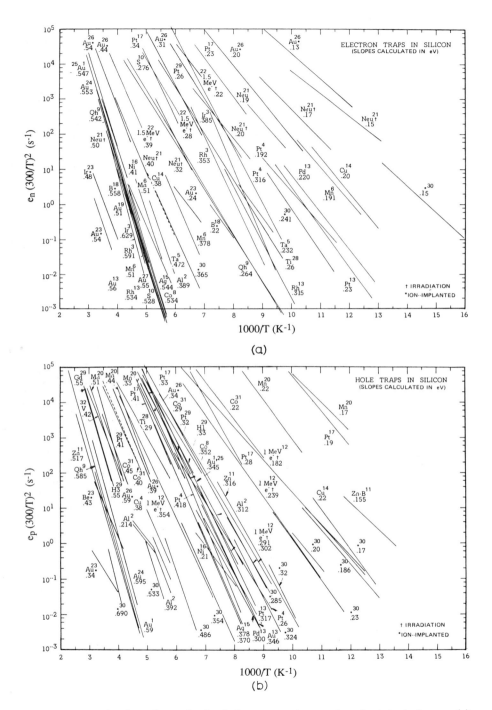

Fig. A7.1 Arrhenius plots obtained from capacitance transient techniques: (a) electron traps in Si, (b) hole traps in Si. The vertical axis is $(300/T^2)e_{n,p}$ instead of $\tau_{n,p}T^2$. Reprinted, with permission, from the Annual Review of Materials Science, Vol. 10, © 1980 by Annual Reviews Inc.

TABLE A7.1 Capacitance Transient Spectral Features for Silicon

Defect	T (K) 1.8 ms	E_T (eV)	σ_{maj} (cm^2)	Anneal	Comments[a]
Ag	286	E (0.51)	1×10^{-16}		Q, *, FZ
	184	H (0.38)	—		Q, *, FZ
Au	288	E (0.53)	2×10^{-16}		Q, *, FZ
	173	H (0.35)	$>1 \times 10^{-15}$		Q, *, FZ
Cu	112	H (0.22)	$>6 \times 10^{-14}$	Out 150°C	Q, *, FZ
	242	H (0.41)	8×10^{-14}		Q, *, FZ
Fe	181	E (0.35)	6×10^{-15}		Q, *, FZ
(Fe–B)	59	H (0.10)	$>4 \times 10^{-15}$	Out >150°C	Q, *, FZ
(Fe$_i$)	267	H (0.46)	—	In >150°C, out >200°C	Q, *, FZ
	208	E (0.21)	—		S, FZ
	299	E (0.46)	—		S, FZ
	184	H (0.23)	—		S, FZ
	170	E (0.35)	5×10^{-15}		Q, CG
	168	H (0.30)	—		Q, CG
	237	H (0.43)	—		Q, CG
	220	H (0.47)	—		Q, CG
Mn	68	E (0.11)	1×10^{-15}		Q, FZ
	216	E (0.41)	$>2 \times 10^{-15}$		Q, FZ
	81	H (0.13)	5×10^{-16}		Q, FZ
Ni	257	E (0.43)	1×10^{-16}		Q, *, FZ
	88	E (0.14)	$>4 \times 10^{-15}$	Out 150°C	Q, *, FZ

Pt	114	E (0.22)	$\sim 10^{-15}$		Q, *, FZ
	174	E (0.30)	$\sim 10^{-15}$		Q, *, FZ
	87	H (0.22)	—		Q, *, FZ
O-donor	Below freezeout	E (0.07)	—	In 400°C, out 600°C	*, CG
Heat treatment	58	E (0.15)	$>3 \times 10^{-15}$	In 400°C, out 600°C	*, CG
	59, 60	E (0.15)	2×10^{-16}	In 900°C	*, CG
	112	E (0.22)	7×10^{-16}	In 900°C	*, CG
	228	E (0.47)	5×10^{-16}	In 900°C	*, CG
Laser donor	115	E (0.19)	5×10^{-19}	Out 550°C	Q, FZ, CG
	200	E (0.33–0.36)	5×10^{-16}	Out 650°C	Q, FZ, CG
	211	H (0.36)	—		Q, *, FZ, CG
Vacancy-O	98	E (0.18)	10^{-14}	In −43°C, out 350°C	1 MeV, CG
Vacancy-vacancy	139	E (0.23)	2×10^{-16}	Out 300°C	1 MeV, CG, FZ
	245	E (0.41)	4×10^{-15}	Out 300°C	1 MeV, CG, FZ
	123	H (0.21)	2×10^{-16}	Out 300°C	1 MeV, CG, FZ
P-vacancy	237	E (0.44)	$>10^{-16}$	Out 150°C	1 MeV, CG, FZ
C_s-C_i	204	H (0.36)	8×10^{-17}	In 43°C	1 MeV, CG, FZ
Dislocation	225	E (0.38)	1.4×10^{-15}		FZ
	206	H (0.35)	$>5 \times 10^{-17}$		FZ
Point defect debris	288	E (0.63–0.68)	$>10^{-15}$	Out 800°C	Fz, cross slip

Source: Benton and Kimerling.[162]

[a] Symbols: Q = quenched material, * = diffused junction, S = slow cool, FZ = float zone growth, CG = crucible growth, and 1 MeV = electron bombardment.

347

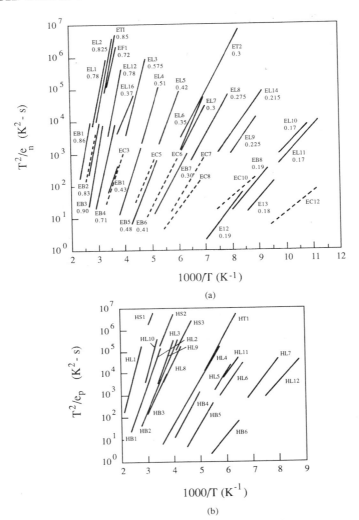

Fig. A7.2 Arrhenius plots obtained from capacitance transient techniques: (a) electron traps in GaAs, (b) hole traps in GaAs. Reprinted with permission after Martin et al.[21] and Mitonneau et al.[22] © Institution of Electrical Engineers.

REFERENCES

[1] A. G. Milnes, *Deep Impurities in Semiconductors*, Wiley-Interscience, New York, 1973.

[2] J. W. Chen and A. G. Milnes, "Energy Levels in Silicon," in *Annual Review of Material Science* (R. A. Huggins, R. H. Bube, and D. A. Vermilyea, eds.), Annual Reviews, Palo Alto, CA, **10**, pp. 157–228, 1980.

[3] A. G. Milnes, "Impurity and Defect Levels (Experimental) in Gallium Arsenide," in *Advances in Electronics and Electron Physics* (P. W. Hawkes, ed.), Academic Press, Orlando, FL, **61**, 63–160, 1983.

[4] M. Jaros, *Deep Levels in Semiconductors*, A. Hilger, Bristol, 1982.

[5] C. T. Sah, L. Forbes, L. L. Rosier, and A. F. Tasch, Jr., "Thermal and Optical Emission and Capture Rates and Cross Sections of Electrons and Holes at Imperfection Centers in Semiconductors from Photo and Dark Junction Current and Capacitance Experiments," *Solid-State Electron.* **13**, 759–788, June 1970.

[6] R. N. Hall, "Electron-Hole Recombination in Germanium," *Phys. Rev.* **87**, 387, July 1952.

[7] W. Shockley and W. T. Read, "Statistics of the Recombinations of Holes and Electrons," *Phys. Rev.* **87**, 835–842, Sept. 1952.

[8] R. Williams, "Determination of Deep Centers in Conducting Gallium Arsenide," *J. Appl. Phys.* **37**, 3411–3416, Aug. 1966.

[9] Y. Furukawa and Y. Ishibashi, "Transient Phenomena in the Capacitance of GaAs Schottky Barrier Diodes," *Japan. J. Appl. Phys.* **5**, 837–838, Sept. 1966.

[10] Y. Furukawa and Y. Ishibashi, "Trapping Effects in Au-*n*-Type GaAs Schottky Barrier Diodes," *Japan. J. Appl. Phys.* **6**, 503–508, April 1967.

[11] R. R. Senechal and J. Basinski, "Capacitance of Junctions on Gold-Doped Silicon," *J. Appl. Phys.* **39**, 3723–3731, July 1968.

[12] R. R. Senechal and J. Basinski, "Capacitance Measurements on Au-GaAs Schottky Barriers," *J. Appl. Phys.* **39**, 4581–4589, Sept. 1968.

[13] M. Bleicher and E. Lange, "Schottky-Barrier Capacitance Measurements for Deep Level Impurity Determination," *Solid-State Electron.* **16**, 375–380, March 1973.

[14] R. F. Pierret, *Advanced Semiconductor Fundamentals*, Addison-Wesley, Reading, MA, 1987, pp. 146–152.

[15] W. Shockley, "Electrons, Holes, and Traps," *Proc. IRE* **46**, 973–990, June 1958.

[16] C. T. Sah, "The Equivalent Circuit Model in Solid-State Electronics—Part I: The Single Energy Level Defect Centers," *Proc. IEEE* **55**, 654–671, May 1967; "The Equivalent Circuit Model in Solid-State Electronics—Part II: The Multiple Energy Level Impurity Centers," *Proc. IEEE* **55**, 672–684, May 1967.

[17] P. A. Martin, B. G. Streetman, and K. Hess, "Electric Field Enhanced Emission from Non-Coulombic Traps in Semiconductors," *J. Appl. Phys.* **52**, 7409–7415, Dec. 1981.

[18] A. F. Tasch, Jr., and C. T. Sah, "Recombination-Generation and Optical Properties of Gold Acceptor in Silicon," *Phys. Rev.* **B1**, 800–809, Jan. 1970.

[19] D. V. Lang, H. G. Grimmeiss, E. Meijer, and M. Jaros, "Complex Nature of Gold-Related Deep Levels in Silicon," *Phys. Rev.* **B22**, 3917–3934, Oct. 1980.

[20] H. D. Barber, "Effective Mass and Intrinsic Concentration in Silicon," *Solid-State Electron.* **10**, 1039–1051, Nov. 1967.

[21] G. M. Martin, A. Mitonneau, and A. Mircea, "Electron Traps in Bulk and Epitaxial GaAs Crystals," *Electron. Lett.* **13**, 191–193, March 1977.

[22] A. Mitonneau, G. M. Martin, and A. Mircea, "Hole Traps in Bulk and Epitaxial GaAs Crystals," *Electron. Lett.* **13**, 666–668, Oct. 1977.

[23] W. B. Leigh, J. S. Blakemore, and R. Y. Koyama, "Interfacial Effects Related to Backgating in Ion-Implanted GaAs MESFET's" *IEEE Trans. Electron Dev.* **ED-32**, 1835–1841, Sept. 1985.

[24] H. Okushi and Y. Tokumaru, "A Modulated DLTS Method for Large Signal Analysis (C^2-DLTS)," *Japan. J. Appl. Phys.* **20**, L45–L47, Jan. 1981.

[25] H. Tomokage, H. Nakashima, and K. Hashimoto, "Note on the Analysis of DLTS and C^2-DLTS," *Japan. J. Appl. Phys.* **21**, 67–70, Jan. 1982.

[26] W. E. Phillips and J. R. Lowney, "Analysis of Nonexponential Transient Capacitance in Silicon Diodes Heavily Doped with Platinum," *J. Appl. Phys.* **54**, 2786–2791, May 1983.

[27] A. C. Wang and C. T. Sah, "Determination of Trapped Charge Emission Rates from Nonexponential Capacitance Transients due to High Trap Densities in Semiconductors," *J. Appl. Phys.* **55**, 565–570, Jan. 1984.

[28] D. Stiévenard, M. Lannoo, and J. C. Bourgoin, "Transient Capacitance Spectroscopy in Heavily Compensated Semiconductors," *Solid-State Electron.* **28**, 485–492, May 1985.

[29] F. D. Auret and M. Nel, "Detection of Minority-Carrier Defects by Deep Level Transient Spectroscopy Using Schottky Barrier Diodes," *J. Appl. Phys.* **61**, 2546–2549, April 1987.

[30] L. Stolt and K. Bohlin, "Deep-Level Transient Spectroscopy Measurements Using High Schottky Barriers," *Solid-State Electron.* **28**, 1215–1221, Dec. 1985.

[31] C. H. Henry, H. Kukimoto, G. L. Miller, and F. R. Merritt, "Photocapacitance Studies of the Oxygen Donor in GaP. II. Capture Cross Sections," *Phys. Rev.* **B7**, 2499–2507, March 1973.

[32] A. C. Wang and C. T. Sah, "New Method for Complete Electrical Characterization of Recombination Properties of Traps in Semiconductors," *J. Appl. Phys.* **57**, 4645–4656, May 1985.

[33] J. A. Borsuk and R. M. Swanson, "Capture-Cross-Section Determination by Transient-Current Trap-Filling Experiments," *J. Appl. Phys.* **52**, 6704–6712, Nov. 1981.

[34] J. A. Pals, "Properties of Au, Pt, Pd and Rh Levels in Silicon Measured with a Constant Capacitance Technique," *Solid-State Electron.* **17**, 1139–1145, Nov. 1974.

[35] S. D. Brotherton and J. Bicknell, "The Electron Capture Cross Section and Energy Level of the Gold Acceptor Center in Silicon," *J. Appl. Phys.* **49**, 667–671, Feb. 1978.

[36] A. Zylbersztejn, "Trap Depth and Electron Capture Cross Section Determination by Trap Refilling Experiments in Schottky Diodes," *Appl. Phys. Lett.* **33**, 200–202, July 1978.

[37] H. Kukimoto, C. H. Henry, and F. R. Merritt, "Photocapacitance Studies of the Oxygen Donor in GaP. I. Optical Cross Sections, Energy Levels, and Concentration," *Phys. Rev.* **B7**, 2486–2499, March 1973.

[38] S. D. Brotherton and J. Bicknell, "Measurement of Minority Carrier Capture Cross Sections and Application to Gold and Platinum in Silicon," *J. Appl. Phys.* **53**, 1543–1553, March 1982.

[39] B. Hamilton, A. R. Peaker, and D. R. Wight, "Deep-State-Controlled Minority-Carrier Lifetime in *n*-Type Gallium Phosphide," *J. Appl. Phys.* **50**, 6373–6385, Oct. 1979.

[40] R. Brunwin, B. Hamilton, P. Jordan, and A. R. Peaker, "Detection of Minority-Carrier Traps Using Transient Spectroscopy," *Electron. Lett.* **15**, 349–350, June 1979.

[41] C. T. Sah, "Bulk and Interface Imperfections in Semiconductors," *Solid-State Electron.* **19**, 975–990, Dec. 1976.

[42] B. W. Wessels, "Determination of Deep Levels in Cu-Doped GaP Using Transient-Current Spectroscopy," *J. Appl. Phys.* **47**, 1131–1133, March 1976.

[43] J. A. Borsuk and R. M. Swanson, "Current Transient Spectroscopy: A High-Sensitivity DLTS System," *IEEE Trans. Electron Dev.* **ED-27**, 2217–2225, Dec. 1980.

[44] Y. Tokuda and A. Usami, "Current DLTS with a Bipolar Rectangular Weighting Function for Neutron-Irradiated *p*-Type Si," *Japan. J. Appl. Phys.* **22**, 371, Feb. 1983.

[45] J. W. Chen, R. J. Ko, D. W. Brzezinski, L. Forbes, and C. J. Dell'Oca, "Bulk Traps in Silicon-on-Sapphire by Conductance DLTS," *IEEE Trans. Electron Dev.* **ED-28**, 299–304, March 1981.

[46] P. K. McLarty, D. E. Ioannou, and H. L. Hughes, "Deep States in Silicon-on-Insulator Substrates Prepared by Oxygen Implantation Using Current Deep Level Transient Spectroscopy," *Appl. Phys. Lett.* **53**, 871–873, Sept. 1988.

[47] M. G. Collet, "An Experimental Method to Analyse Trapping Centres in Silicon at Very Low Concentrations," *Solid-State Electron.* **18**, 1077–1083, Dec. 1975.

[48] I. D. Hawkins and A. R. Peaker, "Capacitance and Conductance Deep Level Transient Spectroscopy in Field-Effect Transistors," *Appl. Phys. Lett.* **48**, 227–229, Jan. 1986.

[49] J. M. Golio, R. J. Trew, G. N. Maracas, and H. Lefèvre, "A Modeling Technique for Characterizing Ion-Implanted Material Using *C-V* and DLTS Data," *Solid-State Electron.* **27**, 367–373, April 1984.

[50] M. J. Kirton and M. J. Uren, "Capture and Emission Kinetics of Individual $Si:SiO_2$ Interface States," *Appl. Phys. Lett.* **48**, 1270–1272, May 1986.

[51] J. W. Farmer, C. D. Lamp, and J. M. Meese, "Charge Transient Spectroscopy," *Appl. Phys. Lett.* **41**, 1063–1065, Dec. 1982.

[52] K. I. Kirov and K. B. Radev, "A Simple Charge-Based DLTS Technique," *Phys. Stat. Sol.* **63a**, 711–716, Feb. 1981.

[53] D. V. Lang, "Deep-Level Transient Spectroscopy: A New Method to Characterize Traps in Semiconductors," *J. Appl. Phys.* **45**, 3023–3032, July 1974.

[54] D. V. Lang, "Fast Capacitance Transient Apparatus: Application to ZnO and O Centers in GaP *p-n* Junctions," *J. Appl. Phys.* **45**, 3014–3022, July 1974.

[55] G. L. Miller, D. V. Lang, and L. C. Kimerling, "Capacitance Transient Spectroscopy," in *Annual Review of Material Science* (R. A. Huggins, R. H.

Bube, and R. W. Roberts, eds.), Annual Reviews, Palo Alto, CA, **7**, 377–448, 1977.

[56] D. V. Lang, "Space-Charge Spectroscopy in Semiconductors," in *Topics in Applied Physics: Thermally Stimulated Relaxation in Solids* (P. Bräunlich, ed.), **37**, Springer, Berlin, 1979, pp. 93–133.

[57] D. S. Day, M. Y. Tsai, B. G. Streetman, and D. V. Lang, "Deep-Level-Transient Spectroscopy: System Effects and Data Analysis," *J. Appl. Phys.* **50**, 5093–5098, Aug. 1979.

[58] S. Misrachi, A. R. Peaker, and B. Hamilton, "A High Sensitivity Bridge for the Measurement of Deep States in Semiconductors," *J. Phys. E: Sci. Instrum.* **13**, 1055–1061, Oct. 1980.

[59] T. I. Chappell and C. M. Ransom, "Modifications to the Boonton 72BD Capacitance Meter for Deep-Level Transient Spectroscopy Applications," *Rev. Sci. Instrum.* **55**, 200–203, Feb. 1984.

[60] H. Lefèvre and M. Schulz, "Double Correlation Technique (DDLTS) for the Analysis of Deep Level Profiles in Semiconductors," *Appl. Phys.* **12**, 45–53, Jan. 1977.

[61] K. Kosai, "External Generation of Gate Delays in a Boxcar Integrator—Application to Deep Level Transient Spectroscopy," *Rev. Sci. Instrum.* **53**, 210–213, Feb. 1982.

[62] G. Goto, S. Yanagisawa, O. Wada, and H. Takanashi, "Determination of Deep-Level Energy and Density Profiles in Inhomogeneous Semi-conductors," *Appl. Phys. Lett.* **23**, 150–151, Aug. 1973.

[63] N. M. Johnson, "Measurement of Semiconductor–Insulator Interface States by Constant-Capacitance, Deep-Level Transient Spectroscopy," *J. Vac. Sci. Technol.* **21**, 303–314, July/Aug. 1982.

[64] G. L. Miller, "A Feedback Method for Investigating Carrier Distributions in Semiconductors," *IEEE Trans. Electron Dev.* **ED-19**, 1103–1108, Oct. 1972.

[65] J. M. Noras, "Thermal Filling Effects on Constant Capacitance Transient Spectroscopy," *Phys. Stat. Sol.* **69a**, K209–K213, Feb. 1982.

[66] M. F. Li and C. T. Sah, "New Techniques of Capacitance-Voltage Measure-ments of Semiconductor Junctions," *Solid-State Electron.* **25**, 95–99, Feb. 1982.

[67] R. Y. DeJule, M. A. Haase, D. S. Ruby, and G. E. Stillman, "Constant Capacitance DLTS Circuit for Measuring High Purity Semiconductors," *Solid-State Electron.* **28**, 639–641, June 1985.

[68] J. J. Shiau, A. L. Fahrenbruch, and R. H. Bube, "A Method to Improve the Speed and Sensitivity of Constant-Capacitance Voltage Transient Measure-ments," *Solid-State Electron.* **30**, 513–518, May 1987.

[69] M. F. Li and C. T. Sah, "A New Method for the Determination of Dopant and Trap Concentration Profiles in Semiconductors," *IEEE Trans. Electron Dev.* **ED-29**, 306–315, Feb. 1982.

[70] N. M. Johnson, D. J. Bartelink, R. B. Gold, and J. F. Gibbons, "Constant-Capacitance DLTS Measurement of Defect-Density Profiles in Semiconduc-tors," *J. Appl. Phys.* **50**, 4828–4833, July 1979.

[71] L. C. Kimerling, "New Developments in Defect Studies in Semiconductors," *IEEE Trans. Nucl. Sci.* **NS-23**, 1497–1505, Dec. 1976.

[72] G. L. Miller, J. V. Ramirez, and D. A. H. Robinson, "A Correlation Method for Semiconductor Transient Signal Measurements," *J. Appl. Phys.* **46**, 2638–2644, June 1975.

[73] M. D. Miller and D. R. Patterson, "Transient Capacitance Deep Level Spectrometry Instrumentation," *Rev. Sci. Instrum.* **48**, 237–239, March 1977.

[74] Y. Tokuda, N. Shimizu, and A. Usami, "Studies of Neutron-Produced Defects in Silicon by Deep-Level Transient Spectroscopy," *Japan. J. Appl. Phys.* **18**, 309–315, Feb. 1979.

[75] A. Rohatgi, J. R. Davis, R. H. Hopkins, and P. G. McMullin, "A Study of Grown-in Impurities in Silicon by Deep-Level Transient Spectroscopy," *Solid-State Electron.* **26**, 1039–1051, Nov. 1983.

[76] G. Couturier, A. Thabti, and A. S. Barrière, "The Baseline Problem in DLTS Technique," *Rev. Phys. Appliqué* **24**, 243–249, Feb. 1989.

[77] J. T. Schott, H. M. DeAngelis, and P. J. Drevinsky, "Capacitance Transient Spectra of Processing- and Radiation-Induced Defects in Silicon Solar Cells," *J. Electron. Mat.* **9**, 419–434, March 1980.

[78] V. I. Strikha, O. V. Tretyak, A. A. Shmatov, and G. M. Mozok, "Problem of Capacitance Relaxation Measurements in Deep Level Spectroscopy," *Sov. Phys. Semicond.* **21**, 400–402, April 1987.

[79] G. Ferenczi and J. Kiss, "Principles of the Optimum Lock-In Averageing in DLTS Measurement," *Acta Phys. Acad. Sci. Hung.* **50**, 285–290, 1981.

[80] P. M. Henry, J. M. Meese, J. W. Farmer, and C. D. Lamp, "Frequency-Scanned Deep-Level Transient Spectroscopy," *J. Appl. Phys.* **57**, 628–630, Jan. 1985.

[81] K. Dmowski and Z. Pióro, "Noise Properties of Analog Correlators with Exponentially Weighted Average," *Rev. Sci. Instrum.* **58**, 2185–2191, Nov. 1987.

[82] M. S. Hodgart, "Optimum Correlation Method for Measurement of Noisy Transients in Solid-State Physics Experiments," *Electron. Lett.* **14**, 388–390, June 1978.

[83] C. R. Crowell and S. Alipanahi, "Transient Distortion and nth Order Filtering in Deep Level Transient Spectroscopy (D^nLTS)," *Solid-State Electron.* **24**, 25–36, Jan. 1981.

[84] E. E. Haller, P. P. Li, G. S. Hubbard, and W. L. Hansen, "Deep Level Transient Spectroscopy of High Purity Germanium Diodes/Detectors," *IEEE Trans. Nucl. Sci.* **NS-26**, 265–270, Feb. 1979.

[85] E. E. Wagner, D. Hiller, and D. E. Mars, "Fast Digital Apparatus for Capacitance Transient Analysis," *Rev. Sci. Instrum.* **51**, 1205–1211, Sept. 1980.

[86] M. D. Jack, R. C. Pack, and J. Henriksen, "A Computer-Controlled Deep-Level Transient Spectroscopy System for Semiconductor Process Control," *IEEE Trans. Electron Dev.* **ED-27**, 2226–2231, Dec. 1980.

[87] T. R. Jervis, W. M. Teter, T. Cole, and D. Dunlavy, "Deep Level Transient Spectroscopy Using CAMAC Components," *Rev. Sci. Instrum.* **53**, 1160–1162, Aug. 1982.

[88] K. Asada and T. Sugano, "Simple Microcomputer-Based Apparatus for

Combined DLTS-*C-V* Measurement," *Rev. Sci. Instrum.* **53**, 1001–1006, July 1982.

[89] C. Y. Chang, W. C. Hsu, C. M. Uang, Y. K. Fang, and W. C. Liu, "A Simple and Low-Cost Personal Computer-Based Automatic Deep-Level Transient Spectroscopy System for Semiconductor Devices Analysis," *IEEE Trans. Instrum. Meas.* **IM-33**, 259–263, Dec. 1984.

[90] K. Hölzlein, G. Pensl, M. Schulz, and P. Stolz, "Fast Computer-Controlled Deep Level Transient Spectroscopy System for Versatile Applications in Semiconductors," *Rev. Sci. Instrum.* **57**, 1373–1377, July 1986.

[91] P. D. Kirchner, W. J. Schaff, G. N. Maracas, L. F. Eastman, T. I. Chappell, and C. M. Ransom, "The Analysis of Exponential and Nonexponential Transients in Deep-Level Transient Spectroscopy," *J. Appl. Phys.* **52**, 6462–6470, Nov. 1981.

[92] K. Ikossi-Anastasiou and K. P. Roenker, "Refinements in the Method of Moments for Analysis of Multiexponential Capacitance Transients in Deep-Level Transient Spectroscopy," *J. Appl. Phys.* **61**, 182–190, Jan. 1987.

[93] K. Ikeda and H. Takaoka, "Deep Level Fourier Spectroscopy for Determination of Deep Level Parameters," *Japan. J. Appl. Phys.* **21**, 462–466, March 1982.

[94] M. Okuyama, H. Takakura, and Y. Hamakawa, "Fourier-Transformation Analysis of Deep Level Transient Signals in Semiconductors," *Solid-State Electron.* **26**, 689–694, July 1983.

[95] S. Weiss and R. Kassing, "Deep Level Transient Fourier Spectroscopy (DLTFS)—A Technique for the Analysis of Deep Level Properties," *Solid-State Electron.* **31**, 1733–1742, Dec. 1988.

[96] J. E. Stannard, H. M. Day, M. L. Bark, and S. H. Lee, "Spectroscopic Line Fitting to DLTS Data," *Solid-State Electron.* **24**, 1009–1013, Nov. 1981.

[97] F. R. Shapiro, S. D. Senturia, and D. Adler, "The Use of Linear Predictive Modeling for the Analysis of Transients from Experiments on Semiconductor Defects," *J. Appl. Phys.* **55**, 3453–3459, May 1984.

[98] M. Henini, B. Tuck, and C. J. Paull, "A Microcomputer-Based Deep Level Transient Spectroscopy (DLTS) System," *J. Phys. E: Sci. Instrum.* **18**, 926–929, Nov. 1985.

[99] R. Langfeld, "A New Method of Analysis of DLTS-Spectra," *Appl. Phys.* **A44**, 107–110, Oct. 1987.

[100] K. L. Wang and A. O. Evwaraye, "Determination of Interface and Bulk-Trap States of IGFET's Using Deep-Level Transient Spectroscopy," *J. Appl. Phys.* **47**, 4574–4577, Oct. 1976.

[101] K. L. Wang, "Determination of Processing-Related Interface States and Their Correlation with Device Properties," in *Semiconductor Silicon 1977* (H. R. Huff and E. Sirtl, eds.), Electrochem. Soc., Princeton, NJ, pp. 404–413.

[102] K. L. Wang, "A Determination of Interface State Energy during the Capture of Electrons and Holes Using DLTS," *IEEE Trans. Electron Dev.* **ED-26**, 819–821, May 1979.

[103] K. L. Wang, "MOS Interface-State Density Measurements Using Transient

Capacitance Spectroscopy," *IEEE Trans. Electron Dev.* **ED-27**, 2231–2239, Dec. 1980.

[104] M. Schulz and N. M. Johnson, "Transient Capacitance Measurements of Hole Emission from Interface States in MOS Structures," *Appl. Phys. Lett.* **31**, 622–625, Nov. 1977.

[105] M. Schulz and N. M. Johnson, "Evidence for Multiphonon Emission from Interface States in MOS Structures," *Solid State Comm.* **25**, 481–484, Feb. 1978; errata: *Solid State Comm.* **26**, 126i, April 1978.

[106] K. Yamasaki, M. Yoshida, and T. Sugano, "Deep Level Transient Spectroscopy of Bulk Traps and Interface States in Si MOS Diodes," *Japan. J. Appl. Phys.* **18**, 113–122, Jan. 1979.

[107] F. Murray, R. Carin, and P. Bogdanski, "Determination of High-Density Interface State Parameters in Metal-Insulator-Semiconductor Structures by Deep-Level Transient Spectroscopy," *J. Appl. Phys.* **60**, 3592–3598, Nov. 1986.

[108] T. J. Tredwell and C. R. Viswanathan, "Determination of Interface-State Parameters in a MOS Capacitor by DLTS," *Solid-State Electron.* **23**, 1171–1178, Nov. 1980.

[109] N. M. Johnson, D. J. Bartelink, and M. Schulz, "Transient Capacitance Measurements of Electronic States at the Si-SiO$_2$ Interface," in *The Physics of SiO$_2$ and Its Interfaces* (S. T. Pantelides, ed.), Pergamon, New York, 1978, pp. 421–427.

[110] T. Katsube, K. Kakimoto, and T. Ikoma, "Temperature and Energy Dependences of Capture Cross Sections at Surface States in Si Metal-Oxide-Semiconductor Diodes Measured by Deep Level Transient Spectroscopy," *J. Appl. Phys.* **52**, 3504–3508, May 1981.

[111] V. Kumar and S. B. Iyer, "Characterization of Surface States in MOS Capacitors," *Phys. Stat. Sol.* **76a**, 637–640, April 1983.

[112] W. D. Eades and R. M. Swanson, "Improvements in the Determination of Interface State Density Using Deep Level Transient Spectroscopy," *J. Appl. Phys.* **56**, 1744–1751, Sept. 1984.

[113] W. D. Eades and R. M. Swanson, "Determination of the Capture Cross Section and Degeneracy Factor of Si-SiO$_2$ Interface States," *Appl. Phys. Lett.* **44**, 988–990, May 1984.

[114] J. Stannard, "Transient Capacitance in GaAs and InP MOS Capacitors," *J. Vac. Sci. Technol.* **15**, 1508–1512, July/Aug. 1978.

[115] B. Monemar and H. G. Grimmeiss, "Optical Characterization of Deep Energy Levels in Semiconductors," *Progr. Cryst. Growth Charact.* **5**, 47–88, Jan. 1982.

[116] H. G. Grimmeiss, "Deep Level Impurities in Semiconductors," in *Annual Review of Material Science* (R. A. Huggins, R. H. Bube, and R. W. Roberts, eds.), Annual Reviews, Palo Alto, CA, **7**, pp. 341–376, 1977.

[117] A. Chantre, G. Vincent, and D. Bois, "Deep-Level Optical Spectroscopy in GaAs," *Phys. Rev.* **B23**, 5335–5339, May 1981.

[118] P. M. Mooney, "Photo-Deep Level Transient Spectroscopy: A Technique to Study Deep Levels in Heavily Compensated Semiconductors," *J. Appl. Phys.* **54**, 208–213, Jan. 1983.

[119] C. T. Sah, L. L. Rosier, and L. Forbes, "Direct Observation of the Multiplicity of Impurity Charge States in Semiconductors from Low-Temperature High-Frequency-Capacitance," *Appl. Phys. Lett.* **15**, 316–318, Nov. 1969.

[120] A. M. White, P. J. Dean, and P. Porteous, "Photocapacitance Effects of Deep Traps in Epitaxial GaAs," *J. Appl. Phys.* **47**, 3230–3239, July 1976.

[121] S. Dhar, P. K. Bhattacharya, F. Y. Juang, W. P. Hong, and R. A. Sadler, "Dependence of Deep-Level Parameters in Ion-Implanted GaAs MESFET's on Material Preparation," *IEEE Trans. Electron Dev.* **ED-33**, 111–118, Jan. 1986.

[122] M. Takikawa and T. Ikoma, "Photo-Excited DLTS: Measurement of Minority-Carrier Traps," *Japan. J. Appl. Phys.* **19**, L436–L438, July 1980.

[123] C. Hurtes, M. Boulou, A. Mitonneau, and D. Bois, "Deep-Level Spectroscopy in High-Resistivity Materials," *Appl. Phys. Lett.* **32**, 821–823, June 1978.

[124] J. K. Rhee and P. K. Bhattacharya, "Photoinduced Current Transient Spectroscopy of Semi-insulating InP:Fe and InP:Cr," *J. Appl. Phys.* **53**, 4247–4249, June 1982.

[125] R. E. Kremer, M. C. Arikan, J. C. Abele, and J. S. Blakemore, "Transient Photoconductivity Measurements in Semi-insulating GaAs. I. An Analog Approach," *J. Appl. Phys.* **62**, 2424–2431, Sept. 1987, and references therein.

[126] J. C. Abele, R. E. Kremer, and J. S. Blakemore, "Transient Photoconductivity Measurements in Semi-insulating GaAs. II. A Digital Approach," *J. Appl. Phys.* **62**, 2432–2438, Sept. 1987.

[127] O. Yoshie and M. Kamihara, "Photo-Induced Current Transient Spectroscopy in High-Resistivity Bulk Material. I. Computer Controlled Multichannel PICTS System with High Resolution," *Japan. J. Appl. Phys.* **22**, 621–628, April 1983.

[128] O. Yoshie and M. Kamihara, "Photo-Induced Current Transient Spectroscopy in High-Resistivity Bulk Material. II. Influence of Non-exponential Transient on Determination of Deep Trap Parameters," *Japan. J. Appl. Phys.* **22**, 629–635, April 1983.

[129] O. Yoshie and M. Kamihara, "Photo-Induced Current Transient Spectroscopy in High-Resistivity Bulk Material. III. Scanning-PICTS System for Imaging Spatial Distributions of Deep Traps in Semi-insulating GaAs Wafer," *Japan. J. Appl. Phys.* **24**, 431–440, April 1985.

[130] D. C. Look, "The Electrical and Photoelectronic Properties of Semi-Insulating GaAs," in *Semiconductors and Semimetals* (R. K. Willardson and A. C. Beer, eds.) Academic Press, Orlando, FL, **19**, pp. 75–170, 1983.

[131] M. R. Burd and R. Braunstein, "Deep Levels in Semi-Insulating Liquid Encapsulated Czochralski-Grown GaAs," *J. Phys. Chem. Sol.* **49**, 731–735, 1988.

[132] J. C. Balland, J. P. Zielinger, C. Noguet, and M. Tapiero, "Investigation of Deep Levels in High-Resistivity Bulk Materials by Photo-Induced Current Transient Spectroscopy: I. Review and Analysis of Some Basic Problems," *J. Phys. D.: Appl. Phys.* **19**, 57–70, Jan. 1986.

[133] J. C. Balland, J. P. Zielinger, M. Tapiero, J. G. Gross, and C. Noguet, "Investigation of Deep Levels in High-Resistivity Bulk Materials by Photo-Induced Current Transient Spectroscopy: II. Evaluation of Various Signal Processing Methods," *J. Phys. D.: Appl. Phys.* **19**, 71–87, Jan. 1986.

[134] C. C. Tin, C. K. Teh, and F. L. Weichman, "Photoinduced Transient Spectroscopy and Photoluminescence Studies of Copper Contaminated Liquid-Encapsulated Czochralski-Grown Semi-insulating GaAs," *J. Appl. Phys.* **62**, 2329–2336, Sept. 1987.

[135] P. M. Petroff and D. V. Lang, "A New Spectroscopic Technique for Imaging the Spatial Distribution of Non-radiative Defects in a Scanning Transmission Electron Microscope," *Appl. Phys. Lett.* **31**, 60–62, July 1977.

[136] O. Breitenstein, "A Capacitance Meter of High Absolute Sensitivity Suitable for Scanning DLTS Application," *Phys. Stat. Sol.* **71a**, 159–167, May 1982.

[137] K. Wada, K. Ikuta, J. Osaka, and N. Inoue, "Analysis of Scanning Deep Level Transient Spectroscopy," *Appl. Phys. Lett.* **51**, 1617–1619, Nov. 1987.

[138] J. Heydenreich and O. Breitenstein, "Characterization of Defects in Semiconductors by Combined Application of SEM (EBIC) and SDLTS," *J. Microsc.* **141**, 129–142, Feb. 1986.

[139] M. C. Chen, D. V. Lang, W. C. Dautremont-Smith, A. M. Sergent, and J. P. Harbison, "Effects of Leakage Current on Deep Level Transient Spectroscopy," *Appl. Phys. Lett.* **44**, 790–793, April 1984.

[140] D. S. Day, M. J. Helix, K. Hess, and B. G. Streetman, "Deep Level Transient Spectroscopy for Diodes with Large Leakage Currents," *Rev. Sci. Instrum.* **50**, 1571–1573, Dec. 1979.

[141] E. V. Astrova, A. A. Lebedev, and A. A. Lebedev, "Influence of Series Resistance of a Diode on Transient Capacitance Measurements of Deep-Level Parameters," *Sov. Phys. Semicond.* **19**, 850–852, Aug. 1985.

[142] A. Broniatowski, A. Blosse, P. C. Srivastava, and J. C. Bourgoin, "Transient Capacitance Measurements on Resistive Samples," *J. Appl. Phys.* **54**, 2907–2910, June 1983.

[143] T. Thurzo and F. Dubecky, "On the Role of the Back Contact in DLTS Experiments with Schottky Diodes," *Phys. Stat. Sol.* **89a**, 693–698, June 1985.

[144] J. H. Zhao, J. C. Lee, Z. Q. Fang, T. E. Schlesinger, and A. G. Milnes, "The Effects of the Nonabrupt Depletion Edge on Deep-Trap Profiles Determined by Deep-Level Transient Spectroscopy," *J. Appl. Phys.* **61**, 5303–5307, June 1987; errata *ibid.* p. 5489.

[145] S. D. Brotherton, "The Width of the Non-steady State Transition Region in Deep Level Impurity Measurements," *Solid-State Electron.* **26**, 987–990, Oct. 1983.

[146] D. Stievenard and D. Vuillaume, "Profiling of Defects Using Deep Level Transient Spectroscopy," *J. Appl. Phys.* **60**, 973–979, Aug. 1986.

[147] L. R. Weisberg and H. Schade, "A Technique for Trap Determination in Low-Resistivity Semiconductors," *J. Appl. Phys.* **39**, 5149–5151, Oct. 1968.

[148] M. C. Driver and G. T. Wright, "Thermal Release of Trapped Space Charge in Solids," *Proc. Phys. Soc. (London)* **81**, 141–147, Jan. 1963.

[149] J. C. Carballes and J. Lebailly, "Trapping Analysis in Gallium Arsenide," *Solid State Commun.* **6**, 167–171, March 1968.

[150] M. G. Buehler and W. E. Phillips, "A Study of the Gold Acceptor in a Silicon p^+n Junction and an n-Type MOS Capacitor by Thermally Stimulated Current and Capacitance Measurements," *Solid-State Electron.* **19**, 777–788, Sept. 1976.

[151] F. Reif, *Fundamentals of Statistical and Thermal Physics*, McGraw-Hill, New York, 1965, pp. 161–166.

[152] O. Engström and A. Alm, "Thermodynamical Analysis of Optimal Recombination Centers in Thyristors," *Solid-State Electron.* **21**, 1571–1576, Nov./Dec. 1978.

[153] O. Engström and A. Alm, "Energy Concepts of Insulator-Semiconductor Interface Traps," *J. Appl. Phys.* **54**, 5240–5244, Sept. 1983.

[154] C. D. Thurmond, "The Standard Thermodynamic Functions for the Formation of Electrons and Holes in Ge, Si, GaAs, and GaP," *J. Electrochem. Soc.* **122**, 1133–1141, Aug. 1975.

[155] J. A. Van Vechten and C. D. Thurmond, "Entropy of Ionization and Temperature Variation of Ionization Levels of Defects in Semiconductors," *Phys. Rev.* **B14**, 3539–3550, Oct. 1976.

[156] A. Mircea, A. Mitonneau, and J. Vannimenus, "Temperature Dependence of Ionization Energies of Deep Bound States in Semiconductors," *J. Physique* **38**, L41–L43, Jan. 1972.

[157] H. Okushi and Y. Takumaru, "Isothermal Capacitance Transient Spectroscopy for Determination of Deep Level Parameters," *Japan. J. Appl. Phys.* **19**, L335–L338, June 1980.

[158] F. Hasegawa, "A New Method (the Three-Point Method) of Determining Transient Time Constants and Its Application to DLTS," *Japan. J. Appl. Phys.* **24**, 1356–1358, Oct. 1985.

[159] J. M. Steele, "Hasegawa's Three Point Method for Determining Transient Time Constant," *Japan. J. Appl. Phys.* **25**, 1136–1137, July 1986.

[160] P. C. Mangelsdorf, Jr., "Convenient Plot for Exponential Functions with Unknown Asymptotes," *J. Appl. Phys.* **30**, 442–443, March 1959.

[161] K. Ikeda and H. Takaoka, "Deep Level Fourier Spectroscopy for Determination of Deep Level Parameters," *Japan. J. Appl. Phys.* **21**, 462–466, March 1982.

[162] J. L. Benton and L. C. Kimerling, "Capacitance Transient Spectroscopy of Trace Contamination in Silicon," *J. Electrochem. Soc.* **129**, 2098–2102, Sept. 1982.

CHAPTER 8

CARRIER LIFETIME

8.1 INTRODUCTION

This chapter is devoted to lifetime measurement techniques. This topic would appear to be straightforward since the concept of electron and hold lifetimes in semiconductors is, in principle, quite simple. However, in practice, there are often as many lifetime values for a given device as there are measurement techniques. A survey of lifetime measurement techniques in 1968 yielded 300 papers published from 1959.[1] There have probably been several hundred papers published since then. In practice, then, "lifetime" is quite a complex concept. Not only does it matter how it is measured, but the operation of the device in which it is measured also plays an important role.

Lifetimes fall into two primary categories: *recombination lifetimes* and *generation lifetimes*. The concept of recombination lifetime τ_r holds when excess carriers, introduced by light or by a forward-biased *pn* junction, decay as a result of recombination. Generation lifetime τ_g applies when there is a paucity of carriers, as in the space-charge region (scr) of a reverse-biased diode, and the device tries to reach equilibrium. During recombination an electron–hole pair (ehp) ceases to exist on average after a time τ_r. The generation lifetime, by analogy, is the time that it takes on average to generate an ehp.

When these recombination and generation events occur in the bulk, they are characterized by τ_r and τ_g. When they occur at the surface, they are characterized by the *surface recombination velocity* s_r and the *surface generation velocity* s_g. Since devices consist of bulk regions and surfaces, both bulk and surface recombination or generation take place simultaneously, and their separation is sometimes quite difficult. Some methods allow this separation, others do not.

Before discussing lifetime measurement techniques, it is instructive to consider τ_r and τ_g in more detail. Those readers not interested in these details can skip these sections and go directly to the measurement methods. Consider a p-type semiconductor in which excess ehp's have been introduced. They may have been generated by photons or particles of energy larger than the band gap or by forward biasing a pn junction. If they are generated by photons or particles, then ehp's are created in equal numbers. Forward biasing injects minority carriers only. The corresponding majority carrier partners are supplied by the ohmic contact. In either case there will be more carriers after the stimulus than before, and the excess carriers try to return to equilibrium by recombination.

A detailed derivation of the relevant equations for optical and electrical excitation is given in Appendix 8.1. Some of those equations are used in this chapter.

8.2 RECOMBINATION LIFETIME/SURFACE RECOMBINATION VELOCITY

The bulk recombination rate U depends non-linearly on the departure of the carrier concentrations from their equilibrium values. We consider a p-type semiconductor throughout this chapter and are chiefly concerned with the behavior of the minority electrons. If we confine ourselves to linear, quadratic, and third-order terms, then U can be written as[2]

$$U = A(n - n_0) + B(pn - p_0 n_0) + C_p(p^2 n - p_0^2 n_0) + C_n(pn^2 - p_0 n_0^2)$$

(8.1)

where $n = n_0 + \Delta n$ and $p = p_0 + \Delta p$. In the absence of trapping, $\Delta n = \Delta p$, allowing Eq. (8.1) to be simplified to

$$U \approx A\Delta n + B(p_0 + n_0 + \Delta n)\Delta n + C_p(p_0^2 + 2p_0\Delta n + \Delta n^2)\Delta n$$
$$+ C_n(n_0^2 + 2n_0\Delta n + \Delta n^2)\Delta n$$

(8.2)

where terms containing n_0 have been dropped because $n_0 \ll p_0$ in a p-type material.

The recombination lifetime is defined as

$$\tau_r = \frac{\Delta n}{U}$$

(8.3)

giving

$$\tau_r = [A + B(p_0 + n_0 + \Delta n) + C_p(p_0^2 + 2p_0\Delta n + \Delta n^2)$$
$$+ C_n(n_0^2 + 2n_0\Delta n + \Delta n^2)]^{-1}$$

(8.4)

Three lifetimes are contained in Eq. (8.4); each is associated with a distinct physical mechanism. The first term represents *multiphonon* or *Shockley-Read-Hall* (SRH) recombination, shown in Fig. 8.1(a). The ehp's recombine through deep-level impurities. The impurity concentration is N_T, and their energy level is E_T. Impurities are furthermore characterized by capture cross sections σ_n and σ_p for electrons and holes, respectively. The energy liberated during the recombination event is dissipated by lattice vibrations or phonons.

The SRH lifetime, $1/A$ in Eq. (8.4), is[3]

$$\tau_{\text{SRH}} = \frac{\tau_p(n_0 + n_1 + \Delta n) + \tau_n(p_0 + p_1 + \Delta n)}{p_0 + n_0 + \Delta n} \cdot \qquad (8.5)$$

with n_1, p_1, τ_n, and τ_p defined as

$$n_1 = n_i \exp\left(\frac{E_T - E_i}{kT}\right), \qquad p_1 = n_i \exp\left(-\frac{E_T - E_i}{kT}\right)$$

$$\tau_p = \frac{1}{\sigma_p v_{\text{th}} N_T}, \qquad \tau_n = \frac{1}{\sigma_n v_{\text{th}} N_T}$$

The second term in Eq. (8.4) is the *radiative* lifetime shown in Fig. 8.1(b).[4] Electron-hole pairs recombine from band to band with the energy carried away by photons. The radiative lifetime is

$$\tau_{\text{rad}} = [B(p_0 + n_0 + \Delta n)]^{-1} \qquad (8.6)$$

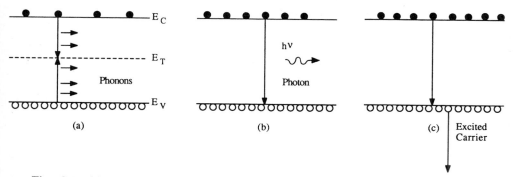

Fig. 8.1 (a) Multiphonon or Shockley-Read-Hall recombination, (b) radiative recombination, (c) Auger recombination.

in which B is the radiative recombination coefficient. Radiative recombination coefficients are given in Table 8.1. The radiative lifetime is inversely proportional to the carrier concentration because in band-to-band processes both electrons and holes must be present simultaneously for a recombination event to take place.

The third term in Eq. (8.4) is the three-carrier *Auger* recombination process with the energy given to a third carrier, as shown in Fig. 8.1(c).[9] Because three carriers are involved in the recombination event, the Auger lifetime is inversely proportional to the square of the carrier concentration

$$\tau_{\text{Auger}} = [C_p(p_0^2 + 2p_0\Delta n + \Delta n^2) + C_n(n_0^2 + 2n_0\Delta n + \Delta n^2)]^{-1} \quad (8.7)$$

C_p and C_n are the Auger recombination coefficients.

Equations (8.5) to (8.7) simplify for both low-level and high-level injection. Low-level injection conditions apply when the *excess minority* carrier concentration is small compared to the *equilibrium majority* carrier concentration, $\Delta n \ll p_0$. Similarly high-level injection holds when $\Delta n \gg p_0$. The injection level is important during lifetime measurements. Low-level injection is used for most methods. The appropriate expressions for *low-level* (*ll*) and the *high-level* (*hl*) injection become

$$\tau_{\text{SRH}}(ll) = \tau_n, \quad \tau_{\text{SRH}}(hl) = \tau_n + \tau_p \quad (8.8)$$

$$\tau_{\text{rad}}(ll) = \frac{1}{Bp_0}, \quad \tau_{\text{rad}}(hl) = \frac{1}{B\Delta n} \quad (8.9)$$

$$\tau_{\text{Auger}}(ll) = \frac{1}{C_p p_0^2} \quad \tau_{\text{Auger}}(hl) = \frac{1}{(C_p + C_n)\Delta n^2} \quad (8.10)$$

TABLE 8.1 Recombination Coefficients

Semicondutor	Temperature (K)	Radiative Recombination Coefficient (B in cm^3/s)	Auger Recombination Coefficient (C in cm^6/s)
Si	300	10^{-14} [5]	$C_n = 2.8 \times 10^{-31}$, $C_p = 10^{-31}$ [10 D/S]
Si	300		$C_n + C_p = 2 \times 10^{-30}$ [10 Y/G]
Si	300		$C_n + C_p = 1.65 \times 10^{-30}$ [10 S/S]
Si	100	10^{-13} [8]	
Si	400	6×10^{-15} [8]	
Ge	300	5.2×10^{-14} [4]	$C_n = 8 \times 10^{-32}$, $C_p = 2.8 \times 10^{-31}$
GaAs	300	2×10^{-10} [6 N/S]	$C_n = 1.8 \times 10^{-31}$, $C_p = 4 \times 10^{-30}$ [6 T]
GaAs	300	1.3×10^{-10} [6 't Hooft]	
GaP	300	5.4×10^{-14} [4]	
InSb	300	4.6×10^{-11} [4]	
InGaAsP	300	4×10^{-10} [7]	$C_n + C_p = 8 \times 10^{-29}$ [7]

An assumption in the SRH expressions is that for low-level injection $p_0 \gg n_1, p_1$ and for high-level injection $\Delta n \gg n_1, p_1$. The conditions are generally satisfied except for high resistivity material where Eq. (8.5) must be used.

When do these three lifetimes apply? Multiphonon or SRH recombination takes place whenever there are impurities or defects in the semiconductor. Since there are always some impurities, this mechanism is always active. It is particularly important for indirect band-gap semiconductors like Si, Ge, and GaP. The SRH lifetime depends inversely on the concentration of the recombination centers and the capture cross sections but does not depend directly on the energy level of the impurity. It does depend indirectly on the energy level because the capture cross section tends to be highest for impurities with energy levels near the middle of the band gap and lowest for E_T near either the conduction or valence band. There is, however, no clear rule—only trends. Band-to-band multiphonon recombination is very unlikely.

Radiative recombination is important in direct band gap materials like GaAs and InP, where the conduction band minimum lies at the same crystal momentum as the valence band maximum. During the recombination event phonons are not required since the energy is dissipated by photons.

Auger recombination is typically observed in either direct or indirect band gap semiconductors when either the majority carrier or excess minority carrier concentration is very high. Similar to the radiative lifetime, the Auger lifetime is independent of any impurity concentration. It is the ultimate recombination mechanism for devices operating at very high injection levels (e.g., laser diodes or power semiconductor devices) or at very high doping levels (e.g., bipolar junction transistor emitter or the shallow n^+ layer in solar cells). It is also a dominant recombination mechanism for narrow band gap semiconductors (e.g., HgCdTe used for infrared detectors). Auger recombination coefficients are given in Table 8.1.

Both radiative and Auger recombination can also occur through intermediate energy levels, as shown in Fig. 8.2. When ehp's recombine radiatively through intermediate levels, as in Fig. 8.2(a), photons are generally not given off for both events. This recombination mechanism is utilized in light-emitting phosphors where semiconductors like CdS are deliberately doped with impurities of a particular kind to generate light of a particular wavelength. A good example of this is the cathode ray tube used in color television receivers where excess carriers in phosphors bombarded by an electron beam emit red, blue, or green light upon recombination. Auger recombination through intermediate energy levels is possible for heavily doped semiconductors.[9]

Experimental values of lifetimes for Si are given in Fig. 8.3. These data suggest that in the high doping limit the lifetimes follow the predicted Auger-limit $1/n_0^2$ or $1/p_0^2$ behavior quite well with $C_p = 10^{-31}$ cm^6/s and $C_n = 2.8 \times 10^{-31}$ cm^6/s. Recent data of very high lifetimes also exhibit the

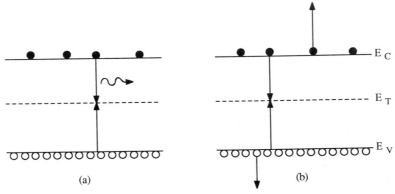

Fig. 8.2 (a) Radiative and (b) Auger recombination through deep-level impurities.

$1/n_0^2$ or the $1/p_0^2$ behavior, but the Auger coefficients are higher as shown in Table 8.1. The inverse square law relationship for Auger recombination does not necessarily hold when the conditions above are not satisfied.[11]

In addition to bulk SRH recombination there is also surface or interface SRH recombination. Where the *bulk SRH recombination rate* is given by[3]

$$U = \frac{\sigma_n \sigma_p v_{th} N_T (pn - n_i^2)}{\sigma_n (n + n_1) + \sigma_p (p + p_1)} = \frac{pn - n_i^2}{\tau_p (n + n_1) + \tau_n (p + p_1)} \quad (8.11)$$

the *surface SRH recombination rate* is

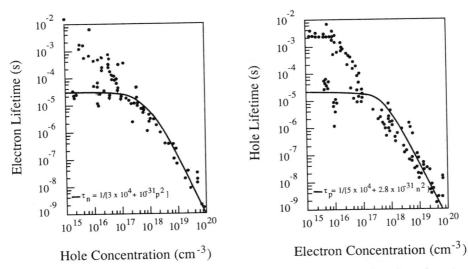

Fig. 8.3 Electron and hole recombination lifetimes in silicon as a function of carrier concentrations.

$$U_s = \frac{\sigma_{ns}\sigma_{ps}v_{th}N_{it}(p_s n_s - n_i^2)}{\sigma_{ns}(n_s + n_{1s}) + \sigma_{ps}(p_s + p_{1s})} = \frac{s_n s_p (p_s n_s - n_i^2)}{s_n(n_s + n_{1s}) + s_p(p_s + p_{1s})}$$

$$(8.12)$$

where

$$s_n = \sigma_{ns}v_{th}N_{it}, \quad s_p = \sigma_{ps}v_{th}N_{it}$$

The subscript "s" refers to the appropriate quantity at the surface; p_s and n_s are the hole and electron concentrations/cm^3 at the surface. The interface trap density N_{it} is assumed constant in Eq. (8.12). If it is not constant, then the interface trap density D_{it} must be integrated over energy. It has been shown that N_{it} in these equations is $N_{it} = kTD_{it}$.[12]

We define the surface recombination velocity as

$$s_r = \frac{U_s}{\Delta n_s} \tag{8.13}$$

From Eq. (8.12)

$$s_r = \frac{s_n s_p (p_{s0} + n_{s0} + \Delta n_s)}{s_n(n_{s0} + n_{1s} + \Delta n_s) + s_p(p_{s0} + p_{1s} + \Delta p_s)} \tag{8.14}$$

The surface recombination velocity for low-level and high-level injection becomes

$$s_r(ll) = s_n, \quad s_r(hl) = \frac{s_n s_p}{s_n + s_p} \tag{8.15}$$

8.3 GENERATION LIFETIME/SURFACE GENERATION VELOCITY

Each of the recombination processes of Fig. 8.1 has a generation counterpart, illustrated in Fig. 8.4. The inverse of multiphonon recombination is the thermal ehp generation in Fig. 8.4(a). From the SRH recombination/generation rate expression in Eq. (8.11), it is obvious that when $pn > n_i^2$, recombination dominates; for $pn < n_i^2$, generation dominates. Furthermore the smaller the pn product, the higher is the generation rate. In the limit of $pn \to 0$, U becomes negative and is then designated the *generation rate G*

$$G = -U = \frac{n_i^2}{\tau_p n_1 + \tau_n p_1} = \frac{n_i}{\tau_g} \tag{8.16}$$

with

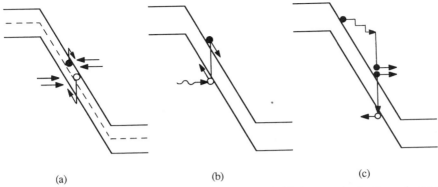

Fig. 8.4 (a) Shockley-Read-Hall, (b) optical, and (c) impact ionization electron-hole pair generation.

$$\tau_g = \tau_p \exp\left(\frac{E_T - E_i}{kT}\right) + \tau_n \exp\left(-\frac{E_T - E_i}{kT}\right) \qquad (8.17)$$

The condition $pn \to 0$ is approximated in the scr of a reverse-biased junction.

The quantity τ_g, defined in Eq. (8.17), is the *generation lifetime*.[13] It is the time to generate one ehp by thermal emission processes, and as evident from Eq. (8.17), it depends inversely on the impurity concentration and the capture cross section for electrons and holes, just as recombination does. In addition, however, it depends exponentially on the energy level E_T. The generation lifetime can be quite large if E_T does not coincide with E_i. Generally, τ_g is higher than τ_r, at least for Si devices, where detailed comparisons have been made and $\tau_g \approx (50 - 100)\tau_r$.[13]

For radiative and Auger recombination, the inverse processes are ehp generation by optical absorption and impact ionization, illustrated in Fig. 8.4(b) and (c). Optical generation is negligible for a device not exposed to light and with negligible blackbody radiation from its surroundings. Impact ionization is usually considered to be negligible if the device is biased sufficiently below its breakdown voltage. Recently, however, it has been shown that impact ionization at low ionization rates can occur at quite low voltages and care must be taken to eliminate this generation mechanism during τ_g lifetime measurements.

When $p_s n_s < n_i^2$ at the surface, we find from Eq. (8.12)

$$G_S = -U_S = \frac{s_n s_p n_i^2}{s_n n_{1s} + s_p p_{1s}} = s_g n_i \qquad (8.18)$$

where s_g is the surface generation velocity, frequently designated as s_0 (see note in Grove [14]), given by

$$s_g = \frac{s_n s_p}{s_n \exp[(E_{it} - E_i)/kT] + s_p \exp[-(E_{it} - E_i)/kT]} \qquad (8.19)$$

For interface trap energies not at E_i, we find $s_r > s_g$ from Eqs. (8.14) and (8.19).

8.4 RECOMBINATION LIFETIME—OPTICAL MEASUREMENT TECHNIQUES

8.4.1 Photoconductive Decay

The *photoconductive decay* (PCD) lifetime technique was developed in 1955.[15] As the name implies, ehp's are created by optical excitation, and their decay is monitored as a function of time following the cessation of the excitation. Other excitation means such as high energy electrons[16] and gamma rays[17] can also be used. The samples have traditionally been contacted, and the current is monitored. More recently, noncontacting techniques have been developed that make the method fast and convenient.

The presence of excess carriers in PCD can be detected in a number of ways. The conductivity σ

$$\sigma = q(\mu_n n + \mu_p p) \qquad (8.20)$$

is usually monitored as a function of time. $n = n_0 + \Delta n$, $p = p_0 + \Delta p$, and we assume both equilibrium and excess carriers to have identical mobilities. This is true under low-level injection when Δn and Δp are small compared to the equilibrium majority carrier concentration. This assumption is no longer true for high optical excitation, however, because carrier–carrier scattering reduces the mobilities.

In some PC decay methods the time-dependent excess carrier concentration is measured directly while in others it is measured indirectly. For insignificant trapping, $\Delta n = \Delta p$, and the excess carrier concentration is related to the conductivity by

$$\Delta n = \frac{\Delta \sigma}{q(\mu_n + \mu_p)} \qquad (8.21)$$

A measure of $\Delta \sigma$ is a measure of Δn, provided the mobilities are constant during the measurement. The time dependence of the carrier decay after an optical pulse is generally a complicated function, as discussed in more detail in Appendix 8.1.[18-21] The carrier decay, governed by the bulk recombination lifetime τ_B and the surface recombination velocity s_r, is a sum of exponentials. The higher-order terms decay more rapidly with time than the first term and may therefore be neglected after an initial transient period. To be safe, one should wait for the transient to have decayed to about half of its maximum value before measuring the time constant.

For $\Delta n(t) = \Delta n(0) \exp(-t/\tau_{\text{eff}})$, the effective lifetime τ_{eff} is given by

$$\frac{1}{\tau_{\text{eff}}} = \frac{1}{\tau_{\text{B}}} + D\beta^2 \tag{8.22}$$

with β found from the relationship

$$\beta \tan\left(\frac{\beta T}{2}\right) = \frac{s_{\text{r}}}{D} \tag{8.23}$$

where τ_{B} is the bulk recombination lifetime, $D = D_p D_n (n + p)/(nD_n + pD_p)$ is the ambipolar diffusion coefficient, and T is the sample thickness. Equation (8.22) holds for any optical absorption depth, provided the excess carrier concentration has ample time to distribute uniformly; that is, $T \ll (Dt)^{1/2}$. The effective lifetime of Eq. (8.22) is plotted in Fig. 8.5 versus T as a function of s_{r}. These figures show the effective lifetime to depend on T and s_{r}. The surface recombination velocity must be known to determine τ_{B} unambiguously unless the sample is sufficiently thick. The surface recombination velocity of a sample is generally not known, but by providing the sample with high s_{r}, with sandblasting, for example, it is possible to determine τ_{B} directly. However, the sample must be extraordinarily thick. For the limiting value of $s_{\text{r}} \approx 10^7$ cm/s, the sample must be about 1 mm for $\tau_{\text{B}} = 10\ \mu\text{s}$, 3–4 mm for $\tau_{\text{B}} = 100\ \mu\text{s}$, and 1 cm thick for $\tau_{\text{B}} = 1000\ \mu\text{s}$ to be able to neglect surface effects.

Two limiting cases are of particular interest. First, for low surface recombination velocity, $s_{\text{r}} \to 0$,

$$\frac{1}{\tau_{\text{eff}}} = \frac{1}{\tau_{\text{B}}} + \frac{2s_{\text{r}}}{T} \tag{8.24}$$

Second, for high s_{r}, $s_{\text{r}} \to \infty$,

$$\frac{1}{\tau_{\text{eff}}} = \frac{1}{\tau_{\text{B}}} + \frac{\pi^2 D}{T^2} \tag{8.25}$$

Writing $1/\tau_{\text{eff}} = 1/\tau_{\text{B}} + 1/\tau_{\text{S}}$, the effective surface lifetime τ_{s} becomes

$$\tau_{\text{S}}(s_{\text{r}} \to 0) = \frac{T}{2s_{\text{r}}} \qquad \tau_{\text{S}}(s_{\text{r}} \to \infty) = \frac{T^2}{\pi^2 D} \tag{8.26}$$

Equations (8.23)–(8.26) hold for samples with one dimension much smaller than the other two dimensions, for example, a wafer. For samples with none of the three dimensions very large, Eq. (8.25) becomes

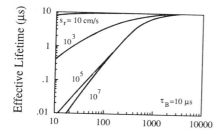

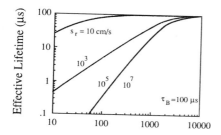

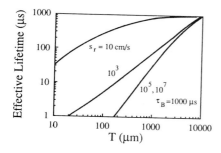

Fig. 8.5 τ_{eff} versus wafer thickness T as a function of surface recombination velocity for $\tau_B = 10$ μs, 100 μs, and 1000 μs. $D = 30$ cm^2/s.

$$\frac{1}{\tau_{eff}} = \frac{1}{\tau_B} + \pi^2 D \left[\frac{1}{a^2} + \frac{1}{b^2} + \frac{1}{c^2} \right] \tag{8.27}$$

where a, b, and c are the sample dimensions. It is recommended that the sample surfaces have high surface recombination velocities, by sandblasting the sample surfaces, for example.[22-23] The recommended dimensions and the maximum bulk lifetimes that can be determined through Eq. (8.27) for Si samples are given in Table 8.2.

A schematic measurement circuit for PC decay is shown in Fig. 8.6, where the photocurrent is measured as a function of time. We follow Ryvkin for the derivation of the appropriate equations.[24] For a sample with dark resistance r_{dk} and steady-state photoresistance r_{ph}, the output voltage change between the dark and the illuminated sample is

$$\Delta V = (i_{ph} - i_{dk}) R \tag{8.28}$$

TABLE 8.2   Recommended Dimensions for PC Decay Samples and Maximum Bulk Lifetime for Silicon

Sample Length (cm)	Sample Width × Height (cm × cm)	Maximum τ_B (μs) n-Si	Maximum τ_B (μs) p-Si
1.5	0.25 × 0.25	240	90
2.5	0.5 × 0.5	950	350
2.5	1 × 1	3600	1340

Source: Proc. IRE[22] and ASTM Standard F28.[23]

where i_{ph}, i_{dk} are the photocurrent and the dark current. With

$$\Delta g = g_{ph} - g_{dk} = (r_{dk} - \Delta r)^{-1} - r_{dk}^{-1} \qquad (8.29)$$

Equation (8.28) becomes

$$\Delta V = \frac{r_{dk}^2 R V_0 \Delta g}{(R + r_{dk})(R + r_{dk} + R r_{dk} \Delta g)} \qquad (8.30)$$

where $\Delta r = r_{dk} - r_{ph}$ and $\Delta g = \Delta \sigma A/L$. According to Eq. (8.30), there is no simple relationship between the time dependence of the measured voltage and the time dependence of the excess carrier concentration.

There are two main versions of the technique in Fig. 8.6: the *constant*

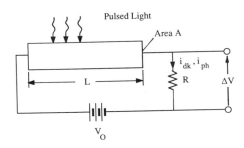

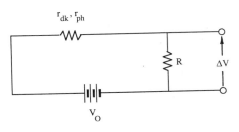

Fig. 8.6   Schematic diagram for photoconductive decay.

voltage method and the *constant current* method. The load resistor R is chosen to be small compared to the sample resistance in the constant voltage method, and Eq. (8.30) becomes

$$\Delta V \approx \frac{\Delta g R V_0}{(1 + \Delta g R)} \approx \Delta g R V_0 \left(1 - \frac{\Delta V}{V_0}\right) \qquad (8.31)$$

For low-level excitation ($\Delta g R \ll 1$ or $\Delta V \ll V_0$) $\Delta V \sim \Delta g \sim \Delta n$; the voltage decay is proportional to the excess carrier concentration. For the constant current case, R is very large, and

$$\Delta V \approx \frac{\Delta g (r_{dk}^2 / R) V_0}{(1 + \Delta g r_{dk})} \approx \Delta g r_{dk} V_0 \left(\frac{r_{dk}}{R} - \frac{\Delta V}{V_0}\right) \qquad (8.32)$$

For $\Delta g r_{dk} \ll 1$ or $\Delta V / V_0 \ll r_{dk} / R$, $\Delta V \sim \Delta g \sim \Delta n$ again.

To determine the time-dependent decay of the excess carriers by the photoconductive voltage decay, the conditions outlined above must be observed. For ΔV to be proportional to Δn, it is necessary that low injection levels be maintained. Otherwise, appropriate corrections must be applied.

The decay of the excess carriers can be monitored in two basic ways. In one, the conductivity or resistivity of the sample is measured. This gives a measure of $\Delta n(t)$ only under the restricted conditions of Eqs. (8.31) and (8.32). The conductivity is determined by providing the sample with contacts and measuring the current. The contacts can be evaporated or plated, as is usually the case for samples in wafer form. The contacts can also be in the form of clamps, often used for samples in the form of uncut ingots. The contacts should not inject minority carriers into the sample, nor should they give rise to sample heating. It is desirable that the illumination be restricted to the non-contacted part of the sample to avoid contact effects or minority carrier sweep-out. The electric field in the sample should be held to a value $\mathscr{E} \leq 0.3 / \sqrt{\mu \tau_r}$ where μ is the minority carrier mobility.[23]

Many circuits have been used for PC decay measurements since its introduction in the 1950s. A recent implementation is shown in Fig. 8.7. The excitation light should be penetrating through the sample. A 1.06-μm laser is suitable for Si. One can also use a light source that is filtered by passing the light through a filter made of the semiconductor to be measured to remove the higher energy light.

The carrier decay can also be monitored without sample contacts. This allows for a fast, nondestructive measure of $\Delta n(t)$. There are several methods to monitor the excess carrier concentration directly. One such technique uses the rf bridge circuit of Fig. 8.8, where the decay time of the photoconductivity produced by a 60-ns burst of near-infrared (0.9-μm) light from a GaAs laser diode array is used to determine the lifetime of Si wafers.[20] The photoconductivity is time resolved with a 100-MHz rf bridge circuit that is capacitively connected to the wafer by two flat electrodes

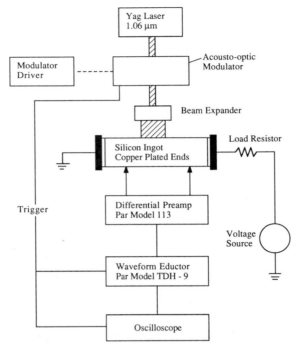

Fig. 8.7 Measurement schematic for photoconductive decay with ohmic contacts. Copyright ASTM. Reprinted with permission after Gerhard and Pearce.[25]

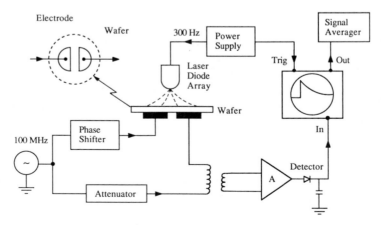

Fig. 8.8 Measurement schematic for contactless photoconductive decay. Reprinted with permission after Tiedje et al.[20]

vacuum pressed against one side of the wafer. The 100-MHz signal is divided in two: one branch going to an attenuator and the other branch containing the sample and a phase shifter. These two branches are combined at the output and amplified. A detailed description of another contactless photoconductive decay circuit is given in Curtis and Verkuil.[26]

In a series of experiments of 19 different Si wafers, Tiedje et al. found good agreement of the lifetimes measured by PCD with lifetimes determined from surface photovoltage measurements.[20] The rf bridge method has proved to be very useful in the development of high-efficiency solar cells.[27–28] A point of interest is the value of $s_r \approx 10^5$ cm/s for wafers etched in conventional HF:NHO$_3$:CH$_3$COOH solutions. This s_r value is neither very high nor very low, making τ_B determination difficult. More recently it has been shown that oxidizing a Si wafer and etching the oxide in HF and then immersing the sample in HF during the PC decay measurement, gives surface recombination velocities as low as 0.25 cm/s.[29] This allows the use of Eq. (8.24), and both τ_B and s_r can be determined by plotting $1/\tau_{eff}$ versus $1/T$. Such a plot has a slope of $2s_r$ and an intercept on the $1/\tau_{eff}$ axis of $1/\tau_B$. The contactless PC decay technique has been extended to lifetime measurements on GaAs by using a Q-switched Nd:YAG laser as the light source.[30] By using inorganic sulfides as passivating layers, surface recombination velocities as low as $s_r \approx 1000$ cm/s were obtained on GaAs samples.

A variation of the contactless technique is the microwave reflection method of Fig. 8.9.[31–33] Excess carriers are created by light pulses and the photoconductivity is monitored by microwave reflection. Microwaves from a Gunn diode at ~10-GHz frequency are directed onto the wafer through a circulator. The microwaves are reflected from the wafer, detected, amplified, and displayed. The change in reflected microwave power ΔP is proportional to the incremental wafer conductivity $\Delta \sigma$[33]

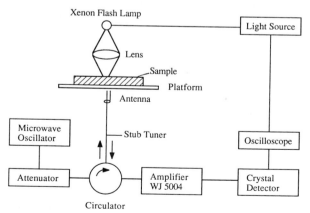

Fig. 8.9 Microwave reflectance contactless photoconductive decay circuit. Reprinted with permission after Mada.[31]

$$\Delta P \sim \Delta \sigma \tag{8.33}$$

The microwaves are not reflected solely from the surface but penetrate a skin depth into the sample. Typical skin depths in Si at 10 GHz range from 350 μm for $\rho = 0.5$ ohm-cm to 2200 μm for $\rho = 10$ ohm-cm. Consequently a good part of the wafer thickness is sampled by the microwaves, and the microwave reflected signal is characteristic of the bulk carrier concentration. The lower limit of τ_r that can be determined depends on the wafer resistivity. Lifetimes as low as 100 ns have been measured.[34]

If a resonant microwave cavity is used in the experimental setup, it is important to ascertain that the signal decay is indeed that of the photoconductor and not that of the measurement apparatus. It has been found that when the cavity is off resonance, the system response is very fast, while an on-resonance cavity resulted in a large increase in the system fall time.[35]

The microwave PC decay method has been combined with Si sample immersion in HF to reduce s_r.[36] The technique has also been used for HgCdTe, where lifetimes of 1 μs have been measured by using GaAs diodes as light sources with 200-ns pulse widths.[37–38] Instead of monitoring the conductivity time decay, one can also measure the steady-state photoconductivity that is related to the lifetime of the substrate. The interpretation becomes more difficult because the mobility is required in lifetime extraction. However, one can make qualitative measurements by mapping the wafer, for example, as has been done for GaAs by comparing the dark conductivity to the photoconductivity.[39]

8.4.2 Short-Circuit Current/Open-Circuit Voltage Decay

Except for the photoconductive decay method, electrical lifetime characterization techniques have been the most popular methods. A few years ago it was recognized that *pn* junction voltage and current decay could also be monitored after *optical* generation of excess carriers.[40–42] More recently the short-circuit current decay method was added to the repertoire of lifetime measuring techniques.[43] The combination *open-circuit voltage decay* (OCVD)/*short-circuit current decay* (SCCD) method was specifically developed for characterizing the lifetime, diffusion length, and surface recombination velocity of solar cells in which the base width is typically on the order of or less than the minority carrier diffusion length, making the determination of these parameters difficult. In fact it is not easy to determine all three recombination parameters under any condition. In contrast to most other methods in which only a single parameter is measured, two measurements—the short-circuit current and the open-circuit voltage—are necessary to determine τ_r and s_r.

The theory is based on a solution of the minority carrier differential equations [Eq. (A8.15)] subject to the boundary conditions[43]

$$\frac{1}{\Delta n(x, t)} \frac{\partial \Delta n(x, t)}{\partial x} = -\frac{s_r}{D_n} \quad \text{for } x = T \tag{8.34a}$$

$$\Delta n(0, t) = 0 \tag{8.34b}$$

for the short-circuit current, and

$$\frac{\partial \Delta n(x, t)}{\partial x} = 0 \quad \text{for } x = 0 \tag{8.35}$$

for the open-circuit voltage method.

So far we have only concerned ourselves with base minority carrier recombination in $n^+ p$ junctions. There is of course also minority carrier recombination in the scr and in the heavily doped emitter. The minority carriers are swept out of the scr by the electric field in times on the order of 10^{-11} s under short-circuit conditions. The emitter lifetime is generally much lower than the base lifetime, and emitter contributions play a role only during the early phase of the current decay.[44] Emitter recombination causes carriers from the base to be injected into the emitter where they recombine at a faster rate. However, the voltage decay is determined by the base recombination parameters for long times.[45–46] If the asymptotic decay rate is measured after the initial transient, then a decay time, representative of base recombination, is observed as discussed by Rose.[44]

An experimental arrangement for the OCVD/SCCD method is shown in Fig. 8.10. Light from a cw 1.06-μm Nd:YAG laser is passed through a polarizer and a Pockel's cell switch. The switch is turned *on* and *off* by a

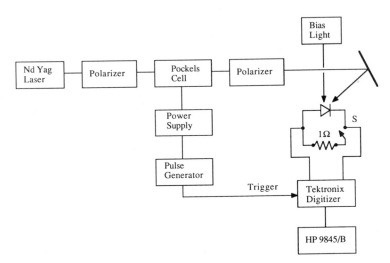

Fig. 8.10 Block diagram for measuring the short-circuit current and open-circuit voltage decay constants. Reprinted with permission after Rose and Weaver.[43]

high voltage pulse to the Pockel's cell. The pulse generator frequency is set to establish steady-state conditions in the solar cell between pulses. Switch S is set to give either open circuit or short circuit. Short circuit is actually a 1-ohm current-sensing resistor. The waveform is digitized and processed by a computer. The lifetime and surface recombination velocity are calculated using lookup tables stored in the computer.

The current decay is found to be exponential with time, with the time constant determined by the time dependence of the excess carrier concentration. This is not necessarily true for the voltage decay. Its decay can be significantly influenced by the junction RC time constant, that can be very large for large-area devices like solar cells. This effect is reduced by measuring the small-signal voltage decay with a steady-state bias light to reduce R.[47] This version of the open-circuit voltage decay technique is discussed further in Section 8.5.2.

One might expect the current and voltage decays to be identical for thick-base devices, when the base thickness is much larger than the minority carrier diffusion length because s_r is no longer important. This is indeed the case. Both have the asymptotic time dependence

$$I_{sc}, V_{oc} \sim \frac{\exp(-t/\tau_B)}{\sqrt{t}} \tag{8.36}$$

which is not a simple exponential as discussed further in Section 8.5.2.

Data reduction difficulties arise for $10^2 \leq s_r \leq 10^4$ cm/s because the method becomes insensitive to s_r and no definite surface recombination velocity can be determined. Lifetime extraction is also difficult for very high lifetimes because the effective lifetime becomes almost completely dominated by surface recombination. It is, however, almost always possible to extract meaningful values when these situations arise.

This method is one of very few allowing *both* the lifetime and surface recombination velocity at the back surface to be determined, by measuring the current and voltage decays of the same device. The method is of course only applicable for diode structures and cannot be used for unprocessed wafers. Being a transient technique, it is subject to higher-order decay time constants and possible trapping. These potential sources of error are considerably reduced by measuring the time constants asymptotically toward the end of the decay and using a bias light. The method also requires a fast turn-off light source and is generally not suitable for lifetimes in the nanosecond range.

8.4.3 Photoluminescence Decay

Photoluminescence (PL) decay is a form of photoconductive decay in the sense that excess carriers are generated by incident photons with energy $hv > E_g$. The excess carrier concentration is monitored not by measuring the change in photoconductivity, but by detecting the time dependence of the light emitted by the recombining electron-hole pairs. The PL signal is larger for efficient light-emitting direct band-gap semiconductors than for indirect band-gap semiconductors for which photoluminescence is quite inefficient.

When excess ehp's recombine radiatively and non-radiatively with lifetimes τ_{rad} and τ_{nrad}, respectively, the PL decay is determined by the lifetime τ_r

$$\tau_r = \frac{\tau_{rad}\tau_{nrad}}{\tau_{rad} + \tau_{nrad}} \qquad (8.37)$$

There may be complications, such as exciton formation, recombination, and thermalization times if the recombination is not band to band but rather impurity level to band or band to impurity level.[48] We do not consider these complications here.

The excess carrier concentration and time decay expressions are those discussed in Section 8.4.1. We expect PL decay to follow those considerations, except that the PL intensity is given by

$$\Phi_{PL}(t) = K \int_0^L \Delta n(x, t) \, dx \qquad (8.38)$$

where K is a constant accounting for the solid angle over which the light is emitted and for the reflectivity of the radiation emitted from the sample.

A complication arises if self-absorption takes place, where some of the photons generated by the recombination radiation are absorbed by the semiconductor before being emitted. Then Eq. (8.38) becomes[49]

$$\Phi_{PL}(t) = K \int_0^L \Delta n(x, t) \, e^{-\alpha x} \, dx \qquad (8.39)$$

with α is the average absorption coefficient. Self-absorption is not important for indirect band-gap semiconductors since α is low for photons with energy near the band gap, but it can be important for direct band-gap semiconductors.

The PL decay lifetime measurement method is usually employed for light-emitting devices with low lifetimes requiring experimental techniques that allow measurements of very short times. One such technique uses carrier excitation by a mode-locked cavity-dumped Ar ion laser with a pulse width of 400 ps and a 10-kHz to 1-MHz repetition rate to measure the lifetime in Si.[50] The emitted radiation is separated from the excitation light with a wide slit, double-prism monochromator and detected by a liquid nitrogen-cooled S1 photomultiplier using photon-counting techniques. The time difference between the start signal, generated by the photomultiplier after detection of the first photon, is converted by a time-to-pulse height converter to a time-proportional voltage pulse. Lifetimes as low as 0.3 ns have been measured. A mode-locked laser with 250-ps wide pulses was utilized for lifetimes of around 10 ns in GaP[51] and a dye laser was used to determine the lifetime in GaAs.[52] PL decay has been used to map the lifetime in Si power devices by scanning the excitation beam across the device.[53]

Instead of optical excitation, electron-beam excitation can also be used; the method is then known as *transient cathodoluminescence* (CL). Using a scanning electron microscope with rise and fall times of 1 ns, lifetimes of 100 ns were determined.[49] Transient CL has also been used to monitor the damage introduced during ion implantation.

8.4.4 Surface Photovoltage

The *surface photovoltage* (SPV) method, also known as the constant-magnitude steady-state surface photovoltage method, is a steady-state technique for determining the *minority carrier diffusion length* using optical excitation. The diffusion length is related to the minority carrier lifetime through the relation $L = (D\tau_r)^{1/2}$. SPV is an attractive technique, because (1) it is nondestructive, (2) sample preparation is simple (no contacts, junctions, or high temperature processing required), and (3) it is a steady-state method relatively immune to the slow trapping and detrapping effects that can influence transient measurements.

The SPV technique was first described in 1957[54] and was used to determine diffusion lengths in Si[55] and GaAs.[56] The sample to be measured is usually in the form of a wafer or part of a wafer. It is assumed to be homogeneous and is of thickness T, as shown in Fig. 8.11. One surface is

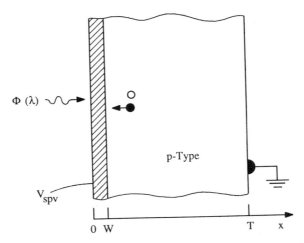

Fig. 8.11 Sample cross section for SPV measurements. The optically transparent, electrically conducting contact allows light to reach the sample.

chemically treated to induce a surface space-charge region (scr) of width W. The scr is the result of surface charges, not due to a bias voltage.

The surface with the induced scr is uniformly illuminated by chopped monochromatic light of energy larger than the band gap, while the back surface is kept in the dark. The wavelength is varied during the measurement. Electron–hole pairs are generated by absorbed photons. Some of the minority carriers diffuse toward the illuminated surface, establishing a surface potential or surface photovoltage voltage V_{SPV} relative to the grounded back surface, much like the open-circuit voltage in a solar cell. V_{SPV} is proportional to the excess minority carrier concentration $\Delta n(W)$ at the edge of the scr. The precise relationship between $\Delta n(W)$ and V_{SPV} need not be known, but it must be a monotonic function. Light reaching the back surface produces an undesirable SPV signal that can be detected by its large amplitude, by a reversal in signal polarity over the SPV wavelength range, or by a signal decrease with increasing illumination at the longer wavelengths.

The excess carrier concentration through the wafer for low-level injection is given by Eq. (A8.4). In principle, it is possible to extract the diffusion length L_n from that expression for arbitrary W, T, and α. In practice, it is very difficult to do, and several constraints are imposed on the system to simplify data extraction. The first requirement is that the undepleted wafer be much thicker than the diffusion length. A second requirement is that the scr width be small compared to L_n. The absorption coefficient should be sufficiently low for $\alpha W \ll 1$, but sufficiently high for $\alpha(T - W) \gg 1$. The assumptions

$$(T - W) \geq 4L_n , \qquad W \ll L_n , \qquad \alpha W \ll 1 , \qquad \alpha(T - W) \gg 1$$

$$(8.40)$$

allow Eq. (A8.4) to be reduced to

$$\Delta n(W) \approx \frac{(1 - R)\Phi}{s_1 + D_n/L_n} \frac{\alpha L_n}{1 + \alpha L_n}$$

$$(8.41)$$

If the surface photovoltage is proportional to $\Delta n(W)$, then V_{SPV} becomes

$$V_{SPV} = C_1 \frac{(1 - R)\Phi L_n}{(s_1 + D_n/L_n)(L_n + 1/\alpha)}$$

$$(8.42)$$

where C_1 is a constant of proportionality. A linear relationship between V_{SPV} and Δn is found for $V_{SPV} < 0.5kT/q$.[57] Typical surface photovoltages are in the low millivolt range, ensuring a linear relationship.

During the SPV measurement, D_n and L_n are assumed to be constant. Furthermore over a restricted wavelength range the reflectivity R can also be considered constant. The surface recombination velocity at the front surface s_1 is usually not known. However, if $\Delta n(W)$ is held constant during the measurement, the surface potential is also constant, and s_1 can be considered reasonably constant. This leaves α and Φ as the only two variables. The SPV is measured by keeping V_{SPV} constant, implying $\Delta n(W)$ is constant. A series of different wavelengths is selected during the measurement with each wavelength providing a different α. The photon flux density Φ is adjusted for each wavelength to hold V_{SPV} constant. This allows Eq. (8.42) to be written as

$$\Phi = C_2\left(L_n + \frac{1}{\alpha}\right) \qquad (8.43)$$

where C_2 is another constant.

Φ is plotted against $1/\alpha$ for each constant-magnitude surface photovoltage. The result is a line whose extrapolated intercept on the negative $1/\alpha$ axis ($\Phi = 0$) is the minority carrier diffusion length L_n. SPV plots for several diffusion lengths in Si are shown in Fig. 8.12. The Φ versus $1/\alpha$ plot is a straight line for well-behaved samples. It has been shown that for moderate trapping the intercept is still the correct diffusion length.[58] Only for strong minority carrier trapping in high resistivity material is the measured diffusion length an effective value that is larger than the true diffusion length.

The condition $W \ll L_n$ is generally satisfied for single-crystal Si samples. However, that may not be true for other semiconductors. For example, the diffusion length in GaAs is often only a few microns. In amorphous Si it is even shorter. In such a situation it has been shown by Moore that the intercept is given by[59–60]

$$\frac{1}{\alpha} = -L_n\left(1 + \frac{(W/L_n)^2}{2(1 + W/L_n)}\right) \qquad (8.44)$$

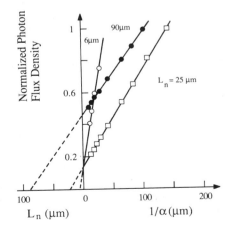

Fig. 8.12 Typical SPV plots of normalized photon flux density versus inverse absorption coefficient. The diffusion lengths are the intercepts on the L_n portion of the horizontal axis.

Equation (8.44) reduces to (8.43) for $W \ll L_n$. For $W \gg L_n$ the $1/\alpha$ intercept is $-W/2$, independent of the diffusion length. For $W = L_n$ the intercept becomes $-5L_n/4$. The scr width can be reduced by shining steady-state light onto the device if the $W \gg L_n$ condition prevails.

In Eqs. (8.43) and (8.44) the photon flux density is plotted against the inverse absorption coefficient. It is not the absorption coefficient, however, but the wavelength that is varied during the measurement. An accurate wavelength–absorption coefficient relationship is therefore very important for SPV measurements. Any error in that relationship leads to incorrect diffusion lengths. Various equations have been proposed. The ASTM Standard recommends[61]

$$\alpha = 5263.67 - \frac{11442.5}{\lambda} + \frac{5853.68}{\lambda^2} + \frac{399.58}{\lambda^3} \qquad (8.45)$$

for stress-relieved Si, that is, heat-treated or chemically-mechanically polished wafers with the wavelength λ in μm and the absorption coefficient α in cm^{-1} units. A more recent equation derived from a fit to published α-λ data for silicon is given by[62]

$$\alpha = \left(\frac{84.732}{\lambda} - 76.417 \right)^2 \qquad (8.46)$$

Both equations are valid for the 0.7 to 1.1 μm wavelength range typically used for Si SPV measurements. For non-stress-relieved Si, the expression is[61]

$$\alpha = -10696.4 + \frac{33498.2}{\lambda} - \frac{36164.9}{\lambda^2} + \frac{13483.1}{\lambda^3} \qquad (8.47)$$

An expression that gives reasonable agreement with experimental GaAs absorption data[63] is given by

$$\alpha = \left(\frac{286.5}{\lambda} - 237.13 \right)^2 \qquad (8.48)$$

for the 0.75- to 0.87-μm wavelength range. For InP we find[64]

$$\alpha = \left(\frac{252.1}{\lambda} - 163.2 \right)^2 \qquad (8.49)$$

to be a reasonable approximation for the 0.8- to 0.9-μm wavelength range.

The reflectance R in Eq. (8.42) is usually considered to be constant. However, there is a weak wavelength dependence for Si, given by[61]

$$(1 - R) = 0.6786 + \frac{0.03565}{\lambda} - \frac{0.03149}{\lambda^2} \qquad (8.50)$$

for $0.7 < \lambda < 1.05$ μm.

Although SPV has been in use since 1961, it has not found wide acceptance in the semiconductor measurement community. This is partly due to its tediousness and time-consuming nature, when the wavelength and light intensity had to be manually adjusted in the past. Another reason is the lack of commercial instruments. Goodman has developed a computer-automated instrument that has proved to be very useful and easy to use.[65] Other instruments that are entirely computer operated have been built by individual groups. Such automated systems make for a rapid and convenient diffusion length characterization technique.

A crucial component of SPV is the surface treatment to create the surface scr. The ASTM method recommends boiling n-Si in water for one hour.[61] For p-Si a one-minute etch in 20-ml concentrated $HF + 80\,ml\ H_2O$ is recommended. Goodman notes that this method works best when care is taken in earlier preparation steps not to produce a stain film by withdrawing the sample from an HF-containing etch directly into air.[65] Otherwise, a low or unstable SPV is likely to result. The stain can be avoided by quenching the HF-containing etch thoroughly with deionized water before withdrawing the sample into air. Another surface treatment for Si samples is a standard Si clean/etch,[66] removing any residual SiO_2 in buffered HF and treating n-Si in an aqueous solution of $KMnO_4$. For p-Si, the $KMnO_4$ step is omitted.

The basic components of an SPV system are a light source, a monochromator, a chopper interrupting the light at typically 100–600 Hz, the sample, a capacitive pick-up probe, and a lock-in amplifier for synchronous detection of the SPV. The monochromator may be replaced by a light source and narrow-band interference filters.

The capacitive pickup probe that measures the surface photovoltage must be properly designed to reduce drifts of the scr capacitance with time. One design, shown in Fig. 8.13, utilizes an electrically conducting, optically

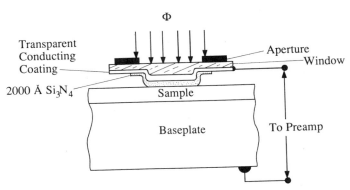

Fig. 8.13 SPV capacitive probe. Reprinted after Goodman[65] by permission of IEEE (© 1980 IEEE).

transparent coating on a glass substrate followed by a silicon nitride coat to prevent direct contact between the wafer surface and the conducting film when the probe is pressed against the sample.[65] Any shunt or stray capacitance must be minimized for good signal-to-noise ratio. A capacitive probe is convenient because it is a temporary contact. A Schottky contact is also suitable, but it must be optically transparent.[67] Aluminum, 125–200 Å thick, is sufficiently transparent to be suitable. Schottky contacts are less convenient due to their permanence, but when the sample surface is not smooth, it is impossible to get a good capacitive contact, and a deposited Schottky contact can generally be made. A Schottky contact allows dc measurements, and the SPV voltage is generally larger than it is for capacitive contacts. It is also possible to use liquid contacts.[68]

SPV has been used for Si,[69] GaAs,[56] InP,[64] and amorphous Si, where diffusion lengths as short as 200 Å were measured.[59] It has also been employed for diffusion length profiling using a scanned optical probe,[70] and for mapping the semiconductor photovoltaic response, using penetrating sub–band-gap radiation.[71] A spatial map is obtained by moving a water electrode probe across the wafer, and the response represents a display of the generation-recombination center distribution.

The size of the optical beam has been found to have an influence on the measured diffusion length. For a beam diameter less than about $30L_n$, the diffusion length is reported to be larger than the true value.[72] Damaged surfaces exhibit a nonlinear Φ-$1/\alpha$ plot.[73] As with all diffusion length measuring techniques, the true diffusion length can only be determined for samples whose thickness is larger than $4L_n$. Effective diffusion lengths are determined for thinner samples.[74] A simulation claims that the measured diffusion length is at least 10% lower than the true value even if $T > 4L_n$.[75] Further complications are introduced if the sample consists of two regions of different diffusion lengths. This is found in Si wafers that have undergone a denuding and oxygen precipitation cycle. The extraction of the diffusion length then becomes quite complicated.[76]

The sample need not be of one conductivity type. It can contain a pn junction. For example, measurements on n^+p solar cells with and without the n^+ region gave identical diffusion lengths, indicative of the p-base dominating the measurement.[77]

8.4.5 Steady-State Short-Circuit Current

The *steady-state short-circuit current* method is related to the SPV method. The sample must be a pn junction or Schottky diode, and the short-circuit current is measured as a function of wavelength. Using the same assumptions as those of the SPV method [Eq. (8.40)] the short-circuit current density of the n^+p junction of Fig. 8.14 is, according to Eq. (A8.11), given by

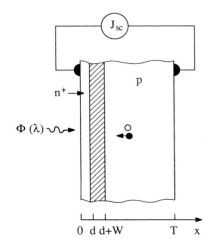

Fig. 8.14 Measurement schematic for the short-circuit current diffusion length measurement method.

$$J_{sc} \approx q(1 - R)\Phi\left(\frac{L_n}{L_n + 1/\alpha} + \frac{L_p}{L_p + 1/\alpha}\right) \tag{8.51}$$

The diffusion length is generally low for heavily doped layers, allowing the second term to be neglected, and the short-circuit current density becomes

$$J_{sc} \approx q(1 - R)\Phi\left(\frac{L_n}{L_n + 1/\alpha}\right) \tag{8.52}$$

Neglecting ehp generation in the n^+ layer and in the space-charge region is permissible if these regions are narrow and if α is not too high. This would not be true for thick n^+ layers and for high α.

Equation (8.52) has been used in two ways to extract the diffusion length. In one technique the current is held constant by adjusting the photon flux density as the wavelength is changed.[78] Equation (8.52) can be written as

$$\Phi = C_1\left(L_n + \frac{1}{\alpha}\right) \tag{8.53}$$

where $C_1 = J_{sc}/q(1 - R)L_n$. L_n is the intercept on the negative $1/\alpha$ axis when Φ is plotted against $1/\alpha$. An example of such a plot is shown in Fig. 8.15. Note the similarity to an SPV plot.

In a second technique Eq. (8.52) is written as

$$\frac{1}{\alpha} = (X - 1)L_n \tag{8.54}$$

where $X = q(1 - R)\Phi/J_{sc}$. Here $1/\alpha$ is plotted against $(X - 1)$ and the diffusion length is given by the slope of this plot, illustrated in Fig. 8.16. A check on the data is provided by the extrapolated lines passing through the

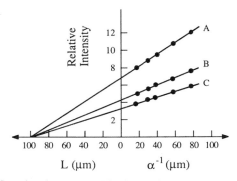

Fig. 8.15 Photon flux density versus $1/\alpha$ for a Si solar cell. A: constant open-circuit voltage, B: constant short-circuit current, and C: capacitively coupled SPV data. Reprinted with permission after Stokes and Chu.[78]

origin. Both methods assume the short-circuit current to be due to minority carrier collection from only the substrate and neglect carrier collection from the n^+ emitter and the scr.

The short-circuit methods are in principle similar to SPV, but they require a junction to collect the minority carriers. In practice, it is easier to measure a current than an open-circuit voltage. However, junction formation may alter the diffusion length, if it requires high temperature process steps. But if the device already contains a junction, then the technique is very attractive. The method in Eq. (8.54) was used in a fairly detailed measurement of the lifetime in Si as a function of doping concentration and temperature.[80] A slightly different implementation, using electrolyte-

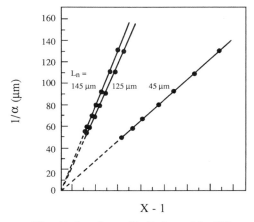

Fig. 8.16 $1/\alpha$ versus $(X-1)$ for three Si diodes with different diffusion lengths. Reprinted with permission after Arora et al.[79]

semiconductor junctions at the front and at the back surface, has been utilized to map the photoresponse which is proportional to the diffusion length.[81]

A different version of the short-circuit current method has been used for negative electron affinity photoemitters. Incident photons generate ehp's; some of the minority carriers are emitted into vacuum and collected as photocurrent. The photoemitter yield is given by

$$Y = (1 - R)P \frac{L_n}{L_n + 1/\alpha} \qquad (8.55)$$

where P is the escape probability. Note the similarity between Eqs. (8.55) and (8.52). Diffusion lengths as small as 0.8 μm have been determined for GaAs.[82]

8.4.6 Free Carrier Absorption

The *free carrier absorption* lifetime method is a non-contacting technique, relying on optical ehp generation and on optical detection using two different wavelengths. As illustrated in Fig. 8.17, a pump beam using photons with energy $h\nu > E_g$ creates ehp's. The readout is based on the dependence of the absorption of photons with $h\nu < E_g$ on the concentration of free carriers. Probe beam photons with $h\nu < E_g$ can be absorbed by free electrons and holes, known as free carrier absorption. The probe beam transmitted photon flux density Φ_t is given by[83]

$$\Phi_t = \frac{(1 - R)^2 \Phi_i \exp(-\alpha_{fc} T)}{1 - R^2 \exp(-2\alpha_{fc} T)} \qquad (8.56)$$

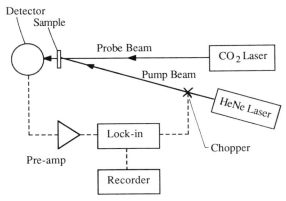

Fig. 8.17 Measurement schematic for free carrier absorption steady-state lifetime measurements. Reprinted after Polla[86] by permission of IEEE (© 1983, IEEE).

α_{fc} is the free carrier absorption coefficient. For n-type semiconductors it is[84]

$$\alpha_{fc} = K_n n \lambda^2 \qquad (8.57)$$

where K_n is a materials constant and λ the wavelength of the probe beam. For n-Si, $K_n \approx 10^{-18}$ cm^2/μm^2, and for p-Si, $K_p \approx (2-2.7) \times 10^{-18}$ cm^2/μm^2.[84-85]

The method can be used in the steady-state and transient mode. A probe beam, for example, a CO_2 laser with $\lambda = 10.6$ μm, is incident on the sample in the steady-state embodiment. The transmitted beam is detected by an infrared detector. The pump beam is chopped at a few hundred Hz for synchronous detection by a lock-in amplifier. In the transient method the pump beam is pulsed, and the time-dependent carrier concentration is detected through the transmitted probe beam.

The change in the transmitted probe beam as a result of a chopped or pulsed pump beam is

$$\Delta\Phi_t \approx -\frac{(1-R)T\Phi_i\Delta\alpha_{fc}}{1+R} \qquad (8.58)$$

using $\exp(-2\alpha_{fc}T) \approx \exp(-\alpha_{fc}T) \approx 1$ in Eq. (8.56) with $\alpha_{fc}T \ll 1$. The change in the absorption coefficient is

$$\Delta\alpha_{fc} = K_n\lambda^2\Delta n = \frac{K_n\lambda^2}{T}\int_0^T \Delta n(x)\,dx \qquad (8.59)$$

Δn in turn is related to the minority carrier lifetime and the surface recombination velocity through Eq. (A8.4). In addition Δn contains the sample reflectivity, the pump beam absorption coefficient, and the photon flux density. The fractional change in transmitted photon flux density, under certain simplifying assumptions, is[86]

$$\frac{\Delta\Phi_t}{\Phi_t} \approx \frac{(1-R)K_n\lambda^2\Phi_i\tau_r(1+s_1/\alpha D_n)}{1+s_1 L_n/D_n} \qquad (8.60)$$

It is obvious that lifetime extraction is not simple, even if the assumptions leading to Eq. (8.60) are satisfied since a number of sample parameters must be known. However, the measurement requires neither high-speed light sources nor detectors because it is a steady-state measurement and is therefore suitable for short lifetime determination. The light chopper is merely used to facilitate the measurement, with the chopper frequency synchronized to that of the narrow bandwidth lock-in amplifier for enhanced signal/noise ratio. The technique has been used for Si,[86] HgCdTe,[87] and GaP, where lifetimes as low as 1 ns were measured.[88] It has also been used to correlate lifetimes in oxygen-precipitated Si with oxygen concentrations.[89]

The transient version is more difficult to implement, but data interpretation is simpler since the transient carrier decay, monitored by the transient absorption coefficient, contains the recombination information. A 3.39-μm HeNe probe beam and a pulsed 1.06-μm Nd:YAG pump beam (150-ns pulse width) was used in one implementation.[90] The lifetime so determined agreed well with the lifetimes measured by open-circuit voltage decay and by photoconductive decay.

8.4.7 Phase Shift

The *steady-state phase-shift* technique relies on a measurement of the phase shift between the excitation source and the detected quantity. Both optical and electrical excitation sources are used, and electrical or optical signals can be detected. The electrical signal is a current due to collected minority carriers, and the optical signal is a photoluminescent output from a material acting as a light emitter.

Optical Signal   The excess minority carrier concentration for a sinusoidal optical excitation is given by Eq. (A8.25). For $T \gg L_n$ and $\alpha T \gg 1$, Eq. (A8.25) simplifies to[91]

$$\Delta n_1(x) \approx \frac{(1-R)\Phi_1 \alpha \tau_r}{\alpha^2 L_n^2 - 1 - j\omega\tau_r} \left(\frac{s_r + \alpha D_n}{s_r + D_n/L_n'} e^{-x/L_n'} - e^{-\alpha x} \right) \quad (8.61)$$

where $L_n' = L_n/(1 + j\omega\tau_r)^{1/2}$ and s_r is the surface recombination velocity at the light-incident side of the sample.

The photoluminescent output P_1 is related to $\Delta n_1(x)$ by

$$P_1 = \frac{\Re}{\tau_{rad}} \int_0^\infty \Delta n_1(x)\, dx \quad (8.62)$$

where $\Re$ is the ratio of external to internal efficiency. Substituting Eq. (8.62) into Eq. (8.61) for $\alpha L_n \gg (1 + j\omega\tau_r)^{1/2}$ gives

$$P_1 \approx \frac{C}{1 + j\omega\tau_r + s_r L_n'/D_n} \quad (8.63)$$

where C is a real number.

Two cases are important:

First, $s_r L_n/D_n \ll 1$. Then

$$P_1 \approx \frac{C}{1 + j\omega\tau_r} \quad (8.64)$$

and

$$\tau_r = \frac{-\tan(\phi)}{\omega} \tag{8.65}$$

with the phase shift $\phi = \arctan(\omega\tau_r)$.
Second, $s_r L_n / D_n \gg 1$, and

$$P_1 \approx \frac{C}{(s_r L_n / D_n)(1 + j\omega\tau_r)^{1/2}} \tag{8.66}$$

The lifetime becomes

$$\tau_r = \frac{-\tan(2\phi)}{\omega} \tag{8.67}$$

with $\phi = -0.5 \arctan(\omega\tau_r)$.

When measurements are made at low fequencies with $\omega \ll 1/\tau_r$, a first-order expansion of $\tan(\phi)$ gives

$$\tau_r(s_r \to 0) \approx -\frac{\phi}{\omega} \qquad \tau_r(s_r \to \infty) \approx -\frac{2\phi}{\omega} \tag{8.68}$$

The lifetime lies between these two values for intermediate surface recombination velocities. A wavelength-dependent phase shift results if reabsorption is not negligible and the lifetime expressions become complicated.[92]

In an experimental circuit a mode-locked laser excites carriers in the sample, and the emitted light is detected with a photodetector. A vector voltmeter measures the phase shift.[92] Care must be taken that the phase angle of either the luminescence or the excitation signal not vary with illumination intensity.

Electrical Signal   In this implementation of the phase shift method, the optically induced excess minority carriers are collected by a junction as a modulated photocurrent, as shown in Fig. 8.18(a). The short-circuit current, calculated from Eq. (A8.25) using $J_{ph}(\omega) = q D_n d(\Delta n_1)/dx$ at $x = 0$, for $T \gg L_n$ and $\alpha L_n \ll 1$ gives

$$J_{ph}(\omega) \approx \frac{q(1 - R)\Phi\alpha L_n}{(1 + j\omega\tau_r)^{1/2}} \tag{8.69}$$

which is similar to Eq. (8.66). The lifetime is given by Eq. (8.67).

When surface recombination is taken into account for $T \leq L_n$, the results can no longer be represented in a simple analytical form. One variation uses a focused, modulated light spot instead of a broad-area beam, as shown in Fig. 8.18(b). This introduces the distance from the beam to the collecting junction r as an additional variable. The diffusion length then becomes[93]

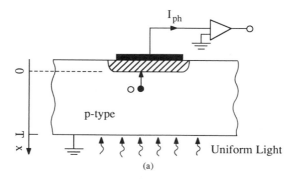

(a)

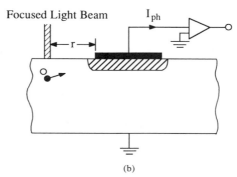

(b)

Fig. 8.18 Modulated light (a) from the back and (b) from the front of the sample. I_{ph} is the modulated photocurrent.

$$L_n = -\frac{2D_n}{\omega}\frac{d\phi}{dr}\qquad(8.70)$$

where $d\phi/dr$ is the slope of the linear phase versus distance plot.

The phase shift is between the excitation beam measured with a photo-diode detector and the sample photocurrent. A 1.15 μm wavelength assures deep penetration into the Si substrate. A mercury probe is used as the collecting junction for lifetime mapping.[94] Special care must be taken to measure the lifetime in thin epitaxial layers to account for substrate recombination.[95] Instead of measuring the ac photocurrent, it is also possible to measure the ac photovoltage.[96]

An electron beam can be used instead of a light beam, with the phase shift between the modulated *e*-beam and the photocurrent being measured. It has been suggested that τ_r and s_r can be determined from the phase shift at two beam-modulation frequencies.[97]

8.4.8 Electron Beam Induced Current

Electron beam induced current (EBIC) is used to measure minority carrier diffusion lengths, minority carrier lifetimes, and defect distributions. In contrast to photons that typically create one ehp pair upon absorption, an absorbed electron of energy E creates

$$N_{\text{eh}} = \left(\frac{E}{E_{\text{eh}}}\right)\left(1 - \frac{\alpha E_{\text{bs}}}{E}\right) \tag{8.71}$$

ehp's.[98] E_{bs} is the mean energy of the backscattered electrons, α is the backscattering coefficient, and E_{eh} is the average energy required to create one ehp ($E_{\text{eh}} \approx 3.2E_{\text{g}}$).[99] The backscattering term $\alpha E_{\text{bs}}/E$ is approximately equal to 0.1 for Si and 0.2–0.25 for GaAs over the 2 to 60 keV electron energy range. The electrons do not follow Lambert-Beer's law of exponential absorption. Rather, the penetration depth or electron range R_{e} is given by[100]

$$R_{\text{e}} = \left(\frac{2.41 \times 10^{-11}}{\rho}\right)E^{1.75} \quad \text{cm} \tag{8.72}$$

where ρ is the semiconductor density (g/cm^3) and E the incident energy (eV). For Si and GaAs, R_{e} (Si) $= 1.04 \times 10^{-11}E^{1.75}$ cm and R_{e}(GaAs) $= 4.53 \times 10^{-12}E^{1.75}$ cm. Equation (8.72) is one of several expressions found in the published literature for the penetration depth.

It is instructive to calculate the ehp concentration generated by an electron beam of energy E and beam current i_{b}. The generation volume tends to be pear shaped, as shown in Fig. 8.19, for atomic numbers $Z < 15$.

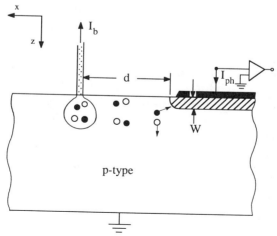

Fig. 8.19 Conventional horizontal-junction EBIC implementation.

For $15 < Z < 40$ it approaches a sphere, and for $Z > 40$ it becomes hemispherical. We approximate it as a sphere of volume $(4/3)\pi(R_e/2)^3$, for simplicity. Combining Eqs. (8.71) and (8.72) gives the generation rate

$$G = \frac{N_{eh}I_b}{(4/3)\pi q(R_e/2)^3} = \frac{8.5 \times 10^{50}\rho^3 I_b}{E_{eh}E^{4.25}} \quad cm^{-3}\cdot s^{-1} \quad (8.73)$$

where the backscattered term in Eq. (8.71) is neglected. For Si with a beam current of 10^{-10} A, $E_{eh} = 3.6$ eV and $E = 10^4$ eV the generation rate is $G = 3 \times 10^{24}$ ehp/cm$^3 \cdot$ s.

The interaction of an electron beam with the semiconductor sample can take place for a variety of geometries. One of these is shown in Fig. 8.19. Changes in the photocurrent I_{ph} collected by the junction can be effected by moving the beam in the x-direction. Changes in the z-direction are produced by changing the beam energy. The e-beam creates ehp's at a distance d from the edge of the scr. Some of the minority carriers diffuse to the junction to be collected, and I_{ph} decreases with increasing d due to bulk and surface recombination.

The photocurrent can be expressed as[101]

$$I_{ph} = \frac{qG'R_eL_n^{1/2}}{(2\pi)^{1/2}d^{3/2}} e^{-d/L_n} \quad (8.74)$$

where $G' = I_bN_{eh}/q$, provided that $s_r \gg D_n/L_n$, $L_n \ll d$, $R_e \ll d$, $R_eL_n \ll d^2$, and low-level injection prevails. A plot of $\ln(I_{ph}d^{3/2})$ versus d should give a straight line of slope $-1/L_n$. A more detailed theory has shown that such a plot can give a diffusion length in error by as much as 25%.[102]

The photocurrent for the configuration in Fig. 8.20 is[103]

$$I_{ph} = I_1\left[\exp\left(-\frac{z}{L_n}\right) - \frac{2s_rF}{\pi}\right] \quad (8.75)$$

where I_1 is a constant and F is a function that depends on s_r and on the ehp generation point. For the configuration of Fig. 8.20(a) the second term in Eq. (8.75) is eliminated for $d \geq L_n$ and L_n is found by recording I_{ph} versus z. Details can be found in van Roosbroeck,[104] Van Opdorp,[105] and Berz and Kuiken.[106] Surface recombination can play an important role in EBIC measurements.[107–109]

Instead of determining the diffusion length from the steady-state photocurrent as a function of lateral motion or beam penetration, it is also possible to use a stationary pulsed beam and extract the minority carrier lifetime from the transient analysis.[110] An approximate expression for the photocurrent, valid for high s_r, is[107]

$$I_{ph}(t) = K_1\left(\frac{\tau_r}{t}\right)^2 \exp\left[\frac{d}{L_n}\left(1 - \frac{\tau_r}{4t}\right) - \frac{t}{\tau_r}\right] \quad (8.76)$$

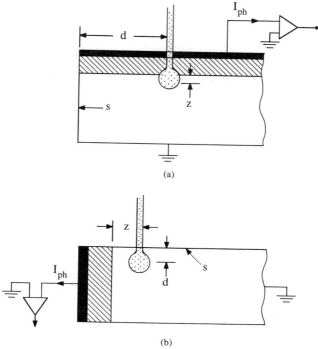

Fig. 8.20   (a) Depth modulation by electron beam energy, (b) vertical junction scan.

valid for $d \gg L_n$ for the configuration of Fig. 8.20(a). Theory predicts that the photocurrent does not decay immediately after the injection has ceased. Instead, there is a delay that is more pronounced the further the beam is from the junction. When optical excitation is used, the technique is known as light beam induced current (LBIC). The considerations are very similar to EBIC except for a different generation expression.[111-115]

Most EBIC measurements are made with the configurations in Figs. 8.19 and 8.20, and the method is fairly straightforward for long diffusion lengths. For short diffusion length measurements, the sample is sometimes bevelled to enlarge the depth.[113] Surface recombination effects can be significantly reduced if the beam penetration is increased. This can be directly tested by plotting $\ln(I_{ph})$ versus d for various beam energies. The plot should approach a straight line for higher energies.[114]

The effect of the beam current on EBIC measurements is illustrated in Fig. 8.21. The beam current for the n-type sample in Fig. 8.21(a) deposits electrons ("1") in the sample and creates ehp's. The minority holes diffuse from the point of generation, and those collected by the junction are measured as a photocurrent, $I_{ph} \approx I_b N_{eh} \exp(-d/L_n)$. The bottom ground lead current is $I_1 = I_b + I_{ph}$. For $d \gg L_n$, $I_{ph} \rightarrow 0$. A different situation arises

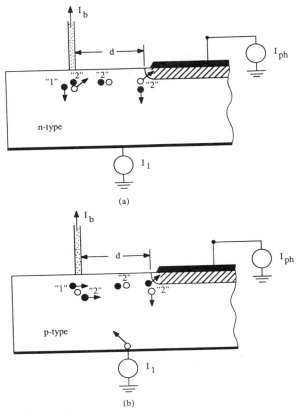

Fig. 8.21 EBIC currents in (a) *n*-type and (b) *p*-type samples.

for the *p*-type sample of Fig. 8.21(b). The photocurrent $I_{ph} \approx I_b(N_{eh} + 1)\exp(-d/L_n)$ is slightly higher than that of the *n*-sample. The ammeter detects those minority carriers collected by the junction, and $I_1 = I_{ph} - I_b$ flows in the ground lead. For $I_{ph} > I_b$, the current direction I_1 is as shown, but for $I_{ph} < I_b$ the current changes direction, as has been observed experimentally.[115] Usually $I_b \ll I_{ph}$ and the ammeter can be placed in either location.

8.5 RECOMBINATION LIFETIME—ELECTRICAL MEASUREMENT TECHNIQUES

8.5.1 Reverse Recovery

The diode *reverse recovery* (RR) method was one of the first electrical lifetime characterization techniques.[116–118] A measurement schematic and typical current-time and voltage-time responses are shown in Fig. 8.22. There are two basic measurement schemes. In Fig. 8.22(b) the current is suddenly switched from forward to reverse current, whereas in 8.22(c) the current is gradually changed. The latter is more typical of power devices in which currents cannot be switched very abruptly.

For a more detailed description of the method, let us consider Fig. 8.22(a) and (b). A forward current I_f flows through the diode for $t < 0$, and the diode voltage is V_d. Excess carriers are injected into the quasi-neutral regions, making the device impedance very small. At $t = 0$ the current is switched from I_f to I_r, with $I_r \approx (V_r - V_d)/R$. The small diode resistance is neglected because the diode remains forward biased during the initial time

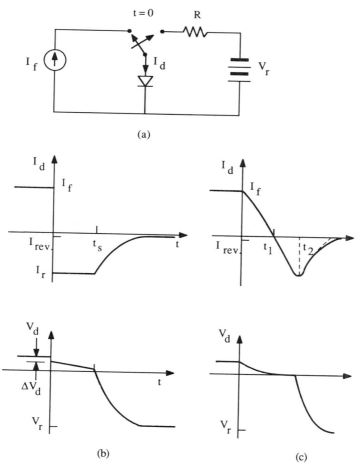

(a)

(b)   (c)

Fig. 8.22   (a) Reverse recovery circuit schematic, (b) current and voltage waveforms for abruptly switched current, (c) current and voltage waveforms for ramped current.

of I_r flow. Currents can be switched very quickly in minority carrier devices because only a change in the *slope* of the minority carrier concentration gradient at the edge of the scr is required. The diode voltage, in contrast, is proportional to the log(excess carrier *concentration*) at the scr edge. The voltage hardly changes during this period and the diode remains forward biased although the current has reversed direction. The voltage step ΔV_d is due to the ohmic voltage drop in the device.[119]

The excess carrier concentration decreases during the reverse current phase for two reasons: some carriers are swept out of the device by the reverse current, and some carriers recombine. The excess minority carrier concentrations at the edges of the scr are approximately zero at $t = t_s$, and the diode becomes zero biased. For $t > t_s$ the excess minority carrier concentrations are depressed below zero, the voltage approaches the reverse-bias voltage V_r and the current becomes the leakage current I_{rev}.

The *I-t* curve is conveniently divided into the constant-current storage phase, $0 \leq t \leq t_s$, and the recovery phase, $t > t_s$. The time t_s is usually well defined and is frequently referred to as the storage time. It is related to the lifetime by the expression[117]

$$\text{erf}\sqrt{\frac{t_s}{\tau_r}} = \frac{1}{1 + I_r/I_f} \tag{8.77}$$

where erf is the error function, defined by

$$\text{erf}(x) = \frac{2}{\sqrt{\pi}} \int_0^x \exp(-z^2)\, dz$$

$$\approx 1 - \left(\frac{0.34802}{1 + 0.4704x} - \frac{0.095879}{(1 + 0.4704x)^2} + \frac{0.74785}{(1 + 0.4704x)^3} \right) e^{-x^2} \tag{8.78}$$

An approximate charge storage analysis that considers the charge Q_s remaining at $t = t_s$ gives[120]

$$t_s = \tau_r \left[\ln\left(1 + \frac{I_f}{I_r}\right) - \ln\left(1 + \frac{Q_s}{I_r \tau_r}\right) \right] \tag{8.79}$$

$Q_s/I_r\tau_r$ can be considered a constant for many cases.

A plot of t_s versus $\ln(1 + I_f/I_r)$ is shown in Fig. 8.23. The lifetime is found from the slope of such a plot; the intercept being $(1 + Q_s/I_r\tau_r)$. The slope is constant only if the second term in Eq. (8.79) is constant. This term is not constant if Q_s depends on I_f or I_r. Various approximations have been derived for this term, and it is found to be approximately constant provided

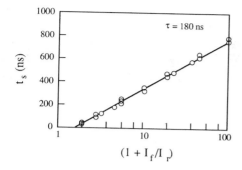

Fig. 8.23 Storage time as a function of $(1 + I_f/I_r)$. Reprinted after Kuno[120] by permission of IEEE (© 1964, IEEE).

$I_r \ll I_f$.[121] The effect of recombination in the heavily doped emitter can be virtually eliminated by keeping $I_r \ll I_f$.[122] The plot of Fig. 8.23 becomes highly curved if these conditions are not met, and a unique lifetime can no longer be extracted. For Fig. 8.22(c) the lifetime is related to t_1 and t_2 by[123]

$$1 - \exp\left(-\frac{t_2}{\tau_r}\right) = \frac{t_2 - t_1}{\tau_r} \qquad (8.80)$$

The junction displacement current $I_j = C_j \, dV_j/dt$ is neglected in all of these expressions because it constitutes only a small fraction of the total current.[118]

What is τ_r in Eqs. (8.77) and (8.79)? To first order it would seem to be the base lifetime in pn junctions. For short-base diodes, τ_r is an effective lifetime representing both bulk and surface recombination.[124–127] An additional problem in forward-biased pn junctions is the existence of excess carriers in *both* quasi-neutral regions and in the scr. Degeneracy and band-gap narrowing add further complications.[128]

The emitter is generally much more heavily doped than the base and the emitter lifetime is much lower than the base lifetime. Hence, one would expect emitter recombination to have a significant influence on the RR transient. This is particularly troublesome for high injection conditions, leading to appreciably reduced lifetimes.[129–131] But the emitter can alter the measured lifetime from its true base value even at low and moderate injection levels. Other potential sources of error are the scr capacitance discharge, scr recombination, and diode series resistance. Generally these are negligible to first order.

The RR method is used in spite of its shortcomings. By measuring coaxially mounted diodes in a matched circuit with a sampling oscilloscope, it is possible to measure lifetimes as low as 1 ns.[121] The constraints on the measuring circuit are relaxed for most Si diodes with their longer lifetimes. An asymmetry in the switching behavior of n^+p compared to p^+n power diodes has been observed with p-base diodes having shorter turnoff times than n-base diodes.[132] This has been attributed to the unequal electron and

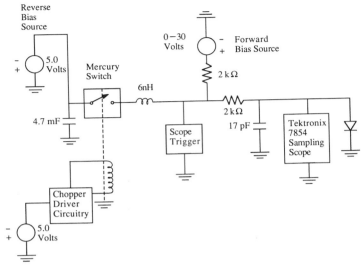

Fig. 8.24   Circuit diagram for reverse-recovery measurements. The mercury switch provides for fast and smooth swtiching. The chopper circuitry allows for repetitive measurements and the 17 pF capacitance is the test equipment capacitance. Reprinted with permission after Liou.[134]

hole mobilities. RR can even be used when the diode contains a drift field in the base, as found in step recovery diodes.[133] A circuit suitable for RR measurements is shown in Fig. 8.24.

8.5.2   Open-Circuit Voltage Decay

The concept of the *open-circuit voltage decay* (OCVD) method was published in 1953 the same year as the reverse recovery technique.[135–136] It is a simple method and has found wide acceptance. The measurement principle is shown in Fig. 8.25(a). A steady-state excess carrier concentration distribution and diode voltage, shown in Fig. 8.25(b), are established by a forward current flow through the diode. At $t = 0$ the diode is open circuited, and the recombination of excess carriers is detected by monitoring the diode's open circuit voltage, shown in Fig. 8.25(c). The voltage step $\Delta V_d = I_f r_s$ is due to the ohmic voltage drop in the diode that is observed when the current flow ceases.[119] ΔV_d can be used to determine the device series resistance. As discussed later in this chapter, emitter recombination also causes a rapid drop in the *V-t* curve near $t = 0$ and may be included in the ohmic drop if the measurement circuit cannot distinguish between the two.

OCVD is similar to the optically excited, open-circuit voltage decay method discussed in Section 8.4.2, the only difference being the method of creating excess minority carriers. The reason is that for optical OCVD ehp's

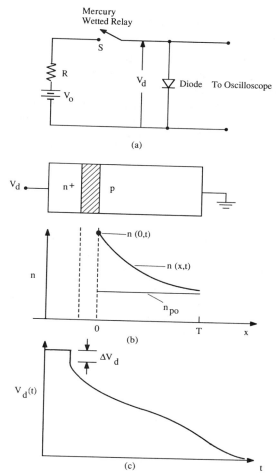

Fig. 8.25 (a) Open-circuit voltage decay schematic, (b) carrier distribution, (c) voltage waveform.

are generated optically, while for electrical OCVD they are injected by a junction. In contrast to RR, in the OCVD method the excess carriers all recombine; they are not swept out of the device by a reverse current. This has certain implications on the time duration of the decay as discussed later.

The external current becomes zero when switch S is opened at $t = 0$. The excess minority carrier concentration at the edge of the scr, $x = 0$ in Fig. 8.25(b), is related to the time-varying junction voltage $V_j(t)$ by

$$\Delta n(0, t) = n_{p0}(e^{qV_j(t)/kT} - 1) \tag{8.81}$$

giving the junction voltage as

$$V_j(t) = \frac{kT}{q} \ln\left(\frac{\Delta n(0,t)}{n_{p0}} + 1\right) \tag{8.82}$$

Equation (8.82) shows the voltage to decrease as Δn is reduced by recombination. Hence a measure of the voltage time dependence is a measure of the excess carrier time dependence.

The diode voltage is $V_d = V_j + V_b$, where V_b is the base voltage, neglecting the voltage across the emitter. You may wonder how there can be a base voltage when there is no current flow during the decay. The base voltage, being the result of unequal electron and hole mobilities, is known as the *Dember voltage*, given by[137]

$$V_b(t) = \frac{kT}{q} \frac{b-1}{b+1} \ln\left(1 + \frac{(b+1)\,\Delta n(0,t)}{n_{p0} + bp_{p0}}\right) \tag{8.83}$$

with $b = \mu_n/\mu_p$. The Dember voltage is negligible for low injection levels, and we will not consider it further. It may have to be considered, however, for high injection levels. We assume that $V_d(t) \approx V_j(t)$, given by Eq. (8.82), and will simply use V for the time-varying device voltage.

We consider the simplest case first and then add second-order effects. For $T \gg L_n$ and low-level injection, it has been shown that[136]

$$V(t) = V(0) - \left(\frac{kT}{q}\right) \ln\left(\mathrm{erfc}\sqrt{\frac{t}{\tau_r}}\right) \tag{8.84}$$

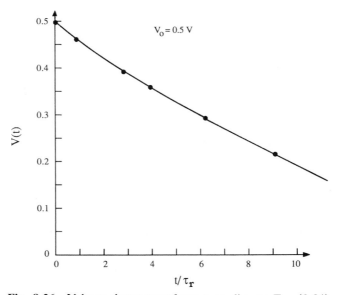

Fig. 8.26 Voltage decay waveform according to Eq. (8.84).

where $V(0)$ is the diode voltage before opening the switch, and $\mathrm{erfc}(x) = 1 - \mathrm{erf}(x)$ is the complementary error function, with $\mathrm{erf}(x)$ defined in Eq. (8.78). Equation (8.84), plotted in Fig. 8.26, obtains for $V(t) \gg kT/q$. The curve has an initial rapid decay followed by a linear region with constant slope. The slope of the $V - t$ curve is given by

$$\frac{dV(t)}{dt} = -\frac{(kT/q)\exp(-t/\tau_r)}{\sqrt{\pi t \tau_r}\,\mathrm{erfc}\sqrt{t/\tau_r}} \tag{8.85}$$

which reduces to

$$\frac{dV(t)}{dt} = -\frac{kT/q}{\tau_r(1 - \tau_r/2t)} \tag{8.86}$$

for $t \geq 4\tau_r$. Equation (8.86) can be further simplified by neglecting the second term in the bracket. For $t \geq 4\tau_r$ the lifetime is determined from the slope according to

$$\boxed{\tau_r = -\frac{kT/q}{dV/dt}} \tag{8.87}$$

This is the most commonly used OCVD equation.

A word of caution regarding Eq. (8.87). The assumption in the derivation leading to this equation is that recombination is dominated by quasi-neutral region recombination with the simple exponential voltage dependence $\exp(qV/kT)$. For scr recombination the dependence becomes $\exp(qV/nkT)$, where the diode ideality factor n lies typically between 1 and 2. Equation (8.87) should correctly contain n as a prefactor. Of course as the diode voltage drops from $V(0) \approx 0.6$ V or so to zero, n is likely to vary from 1 to a value closer to 2, and since one usually does not know what n is, it is generally taken to be unity.

As with all junction lifetime measuring techniques, τ_r is an effective lifetime influenced by emitter recombination and by back surface recombination for short-base diodes. Reliable lifetime extraction becomes difficult if the voltage decay is significantly influenced by surface recombination. OCVD must then be augmented by other techniques, such as the short-circuit current method, discussed in Section 8.4.2.

With the emitter lifetime generally much lower than the base lifetime, excess emitter carriers recombine more rapidly than excess base carriers causing carriers from the base to be injected into the emitter during the voltage decay. Obviously emitter recombination influences base recombination, and the voltage decay time is reduced. Fortunately this effect becomes negligible for $t \geq 2.5\tau_b$ where τ_b is the base lifetime, and the V-t curve becomes linear with slope $(kT/q\tau_b)$ regardless of emitter recombination or band-gap narrowing.[138]

Recombination in the n^+ and p^+ end regions of power p^+in^+ diodes is important when the i region width is less than the ambipolar diffusion length. The base lifetime can then only be determined if the recombination properties of the end regions are known.[139–140] Under high-level injection, the lifetime is given by[141–142]

$$\tau_r = -\frac{2kT/q}{dV/dt} \tag{8.88}$$

subject to the restrictions: The excess carrier concentration in the base is uniform, end region recombination is negligible, and the base excess carrier concentration is higher than the base doping concentration. The 2 accounts for high injection effects. The high injection level V-t curve frequently exhibits two distinct slopes shown in Fig. 8.27.

Unusual V-t responses, shown in Fig. 8.28, are sometimes observed when the diode capacitance is not negligible or when the junction shunt resistance is low. Capacitance tends to extend the V-t curve, giving the curve a smaller slope.[143] The lifetime is too high if this slope is used to determine τ_r. Space-charge region recombination and shunt resistance cause the V-t curve to drop faster than observed for quasi-neutral bulk recombination only. A variation of the OCVD method that has been found to be useful for devices exhibiting such decay curves is one in which an external resistor and capacitor is switched into the measurement circuit and the curve is differentiated to extract the lifetime.[144] Another possible anomaly is a peak in the V-t curve near $t = 0$ that has been attributed to emitter recombination.[145] It comes about by a time-dependent emitter-base coupling parameter invoked to account for injection of carriers from the base into the emitter.

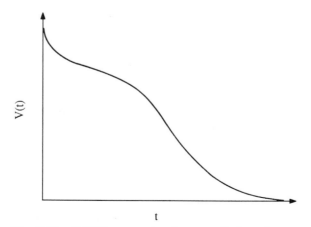

Fig. 8.27 OCVD curve showing two distinct slopes.

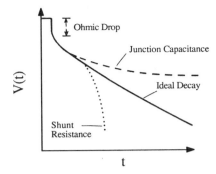

Fig. 8.28   Effects of junction capacitance and shunt resistance on voltage decay.

A variation of OCVD is the *small-signal OCVD* method in which the diode is biased to a steady-state voltage by illuminating the device, and a small electrical pulse is superimposed on the "optical" bias. With the pulse "on," additional carriers are injected, and with it "off," these additional carriers recombine. This method is used to measure τ_r under bias conditions and also to reduce effects of capacitance and shunt resistance. The small-signal voltage decay, monitored with the circuit shown in Fig. 8.29, is approximately given by[47,146]

$$\Delta V(t) = V(t) - V_{ss} = \frac{kT}{q} \ln\left(\frac{\Delta n}{n_{ss}} \operatorname{erfc}\sqrt{\frac{t}{\tau_r}} + 1 \right) \approx \frac{kT}{q} \frac{\Delta n}{n_{ss}} \frac{e^{-t/\tau_r}}{\sqrt{\pi t/\tau_r}}$$

(8.89)

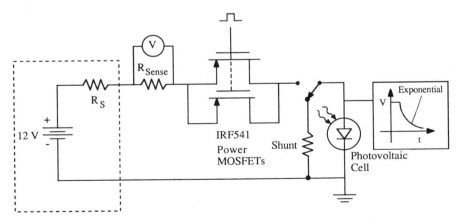

Fig. 8.29   Circuit for OCVD and small-signal OCVD. The dashed block represents a 12 V battery with its internal resistance R_s, $R_{sense} = 0.4\,\Omega$, shunt resistance = $0.01\,\Omega$. The switch is a power MOSFET. This circuit can supply several amps of current. It can also be used without steady-state illumination for conventional OCVD measurements.

where V_{ss} is the steady-state voltage and n_{ss} the steady-state carrier concentration determined by the steady-state illumination. It has been shown that Eq. (8.89) is an approximation that holds only for $[(\Delta n / n_{ss}) \, \text{erfc}(t/\tau_r)^{1/2}] < 0.23$.[146] The voltage decay is nearly exponential if that condition is satisfied, and the lifetime can be extracted from the slope of a $\ln(V)$-t plot.

A comparison of the RR and the OCVD techniques favored OCVD for its ease and accuracy.[129] In OCVD the lifetime can be extracted for that part of the V-t curve where base recombination dominates, whereas in RR storage time measurements there is some averaging over a voltage range that includes at the lower current a component where scr recombination may be important. During OCVD the experimental considerations are relaxed because the carriers decay by recombination, and the voltage decay time takes many lifetimes since excess carriers decay by a factor of $1/e$ in one lifetime. First-order considerations give the excess carrier concentration as

$$\Delta n(t) = n_{p0} \, e^{qV/kT} = \Delta n(0) \, e^{-t/\tau_r} \tag{8.90}$$

leading to

$$t = \frac{V_0 - V}{kT/q} \, \tau_r \tag{8.91}$$

using $V_0 = (kT/q) \ln(\Delta n(0)/n_{p0})$, the initial bias voltage. The voltage decay time can be thought of as the time when $V \to 0$

$$t_{OCVD} \approx \frac{V_0}{kT/q} \, \tau_r \tag{8.92}$$

With V_0 around $20kT/q$ or so, we find the decay time to be around $20\tau_r$.

8.5.3 Pulsed MOS Capacitor

The principle of the *pulsed MOS capacitor* (MOS-C) recombination lifetime measurement technique is quite different from the other methods described in this chapter. The MOS-C technique in turn is divided into two very different methods. In the first of these an MOS-C is biased into strong inversion shown in Fig. 8.30(a) and by point A in 8.30(d). The inversion charge is

$$Q_{N1} = (V_{G1} - V_T)C_{ox} \tag{8.93}$$

where V_T is the threshold voltage. A voltage pulse of amplitude $-\Delta V_G$ and pulse width t_p is superimposed on V_{G1}, reducing the gate voltage during the pulse period to $V_{G2} = V_{G1} - \Delta V_G$ shown in Fig. 8.30(b) and by point B in Fig. 8.30(d). The corresponding inversion charge is

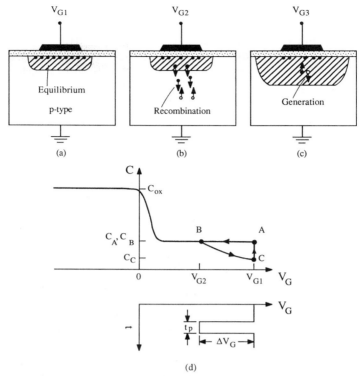

Fig. 8.30 Pulsed MOS capacitor recombination lifetime measurement technique. The device behavior at various voltages and times is shown in (a), (b), and (c) and the $C - V_G$ and $V_G - t$ curves are shown in (d).

$$Q_{N2} = (V_{G2} - V_T)C_{ox} < Q_{N1} \qquad (8.94)$$

The charge difference $\Delta Q_N = (Q_{N1} - Q_{N2})$ is injected into the substrate as indicated in Fig. 8.30(b).

What happens to ΔQ_N? Minority carriers in an inversion layer do not recombine with majority carriers because they are separated by the electric field of the scr. However, when the minority carriers are injected into the substrate, they find themselves surrounded by holes and are able to recombine.

Let us now consider two extrema. First, the pulse is sufficiently wide $(t_p > \tau_r)$ for the injected minority carriers to have sufficient time to recombine. When the gate voltage returns to V_{G1} only Q_{N2} is available, and the MOS-C is driven into partial depletion, shown in Fig. 8.30(c) and by point C in 8.30(d). Thermal ehp generation subsequently returns the device to equilibrium, point A, in Fig. 8.30(d). Second, the pulse is narrow $(t_p \ll \tau_r)$. The device goes through similar stages as in the first case with one major exception: The injected minority carriers have insufficient time to recombine because the pulse width is much less than the recombination lifetime. Consequently the capacitance sequence in Fig. 8.30(d) is $C_A \rightarrow C_B \rightarrow C_A$. For intermediate pulse widths the capacitance lies between C_C and C_A.

The capacitance at the end of the injection pulse is a measure of how many minority carriers have recombined during the pulse period. If the recombination of injected carriers follows the simple exponential decay

$$\Delta Q_N(t) = \Delta Q_N(0) \exp\left(-\frac{t}{\tau_r}\right) \tag{8.95}$$

then it can be shown that[147]

$$\Delta Q_N(t_p) = \Delta Q_N(0) \exp\left(-\frac{t_p}{\tau_r}\right) = K\left(C_A^{-2} - C_C^{-2}\right) \tag{8.96}$$

where K is a constant. To determine the lifetime, the pulse width is varied and the capacitance C_C is measured for each pulse width. Then $\ln(1/C_A^2 - 1/C_C^2)$ is plotted against t_p. τ_r is obtained from the slope of this plot. A more detailed theory shows the exponential time decay of the carriers in Eq. (8.95) to be too simplistic because minority carriers recombine not only in the quasi-neutral substrate but also in the scr and at the surface.[148]

The pulsed MOS-C recombination lifetime measurement method has not found wide acceptance despite the simple and ubiquitous MOS-C found on many test structures. Most capacitance meters and bridges cannot be used directly because the instruments are unable to pass the required narrow pulses undistorted. For lifetimes in the microsecond range submicrosecond pulses are required. It is easier to modify the experimental arrangement by coupling the device to the capacitance meter through a pulse transformer at its input terminals and apply the pulse there instead of trying to feed it through the meter. One experimental arrangement is given in Wang and Sah.[149]

A variation of the pulsed MOS-C technique, based on charge pumping, has been proposed for MOSFET's.[150] When a MOSFET is pulsed from inversion into accumulation, most of the inversion charge leaves the device through the source and the drain. However, a small fraction of the charge is unable to reach either source or drain and recombines with majority carriers. This fraction, proportional to the pulse frequency, is detected as a substrate current. As the frequency increases to the point where the time between successive pulses is on the order of τ_r, the substrate current-pulse frequency relationship becomes non-linear, and τ_r can be extracted from the current. Recombination lifetimes from 100 ns to 100 ms can be determined with this method.

The second pulsed MOS-C method is based on an entirely different principle—a measurement of the relaxation time of an MOS-C pulsed into deep depletion. The device may originally be biased in accumulation, depletion, or inversion. We assume that prior to the depletion gate voltage pulse, the device is in equilibrium and illustrate the technique in Fig. 8.31, where the MOS-C capacitance is driven from A to B by a depleting voltage step. Thermal generation returns the device to equilibrium, shown by the path B to C, in Fig. 8.31(a). The return to equilibrium on the C-t diagram is typically as shown in Fig. 8.31(b). The recovery time t_f is determined by the thermal ehp generation properties of the bulk semiconductor and the oxide–semiconductor interface.

Majority carriers are repelled over the depth of the depleted scr by the depleting voltage pulse. This occurs in a time around 10^{-10} s, much shorter than typical C-t responses which last on the order of seconds or minutes. Consequently the capacitance decreases very quickly. In fact it is the capacitance measurement instrument that is the time-limiting element during the capacitance decrease. Device equilibrium is restored through thermal generation of ehp's, provided that there are no other sources of minority carriers. There should be no *pn* junctions nearby to inject minority carriers. For example, if the measurement is made on a MOSFET, both the source and drain can inject minority carriers and the C-t relaxation time becomes a measure of the injection efficiency of the source and drain, but not of the generation parameters of the device. Similarly it should be evident that the measurement must be performed in the dark. Otherwise, photon-generated ehp's will contribute to the capacitor discharge.

The thermal generation components, shown in Fig. 8.32, are (1) bulk scr generation characterized by the generation lifetime τ_g, (2) lateral surface scr generation characterized by the surface generation velocity s_g, (3) surface scr generation under the gate characterized by the surface generation velocity s, (4) quasi-neutral bulk generation characterized by the minority carrier diffusion length L_n', and (5) back surface generation characterized by the generation velocity s_c. Components (1) and (2) depend on the scr width

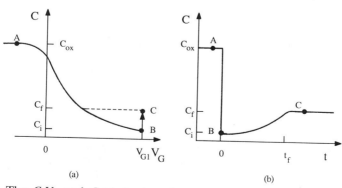

(a)   (b)

Fig. 8.31 The C-V_G and C-t behavior of an MOS-C pulsed into deep depletion.

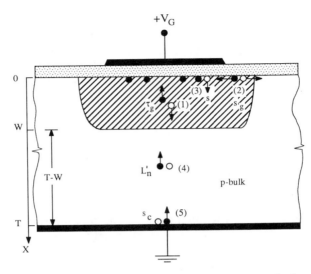

Fig. 8.32 The thermal generation components of a deep-depleted MOS-C.

and are discussed in more detail in Section 8.6.2. Components (3)–(5) are independent of the scr width and are the subject of this section.

Surface generation considerations hold when the MOS capacitor is pulsed from accumulation into deep depletion because the surface is depleted at $t = 0^+$. However, the area under the gate is already inverted at $t = 0$ when the device is initially biased in inversion and then pulsed into deep depletion. The two surface generation components, characterized by s_g and s, are never equal in that case even at $t = 0$. Pulsing from inversion instead of from accumulation is commonly used for low surface generation.

The capacitance depends on the gate voltage and on the inversion charge Q_N as

$$C = \frac{C_{ox}}{\sqrt{1 + 2(V_G' + Q_N/C_{ox})/V_0}} \qquad (8.97)$$

where $V_G' = V_G - V_{FB}$ and $V_0 = qK_s\varepsilon_0 N_A/C_{ox}^2$. Solving Eq. (8.97) for V_G' and differentiating with respect to t gives

$$\frac{dV_G}{dt} = -\frac{1}{C_{ox}}\frac{dQ_N}{dt} - \frac{qK_s\varepsilon_0 N_A}{C^3}\frac{dC}{dt} \qquad (8.98)$$

with $dV_{FB}/dt = 0$.

Equation (8.98) is an important equation relating the gate voltage rate of change with time to inversion charge and capacitance rate of change with time. For the *pulsed* capacitor, V_G is constant, $dV_G/dt = 0$, and Eq. (8.98) solved for dQ_N/dt becomes

$$\frac{dQ_N}{dt} = -\frac{qK_s\varepsilon_0 C_{ox}N_A}{C^3}\frac{dC}{dt} \qquad (8.99)$$

dQ_N/dt represents the thermal generation components in Fig. 8.32, which can be written as[151-152]

$$\frac{dQ_N}{dt} = -\frac{qn_iW}{\tau_g} - \frac{qn_i s_g A_S}{A_G} - qn_i s - \frac{qn_i^2 D_n}{N_A L_n'} \qquad (8.100)$$

where $A_S = 2\pi rW$ is the area of the lateral scr (assuming the lateral scr width to be identical to the vertical scr width) and $A_G = \pi r^2$ is the gate area. The first two terms in Eq. (8.100) are scr width-dependent generation components. Here we consider the last three scr width-independent components.

The five generation components are plotted in Fig. 8.33. The components with the n_i dependence all exhibit identical temperature dependence. However, G_4 has a n_i^2 dependence causing it to increase faster with temperature. This is the basis for the recombination lifetime measurement. G_4 dominates for temperatures above about 75°C. When $dQ_N/dt = -qn_i^2 D_n/N_A Ln'$ is substituted into Eq. (8.99), we find[152]

$$C = \frac{C_i}{\sqrt{1 - t/t_1}} \qquad (8.101)$$

where C_i is the capacitance at $t = 0$ (see Fig. 8.31) and $t_1 = (K_s/K_{ox})(C_{ox}/C_i)^2(N_A/n_i)^2(W_{ox}/2)(L_n'/D_n)$.

The measurement consists of a C-t plot when quasi-neutral region generation dominates. Then $1 - (C_i/C)^2$ is plotted against t, with the slope

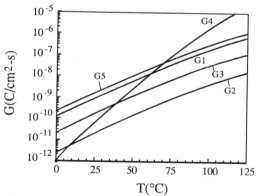

Fig. 8.33 Generation components as a function of temperature for Si with $\tau_g = 500$ μs, $s_g = 2$ cm/s, $s = 0.1$ cm/s, $L_n' = 100$ μm, $s_c = 1$ cm/s, $(W - W_t) = 3$ μm, $r = 0.5$ mm, $N_A = 10^{15}$ cm^{-3}, $D_n = 37(300/T)^{1.2}$ cm^2/s, $n_i = 3.87 \times 10^{16} \, T^{1.5}$ exp$(-0.605/kT)$ cm^{-3}.

being $1/t_1$. The diffusion length is determined through the definition of t_1. To ensure that qn-region generation dominates, the $1 - (C_i/C)^2$ versus t curve should be straight line. If it is not, then the measurement temperature is probably too low. L_n' is an effective diffusion length that takes bulk and back surface generation into account, and is given by[152]

$$L_n' = L_n \frac{\cosh(\zeta) + (s_c L_n/D_n)\sinh(\zeta)}{(s_c L_n/D_n)\cosh(\zeta) + \sinh(\zeta)} \qquad (8.102)$$

where $\zeta = W_B/L_n$ with $W_B = T - W$. For $W_B \gg L_n$ we find $L_n' = L_n$, and for $W_B \ll L_n$, $L_n' = (s_c + D_n/W_B)/(s_c L_n/W_B + D_n/L_n)$. The surface generation velocity s_c depends on the type of back contact. If it is a p-semiconductor/metal contact, its surface generation velocity is very high. If, however, a p-p^+ semiconductor/metal contact is formed, it has low s_c because the low-high p-p^+ contact represents a barrier for minority carriers. It should be mentioned that though the back ohmic contact can generate ehp's by virtue of surface generation, it cannot inject minority carriers. That can only be done by a pn junction, not by an ohmic contact. The dependence of the effective diffusion length on the back surface recombination velocity has been experimentaly verified.[153]

An implicit assumption in Eq. (8.100) is that the minority carrier collection area from the quasi-neutral region is the gate area. For gate diameters of 1 mm and larger and diffusion lengths of 100 μm or less, this is a reasonable assumption. But for high diffusion length devices, it is necessary to consider lateral minority carrier collection. This is particularly important for MOS-C's on thin epitaxial layers on heavily doped substances.[154]

8.5.4   Other Techniques

Short-Circuit Current Decay   In the reverse-recovery method the diode current is switched from forward to reverse; for open-circuit voltage it is switched from forward to zero current. In the *short-circuit current decay method* (SCCD) the current is switched from forward current to short-current or zero voltage. Emitter minority carriers play a relatively minor role in this measurement because they recombine very quickly when the diode is short-circuited. The method is relatively simple to implement and in combination with RR or OCVD allows extraction of both base lifetime and surface recombination velocity.[155-157]

Conductivity Modulation   The *conductivity modulation* technique was developed to measure the recombination lifetime in epitaxial layers with thicknesses less than the minority carrier diffusion length. The measured lifetime is neither affected by substrate recombination nor by recombination in heavily doped regions. The structure consists of alternate n^+ and p^+

stripes diffused or implanted into a p-epi layer on a p^+ substrate. All p^+ stripes are connected to each other, and all n^+ stripes are connected to each other forming a lateral n^+pp^+ diode. The spacing between stripes is smaller than the minority carrier diffusion length.

The lateral diode is forward biased to a constant dc voltage. A small ac voltage is superimposed on the dc bias, and the ac current resulting from recombination in the epitaxial layer is measured and related to the recombination lifetime.[158] By applying a dc voltage to the p^+ substrate, it is possible to profile the lifetime.

8.6 GENERATION LIFETIME

8.6.1 Gate-Controlled Diode

The generation lifetime is an important device parameter because it determines the leakage current of pn junctions and the storage time of MOS capacitors. MOS-C's are the basic elements of charge-coupled devices and dynamic random access memories, and τ_g is determined from measurements of junction leakage current and MOS capacitor storage times.

The thermal generation rate of ehp's in a reverse-biased junction is given by Eq. (8.100). Space-charge region generation components 1 and 2 dominate for semiconductors like Si and GaAs at room temperature, and the scr current can be written as

$$I_{scr} = \frac{qn_i A_J W}{\tau_g}\left(1 + \frac{2s_g\tau_g}{r}\right) = \frac{qn_i A_J W}{\tau_g'} \tag{8.103}$$

where A_J is the junction diode area. An implicit assumption in Eq. (8.103) is that the lateral scr width is identical to the vertical scr width. That may not be strictly correct, but it is a reasonable assumption in view of the uncertainty of the lateral scr shape. The effective generation lifetime τ_g' contains both τ_g and s_g, and a measure of I_{scr} does not allow a separation of the two. Hence a simple diode does not lend itself to a measurement of both bulk and surface generation parameters. However, a gate-controlled diode does allow surface and volume generation to be easily separated.

The three-terminal *gate-controlled diode*, consisting of a p substrate, an n^+ region (D), a circular gate (G) surrounding the n^+ region, and a circular guard ring (GR) surrounding the gate, is shown in Fig. 8.34. The gate is sometimes located in the center surrounded by the circular n^+ region. The gate should overlap the n^+ region slightly because potential barriers can develop if there are gaps between the two. The guard ring is preferred but is not always necessary. It should be close to the gate and be biased for the semiconductor under it to be accumulated to isolate the gate-controlled diode from the rest of the wafer. For example, moderately doped and

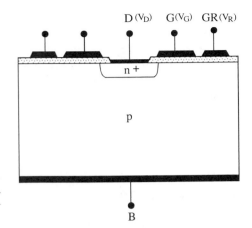

Fig. 8.34 The gate-controlled diode. D is the n^+p diode, G is the gate, GR is the guard ring, and B is the substrate.

thermally oxidized p-type silicon substrates are easily inverted by the positive fixed oxide charge in thermally grown SiO_2. The guard ring prevents coupling adjacent devices by such an inversion layer. Devices can also be decoupled by doping the semiconductor between the devices more heavily, making inversion layer formation more difficult.

A detailed discussion of gate-controlled diodes is found in Schroder.[151] We give here the necessary background for lifetime and surface generation velocity measurements. Let us first consider the semiconductor under the gate with the diode short-circuited to the substrate ($V_D = 0$). The surface is accumulated for $V_G' < 0$ ($V_G' = V_G - V_{FB}$), it is at flat band for $V_G' = 0$, and for $V_G' > 0$ the surface is depleted if $0 < \phi_s < 2\phi_F$ and inverted if $\phi_s \geq 2\phi_F$, where $\phi_F = (kT/q) \ln(N_A/n_i)$. Accumulation and flatband conditions remain unchanged when the diode is reversed biased, but depletion and inversion conditions change. For diode voltage $V_D \neq 0$, depletion holds for $0 < \phi_s < (V_D + 2\phi_F)$ and inversion for $\phi_s \geq (V_D + 2\phi_F)$, with the surface potential ϕ_s related to the gate voltage by Eq. (6.14).

The diode is biased at a constant voltage, and the gate voltage is varied during the gate-controlled diode measurement. Let the diode be reverse biased to $V_D = V_{D1}$. The surface under the gate is accumulated for negative gate voltage $-V_{G1}$, and the junction scr is slightly pulled in at the surface, illustrated in Fig. 8.35(a). The measured current is the diode scr generated current I_J shown in Fig. 8.35(a) and by point A in Fig. 8.35d. The current increase for more negative gate voltages has been attributed to weak breakdown of the gate-induced n^+-p^+ junction at the surface. At $V_G = V_{FG}$ the semiconductor is in flat band, the diode scr width is the same at the surface as in the bulk, and the current is slightly increased as a result of the slightly larger generation volume. The surface under the gate depletes for $V_G > V_{FB}$, and the current increases rapidly. This abrupt increase is due to the surface generation current I_S and the gate-induced scr bulk current I_{GIJ}

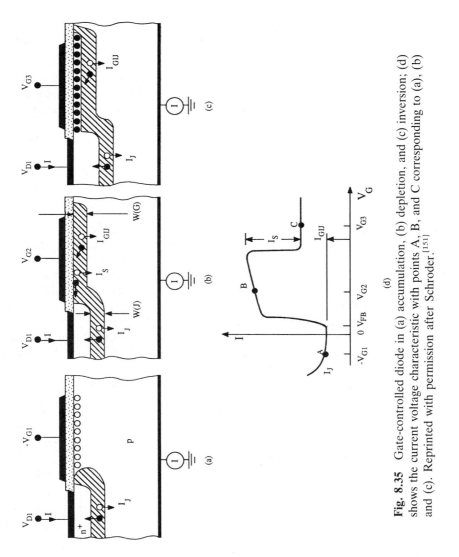

Fig. 8.35 Gate-controlled diode in (a) accumulation, (b) depletion, and (c) inversion; (d) shows the current voltage characteristic with points A, B, and C corresponding to (a), (b) and (c). Reprinted with permission after Schroder.[151]

413

in Fig. 8.35(b). Higher gate voltages lead to a more gradual current increase because the scr width under the gate widens with gate voltage. Gate voltage V_{G2} [point B in Fig. 8.35(d)] is characteristic of this part of the current-voltage curve. The surface potential lies in the range $0 < \phi_s < V_{D1} + 2\phi_F$, and the scr width under the gate is given by[151]

$$W(G) = \left(\frac{K_s W_{ox}}{K_{ox}} \right)\left(\sqrt{1 + \frac{2V'_G}{V_0}} - 1 \right) \qquad (8.104)$$

assuming that there is no inversion charge. The generation regions in Fig. 8.35(b) consist of three components: (1) the diode scr, (2) the gate-induced scr, and (3) the depleted surface under the gate. The total current is $I_J + I_S + I_{GIJ}$.

The surface inverts for surface potentials $\phi_s \geq V_{D1} + 2\phi_F$, and the gate scr width pins to

$$W(G) = \sqrt{\frac{2K_s\varepsilon_0(V_{D1} + 2\phi_F)}{qN_A}} \qquad (8.105)$$

Surface generation drops precipitously and generation component (3) effectively dissappears, as shown in Fig. 8.35(c). Further gate voltage increases beyond the inversion voltage give no further current changes. This is also evident from Eq. (8.12), which shows surface generation to be a maximum for a depleted surface when p_s and n_s are small. Thermal generation is reduced for heavily inverted surfaces when n_s is very high, and most interface traps are occupied by electrons. To first order, we assume that surface generation becomes zero. The current is due to components (1) and (2), shown in Fig. 8.35(c), and by C in Fig. 8.35(d). Experimental current-voltage curves are shown in Fig. 8.36. The current rise near zero volts is independent of the diode voltage, but the current drop is not.

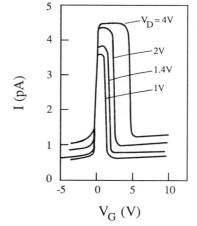

Fig. 8.36 Experimental current-voltage characteristics for various diode voltages. Gate length 20 μm, $\tau_g = 5.5$ ms and $s_g = 1.5$ cm/s; $T = 23°C$. Reprinted with permission after van der Spiegel and Declerck.[164]

For a quantitative description, we make several assumptions. The first is a sufficiently high reverse bias for the mobile carrier concentration to be negligibly small in the scr and at the depleted surface. The bulk and surface generation rates are given by Eqs. (8.16) and (8.18), respectively. The bulk scr generation current is $I_B = qG \times$ volume and the surface current is $I_S = qG_S \times$ area, where the volume and area are the thermal generation volume and area.

The generation current is $I_g = I_J + I_{GIJ} + I_S$ with

$$I_J = \frac{qn_iW(J)A_J}{\tau_g(J)} \qquad I_{GIJ} = \frac{qn_iW(G)A_G}{\tau_g(G)} \qquad I_S = qn_is_gA_G$$

(8.106)

where J and G stand for the n^+p junction and the gate, respectively. $W(G)$ is given either by Eq. (8.104) or (8.105), and $W(J)$ is

$$W(J) = \sqrt{\frac{2K_s\varepsilon_0(V_{D1} + V_{bi})}{qN_A}} \qquad (8.107)$$

with $V_{bi} = (kT/q)\ln(N_AN_D/n_i^2)$ the built-in or diffusion potential.

To extract the generation parameters from a gate-controlled diode, one must be able to measure or calculate the various scr widths in order to extract the generation parameters. The widths can be experimentally determined from capacitance measurements, but it is usually more convenient to calculate them using Eqs. (8.104), (8.105), and (8.107). Note the close relationship between Eqs. (8.105) and (8.107). The only variation between them is the built-in voltage V_{bi} in the n^+p junction and the Fermi potential $2\phi_F$ in the gate-induced junction. To determine the surface generation velocity, one usually makes I-V_G measurements at low diode voltages ($V_D \approx 0.5 - 1$ V), thereby increasing the importance of surface current relative to bulk current. It is also possible of course to determine the generation lifetime under the gate separately from the generation lifetime under the n^+ diffusion and to profile both as a function of depth.[160]

The theory discussed so far was originally proposed by Grove and Fitzgerald.[161–162] It is based on several simplifying assumptions. It assumes the current to be due to scr-generated current only. This is a reasonable assumption for Si devices at room temperature, but the quasi-neutral current component may not be totally negligible for high lifetime devices. The ratio of the bulk scr current to the bulk quasi-neutral region current, that is, component 1 to component 4 in Eq. (8.100), becomes

$$\frac{N_A}{n_i}\frac{W}{L_n}\frac{\tau_r}{\tau_g} = \frac{N_A}{n_i}\frac{W}{\sqrt{D_n}}\frac{\sqrt{\tau_r}}{\tau_g} \approx 3.6\frac{\sqrt{\tau_r}}{\tau_g} \qquad (8.108)$$

for $N_A/n_i = 10^5$, $W = 2\ \mu m$, $D_n = 30\ cm^2/s$, and $L_n = (D_n \tau_r)^{1/2}$. For $\tau_g = \tau_r = 1\ \mu s$ we find the ratio to be 3600, and clearly scr current dominates. But for $\tau_g = 1\ ms$ and $\tau_r = 100\ \mu s$, the ratio becomes 36. The ratio approaches unity for temperatures above room temperature. For narrow-band-gap semiconductors quasi-neutral current dominates, and the theory must be suitably modified.

A second assumption is total depletion of the surface before inversion. This has been shown not to be the case.[163] The lateral surface current inverts the surface weakly for all but a very small fraction of the channel for gate biases far below the gate voltage required to invert the surface strongly. It follows that s_g deduced from the simple equations will be lower than the true s_g. Experiments with long gates (500–1000 μm) gave $s_g = 0.3\ cm/s$, whereas short-gate devices (20 μm) gave 1.5 cm/s.[164]

A further assumption is that thermal scr generation proceeds over the entire scr width. That is also generally not true since the scr is not entirely depleted of mobile carriers. Toward the edge of the scr at $x = W$, the majority carriers have a smeared-out tail, causing the active generation width to be less than W. Furthermore, as the inversion layer forms, the scr under the gate becomes inverted, and generation in that part of the scr diminishes too. Hence the actual generation width is less than W. The exact shape of the generation width is difficult to determine, but by using an effective scr width $W_{eff} = (W - W_a)$ instead of W, a more realistic generation width ensues and simple solutions are obtained that agree well with experiment. W_a is a function of the energy level of the G-R centers. W_{eff} approaches W for high reverse bias but is considerably less for low voltages.[159] If W_{eff} is used instead of W, then τ_g cannot be extracted from one discrete current measurement, but should be determined from the slope of the $I_g - W$ characteristic.[164]

Active devices are frequently surrounded by implantation-doped/thick-oxide channel stops. These channel-stop sidewalls contribute additional current, and the gate-controlled diode is an effective test structure to measure this current contribution.[165]

8.6.2 Pulsed MOS Capacitor

The pulsed MOS capacitor lifetime measuring technique is one of the most popular techniques due to its simplicity and the ubiquitous MOS-C found on many test structures. The method is generally used to determine τ_g, but as shown in Section 8.5.3, it can also be used to determine τ_r. Many papers have been written on the basic method, first proposed by Zerbst in 1966,[166] and on subsequent variations. A review of the various methods can be found in Kang and Schroder.[167] We give the most relevant concepts and equations here for three popular versions of this method, leaving many of the details to the published literature.

Zerbst Plot   The MOS capacitor is pulsed into deep depletion, and the capacitance-time curve is measured, as shown in Fig. 8.31. An experimental room-temperature *C-t* curve is shown in Fig. 8.37(a). The capacitance relaxation is determined by thermal ehp generation. The ehp generation rate, given by Eq. (8.100), consist of scr and quasi-neutral region generations components

$$\frac{dQ_N}{dt} = -\frac{qn_i(W - W_f)}{\tau_g} - \frac{qn_i s_g A_S}{A_G} - qn_i s_{eff} = -\frac{qn_i(W - W_f)}{\tau_g'} - qn_i s_{eff} \tag{8.109}$$

where $W_f = (4K_s \varepsilon_0 \phi_F / qN_A)^{1/2}$ and τ_g' is defined in Eq. (8.103). The effective scr width $(W - W_f)$ approximates the actual generation width and ensures that at the end of the *C-t* transient the scr generation becomes zero. The major uncertainty in estimating the true generation width arises from a lack of knowledge of the quasi-Fermi levels during the transient. Expressions

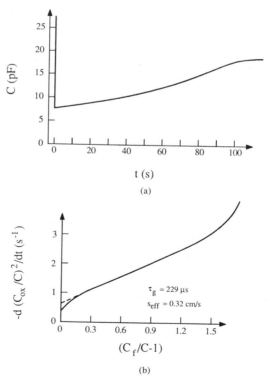

(a)

(b)

Fig. 8.37   (a) Experimental *C-t* response of an MOS-C, (b) its Zerbst plot. Reprinted with permission after Kang and Schroder.[167]

other than W_f have been derived, but they are usually more complicated and have not found wide acceptance.[159,168]

The term $qn_i s_{eff}$ accounts for the scr width-independent generation components (surface generation under the gate and in the quasi-neutral region) with

$$s_{eff} = s + \frac{n_i D_n}{N_A L'_n} \qquad (8.110)$$

The scr width is related to the capacitance C through[151]

$$W = \frac{K_s \varepsilon_0 (C_{ox} - C)}{C_{ox} C} \qquad (8.111)$$

Combining Eqs. (8.99), (8.109), and (8.111) gives

$$\boxed{-\frac{d}{dt}\left(\frac{C_{ox}}{C}\right)^2 = \frac{2n_i}{\tau'_g N_A}\frac{C_{ox}}{C_f}\left(\frac{C_f}{C} - 1\right) + \frac{K_{ox}}{K_s}\frac{2n_i s_{eff}}{W_{ox} N_A}} \qquad (8.112)$$

using the identity $(2/C^3)\,dC/dt = -[d(1/C)^2/dt]$.

Equation (8.112) is the basis of the well-known *Zerbst plot*, $-d(C_{ox}/C)^2/dt$ versus $(C_f/C - 1)$, shown in Fig. 8.37(b). The most useful part of this plot is the linear portion. The curved portion near the origin is when the device approaches equilibrium and the curvature at the other end of the straight line has been attributed to field-enhanced emission from interface and/or bulk traps.[169–170] The slope of the straight line is given by $2n_i C_{ox}/N_A C_f \tau'_g$, and its extrapolated intercept on the vertical axis is $2n_i K_{ox} s_{eff}/K_s W_{ox} N_A$. Clearly the slope is a measure of the scr generation parameters τ_g and s_g, whereas the intercept is related to the scr width-independent generation parameters s, L_n, and s_c. s_{eff} obtained from the intercept should not be interpreted as the surface generation velocity as is sometimes done. First, it includes the quasi-neutral bulk generation components. Second, a more detailed analysis of the C-t response shows that the inherent inaccuracy of the $(W - W_f)$ approximation for the generation width can lead to a non-zero Zerbst plot intercept even if $s_{eff} = 0$.[164]

It is instructive to examine the two axes of the *Zerbst plot* in more detail for a better insight into the physical meaning of such a plot. For the identity that leads to Eq. (8.112), we find from Eq. (8.99)

$$-\frac{d}{dt}\left(\frac{C_{ox}}{C}\right)^2 \sim \frac{dQ_N}{dt} \qquad (8.113)$$

The Zerbst plot vertical axis is proportional to the total ehp carrier generation rate or to the generation current. From Eq. (8.111)

$$\left(\frac{C_f}{C} - 1\right) \sim (W - W_f) \qquad (8.114)$$

The horizontal axis is proportional to the scr generation width. So we find this rather complicated plot to be nothing more than a plot of generation current versus scr generation width. The current can of course be measured directly. The generation width cannot be measured directly and is most easily extracted from the capacitance.

The MOS-C pulsed C-t technique has found wide acceptance because it is easily implemented with commercially available capacitance meters, a dc power supply, a micro switch, or very low frequency function generator, and an X-Y recorder. No fast response circuitry is required. For short C-t response times, the output is displayed on an oscilloscope instead of an X-Y recorder. Computer-controlled implementations are commonly used.

The measured C-t transient times are usually quite long with times of tens of seconds to minutes being common. The relaxation time t_f is related to τ'_g by[171–172]

$$t_f \approx 10\left(\frac{N_A}{n_i}\right)\tau'_g \qquad (8.115)$$

This equation brings out a very important feature of the pulsed MOS-C technique, which is the magnification factor N_A/n_i built into the measurement. Values of τ'_g range over many orders of magnitude, but representative values for high quality silicon devices lie in the range of 10^{-5} to 10^{-3} s. Equation (8.115) predicts the actual C-t transient time to be 1 to 1000 s. These long times point out the great virtue of the pulsed MOS capacitor lifetime measurement technique. To measure lifetimes in the microsecond range, it is only necessary to measure capacitance recovery times on the order of seconds. The circuit implementation to do this is very simple, accounting for the method's popularity. Equation (8.115) can be used as a first-order calculation of the generation lifetime.

The time magnification factor in Eq. (8.115) is also a disadvantage. For $N_A/n_i \approx 10^5$ and $\tau_g = 1$ ms, we find $t_f \approx 1000$ s. Such long measurement times preclude mapping of large number of devices even for automated equipment. The measurement bottleneck is the data acquisition time. Several approaches have been proposed to reduce the measurement time. Equations (8.109) and (8.110) show that $s_{eff} \sim n_i^2$. As the temperature is raised, this scr width-independent term dominates and the relaxation time is considerably reduced. The Zerbst plot is shifted vertically retaining the slope determined by τ'_g.[172] As pointed out in Section 8.5.3, the temperature should not be so high that quasi-neutral generation dominates, for then it is impossible to extract τ'_g. A t_f reduction and similar vertical shift is also attained by illuminating the sample, adding a constant optical generation term to Eq. (8.109).[173–174]

In another version the MOS-C is driven into deep depletion by a voltage pulse and then into inversion by a light pulse. Subsequently a series of small pulses of opposite polarity and varying amplitudes are superimposed on the depleting voltage, driving the device into weaker inversion and then into depletion. C and dC/dt are determined after each pulse to construct a Zerbst plot. The total measurement time can be reduced by as much as a factor of 10 and τ'_g agrees with the conventional method to within about 10%.[175] In yet another simplification the scr width is calculated from the C-t response, and $\ln(W)$ is plotted against time.[176] Such a plot is nearly linear. The line can be extrapolated to t_f without recording the entire curve by using only the initial portion of the C-t response.

The generation lifetime determined from pulsed MOS-C's agrees within 5% with values determined from gate-controlled diode measurements.[177] The generation lifetime is frequently constant over the measurement depth. However, there are cases when that is not so. For example, many modern wafers contain a denuded zone and a precipitated interior formed by appropriate oxygen outdiffusion and precipitation processes. Recombination lifetime measurement methods have difficulty in probing lifetime variations into the wafer. However, pulsed MOS-C methods lend themselves very nicely to such measurements because the scr width is under the operator's control. Using successively higher gate voltage pulses allows τ'_g to be determined as a function of depth.[172,178] A similar approach has been used to determine the spatial lifetime variation in epitaxial Si layers with misfit dislocations introduced by incorporating a small fraction of Ge into the Si.[179]

Current-Capacitance   The *Zerbst* technique requires differentiation of the experimental data and a knowledge of N_A. The *current-capacitance* technique, in which both current and capacitance are measured, requires neither N_A nor differentiation. However, it does require a measure of both the current and the capacitance of a pulsed MOS-C. The current is given by

$$I = A_G\left(\frac{dQ_N}{dt} + qN_A\frac{dW}{dt}\right) \tag{8.116}$$

where the first term is the conduction and the second term the displacement current. The current-capacitance relationship is[169,180]

$$\frac{I}{(1 - C/C_{ox})} = \frac{qK_s\varepsilon_0 n_i A_G^2}{\tau'_g}\left(\frac{1}{C} - \frac{1}{C_f}\right) + qn_i s_{eff}A_G \tag{8.117}$$

From the C-t and the I-t curve one plots $I/(1 - C/C_{ox})$ versus $(1/C - 1/C_f)$. The slope of this curve gives τ'_g and the intercept gives s_{eff}, as shown in Fig. 8.38.

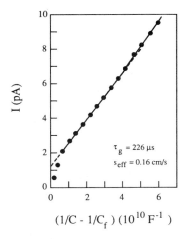

$(1/C - 1/C_f) (10^{10} F^{-1})$

Fig. 8.38 Experimental current versus inverse capacitance curve. Reprinted with permission after Kang and Schroder.[167]

For a profile of the generation lifetime, Eq. (8.117) can be rewritten as

$$\tau'_g = \frac{qK_s\varepsilon_0 n_i A_G^2}{C_{ox}} \frac{d[C_{ox}/C]/dt}{d[I/(1 - C/C_{ox})]/dt} \tag{8.118}$$

By measuring current and capacitance simultaneously, forming the proper combinations given in Eq. (8.118), and differentiating the data, it is possible to plot a profile of τ'_g directly without knowing the doping profile. The computer-controlled instrumentation for this is described in Fahrner et al.[172]

Linear Sweep The *linear sweep* technique has some advantages over the pulsed method. For example, for a complete characterization of the *C-t* transient in the pulsed method, it is necessary to wait for the entire response, which may be many minutes for high lifetime devices. The linear sweep technique has the potential of reduced measurement time—a decided plus when many devices need to be characterized.

Consider a linearly varying voltage applied to the gate of an MOS-C of a polarity to drive the device into depletion. We show in Chapter 6 that for sufficiently slow sweep rates, the equilibrium C-V_G curve is traced out. We also know that when the sweep rate is so high that it can be considered a voltage step, the pulsed MOS-C deep-depletion curve is obtained. For intermediate sweep rates, an intermediate trace is swept out, shown in Fig. 8.39, lying between the deep-depletion and the equilibrium curves.

The interesting point about this curve is its saturation characteristic.[181] It comes about for the following reason. Assume the voltage sweeps from point A in Fig. 8.39 to the right. For voltages more positive than V_B, the device enters deep depletion and the scr width widens beyond W_f, with the capacitance being driven below C_f. Electron–hole pair generation attempts

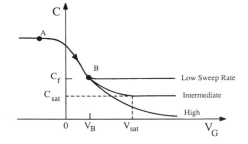

Fig. 8.39 Inversion, saturation, and deep-depletion curves of an MOS-C. Reprinted with permission after Schroder.[151]

to reestablish equilibrium. However, the gate voltage continues to drive the device into deep depletion, further increasing W. This in turn enhances the generation rate because it is proportional to W. At the voltage V_{sat} the attempt by the linearly varying gate voltage to drive the device into deeper depletion is exactly balanced by the generation rate holding it at the capacitance. The capacitance-voltage curve saturates at C_{sat}. For other sweep rates the C-V_G curves will saturate at other C_{sat} values. There will always be saturation, unless the sweep rate is so high that the device will be driven along the "high sweep rate" true deep-depletion curve.

For a constant sweep rate, $dV_G/dt = R$, Eq. (8.98) becomes

$$\frac{dQ_N}{dt} = -\frac{qK_s\varepsilon_0 C_{ox} N_A}{C^3}\frac{dC}{dt} - C_{ox}R \tag{8.119}$$

Using the generation rate expression Eq. (8.109), leads to

$$-\frac{d}{dt}\left(\frac{C_{ox}}{C}\right)^2 = \frac{2}{V_0}\left(\frac{qK_s\varepsilon_0 n_i(C_f/C - 1)}{C_f C_{ox}\tau_g'} + \frac{qn_i s_{eff}}{C_{ox}} - R\right) \tag{8.120}$$

When the device enters the saturation regime, capacitance C_{sat} changes neither with voltage nor with time. Hence the left side of Eq. (8.120) becomes zero, and

$$R = \frac{qK_s\varepsilon_0 n_i(C_f/C_{sat} - 1)}{C_f C_{ox}\tau_g'} + \frac{qn_i s_{eff}}{C_{ox}} \tag{8.121}$$

Equation (8.121) gives the relationship between the linear sweep rate R and the generation parameters τ_g' and s_{eff}. In the experiment, a series of C-V_G curves at different linear sweeps rates are plotted. The C_{sat} values are taken from these curves, and a plot of R versus $(C_f/C_{sat} - 1)$ yields a straight line of slope $qK_s\varepsilon_0 n_i/C_f C_{ox}\tau_g'$ and intercept $qn_i s_{eff}/C_{ox}$. Similar to the *Zerbst plot*, τ_g' is obtained from the slope and s_{eff} from the intercept.

It may appear from Eq. (8.121) that a single value of R and $(C_f/C_{sat} - 1)$ suffice to determine the two generation parameters. That is incorrect because the R versus $(C_f/C_{sat} - 1)$ line does not pass through the origin, and a slope of that plot is not the same as the slope of a line from the origin to any one point on the line. This is important to keep in mind. It is more accurate to determine a device parameter from the slope of several data points than it is to use a single datum point if the line does not pass through the origin.

Experimental data for the linear sweep method are shown in Fig. 8.40 for the device whose *Zerbst plot* is shown in Fig. 8.37. Note the good agreement between the experimentally determined values for τ'_g and s_{eff}. The linear sweep technique does not require the acquisition of an entire *C-t* curve, nor the differentiation of the experimental data. It does, however, require multiple saturating *C-V_G* curves. For those devices with high lifetimes and consequent long *C-t* transients, it is found that very low sweep rates are required, with resultant long data acquisition times. The use of a feedback circuit, with the capacitance preset to a certain value and the linear sweep rate adjusting itself through feedback to maintain this preset value, reduces the data acquisition time.[182] Computer automation of the linear sweep technique has also been developed.[183]

All three MOS-C methods have found applications. The choice of a particular technique is largely determined by the experience and preference of the experimenter and the availability of commercial instrumentation. We will no doubt see further variations and automatization of the methods. They are clearly among the more powerful measurement methods for device as well as a process characterization. Frequently, a complete analysis of the experimental data is not necessary, and only the *C-t* data of devices made on different days or with different processes are compared. Significant variances are indicative of process problems. For a detailed review of the many

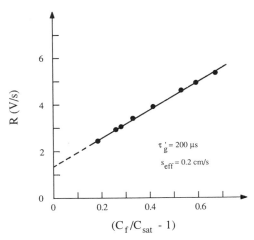

Fig. 8.40   Linear ramp plot for the device whose Zerbst plot is shown in Fig. 8.37. Reprinted with permission after Schroder.[151]

lifetime extraction methods based in one form or another on the deep-depletion MOS-C, the reader is referred to Kang and Schroder.[167]

8.7 STRENGTHS AND WEAKNESSES

- *Recombination Lifetime* Among the optical recombination lifetime measuring methods, the photoconductive decay technique is most commonly used. Originally it was implemented with ohmic contacts to the device; more recently contactless methods have been implemented using either microwave reflection or inductive coupling. This allows for rapid measurements. Its major strength is the noncontacting nature and rapid measurement. Its major weakness is the unknown surface recombination velocity. If one is satisfied with an effective lifetime, only one sample need be measured. If more then one sample is available, or if the sample thickness can be changed by etching, then both the bulk lifetime and the surface recombination velocity can be extracted. The open-circuit voltage decay method is the most common electrical recombination lifetime method. It is easy to interpret, but a junction diode is required.
- *Generation Lifetime* The generation lifetime is most often used as a process monitor using the pulsed MOS capacitor. The device is available on many test structures, and the measurement is easy to do. The Zerbst plot implementation is the most common, but the current versus inverse capacitance is easier to interpret since the doping concentration of the sample need not be known.

APPENDIX 8.1 OPTICAL EXCITATION

Steady State Optical excitation of semiconductors is used in a variety of lifetime or diffusion length measurement techniques. We include in this appendix excitation by non-optical methods, such as electron beam and X-ray techniques, as well because all of them create excess ehp's. Once excess ehp's have been created, there are two basic methods of measuring recombination properties. (1) The excitation source is abruptly terminated and the excess carrier decay rate is measured. An example of this is the photoconductive decay technique where the conductivity of the sample is monitored as a function of time. (2) The steady-state excess carrier concentration due to continuous generation and recombination, sets up an open-circuit voltage or short-circuit current which is measured. Being a steady-state method, it does not have the problems that transient methods are prone to, such as trapping effects.

We consider the *p*-type semiconductor of Fig. A8.1. It is a wafer of thickness T, reflectivity R, minority carrier lifetime τ, minority carrier diffusion coefficient D, minority carrier diffusion length L, the surface

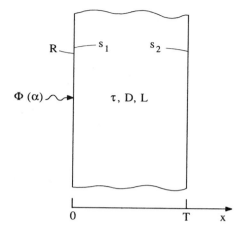

Fig. A8.1 Homogeneous p-type sample geometry for optical excitation.

recombination velocities s_1 and s_2 at the two surfaces. Monochromatic light of photon flux density Φ, wavelength λ, and absorption coefficient α, is incident on one side of this wafer. We assume that carriers generated by absorbed photons diffuse in the x-direction and that the wafer has infinite extent in the y-z plane. This allows the neglect of edge effects. The steady-state, small-signal excess minority carrier concentration $\Delta n(x)$ is obtained from a solution of the one-dimensional continuity equation

$$D \frac{d^2\, \Delta n(x)}{dx^2} - \frac{\Delta n(x)}{\tau} + G(x) = 0 \qquad (A8.1)$$

subject to the boundary conditions

$$\frac{d\, \Delta n(x)}{dx} = s_1 \frac{\Delta n(0)}{D} \quad \text{at } x = 0 \qquad (A8.2a)$$

$$\frac{d\, \Delta n(x)}{dx} = -s_2 \frac{\Delta n(T)}{D} \quad \text{at } x = T \qquad (A8.2b)$$

The generation rate is given by

$$G(x, \lambda) = \Phi(\lambda)\alpha(\lambda)[1 - R(\lambda)] \exp[-\alpha(\lambda)x] \qquad (A8.3)$$

An implicit assumption in this expression is that each *absorbed* photon generates one ehp. This is usually a good assumption for semiconductors. The solution to Eq. (A8.1) using (A8.2) and (A8.3) is[184]

$$\Delta n(x) = \frac{(1 - R)\Phi\alpha\tau}{(\alpha^2 L^2 - 1)} \left[\frac{A_1}{D_1} + e^{-\alpha T} \frac{B_1}{D_1} - e^{-\alpha x} \right] \qquad (A8.4)$$

where

$A_1 = (s_1 s_2 L/D + s_2 \alpha L) \sinh[(T - x)/L] + (s_1 + \alpha D) \cosh[(T - x)/L]$
$B_1 = (s_1 s_2 L/D - s_1 \alpha L) \sinh(x/L) + (s_2 - \alpha D) \cosh(x/L)$
$D_1 = (s_1 s_2 L/D + D/L) \sinh(T/L) + (s_1 + s_2) \cosh(T/L)$

For some measurement methods the excess carrier concentration is required; for others, the current density.

In the derivation of Eq. (A8.4) only diffusion was considered. The electric fields are assumed to be sufficiently small that drift is negligible. The diffusion current density is given by

$$J_n(x) = qD \frac{d\,\Delta n(x)}{dx} \tag{A8.5}$$

From Eq. (A8.4), $J_n(x)$ can be written as

$$J_n(x) = \frac{q(1 - R)\Phi\alpha L}{(\alpha^2 L^2 - 1)} \left[\frac{A_2}{D_1} - e^{-\alpha T} \frac{B_2}{D_1} - \alpha L\, e^{-\alpha x} \right] \tag{A8.6}$$

where

$A_2 = (s_1 s_2 L/D + s_2 \alpha L) \cosh[(T - x)/L] + (s_1 + \alpha D) \sinh[(T - x)/L]$
$B_2 = (s_1 s_2 L/D - s_1 \alpha L) \cosh(x/L) + (s_2 - \alpha D) \sinh(x/L)$

Two special cases are the current density at the left and right boundaries. At the left boundary

$$J_n(0) = \frac{q(1 - R)\Phi\alpha L}{(\alpha^2 L^2 - 1)} \left[\frac{A_3}{D_1} - e^{-\alpha T} \frac{(s_1 s_2 L/D - s_1 \alpha L)}{D_1} - \alpha L \right] \tag{A8.7}$$

where

$$A_3 = (s_1 s_2 L/D + s_2 \alpha L) \cosh(T/L) + (s_1 + \alpha D) \sinh(T/L)$$

Similarly for the right boundary

$$J_n(T) = \frac{q(1 - R)\Phi\alpha L}{(\alpha^2 L^2 - 1)} \left[\frac{(s_1 s_2 L/D + s_2 \alpha L)}{D_1} - e^{-\alpha T} \frac{B_3}{D_1} - \alpha L\, e^{-\alpha T} \right] \tag{A8.8}$$

where

$$B_3 = (s_1 s_2 L/D - s_1 \alpha L) \cosh(T/L) + (s_2 - \alpha D) \sinh(T/L)$$

Figure A8.1 serves as a starting point for the derivation of the excess carrier concentration and the current density, but a more realistic structure is the n^+p junction of Fig. A8.2. For this structure we can derive the excess carrier concentration and the current density expressions by some modification to Eqs. (A8.4) and (A8.6).[185]

For the n^+ layer we are concerned with the thin top layer of thickness d. Hence in Eq. (A8.4) we make the following replacements: $T \to d$ and $s_1 \to s_p$. We are especially interested in the excess carrier concentrations under short-circuit current conditions, where that the excess carrier concentration is zero at the edge of the space-charge region ($x = d$). From a surface recombination point of view, this means $s_2 = \infty$, resulting in

$$\Delta p(x) = \frac{(1-R)\Phi\alpha\tau_p}{(\alpha^2 L_p^2 - 1)} \left[\frac{A_4}{D_4} + e^{-\alpha d}\frac{B_4}{D_4} - e^{-\alpha x} \right] \tag{A8.9}$$

with

$$A_4 = (s_p L_p/D_p + \alpha L_p)\sinh[(d-x)/L_p]$$
$$B_4 = (s_p L_p/D_p)\sinh(x/L_p) + \cosh(x/L_p)$$
$$D_4 = (s_p L_p/D_p)\sinh(d/L_p) + \cosh(d/L_p)$$

Similar arguments for the p-substrate, using $x' = (x - d - W)$, $T' = (T - d - W)$, and $s_1 = \infty$, give

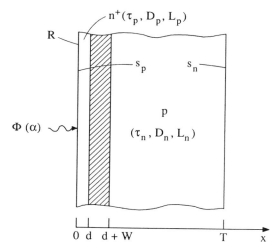

Fig. A8.2 Junction geometry for optical excitation.

$$\Delta n(x') = \frac{(1-R)\Phi\alpha\tau_n}{(\alpha^2 L_n^2 - 1)}\left[\frac{A_5}{D_5} + e^{-\alpha T'}\frac{B_5}{D_5} - e^{-\alpha x'}\right]e^{-\alpha(d+W)}$$

(A8.10)

with

$$A_5 = (s_n L_n/D_n)\sinh[(T-x)/L_n] + \cosh[(T-x)/L_n]$$
$$B_5 = (s_n L_n/D_n - \alpha L_n)\sinh(x'/L_n)$$
$$D_5 = (s_n L_n/D_n)\sinh(T'/L_n) + \cosh(T'/L_n)$$

The additional term $\exp[-\alpha(d+W)]$ in Eq. (A8.10) accounts for the carrier generation beyond $x = d + W$. The absorbed photon flux density has already diminished by this factor when the photons enter the p-substrate. The excess carrier concentration in the scr is considered to be zero. Of course ehp's are generated in the scr and do give rise to a photocurrent, but they are considered to be swept out of that region as soon as they are generated.

The current density for the short-circuited structure of Fig. A8.2 is obtained by considering the diffusion current only, as in Eq. (A8.5). An implicit assumption is that there are no voltage drops across the n^+ and p regions and that drift currents are negligible in these two regions. In the scr the electric field is dominant, and recombination is negligible. With these assumptions the short-circuit current density is

$$J_{sc} = J_p + J_n + J_{scr}$$

(A8.11)

The *hole* current density is

$$J_p = \frac{q(1-R)\Phi\alpha L_p}{(\alpha^2 L_p^2 - 1)}\left[\frac{A_6}{D_6} - e^{-\alpha d}\frac{B_6}{D_6} - \alpha L_p e^{-\alpha d}\right]$$

(A8.12)

where

$$A_6 = s_p L_p/D_p + \alpha L_p$$
$$B_6 = (s_p L_p/D_p)\cosh(d/L_p) + \sinh(d/L_p)$$
$$D_6 = (s_p L_p/D_p)\sinh(d/L_p) + \cosh(d/L_p)$$

The *electron* current density is

$$J_n = \frac{q(1-R)\Phi\alpha L_n}{(\alpha^2 L_n^2 - 1)}\left[-\frac{A_7}{D_7} + e^{-\alpha T'}\frac{B_7}{D_7} + \alpha L_n\right]e^{-\alpha(d+W)}$$

(A8.13)

where

$$A_7 = (s_n L_n / D_n) \cosh(T'/L_n) + \sinh(T'/L_n)$$
$$B_7 = (s_n L_n / D_n - \alpha L_n)$$
$$D_7 = (s_n L_n / D_n) \sinh(T'/L_n) + \cosh(T'/L_n)$$

and the *space-charge region* current density is

$$J_{scr} = q(1 - R)\Phi \, e^{-\alpha d}(1 - e^{-\alpha W}) \qquad (A8.14)$$

Transient For the transient case the one-dimensional continuity equation for the sample geometry of Fig. A8.1 becomes

$$\frac{\partial \Delta n(x, t)}{\partial t} = D \, \frac{\partial^2 \Delta n(x, t)}{\partial x^2} - \frac{\Delta n(x, t)}{\tau} + G(x, t) \qquad (A8.15)$$

Generally during transient measurements, the carrier decay is monitored after the excitation source is turned off; that is, $G(x, t) = 0$ during the measurement.

A solution of Eq. (A8.15) with $G(x, t) = 0$ and the boundary conditions

$$\frac{\partial \Delta n(x, t)}{\partial x} = s_1 \, \frac{\Delta n(0, t)}{D} \quad \text{at } x = 0 \qquad (A8.16a)$$

$$\frac{\partial \Delta n(x, t)}{\partial x} = -s_2 \, \frac{\Delta n(T, t)}{D} \quad \text{at } x = T \qquad (A8.16b)$$

gives the general solution[186]

$$\Delta n(x, t) = \sum_{m=1}^{\infty} A_m(x) \exp\left(-\frac{t}{\tau_m}\right) \qquad (A8.17)$$

where the coefficients $A_m(x)$ depend upon the initial conditions $\Delta n(x, 0)$ and the decay time constants τ_m given by

$$\frac{1}{\tau_m} = \frac{1}{\tau} + D\beta_m^2 \qquad (A8.18a)$$

with β_m being the mth root of

$$\tan(\beta T) = \frac{\beta(s_1 + s_2)D}{(\beta^2 D^2 - s_1 s_2)} \qquad (A8.18b)$$

The excess carrier decay curve is a sum of exponentials in which the higher-order solutions decay more rapidly with time than the fundamental and may be neglected after an initial transient period as shown in Fig. A8.3. The dominant mode decays exponentially with a time constant, denoted as τ_{eff}, given by

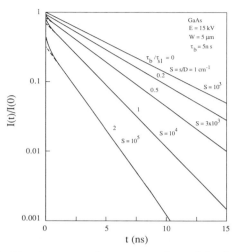

Fig. A8.3 Calculated luminescence decay curves for a delta function electron beam pulse in GaAs as a function of surface recombination velocity. Reprinted with permission after Boulou and Bois.[19]

$$\frac{1}{\tau_{eff}} = \frac{1}{\tau} + D\beta_1^2 \qquad (A8.19)$$

β_1 being the first real root of Eq. (A8.18b).

Two limiting cases are of particular interest. For low surface recombination velocity ($s_1 = s_2 = s \to 0$)

$$\frac{1}{\tau_{eff}} = \frac{1}{\tau} + \frac{2s}{T} \qquad (A8.20)$$

while for high $s(s \to \infty)$

$$\frac{1}{\tau_{eff}} = \frac{1}{\tau} + \frac{\pi^2 D}{T^2} \qquad (A8.21)$$

If we write $1/\tau_{eff} = 1/\tau + 1/\tau_s$, where τ_s is the surface lifetime, then

$$\tau_s(s \to 0) = \frac{T}{2s}$$

$$\tau_s(s \to \infty) = \frac{T^2}{\pi^2 D} \qquad (A8.22)$$

The measured lifetime is always less than the true recombination lifetime from Eqs. (A8.20) and (A8.21). The discrepancy of the measured lifetime from the true lifetime depends on s, τ, and T. A more detailed discussion of the decay rate is given in Boulou and Bois[187] and in Luke and Cheng.[188]

The discussion so far has centered on the shape of the decay curve when recombination is the result of a single mechanism, characterized by a single lifetime τ. Even then, the decay is not purely exponential, owing to surface recombination. A more complex decay is observed if more than one bulk recombination mechanism is active. This might happen if there are several impurities in the sample with different concentrations and capture cross sections. It can also happen if different mechanisms like SRH, radiative, or Auger recombination are active.

All of the above theories are valid for low-level injection, where the SRH, radiative, and Auger lifetimes can be treated as constants and, aside from surface effects, the transient decay can be considered to be of an exponential form. This is no longer true for higher-level injection, especially for radiative and Auger recombination, because the lifetimes themselves are functions of the excess carrier concentrations and the decay is no longer exponential. The equations become very complex, and a detailed discussion is given by Blakemore.[189]

In some measurement techniques a phase shift between the optical excitation source and the detected parameter is measured. For a sinusoidally varying generation rate

$$G(x, t) = [G_0 + G_1 \exp(j\omega t)] \exp(-\alpha x)$$

$$= [\Phi_0 + \Phi_1 \exp(j\omega t)]\alpha(1 - R) \exp(-\alpha x) \quad \text{(A8.23)}$$

the fundamental component of the variation of the excess minority carrier concentration $\Delta n_1(x) \exp(j\omega t)$ is determined from the equation

$$D \frac{d^2 \Delta n_1(x)}{dx^2} - \frac{\Delta n_1(x)}{\tau} + G_1 e^{-\alpha x} = j\omega \Delta n_1(x) \quad \text{(A8.24)}$$

The solution to this equation, subject to the same boundary conditions as Eq. (A8.1), is

$$\Delta n_1(x) = \frac{(1 - R)\Phi_1 \alpha \tau}{\alpha^2 L^2 - 1 - j\omega\tau} \left[\frac{A'}{D'} + e^{-\alpha T} \frac{B'}{D'} - e^{-\alpha x} \right] \quad \text{(A8.25)}$$

where A', B', and D' are similar to A, B, and D in Eq. (A8.4), except that in those equations L is replaced by $L/(1 + j\omega t)^{1/2}$, the frequency-dependent diffusion length.

Trapping   If carrier injection is kept at low level and if the recombination center concentration is low ($N_T \ll N_A$) for p-type material, then the above analysis holds. For high N_T, $\Delta n \neq \Delta p$, and the transient decay is not a simple exponential. There may in addition be trapping centers which capture the carriers and then release them back to the band from which they were captured. This is illustrated in Fig. A8.4. An excess ehp is introduced into

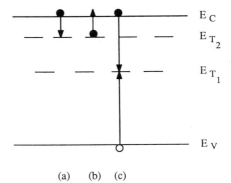

Fig. A8.4 Band diagram showing trapping and recombination.

(a) (b) (c)

the semiconductor. Instead of recombining directly, the electron is temporarily captured or trapped onto level E_{T2} [Fig. A8.4(a)]. It is subsequently re-emitted into the conduction band [Fig. A8.4(b)], and finally it recombines with the hole, either through the impurity with energy level E_{T1} [Fig. A8.4(c)] or by a radiative or Auger process. Clearly the electron "lives" longer in this case by the length of time that it is trapped, and a lifetime measurement would give an erroneously high value.

The resultant effective lifetime has been shown to be

$$\tau'_n = \frac{\tau_n(1 + b + b\tau_2/\tau_1)}{1 + b} \qquad (A8.26)$$

where $b = \mu_n/\mu_p$, τ_1 is the average time that a minority electron spends in the conduction band before it is trapped by a trapping center, and τ_2 is the mean time that the electron spends in the trap before being emitted back into the conduction band. With no trapping $\tau'_n = \tau_n$; with trapping $\tau'_n > \tau_n$, and τ'_n can be very long. For example, certain wideband-gap phosphors exhibit afterglow effects lasting minutes following cessation of the excitation caused by trapping effects in these materials.

It has been found that trapping can be much reduced by illuminating the sample with a steady-state light. The light continually creates ehp's that keep the traps filled, and any additional ehp's created by a light flash will tend to recombine with reduced trapping. Another alternative is to use a very short, intense light pulse. If the pulse width is much less than τ_1, the trap concentration will not change appreciably during the pulse and will play a negligible part during the carrier decay.

Reabsorbed Photons (Photon Recycling) Some ehp's recombine radiatively in direct band-gap semiconductors. It is a property of direct band-gap materials that in addition to radiative recombination, their absorption coefficient is quite high. This follows from the van Roosbroeck and Shockley detailed balance argument.[190] A consequence of this is that the photons generated by a radiative recombination event have a high probability of being reabsorbed to generate electron-hole pairs. The effect of re-absorption is in general a complicated problem with two major effects on the device: the internal quantum efficiency of light emitting devices is enhanced, and the recombination lifetime is increased.[191]

A structure of considerable interest is a double heterostructure of a GaAs layer sandwiched between two AlGaAs layers. The AlGaAs/GaAs interfaces serve as low surface recombination velocity surfaces as well as potential barriers to minority carriers, thereby confining them to the GaAs layer. A photon generated by radiative recombination within the GaAs layer at a distance x from the interface and propagating at an angle θ from the interface normal has a probability of reaching the interface given by[192]

$$p_1 = \exp\left(-\frac{\alpha x}{\cos\theta}\right) \qquad (A8.27)$$

where α is the absorption coefficient in the active region for a photon generated by the recombination event.

The probability for the photon to generate a new ehp is

$$p_2 = 1 - p_1 \qquad (A8.28)$$

It is one minus the probability of escaping. The net fraction F of the emitted photons that generate ehp's is found by averaging p_2 over x, θ, and photon energy E. This averaging process has been discussed in Asbeck,[192] with the result that the net lifetime τ can be written as

$$\frac{1}{\tau} = \frac{1}{\tau_{nrad}} + \frac{1-F}{\tau_{rad}} \qquad (A8.29)$$

where τ_{nrad} is the non-radiative lifetime due to non-radiative processes as well as radiative processes that do not contribute to near band edge luminescence.

For negligible photon recycling, $F = 0$. The lifetime ratio considering photon recycling and zero recycling is

$$\frac{\tau(F \neq 0)}{\tau(F = 0)} = \frac{1 + (\tau_{nrad}/\tau_{rad})}{1 + (\tau_{nrad}/\tau_{rad})(1 - F)} \qquad (A8.30)$$

Typically for $d \leq 0.5\ \mu m$, $F \approx 0$, and for $d \approx 10\ \mu m$, $F \approx 0.9$.[192–193] As the layer becomes thicker and $F \rightarrow 1$, the lifetime with photon recycling taking into account becomes larger than that without recycling by the factor $1 + \tau_{nrad}/\tau_{rad}$. For a device in which $\tau_{rad} \ll \tau_{nrad}$, as is required for a high quality light-emitting device, this factor can become quite large. Hence care should be exercised in the interpretation of lifetime data for thick layers with low surface recombination velocities.

APPENDIX 8.2   ELECTRICAL EXCITATION

Optical excitation as a means to create ehp's in semiconductors for lifetime measuring is non-contacting. The non-uniform carrier excitation due to the exponential absorption of the light can be a disadvantage, or it can be an advantage, especially for low α where ehp generation can be quite uniform throughout the sample. The α-λ relationship must be accurately known for some methods.

Electrical injection is easier to control, and it is a planar source of minority carriers injected at the edges of the scr in a pn junction. The main disadvantage is the requirement of a junction as the source of minority carriers. In most electrical lifetime methods, a junction is forward biased to inject minority carriers into both quasi-neutral regions, illustrated in Fig. A8.5. The injection can be thought of as proceeding from a plane located at the edge of the scr. Consider the p-base of the n^+p junction of Fig. A8.5. The spatial distribution of electrons injected from $x = 0$ into the base is given by

$$\Delta n(x) = n_{p0}(e^{qV_f/kT} - 1) \frac{A}{B} \tag{A8.31}$$

where

$$A = (s_n L_n / D_n) \sinh[(T - x)/L_n] + \cosh[(T - x)/L_n]$$
$$B = (s_n L_n / D_n) \sinh(T/L_n) + \cosh(T/L_n)$$

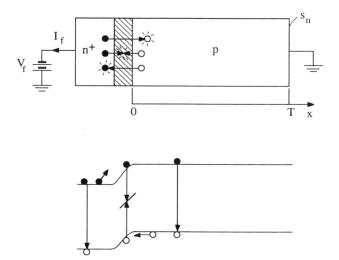

Fig. A8.5   Junction geometry for electrical injection.

Equation (A8.31) resembles Eq. (A8.4) if in the latter we let $\alpha \to \infty$, which is similar to confining the optical carrier generation to the plane at $x = 0$ in Fig. A8.5.

One of the key differences between optical and electrical injection is that during optical injection excess carriers are generated in the sample volume, with the generation depth controlled by the absorption coefficient. Electrical injection proceeds from a plane. Excess carriers exist beyond that plane because they diffuse there, *not* because they are generated there.

REFERENCES

[1] W. M. Bullis, "Measurement of Carrier Lifetime in Semiconductors—An Annotated Bibliography Covering the Period 1949–1967," National Bureau of Stand. Techn. Note 465, Nov. 1968.

[2] R. N. Hall, "Recombination Processes in Semiconductors," *Proc. IEE* **106B**, 923–931, March 1960.

[3] W. Shockley and W. T. Read, "Statistics of the Recombinations of Holes and Electrons," *Phys. Rev.* **87**, 835–842, Sept. 1952.

[4] Y. P Varshni, "Band-to-Band Radiative Recombination in Groups IV, VI and III-V Semiconductors (I) and (II)," *Phys. Stat. Sol.* **19**, 459–514, Feb. 1967; ibid. **20**, 9–36, March 1967.

[5] W. Gerlach, H. Schlangenotto, and H. Mäder, "On the Radiative Recombination Rate in Silicon," *Phys. Stat. Sol.* **13a**, 277–283, Sept. 1972.

[6] R. J. Nelson and R. G. Sobers, "Minority-Carrier Lifetime and Internal Quantum Efficiency of Surface-Free GaAs," *J. Appl. Phys.* **49**, 6103–6108, Dec. 1978; G. W. 't Hooft, "The Radiative Recombination Coefficient of GaAs from Laser Delay Measurements and Effective Nonradiative Lifetimes," *Appl. Phys. Lett.* **39**, 389–390, Sept. 1981, M. Takeshima, "Effect of Auger Recombination on Laser Operation in $Ga_{1-x}Al_xAs$," *J. Appl. Phys.* **58**, 3846–3850, Nov. 1985.

[7] J. Pietzsch and T. Kamiya, "Determination of Carrier Density Dependent Lifetime and Quantum Efficiency in Semiconductors with a Photoluminescence Method (Application to InGaAsP/InP Heterostructures)," *Appl. Phys.* **A42**, 91–102, Jan. 1987.

[8] H. Schlangenotto, H. Mäder, and W. Gerlach, "Temperature Dependence of the Radiative Recombination Coefficient in Silicon," *Phys. Stat. Sol.* **21a**, 357–367, Jan. 1974.

[9] J. G. Fossum, R. P. Mertens, D. S. Lee, and J. F. Nijs, "Carrier Recombination and Lifetime in Highly Doped Silicon," *Solid-State Electron.* **26**, 569–576, June 1983.

[10] D. K. Schroder, "Carrier Lifetimes in Silicon," in *Handbook of Silicon Technology* (W. C. O'Mara and R. B. Herring, eds.) Noyes Publ., Park Ridge, NJ, 1990; J. Burtscher, F. Dannhäuser, and J. Krausse, "The Recombination in Thyristors and Rectifiers in Silicon: Its Influence on the Forward-bias Characteristic and the Turn-off Time," (in German) *Solid-State Electron.* **18**,

35–63, Jan. 1975; J. Dziewior and W. Schmid, "Auger Coefficients for Highly Doped and Highly Excited Silicon," *Appl. Phys. Lett.* **31**, 346–348, Sept. 1977; I. V. Grekhov and L. A. Delimova, "Auger Recombination in Silicon," *Sov. Phys. Semicond.* **14**, 529–532, May 1980; L. A. Delimova, "Auger Recombination in Silicon at Low Temperatures," *Sov. Phys. Semicond.* **15**, 778–780, July 1981; L. Passari and E. Susi, "Recombination Mechanisms and Doping Density in Silicon," *J. Appl. Phys.* **54**, 3935–3937, July 1983; D. Huber, A. Bachmeier, R. Wahlich, and H. Herzer, "Minority Carrier Diffusion Length and Doping Density in Nondegenerate Silicon," in *Semiconductor Silicon/1986* (H. R. Huff, T. Abe, and B. Kolbesen, eds.) Electrochem. Soc., Pennington, NJ, 1986, pp. 1022–1032; E. Yablonovitch and T. Gmitter, "Auger Recombination in Silicon at Low Carrier Densities," *Appl. Phys. Lett.* **49**, 587–589, Sept. 1986; R. A. Sinton and R. M. Swanson, "Recombination in Highly Injected Silicon," *IEEE Trans. Electron Dev.* **ED-34**, 1380–1389, June 1987; T. F. Ciszek, T. Wang, T. Schuyler, and A. Rohatgi, "Some Effects of Crystal Growth Parameters on Minority Carrier Lifetime in Float-Zoned Silicon," *J. Electrochem. Soc.* **136**, 230–234, Jan. 1989, and citations in these references.

[11] A. Haug, "Remarks on the Carrier Density Dependence of Auger Recombination," *J. Phys. C: Solid State Phys.* **21**, L287–L290, March 1988.

[12] D. J. Fitzgerald and A. S. Grove, "Surface Recombination in Semiconductors," *Surf. Sci.* **9**, 347–369, Feb. 1968.

[13] D. K. Schroder, "The Concept of Generation and Recombination Lifetimes in Semiconductors," *IEEE Trans. Electron Dev.* **ED-29**, 1336–1338, Aug. 1982.

[14] A. S. Grove in *Physics and Technology of Semiconductor Devices* (J. Wiley, 1967) introduced τ_0 and s_0 as bulk and surface generation parameters. He assumes that $\sigma_n = \sigma_p$ and $E_T = E_i$ and finds $\tau_0 = \tau_n = \tau_p$ and $G = n_i/2\tau_0$. This places undue restrictions on τ_0. I prefer the more general definition of Eq. (8.16) which requires no assumptions regarding τ_g. By similar arguments Grove defines the surface generation rate as $G_s = n_i(s_0/2)$. Again, I prefer the more general definition of Eq. (8.18) with no assumptions.

[15] D. T. Stevenson and R. J. Keyes, "Measurement of Carrier Lifetimes in Germanium and Silicon," *J. Appl. Phys.* **26**, 190–195, Feb. 1955.

[16] G. K. Wertheim and W. M. Augustyniak, "Measurement of Short Carrier Lifetimes," *Rev. Sci. Instrum.* **27**, 1062–1064, Dec. 1956.

[17] R. Gremmelmaier, "Irradiation of PN Junctions with Gamma Rays: A Method for Measuring Diffusion Lengths," *Proc. IRE*, **46**, 1045–1049, June 1958.

[18] J. H. Reynolds and A. Meulenberg, Jr., "Measurement of Diffusion Length in Solar Cells," *J. Appl. Phys.* **45**, 2582–2592, June 1974.

[19] M. Boulou and D. Bois, "Cathodoluminescence Measurements of the Minority-Carrier Lifetime in Semiconductors," *J. Appl. Phys.* **48**, 4713–4721, Nov. 1977.

[20] T. Tiedje, J. I. Haberman, R. W. Francis, and A. K. Ghosh, "An RF Bridge Technique for Contactless Measurement of the Carrier Lifetime in Silicon Wafers," *J. Appl. Phys.* **54**, 2499–2503, May 1983.

[21] K. L. Luke and L. J. Cheng, "Analysis of the Intraction of a Laser Pulse with

a Silicon Wafer: Determination of ~~Bulk Lifetime~~ and Surface Recombination Velocity," *J. Appl. Phys.* **61**, 2282–2293, March 1987.

[22] "IRE Standard on Solid-State Devices: Measurement of Minority-Carrier Lifetime in Germanium and Silicon by the Method of Photoconductive Decay," *Proc. IRE* **49**, 1292–1299, Aug. 1961.

[23] ASTM Standard F28 (Reapproved 1981), "Standard Method for Measuring the Minority-Carrier Lifetime in Bulk Germanium and Silicon," *1985 Annual Book of ASTM Standards*, Am. Soc. Test. Mat., Philadelphia, 1985.

[24] S. M. Ryvkin, *Photoelectric Effects in Semiconductors*, Consultants Bureau, New York, 1964, pp. 19–22.

[25] A. R. Gerhard and C. W. Pearce, "Measurement of Minority Carrier Lifetime in Silicon Crystals by the Photoconductive Decay Technique," in *Lifetime Factors in Silicon* (R. D. Westbrook, ed.), Am. Soc. Test. Mat., Philadelphia, PA, 1980, pp. 161–170.

[26] H. W. Curtis and R. L. Verkuil, "A High Signal-to-Noise Oscillator for Contactless Measurement of Photoinduced Carrier Lifetimes," in *Lifetime Factors in Silicon* (R.D. Westbrook, ed.), Am. Soc. Test. Mat., Philadelphia, PA, 1980, pp. 210–224.

[27] D. E. Kane and R. M. Swanson, "Measurement of the Emitter Saturation Current by a Contactless Photoconductivity Decay Method," *18th IEEE Photovoltaic Spec. Conf.*, Las Vegas, NV, 1985, pp. 578–583.

[28] E. Yablonovitch, R. M. Swanson, W. D. Eades, and B. R. Weinberger, "Electron-Hole Recombination at the Si-SiO$_2$ Interface," *Appl. Phys. Lett.* **48**, 245–247, Jan. 1986.

[29] E. Yablonovitch, D. L. Allara, C. C. Chang, T. Gmitter, and T. B. Bright, "Unusually Low Surface-Recombination Velocity on Silicon and Germanium Surfaces," *Phys. Rev. Lett.* **57**, 249–252, July 1986.

[30] E. Yablonovitch, C. J. Sandroff, R. Bhat, and T. Gmitter, "Nearly Ideal Electronic Properties of Sulfide Coated GaAs Surfaces," *Appl. Phys. Lett.* **51**, 439–441, Aug. 1987.

[31] Y. Mada, "A Nondestructive Method for Measuring the Spatial Distribution of Minority Carrier Lifetime in Si Wafer," *Japan. J. Appl. Phys.* **18**, 2171–2172, Nov. 1979.

[32] T. Kato, H. Morita, H. Washida, and A. Onoe, "Some Investigations on Photovoltaic Mechanisms in Silicon Ribbon," *J. Electron. Mat.* **12**, 11–28, Jan. 1983.

[33] M. Kunst and G. Beck, "The Study of Charge Carrier Kinetics in Semiconductors by Microwave Conductivity Measurements," *J. Appl. Phys.* **60**, 3558–3566, Nov. 1986; J. M. Borrego, R. J. Gutmann, N. Jensen, and O. Paz, "Non-destructive Lifetime Measurement in Silicon Wafers by Microwave Reflection," *Solid-State Electron.* **30**, 195–203, Feb. 1987.

[34] T. Yamazaki, Y. I. Ogita, Y. Ikegami, H. Onaka, E. Ohta, and M. Sakata, "Contactless Measurement of Short Carrier Lifetime in Heat-Treated *n*-Type Silicon," *Japan. J. Appl. Phys.* **23**, 322–325, March 1984.

[35] R. J. Deri and J. P. Spoonhower, "Microwave Photoconductivity Lifetime Measurements: Experimental Limitations," *Rev. Sci. Instrum.* **55**, 1343–1347, Aug. 1984.

[36] K. L. Luke and L. J. Cheng, "A Chemical/Microwave Technique for the Measurement of Bulk Minority Carrier Lifetime in Silicon Wafers," *J. Electrochem. Soc.* **135**, 957–961, April 1988.

[37] R. G. Pratt, J. Hewett, P. Capper, C. L. Jones, and M. J. Quelch, "Minority Carrier Lifetime in *n*-type Bridgman Grown HgCdTe," *J. Appl. Phys.* **54**, 5152–5157, Sept. 1983.

[38] M. C. Chen, "Photoconductivity Lifetime Measurements on HgCdTe Using a Contactless Microwave Technique," *J. Appl. Phys.* **64**, 945–947, July 1988.

[39] K. D. Cummings, S. J. Pearton, and G. P. Vella-Coleiro, "Characterization of GaAs and Si by a Microwave Photoconductance Technique," *J. Appl. Phys.* **60**, 1676–1680, Sept. 1986.

[40] J. E. Mahan, T. W. Ekstedt, R. I. Frank, and R. Kaplow, "Measurement of Minority Carrier Lifetime in Solar Cells from Photo-Induced Open Circuit Voltage Decay," *IEEE Trans. Electron Dev.* **ED-26**, 733–739, May 1979.

[41] S. R. Dhariwal and N. K. Vasu, "Mathematical Formulation for the Photo-Induced Open Circuit Voltage Decay Method for Measurement of Minority Carrier Lifetime in Solar Cells," *IEEE Electron Dev. Lett.* **EDL-2**, 53–55, Feb. 1981.

[42] O. von Roos, "Analysis of the Photon Voltage Decay (PVD) Method for Measuring Minority Carrier Lifetimes in *pn* Junction Solar Cells," *J. Appl. Phys.* **52**, 5833–5837, Sept. 1981.

[43] B. H. Rose and H. T. Weaver, "Determination of Effective Surface Recombination Velocity and Minority-Carrier Lifetime in High-Efficiency Si Solar Cells," *J. Appl. Phys.* **54**, 238–247, Jan. 1983; Corrections *J. Appl. Phys.* **55**, 607, Jan. 1984.

[44] B. H. Rose, "Minority-Carrier Lifetime Measurements on Si Solar Cells Using I_{sc} and V_{oc} Transient Decay," *IEEE Trans. Electron Dev.* **ED-31**, 559–565, May 1984.

[45] S. C. Jain, "Theory of Photo Induced Open Circuit Voltage Decay in a Solar Cell," *Solid-State Electron.* **24**, 179–183, Feb. 1981.

[46] S. C. Jain and U. C. Ray, "Photovoltage Decay in *pn* Junction Solar Cells Including the Effects of Recombination in the Emitter," *J. Appl. Phys.* **54**, 2079–2085, April 1983.

[47] A. R. Moore, "Carrier Lifetime in Photovoltaic Solar Concentrator Cells by the Small Signal Open Circuit Decay Method," *RCA Rev.* **40**, 549–562, Dec. 1980.

[48] R. Z. Bachrach and O. G. Lorimor, "Measurement of the Extrinsic Room-Temperature Minority Carrier Lifetime in GaP," *J. Appl. Phys.* **43**, 500–507, Feb. 1972; P. D. Dapkus, W. H. Hackett, O. G. Lorimor, and R. Z. Bachrach, "Kinetics of Recombination in Nitrogen-Doped GaP," *J. Appl. Phys.* **45**, 4920–4930, Nov. 1974.

[49] M. L. Cone and R. L. Hengehold, "Characterization of Ion-Implanted GaAs Using Cathodoluminescence," *J. Appl. Phys.* **54**, 6346–6351, Nov. 1983.

[50] J. Dziewior and W. Schmid, "Auger Coefficients for Highly Doped and Highly Excited Silicon," *Appl. Phys. Lett.* **31**, 346–348, Sept. 1977.

[51] R. Z. Bachrach, "A Photon Counting Apparatus for Kinetic and Spectral Measurements," *Rev. Sci. Instrum.* **43**, 734–737, May 1972.

[52] R. K. Ahrenkiel, D. J. Dunlavy, J. Benner, R. P. Gale, R. W. McClelland, J. V. Gormley, and B. D. King, "Minority-Carrier Lifetime in GaAs Thin Films," *Appl. Phys. Lett.* **53**, 598–599, Aug. 1988.

[53] G. Bohnert, R. Häcker, and A. Hangleiter, "Position Resolved Carrier Lifetime Measurement in Silicon Power Devices by Time Resolved Photoluminescence Spectroscopy," *J. Physique* **C4**, 617–620, Sept. 1988.

[54] E. O. Johnson, "Measurement of Minority Carrier Lifetime with the Surface Photovoltage," *J. Appl. Phys.* **28**, 1349–1353, Nov. 1957.

[55] A. Quilliet and P. Gosar, "The Surface Photovoltaic Effect in Silicon and Its Application to Measure the Minority Carrier Lifetime (in French)," *J. Phys. Rad.* **21**, 575–580, July 1960.

[56] A. M. Goodman, "A Method for the Measurement of Short Minority Carrier Diffusion Lengths in Semiconductors," *J. Apply. Phys.* **32**, 2550–2552, Dec. 1961.

[57] M. Saritas and H. D. McKell, "Diffusion Length Studies in Silicon by the Surface Photovoltage Method," *Solid-State Electron.* **31**, 835–842, May 1988.

[58] S. C. Choo and A. C. Sanderson, "Bulk Trapping Effect on Carrier Diffusion Length as Determined by the Surface Photovoltage Method: Theory," *Solid-State Electron.* **13**, 609–617, May 1970.

[59] A. R. Moore, "Theory and Experiment on the Surface-Photovoltage Diffusion-Length Measurement as Applied to Amorphous Silicon," *J. Appl. Phys.* **54**, 222–228, Jan. 1983; C. L. Chiang, R. Schwarz, D. E. Slobodin, J. Kolodzey, and S. Wagner, "Measurement of the Minority-Carrier Diffusion Length in Thin Semiconductor Films," *IEEE Trans. Electron Dev.* **ED-33**, 1587–1592, Oct. 1986.

[60] C. L. Chiang and S. Wagner, "On the Theoretical Basis of the Surface Photovoltage Technique," *IEEE Trans. Electron Dev.* **ED-32**, 1722–1726, Sept. 1985.

[61] ASTM Standard F391-87, "Standard Test Method for Minority-Carrier Diffusion Length in Silicon by Measurement of Steady-State Surface Photovoltage," *1987 Annual Book of ASTM Standards*, Am. Soc. Test. Mat., Philadelphia, 1987.

[62] E. S. Nartowitz and A. M. Goodman, "Evaluation of Si Optical Absorption Data for Use in Minority-Carrier Diffusion Length Measurements by the SPV Method," *J. Electrochem. Soc.* **132**, 2992–2997, Dec. 1985.

[63] M. D. Sturge, "Optical Absorption of Gallium Arsenide Between 0.6 and 2.75 eV," *Phys. Rev.* **127**, 768–773, Aug. 1962; D. D. Sell and H. C. Casey, Jr., "Optical Absorption and Photoluminescence Studies of Thin GaAs Layers in GaAs-AlGaAs Double Heterostructures," *J. Appl. Phys.* **45**, 800–807, Feb. 1974; D. E. Aspnes and A. A. Studna, "Dielectric Functions and Optical Parameters of Si, Ge, GaP, GaAs, GaSb, InP, InAs and InSb from 1.5 to 6 eV," *Phys. Rev.* **B27**, 985–1009, Jan. 1983.

[64] S. S. Li, "Determination of Minority-Carrier Diffusion Length in Indium Phosphide by Surface Photovoltage Measurement," *Appl. Phys. Lett.* **29**, 126–127, July 1976; H. Burkhard, H. W. Dinges, and E. Kuphal, "Optical Properties of InGaPAs, InP, GaAs, and GaP Determined by Ellipsometry," *J. Appl. Phys.* **53**, 655–662, Jan. 1982.

[65] A. M. Goodman, "Improvements in Method and Apparatus for Determining Minority Carrier Diffusion Length," *IEEE Int. Electron Dev. Meet.*, Washington, DC, 1980, pp. 231–234.

[66] W. Kern and D. A. Puotinen, "Cleaning Solutions Based on Hydrogen Peroxide for Use in Silicon Semiconductor Technology," *RCA Rev.* **31**, 187–206, June 1970.

[67] R. O. Bell and G. M. Freedman, "Minority Carrier Diffusion Length from Spectral Response Measurements," *Proc. of the 13th Photovoltaic Spec. Conf.*, Washington, DC, 1978, pp. 89–94.

[68] R. H. Micheels and R. D. Rauh, "Use of a Liquid Electrolyte Junction for the Measurement of Diffusion Length in Silicon Ribbon," *J. Electrochem. Soc.* **131**, 217–219, Jan 1984; A. R. Moore and H. S. Lin, "Improvement in the Surface Photovoltage Method of Determining Diffusion Length in thin Films of Hydrogenated Amorphous Silicon," *J. Appl. Phys.* **61**, 4816–4819, May 1987.

[69] A. M. Goodman, L. A. Goodman, and H. F. Gossenberger, "Silicon-Wafer Process Evaluation Using Minority-Carrier Diffusion Length Measurements by the SPV Method," *RCA Rev.* **44**, 326–341, June 1983.

[70] D. L. Lile and N. M. Davis, "Semiconductor Profiling Using an Optical Probe," *Solid-State Electron.* **18**, 699–704, July-Aug. 1975.

[71] J. I. Pankove and J. E. Berkeyheiser, "Mapping the Quality of Semiconductor Wafers," *Rev. Sci. Instrum.* **57**, 674–679, April 1986.

[72] B. L. Sopori, R. W. Gurtler, and I. A. Lesk, "Effects of Optical Beam Size on Diffusion Length Measured by the Surface Photovoltage Method," *Solid-State Electron.* **23**, 139–142, Feb. 1980.

[73] A. M. Goodman, "Silicon Wafer Surface Damage Revealed by Surface Photovoltage Measurements," *J. Appl. Phys.* **53**, 7561–7565, Nov. 1982.

[74] W. E. Phillips, "Interpretation of Steady-State Surface Photovoltage Measurements in Epitaxial Semiconductor Layers," *Solid-State Electron.* **15**, 1097–1102, Oct. 1972.

[75] A. K. Ghosh, J. I. Haberman, and T. Feng, "Limitations of Surface Photovoltage Measurements," *J. Apply. Phys.* **55**, 280–281, Jan. 1984.

[76] T. I. Chappell, P. W. Chye, and M. A. Tavel, "Determination of the Oxygen Precipitate-Free Zone Width in Silicon Wafers from Surface Photovoltage Measurements," *Solid-State Electron.* **26**, 33–36, Jan. 1983.

[77] E. Y. Wang, C. R. Baraona, and H. W. Brandhorst, Jr., "Surface Photovoltage Method Extended to Silicon Solar Cell Junction," *J. Electrochem. Soc.* **121**, 973–975, July 1974.

[78] E. D. Stokes and T. L. Chu, "Diffusion Lengths in Solar Cells from Short-Circuit Current Measurements," *Appl. Phys. Lett.* **30**, 425–426, April 1977.

[79] N. D. Arora, S. G. Chamberlain, and D. J. Roulston, "Diffusion Length Determination in *pn* Junction Diodes and Solar Cells," *Appl. Phys. Lett.* **37**, 325–327, Aug. 1980.

[80] C. Y. Wu and J. F. Chen, "Doping and Temperature Dependences of Minority Carrier Diffusion Length and Lifetime Deduced from the Spectral Response Measurements of *pn* Junction Solar Cells," *Solid-State Electron.* **25**, 679–682, July 1982.

[81] V. Lehmann and H. Föll, "Minority Carrier Diffusion Length Mapping in Silicon Wafers Using a Si-Electrolyte-Contact," *J. Electrochem. Soc.* **135**, 2831–2835, Nov. 1988.

[82] M. Allenson and S. J. Bass, "GaAs Reflection Photocathodes Grown by Metal Alkyl Vapor Phase Epitaxy," *Appl. Phys. Lett.* **28**, 113–115, Feb. 1976.

[83] M. Born and E. Wolf, *Principles of Optics*, 6th ed., Pergamon, Oxford, 1960.

[84] D. K. Schroder, R. N. Thomas, and J. C. Swartz, "Free Carrier Absorption in Silicon," *IEEE Trans Electron Dev.* **ED-25**, 254–261, Feb. 1978.

[85] L. Jastrzebski, J. Lagowski, and H. C. Gatos, "Quantitative Determination of the Carrier Concentration Distribution in Semiconductors by Scanning Infrared Absorption: Si," *J. Electrochem. Soc.* **126**, 260–263, Feb. 1979.

[86] D. L. Polla, "Determination of Carrier Lifetime in Silicon by Optical Modulation," *IEEE Electron Dev. Lett.* **EDL-4**, 185–187, June 1983.

[87] J. A. Mroczkowski, J. F. Shanley, M. B. Reine, P. LoVecchio, and D. L. Polla, "Lifetime Measurements in $Hg_{0.7}Cd_{0.3}Te$ by Population Modulation," *Appl. Phys. Lett.* **38**, 261–263, Feb. 1981.

[88] M. A. Afromowitz and M. DiDomenico, "Measurement of Free-Carrier Lifetimes in GaP by Photoinduced Modulation of Infrared Absorption," *J. Appl. Phys.* **42**, 3205–3208, July 1971.

[89] K. Nauka, H. C. Gatos, and J. Lagowski, "Oxygen-Induced Recombination Centers in As-Grown Czochralski Silicon Crystals," *Appl. Phys. Lett.* **43**, 241–243, Aug. 1983.

[90] J. Waldmeyer, "A Contactless Method for Determination of Carrier Lifetime, Surface Recombination Velocity, and Diffusion Constant in Semiconductors," *J. Appl. Phys.* **63**, 1977–1983, March 1988.

[91] G. A. Acket, W. Nijman, and H. 't Lam, "Electron Lifetime and Diffusion Constant in Germanium-Doped Gallium Arsenide," *J. Appl. Phys.* **45**, 3033–3040, July 1974.

[92] C. J. Hwang, "Doping Dependence of Hole Lifetime in *n*-Type GaAs," *J. Appl. Phys.* **42**, 4408–4413, Oct. 1971.

[93] G. Schwab, "The Measurement of Minority Carrier Lifetime and Surface Recombination Velocity in Silicon by Means of Photocurrent Technique," in *Semiconductor Silicon/1977* (H. R. Huff and E. Sirtl, eds.), Electrochem. Soc., Princeton, NJ, 1977, pp. 481–490.

[94] G. Schwab, H. Bernt, and H. Reichl, "Measurement of Minority Carrier Lifetime Profiles in Silicon," *Solid-State Electron.* **20**, 91–94, Feb. 1977.

[95] H. Reichl and H. Bernt, "Lifetime Measurements in Silicon Epitaxial Material," *Solid-State Electron.* **18**, 453–458, May 1975.

[96] N. Honma, C. Munakata, H. Itoh, and T. Warabisaka, "Nondestructive Measurement of Minority Carrier Lifetimes in Si Wafers Using Frequency Dependence of ac Photovoltages," *Japan. J. Appl. Phys.* **25**, 743–749, May 1986.

[97] O. von Roos, "Recombination Lifetimes and Surface Recombination Velocities of Minority Carriers in *np* Junctions. A New Method for Their Determination by Means of a Stationary Amplitude-Modulated Electron Beam," *J. Appl. Phys.* **50**, 3738–3742, May 1979.

[98] J. F. Bresse, "Quantitative Investigations in Semiconductor Devices by Electron Beam Induced Current Mode: A Review," in *Scanning Electron Microscopy* 1, 717–725, 1978.

[99] C. A. Klein, "Band Gap Dependence and Related Features of Radiation Ionization Energies in Semiconductors," *J. Appl. Phys.* **39**, 2029–2038, March 1968.

[100] H. J. Leamy, "Charge Collection Scanning Electron Microscopy," *J. Appl. Phys.* **53**, R51–R80, June 1982.

[101] D. E. Ioannou and C. A. Dimitriadis, "A SEM-EBIC Minority Carrier Diffusion Length Measurement Technique," *IEEE Trans. Electron Dev.* **ED-29**, 445–450, March 1982.

[102] H. K. Kuiken and C. van Opdorp, "Evaluation of Diffusion Length and Surface-Recombination Velocity from a Planar-Collector-Geometry Electron-Beam-Induced Current Scan," *J. Appl. Phys.* **57**, 2077–2090, March 1985.

[103] J. D. Zook, "Theory of Beam-Induced Currents in Semiconductors," *Appl. Phys. Lett.* **42**, 602–604, April 1983.

[104] W. van Roosbroeck, "Injected Current Carrier Transport in a Semi-infinite Semiconductor and the Determination of Lifetimes and Surface Recombination Velocities," *J. Appl. Phys.* **26**, 380–391, April 1955.

[105] C. Van Opdorp, "Methods of Evaluating Diffusion Lengths and Near-Junction Luminescence-Efficiency Profiles from SEM Scans," *Phil. Res. Rep.* **32**, 192–249, 1977.

[106] F. Berz and H. K. Kuiken, "Theory of Lifetime Measurements with the Scanning Electron Microscope: Steady State," *Solid-State Electron.* **19**, 437–445, June 1976.

[107] H. K. Kuiken, "Theory of Lifetime Measurements with the Scanning Electron Microscope: Transient Analysis," *Solid-State Electron.* **19**, 447–450, June 1976.

[108] M. Watanabe, G. Actor, and H. C. Gatos, "Determination of Minority-Carrier Lifetime and Surface Recombination Velocity with High Spatial Resolution," *IEEE Trans. Electron Dev.* **ED-24**, 1172–1177, Sept. 1977.

[109] C. H. Seager, "The Determination of Grain-Boundary Recombination Rates by Scanned Spot Excitation Methods," *J. Appl. Phys.* **53**, 5968–5971, Aug. 1982.

[110] W. Zimmermann, "Measurement of Spatial Variations of the Carrier Lifetime in Silicon Power Devices," *Phys. Stat. Sol.* **12a**, 671–678, Aug. 1972.

[111] M. Ettenberg, H. Kressel, and S. L. Gilbert, "Minority Carrier Diffusion Length and Recombination Lifetime in GaAs:Ge Prepared by Liquid-Phase Epitaxy," *J. Appl. Phys.* **44**, 827–831, Feb. 1973.

[112] C. M. Hu and C. Drowley, "Determination of Diffusion Length and Surface Recombination Velocity by Light Excitation," *Solid-State Electron.* **21**, 965–968, July 1978.

[113] W. H. Hackett, "Electron-Beam Excited Minority-Carrier Diffusion Profiles in Semiconductors," *J. Appl. Phys.* **43**, 1649–1654, April 1972.

[114] W. H. Hackett, R. H. Saul, R. W. Dixon, and G. W. Kammlott, "Scanning Electron Microscope Characterization of GaP Red-Emitting Diodes," *J. Appl. Phys.* **43**, 2857–2868, June 1972.

[115] J. D. Zook, "Effects of Grain Boundaries in Polycrystalline Solar Cells," *Appl. Phys. Lett.* **37**, 223–226, July 1980.

[116] E. M. Pell, "Recombination Rate in Germanium by Observation of Pulsed Reverse Characteristic," *Phys. Rev.* **90**, 278–279, April 1953.

[117] R. H. Kingston, "Switching Time in Junction Diodes and Junction Transistors," *Proc. IRE* **42**, 829–834, May 1954.

[118] B. Lax and S. F. Neustadter, "Transient Response of a *pn* Junction," *J. Appl. Phys.* **25**, 1148–1154, Sept. 1954.

[119] K. Schuster and E. Spenke, "The Voltage Step at the Switching of Alloyed PIN Rectifiers," *Solid-State Electron.* **8**, 881–882, Nov. 1965.

[120] H. J. Kuno, "Analysis and Characterization of *pn* Junction Diode Switching," *IEEE Trans. Electron Dev.* **ED-11**, 8–14, Jan. 1964.

[121] R. H. Dean and C. J. Nuese, "A Refined Step-Recovery Technique for Measuring Minority Carrier Lifetimes and Related Parameters in Asymmetric *pn* Junction Diodes," *IEEE Trans. Electron Dev.* **ED-18**, 151–158, March 1971.

[122] S. C. Jain and R. Van Overstraeten, "The Influence of Heavy Doping Effects on the Reverse Recovery Storage Time of a Diode," *Solid-State Electron.* **26**, 473–481, May 1983.

[123] Y. C. Kao and J. R. Davis, "Correlation Between Reverse Recovery Time and Lifetime of *pn* Junction Driven by a Current Ramp," *IEEE Trans. Electron Dev.* **ED-17**, 652–657, Sept. 1970.

[124] L. De Smet and R. Van Overstraeten, "Calculation of the Switching Time in Junction Diodes," *Solid-State Electron.* **18**, 557–562, June 1975.

[125] F. Berz, "Step Recovery of pin Diodes," *Solid-State Electron.* **22**, 927–932, Nov. 1979.

[126] A. S. Grove and C. T. Sah, "Simple Analytical Approximation to the Switching Times in Narrow Base Diodes," *Solid-State Electron.* **7**, 107–110, Jan. 1964.

[127] L. A. Davidson, "Simple Expressions for Storage Time of Arbitrary Base Diodes," *Solid-State Electron.* **9**, 1145–1147, Nov./Dec. 1966.

[128] R. P. Mertens and R. J. Van Overstraeten, "Heavy Doping Effects in Silicon," in *Advances in Electronics and Electron Physics* (L. Marton, ed.), Academic Press, New York, 1981, pp. 77–117.

[129] M. Derdouri, P. Leturcq, and A. Muñoz-Yague, "A Comparative Study of Methods of Measuring Carrier Lifetime in pin Devices," *IEEE Trans. Electron Dev.* **ED-27**, 2097–2101, Nov. 1980.

[130] J. A. G. Slatter and J. P. Whelan, "PIN Diode Recovery Storage Time," *Solid-State Electron.* **23**, 1235–1242, Dec. 1980.

[131] S. C. Jain, S. K. Agarwal, and Harsh, "Importance of Emitter Recombination in Interpretation of Reverse-Recovery Experiments at High Injections, "*J. Appl. Phys.* **54**, 3618–3619, June 1983.

[132] H. J. Benda and E. Spenke, "Reverse Recovery Processes in Silicon Power Rectifiers," *Proc. IEEE* **55**, 1331–1354, Aug. 1967.

[133] J. L. Moll, U. C. Ray, and S. C. Jain, "Reverse Recovery in *pn* Junction Diodes with Built-In Drift Fields," *Solid-State Electron.* **26**, 1077–1081, Nov. 1983.

[134] J. J. Liou, H. K. Brown, and M. S. Clamme, "Characterization of Reverse Recovery Transient Behavior of Bipolar Transistors for Emitter Parameters Determination," *Solid-State Electron.* **31**, 1595–1601, Nov. 1988.

[135] B. R. Gossick, "Post-injection Barrier Electromotive Force of *pn* Junctions," *Phys. Rev.* **91**, 1012–1013, Aug. 1953; "On the Transient Behavior of Semiconductor Rectifiers," *J. Appl. Phys.* **26**, 1356–1365, Nov. 1955.

[136] S. R. Lederhandler and L. J. Giacoletto, "Measurement of Minority Carrier Lifetime and Surface Effects in Junction Devices," *Proc. IRE* **43**, 477–483, April 1955.

[137] S. C. Choo and R. G. Mazur, "Open Circuit Voltage Decay Behavior of Junction Devices," *Solid-State Electron.* **13**, 553–564, May 1970.

[138] S. C. Jain and R. Muralidharan, "Effect of Emitter Recombinations on the Open Circuit Voltage Decay of a Junction Diode," *Solid-State Electron.* **24**, 1147–1154, Dec. 1981.

[139] F. Berz and J. A. G. Slatter, "Effect of Linear Emitter Recombination on OCVD Determination of Lifetime in pin Diodes," *Solid-State Electron.* **25**, 693–697, Aug. 1982.

[140] H. Schlangenotto and W. Gerlach, "On the Post-Injection Voltage Decay of psn Rectifiers at High Injection Levels," *Solid-State Electron.* **15**, 393–402, April 1972.

[141] R. J. Basset, W. Fulop, and C. A. Hogarth, "Determination of the Bulk Carrier Lifetime in Low-Doped Region of a Silicon Power Diode by the Method of Open Circuit Voltage Decay," *Int. J. Electron.* **35**, 177–192, Aug. 1973.

[142] P. G. Wilson, "Recombination in Silicon p-π-n Diodes," *Solid-State Electron.* **10**, 145–154, Feb. 1967.

[143] J. E. Mahan and D. L. Barnes, "Depletion Layer Effects in the Open-Circuit-Voltage-Decay Lifetime Measurement," *Solid-State Electron.* **24**, 989–994, Oct. 1981.

[144] M. A. Green, "Minority Carrier Lifetimes Using Compensated Differential Open Circuit Voltage Decay," *Solid-State Electron.* **26**, 1117–1122, Nov. 1983; "Solar Cell Minority Carrier Lifetime Using Open-Circuit Voltage Decay," *Solar Cells* **11**, 147–161, March 1984.

[145] D. H. J. Totterdell, J. W. Leake, and S. C. Jain, "High-Injection Open-Circuit Voltage Decay in *pn*-Junction Diodes with Lightly Doped Bases," *IEE Proc. Pt I* **133**, 181–184, Oct. 1986.

[146] K. Joardar, R. C. Dondero, and D. K. Schroder, "A Critical Analysis of the Small-Signal Voltage Decay Technique for Minority-Carrier Lifetime Measurement in Solar Cells," *Solid-State Electron.* **32**, 479–483, June 1989.

[147] P. Tomanek, "Measuring the Lifetime of Minority Carriers in MIS Structures," *Solid-State Electron.* **12**, 301–303, April 1969.

[148] J. Müller and B. Schiek, "Transient Responses of a Pulsed MIS-Capacitor," *Solid-State Electron.* **13**, 1319–1332, Oct. 1970.

[149] A. C. Wang and C. T. Sah, "New Method for Complete Electrical Characterization of Recombination Properties of Traps in Semiconductors," *J. Appl. Phys.* **57**, 4645–4656, May 1985.

[150] E. Soutschek, W. Müller, and G. Dorda, "Determination of Recombination Lifetime in MOSFET's," *Appl. Phys. Lett.* **36**, 437–438, March 1980.

[151] D. K. Schroder, *Advanced MOS Devices*, Addison-Wesley, Reading, MA, 1987, ch. 2.

[152] D. K. Schroder, J. D. Whitfield, and C. J. Varker, "Recombination Lifetime Using the Pulsed MOS Capacitor," *IEEE Trans. Electron Dev.* **ED-31**, 462–467, April 1984.

[153] D. K. Schroder, "Bulk and Optical Generation Parameters Measured with the Pulsed MOS Capacitor," *IEEE Trans. Electron Dev.* **ED-19**, 1018–1023, Sept. 1972.

[154] M. Aminzadeh and L. Forbes, "Recombination Lifetime of Short-Base-Width Devices Using the Pulsed MOS Capacitor Technique," *IEEE Trans. Electron Dev.* **ED-35**, 518–521, April 1988.

[155] T. W. Jung, F. A. Lindholm, and A. Neugroschel, "Unifying View of Transient Responses for Determining Lifetime and Surface Recombination Velocity in Silicon Diodes and Back-Surface-Field Solar Cells," *IEEE Trans. Electron Dev.* **ED-31**, 588–595, May 1984.

[156] T. W. Jung, F. A. Lindholm, and A. Neugroschel, "Variations in the Electrical Short-Circuit Current Decay for Recombination Lifetime and Velocity Measurements," *Solar Cells* **22**, 81–96, Oct. 1987.

[157] A. Zondervan, L. A. Verhoef, and F. A. Lindholm, "Measurement Circuits for Silicon-Diode and Solar-Cell Lifetime and Surface Recombination Velocity by Electrical Short-Circuit Current Delay," *IEEE Trans. Electron Dev.* **ED-35**, 85–88, Jan. 1988.

[158] P. Spirito and G. Cocorullo, "Measurement of Recombination Lifetime Profiles in Epilayers Using a Conductivity Modulation Technique," *IEEE Trans. Electron Dev.* **ED-32**, 1708–1713, Sept. 1985; P. Spirito, S. Bellone, C. M. Ransom, G. Busatto, and G. Cocorullo, "Recombination Lifetime Profiling in Very Thin Si Epitaxial Layers Used for Bipolar VLSI," *IEEE Electron Dev. Lett.* **EDL-10**, 23–24, Jan. 1989.

[159] P. U. Calzolari and S. Graffi, "A Theoretical Investigation on the Generation Current in Silicon *p-n* Junctions under Reverse Bias," *Solid-State Electron.* **15**, 1003–1011, Sept. 1972.

[160] P. C. T. Roberts and J. D. E. Beynon, "An Experimental Determination of the Carrier Lifetime near the Si-SiO$_2$ Interface," *Solid-State Electron.* **16**, 221–227, Feb. 1973; "Effect of a Modified Theory of Generation Currents on an Experimental Determination of Carrier Lifetime," *Solid-State Electron.* **17**, 403–404, April 1974.

[161] A. S. Grove and D. J. Fitzgerald, "Surface Effects on *pn* Junctions: Characteristics of Surface Space-Charge Regions under Non-equilibrium Conditions," *Solid-State Electron.* **9**, 783–806, Aug. 1966.

[162] D. J. Fitzgerald and A. S. Grove, "Surface Recombination in Semiconductors," *Surf. Sci.* **9**, 347–369, Feb. 1968.

[163] R. F. Pierret, "The Gate-Controlled Diode s_0 Measurement and Steady-State Lateral Current Flow in Deeply Depleted MOS Structures," *Solid-State Electron.* **17**, 1257–1269, Dec. 1974.

[164] J. van der Spiegel and G. J. Declerck, "Theoretical and Practical Investigation of the Thermal Generation in Gate Controlled Diodes," *Solid-State Electron.* **24**, 869–877, Sept. 1981.

[165] G. A. Hawkins, E. A. Trabka, R. L. Nielsen, and B. C. Burkey, "Characterization of Generation Currents in Solid-State Imagers," *IEEE Trans. Electron Dev.* **ED-32**, 1806–1816, Sept. 1985; G. A. Hawkins, "Generation Currents from Interface States in Selectively Implanted MOS Structures," *Solid-State Electron.* **31**, 181–196, Feb. 1988.

[166] M. Zerbst, "Relaxation Effects at Semiconductor-Insulator Interfaces" (in German), *Z. Angew. Phys.* **22**, 30–33, May 1966.

[167] J. S. Kang and D. K. Schroder, "The Pulsed MIS Capacitor—A Critical Review," *Phys. Stat. Sol.* **89a**, 13–43, May 1985.

[168] K. S. Rabbani and D. R. Lamb, "On the Analysis of Pulsed MOS Capacitance Measurements," *Solid-State Electron.* **21**, 1171–1173, Sept. 1978.

[169] P. U. Calzolari, S. Graffi, and C. Morandi, "Field-Enhanced Generation in MOS Capacitors," *Solid-State Electron.* **17**, 1001–1011, Oct. 1974.

[170] K. S. Rabbani, "Investigations on Field Enhanced Generation in Semiconductors," *Solid-State Electron.* **30**, 607–613, June 1987.

[171] D. K. Schroder and J. Guldberg, "Interpretation of Surface and Bulk Effects Using the Pulsed MIS Capacitor," *Solid-State Electron.* **14**, 1285–1297, Dec. 1971.

[172] W. R. Fahrner, D. Braeunig, C. P. Schneider, and M. Briere, "Reduction of Measurement Time of Lifetime Profiles by Applying High Temperatures," *J. Electrochem. Soc.* **134**, 1291–1296, May 1987.

[173] D. K. Schroder, "Bulk and Optical Generation Parameters Measured with the Pulsed MOS Capacitor," *IEEE Trans. Electron Dev.* **ED-19**, 1018–1023, Sept. 1972.

[174] R. F. Pierret and W. M. Au, "Photo-Accelerated MOS-C *C-t* Transient Measurements," *Solid-State Electron.* **30**, 983–984, Sept. 1987.

[175] W. W. Keller, "The Rapid Measurement of Generation Lifetime in MOS Capacitors with Long Relaxation Times," *IEEE Trans. Electron Dev.* **ED-34**, 1141–1146, May 1987.

[176] C. S. Yue, H. Vyas, M. Holt, and J. Borowick, "A Fast Extrapolation Technique for Measuring Minority-Carrier Generation Lifetime," *Solid-State Electron.* **28**, 403–406, April 1985.

[177] K. S. Rabbani and D. R. Lamb, "Direct Correlation of Generation Lifetimes Obtained from Pulsed MOS Capacitance and Gated Diode Measurements," *Solid-State Electron.* **26**, 161–168, Feb. 1983.

[178] O. Paz and C. P. Schneider, "Determination of the Denuded Zone in Czochralski-Grown Silicon Wafers through MOS Lifetime Profiling," *IEEE Trans. Electron Dev.* **ED-32**, 2830–2838, Dec. 1985.

[179] Z. Radzimski, J. Honeycutt, and G. A. Rozgonyi, "Minority-Carrier Lifetime Analysis of Silicon Epitaxy and Bulk Crystals with Nonuniformly Distributed Defects," *IEEE Trans. Electron Dev.* **ED-35**, 80–84, Jan 1988.

[180] P. U. Calzolari, S. Graffi, A. M. Mazzone, and C. Morandi, "Bulk Lifetime Determination from Current and Capacitance Transient Response of MOS Capacitors," *Alta Freq.* **41**, 848–853, Nov. 1972.

[181] R. F. Pierret, "A. Linear Sweep MOS-C Technique for Determining Minority Carrier Lifetimes," *IEEE Trans. Electron Dev.* **ED-19**, 869–873, July 1972.

[182] R. F. Pierret and D. W. Small, "A Modified Linear Sweep Technique for MOS-C Generation Rate Measurements," *IEEE Trans. Electron Dev.* **ED-22**, 1051–1052, Nov. 1975.

[183] W. D. Eades, J. D. Shott, and R. M. Swanson, "Refinements in the Measurement of Depleted Generation Lifetime," *IEEE Trans. Electron Dev.* **ED-30**, 1274–1277, Oct. 1983.

[184] G. Duggan and G. B. Scott, "The Efficiency of Photoluminescence of Thin Epitaxial Semiconductors," *J. Appl. Phys.* **52**, 407–411, Jan. 1981.

[185] H. J. Hovel, "Solar Cells," in *Semiconductors and Semimetals* (R. K. Willardson and A. C. Beer, eds.) **11**, Academic Press, New York, 1975, pp. 17–20.

[186] H. S. Carslaw and J. C. Jaeger, *Conduction of Heat in Solids*, Oxford University Press, Oxford, 1959.

[187] M. Boulou and D. Bois, "Cathodoluminescence Measurements of the Minority-Carrier Lifetime in Semiconductors," *J. Appl. Phys.* **48**, 4713–4721, Nov. 1977.

[188] K. L. Luke and L. J. Cheng, "Analysis of the Interaction of a Laser Pulse with a Silicon Wafer: Determination of Bulk Lifetime and Surface Recombination Velocity," *J. Appl. Phys.* **61**, 2282–2293, March 1987.

[189] J. S. Blakemore, *Semiconductor Statistics*, Pergamon, New York, 1962.

[190] W. van Roosbroeck and W. Shockley, "Photon-Radiative Recombination of Electrons and Holes in Germanium," *Phys. Rev.* **94**, 1558–1560, 1954.

[191] T. Kuriyama, T. Kamiya, and H. Yanai, "Effect of Photon Recycling on Diffusion Length and Internal Quantum Efficiency in AlGaAs-GaAs Heterostructures," *Japan. J. Appl. Phys.* **16**, 465–477, March 1977.

[192] P. Asbeck, "Self-absorption Effects on the Radiative Lifetime in GaAs-AlGaAs Double Heterostructures," *J. Appl. Phys.* **48**, 820–822, Feb. 1977.

[193] R. J. Nelson and R. G. Sobers, "Minority Carrier Lifetime and Internal Quantum Efficiency of Surface-Free GaAs," *J. Appl. Phys.* **49**, 6103–6108, Dec. 1978.

CHAPTER 9

OPTICAL CHARACTERIZATION

9.1 INTRODUCTION

In this chapter we discuss those optical characterization techniques most commonly used in the semiconductor industry. Optical measurements are becoming more popular. They are almost always non-contacting, which is a big advantage when contact formation is detrimental. The instrumentation for many optical techniques is commercially available and has become easier to use, and is often automated. The measurements can be used for a large class of characterization and can have very high sensitivity. The main concepts are discussed in this chapter, with some of the details left to the published literature.

Optical measurements fall into three broad categories: *photometric* measurements (amplitude of reflected or transmitted light is measured), *interference* measurements (phase of reflected or transmitted light is measured), and *polarization* measurements (ellipticity of reflected light is measured). The main optical techniques are summarized in the diagram in Fig. 9.1. Incident light is reflected, absorbed, emitted, or transmitted. Most of the techniques in that figure are discussed here; some have been discussed in earlier chapters (e.g., photoconductivity), and some are not discussed at all (e.g., UV photoelectron spectroscopy). For completeness we also mention several non-optical methods for film thickness and line-width determination.

9.2 OPTICAL MICROSCOPY

The compound optical microscope is one of the most versatile and useful instruments in a semiconductor laboratory. Most of the features of integ-

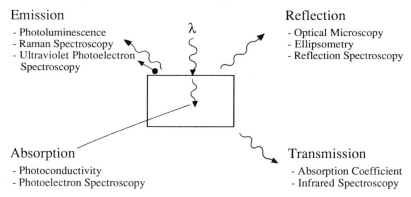

Emission

- Photoluminescence
- Raman Spectroscopy
- Ultraviolet Photoelectron
  Spectroscopy

λ

Reflection

- Optical Microscopy
- Ellipsometry
- Reflection Spectroscopy

Absorption

- Photoconductivity
- Photoelectron Spectroscopy

Transmission

- Absorption Coefficient
- Infrared Spectroscopy

Fig. 9.1   Optical characterization techniques.

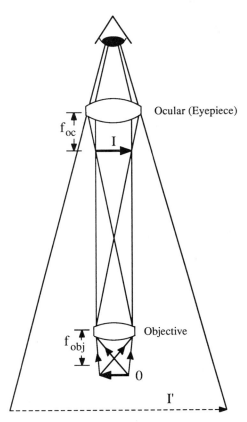

Ocular (Eyepiece)

f_{oc}

I

Objective

f_{obj}

0

I'

Fig. 9.2   Simplified representation of a compound microscope's optical paths.

rated circuits and other semiconductor devices are sufficiently gross to be seen through such a microscope. However, optical microscopy loses its usefulness as feature sizes shrink to the submicron regime. Electron beam microscopes may be used then. The basic optical microscope can be enhanced by adding phase and differential interference contrast as well as polarizing filters. Optical microscopy is not only used to view the features of integrated circuits; it is also useful for analyzing particles found on such circuits. To identify and analyze particles requires a skilled and practiced microscopist. The technique is most useful for particles larger than one micron, and the analysis depends on matching the unknown with data on known particles. Particle atlases are available to aid in identification.[1,2]

The essential elements of a compound optical microscope are illustrated in Fig. 9.2. Its optical elements, the *objective* and the *ocular* or *eyepiece*, are shown as simple lenses; in modern microscopes they consist of six or more highly corrected compound lenses. Object O is placed just beyond the first focal point f_{obj} of the objective lens that forms a real and enlarged image I. This image lies just within the first focal point f_{oc} of the ocular, forming a virtual image of I at I'. The position of I' may lie anywhere between near and far points of the eye. The objective merely forms an enlarged real image which is examined by the eye looking through the ocular. The overall magnification M is a product of the lateral magnification of the objective and the angular magnification of the ocular. The simplest microscope is the monocular microscope, which has only one eyepiece. The binocular instrument has two eyepieces chiefly to make viewing of the sample more convenient. When one objective is used with a binocular microscope, the observed image is generally not stereoscopic.

9.2.1 Resolution, Magnification, Contrast

Light can be thought of as waves as well as particles. To explain some experimental results, it is easier to use the wave concept, but for others the particle concept is more useful. Waves interfere with one another, placing certain limits on the performance of microscopes. Airy first computed the diffracted image and showed in 1834 that for diffraction at a circular aperture of diameter d, the angular position of the first minimum (measured from the center) is given by [see Fig. 9.3(a)][5]

$$\sin(\alpha) = \frac{1.22\lambda}{d} \tag{9.1}$$

where λ is the wavelength of light in free space. The central spot that contains most of the light is called the *Airy* or *diffraction disc*. You can do your own experiment by looking at a bright point source at a distance of several meters, for example, at a lamp, through a small pinhole in a cardboard sheet. The same kind of pattern is formed when a point object is

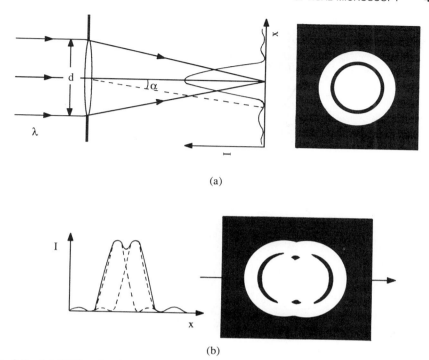

Fig. 9.3 (a) Diffraction at the aperture of a lens showing the Airy disc and (b) the Raleigh criterion for resolution. *I* represents the intensity. Reprinted with permission after Spencer.[3]

imaged by a microscope. There is no lower limit to the size of an object that can be detected *in isolation* under adequate illumination conditions.

Generally one is not interested in detecting a point object, but a two- or three-dimensional object. Two point objects, a distance *s* apart, produce overlapping images, as shown in Fig. 9.3(b). If they are too close together, it is impossible to resolve them. As they are separated, it becomes possible to tell that there are two objects. The definition of this spacing is somewhat arbitrary. Raleigh suggested that two objects can be distinguished when the central maximum of one coincides with the first minimum of the other. The intensity between the two peaks then decreases to 80% of the peak height.

This is shown in Fig. 9.4 for a microscope. The equation

$$s = \frac{0.61\lambda}{n \sin(\theta)} = \frac{0.61\lambda}{NA} \tag{9.2a}$$

gives the *resolution* (the minimum distance between points or parts of an

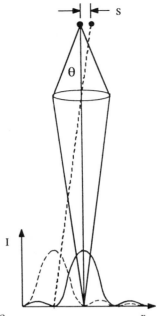

Fig. 9.4   The resolution limit of an optical microscope.

object) and satisfies Raleigh's criterion.[3] In the equation n is the refractive index of the medium separating the object from the objective, and θ is the half angle subtended by the lens at the object. The *numerical aperture* (NA), usually engraved on the objective mount, is a number that expresses the resolving power of the lens and the brightness of the image it forms. The higher the NA of a lens, the higher is the quality of the lens. For high resolution, that is, small s, NA should be made as large as possible. However, high NA corresponds not only to high resolution but also to shallow depth of field and shallow working distance—the distance from the focus point of the object plane to the front surface of the objective. The depth of focus D_{focus}—the thickness of the image space that is simultaneously in focus—is given by

$$D_{focus} = \frac{\lambda}{4(NA)^2} \qquad (9.2b)$$

D_{focus} is insufficient for both the top and bottom surfaces of an integrated circuit to be in focus simultaneously at 200× magnification. The depth of field D_{field}—the thickness of the object space that is simultaneously in focus—is given by

$$D_{field} = \frac{\sqrt{n^2 - (NA)^2}}{(NA)^2} \lambda \qquad (9.2c)$$

Both D_{focus} and D_{field} decrease with increasing NA, but the resolution increases.

It is possible to purchase long working distance objectives that allow greater clearance between the objective lens and the sample. Whereas normal objectives may have working distances of 3 mm ($20\times$ magnification, 0.46 NA) or 0.5 mm ($50\times$, 0.8 NA), long working distance objectives have 11 mm ($20\times$, 0.4 NA) or 8 mm ($50\times$, 0.55 NA). However, they generally have lower NA and require higher illumination.

According to Eq. (9.2a) three variables may be adjusted to reduce s or increase the resolution. The wavelength may be reduced by using selective filters. Blue light has higher resolution than red light. One frequently uses a green filter with its transmission peak at or near the wavelength for which the objective is chromatically corrected and the eye is most sensitive. The resolution may be improved by increasing the angle θ_0 toward the theoretical maximum of $90°$. $NA \approx 0.95$ is the upper practical limit. Beyond this, further gain in resolution is achieved by use of immersion objectives in which a fluid with higher index of refraction than air is used between the sample and the front lens of the objective. With air as the immersion medium, the numerical aperture is sometimes referred to as "dry" NA. Immersion fluids can be oil ($n = 1.52$) and water ($n = 1.33$), but mono-bromonaphthalene ($n = 1.66$) has also been used. Oil is not suitable for semiconductors, but distilled water is sometimes used. Practical limits of $NA \approx 1.3$–1.4 for oil-immersion optics limit the resolution to $s \approx 0.25$ μm for green light with $\lambda = 0.5$ μm.

Magnification M is related to the resolving power of the microscope objective and the eye. However, the image must be magnified sufficiently for detail to be visible to the eye. The *resolving power* is the ability to reveal detail in an object by means of the eye, microscope, camera, or photograph. An approximate relationship for the magnification is[4]

$$M = \frac{\text{maximum NA (microscope)}}{\text{minimum NA (eye)}} = \frac{1.4}{0.002} = 700\times \qquad (9.3)$$

Magnification is sometimes expressed as the ratio of the resolution limits

$$M = \frac{\text{limit of resolution (eye)}}{\text{limit of resolution (microscope)}} = \frac{0.15 \text{ mm}}{0.0002 \text{ mm}} = 750\times \qquad (9.4)$$

where the eye resolution is set by the distance between the rod and cone receptors on the retina of the eye. The maximum magnification of a microscope when the image is viewed by the eye is around $750\times$. Magnification above this is *empty magnification*; it gives no additional information. $M \approx 400$–$500\times$ with air as the immersion medium.

The eye fatigues easily if used at its limits of resolving power, and it is desirable to supply more magnification than the minimum required for convenience. A reasonable rule is to make the magnification about 750 NA, but one should always use the lowest magnification that permits comfortable viewing. Excessive magnification produces images of lower brilliance and poorer definition, with the result that the amount of object detail that can be seen is frequently reduced. Recent objectives have been made with magnifications up to 150 × with a "dry" NA of 0.95.

Contrast—the ability to distinguish between parts of an object—depends on many factors. Dirty eyepieces or objectives degrade image quality. Glare will reduce contrast, especially if the sample is highly reflecting. It is most serious when viewing samples with little contrast and can be controlled to some extent by controlling the field diaphragm, the opening that controls the area of the lighted region. The diaphragm should never be open more than just enough to illuminate the complete field of the microscope. For critical cases it may be reduced to illuminate only a small portion of the normal field.

Contrast of surface-height variation can be considerably improved by *dark-field microscopy*. Instead of the light impinging vertically onto the sample, it is deflected by an appropriate condenser lens and impinges on the sample at an angle. Only the scattered light reaches the objective, allowing small height variations to be seen. Flat portions of the sample do not reflect the light and appear black. The effect is not unlike the observation of dust particles in air that appear dramatically when illuminated by sunlight streaming through a window. The eye sees the light scattered from the dust particles.

9.2.2 Differential Interference Contrast

Differential interference contrast microscopes or *Nomarski* microscopes are extremely useful to accentuate small sample height variations. The basic technique is due to Nomarski.[5-6] For simplicity we will describe the method for a transmission microscope and follow the discussion of Spencer.[3] In the optical system of Fig. 9.5 the first step is to produce plane-polarized light (polarization is discussed in more detail in Section 9.3), which strikes a beam splitter or modified Wollastan prism and a condenser lens. The left part of Fig. 9.5 shows the relative polarizations, the central part shows the optical paths, and the right side shows the optical path variations. The prism splits the polarized light into two wavefronts that follow different paths and are polarized along directions at right angles to one another. By arranging these two directions to make angles of 45° with the vibration axis of the polarizer, two broad beams of light with a lateral displacement or *shear* are produced, shown greatly exaggerated in Fig. 9.5.

Placed above the objective is a second double prism or beam combiner. This recombines the beam, and in some instruments it also introduces a constant optical path difference (opd) δ between the beams that can be varied by moving the combiner. The planes of polarization are shown by the

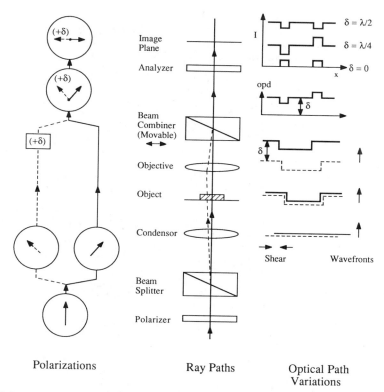

Polarizations   Ray Paths   Optical Path
                                              Variations

Fig. 9.5 The optical system for Nomarski interference contrast microscopy. Reprinted with permission after Spencer.[3]

arrows on the left side of Fig. 9.5. For $\delta = 0$ the background is dark, whereas for $\delta = \lambda/2$ (equivalent to reversal of the dashed analyzer component) maximum brightness for light of wavelength λ is observed. In white light the background shows a strong color that can be changed by small movements of the beam combiner.

The right side of Fig. 9.5 shows the result of the beam-splitting operation. The sample introduces a "trough" of optical path difference into both beams. When the shear is removed by the objective–beam combiner, these troughs no longer overlap exactly. For the idealized uniform sample illustrated here, there is an extra "blip" of optical path difference corresponding to each *edge* of the sample, positive at one edge and negative at the other. With $\delta = 0$ both edges appear bright against a dark background. For $\delta = \lambda/2$ the converse is true, with dark edges against a light background. For white light, edges of one color appear against a different colored background. Other effects are produced for intermediate δ values; one edge is then brighter and the other darker than the background, giving an interesting but

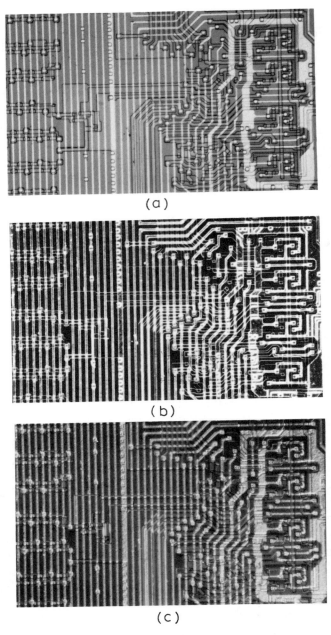

Fig. 9.6 Micrograph for (a) reflected light bright field, (b) reflected light dark field, and (c) reflected light differential interference contrast conditions. Courtesy of I. Toplin, Carl Zeiss, Inc.

strictly spurious impression of "shadowing." Step heights as a small as 30 Å can be observed, making this technique suitable for measurements of planarity of wafer surface and etch pit studies.[6] Figure 9.6 shows a comparison of bright-field, dark-field, and differential interference contrast micrographs. Hilton gives a good, simple discussion of microscopy, including differential interference contrast, with many examples.[7]

9.2.3 Defect Etches

Optical microscopy is frequently used to look at and measure defect size, type, and density in semiconductors delineated by particular defect etches.

TABLE 9.1   Etches for Semiconductor Defect Delineation

Semiconductor	Etch	Chemical Composition	Application
Si	Sirtl[8]	Dissolve 50 g CrO_3 in 100 ml H_2O, immediately before using add 1 part HF to 1 part of the solution by volume.	Best applicable to {111}-oriented surfaces.
Si	Dash[9]	$HF:HNO_3:CH_3COOH$ 1:3:10	Generally applicable for both n-Si and p-Si of {111} and {100} orientation, but works best for p-Si.
Si	Secco[10]	Dissolve 55 g $CuSO_4$ 5 H_2O in 950 ml H_2O, add 50 ml HF. $HF:K_2Cr_2O_7$ (0.15M) (11 g $K_2Cr_2O_7$ in 250 ml H_2O) i.e. 2:1 or $HF:CrO_3$ (0.15M) 2:1	Cu displacement etch; delineates defects by Cu decoration. Generally applicable, but is particularly suitable for {100} orientation.
Si	Schimmel[11]	Add 75 g CrO_3 to H_2O to make 1000 ml solution (0.75M solution)	For n-Si, p-Si, {100}, {111} For $\rho > 0.2$ Ω-cm add 2 parts HF to 1 part solution. For $\rho < 0.2$ Ω-cm add 2 parts HF to 1 part solution and 1.5 parts H_2O
Si	Wright[12]	$HF:HNO_3:5MCrO_3:Cu(NO_3)_2 \cdot 3H_2O; CH_3COOH:H_2O$ 2:1:1:2 g:2:2. Best results obtained by first dissolving the $Cu(NO_3)_2$ in the H_2O; otherwise, order of mixing not critical.	For n-Si and p-Si, {100} and {111}; defect-free regions are not roughened following etching.
Si	Yang[13]	Add 150 g CrO_3 to 1000 ml H_2O (1.5M); add 1 part solution to 1 part HF.	Delineates various defects on {100}, {111}, and {110} surfaces without agitation.
Si	Seiter[14]	9 parts by volume of a solution of 120 g CrO_3 in 100 ml H_2O and 1 part HF (49%).	Etches {100} planes 0.5–1 μm/min; 20–60 s etch time; delineates dislocations, stacking faults, swirl defects.
GaAs	KOH[15]	Molten KOH	Sample immersed in molten KOH for 3 h at 350°C in covered Ni crucible.
InP	Huo et al.[16]	$HBr:H_2O_2:H_2O:HCl$ 20:2:20:20	Reveals dislocations on {100} and {111} surfaces.

Note: CH_3COOH is glacial acetic acid.

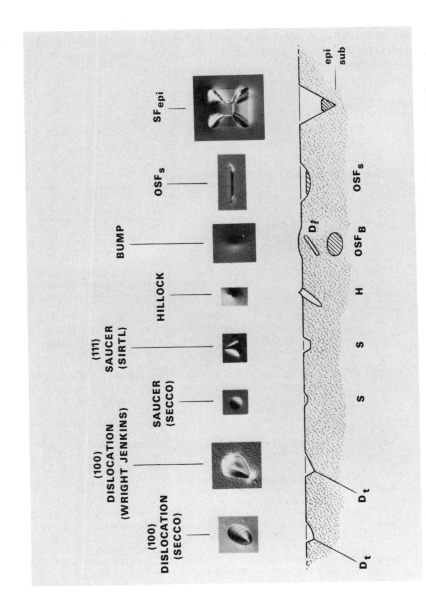

Fig. 9.7 Some common etch patterns found in silicon when etched with some of the etches in Table 9.1. Reprinted with permission after Miller and Rozgonyi.[6]

The sample is etched in an etchant that renders particular defects visible through etch pits of particular shapes. Table 9.1 lists a number of such etches. Detailed instructions on using some of these etches are given in ASTM standards F47 and F26. To count the defects, recommended procedure is to use an optical microscope with $100\times$ magnification and count the number of defects within a known area.[17–18] This should be done on nine locations on the wafer and the readings should be averaged. A very detailed series of photographs showing many examples of defects and how they appear when viewed through a microscope are given in ASTM standard F154.[19] A cross section and top view of some defects in silicon are shown in Fig. 9.7.

9.3 ELLIPSOMETRY

9.3.1 Theory

Ellipsometry is used predominantly to measure the thickness of thin dielectric films on highly absorbing substrates but can also be used to determine the optical constants of films or substrates. It allows thickness measurements at least an order of magnitude smaller than interferometric methods. Ellipsometry is based on measuring the state of polarization of polarized light. Before going into the details of ellipsometry, it is important to understand the properties of polarized light. When light is reflected from a single surface, it will generally be reduced in amplitude and shifted in phase. For multiple reflecting surfaces the various reflecting beams will further interact and give maxima and minima as a function of wavelength or incident angle. These interference effects are discussed in Section 9.5. Since ellipsometry depends on angle measurements, optical variables can be measured with great precision that is independent of light intensity, total reflectance, and detector-amplitude sensitivity.

Consider the plane-polarized light incident on a plane surface illustrated in Fig. 9.8. The incident polarized light can be resolved into a component p, parallel to the plane of incidence and a component s ("s" is the first letter of the German work *senkrecht* meaning vertical) perpendicular to the plane of incidence. If the material from which the wave is reflected has zero absorption, then only the amplitude of the reflected wave is affected. Linearly polarized light is reflected as linearly polarized light. However, the two components experience different amplitudes *and* phase shifts upon reflection from absorbing materials and for multiple reflections in a thin layer between air and the substrate. The parallel component reflectance is always less than the vertical component reflectance for angles of incidence other than $0°$ and $90°$. The two are equal at those two angles. The phase-shift difference introduces an additional component polarized $90°$ to the incident beam rendering the reflected light *elliptically* polarized. Project-

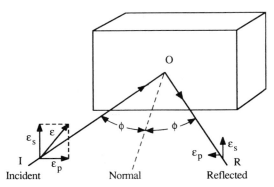

Fig. 9.8   Schematic of polarized light reflection from a plane surface. Lines *I–O* and *O–R* define the plane of incidence, and φ is the angle of incidence.

ed onto a plane perpendicular to the reflected beam, the resultant electric field vector of the elliptically polarized light traces out an ellipse. The key property of polarized light for ellipsometry is the change of plane-polarized light into elliptically polarized light or elliptically polarized light into plane-polarized light upon reflection.

Light propagates as a fluctuation in electric and magnetic fields at right angles to the direction of propagation. The total electric field vector is given by

$$\mathscr{E} = \mathscr{E}_p \hat{\mathbf{p}} + \mathscr{E}_s \hat{\mathbf{s}} \tag{9.5}$$

where $\mathscr{E}_p$ and $\mathscr{E}_s$ are the parallel and perpendicular components of the electric field. The "hatted" symbols represent unit vectors. The reflection coefficients

$$R_p = \frac{\mathscr{E}_p(\text{reflected})}{\mathscr{E}_p(\text{incident})} \qquad R_s = \frac{\mathscr{E}_s(\text{reflected})}{\mathscr{E}_s(\text{incident})} \tag{9.6}$$

are not separately measurable. However, the complex reflection ratio ρ defined in terms of the reflection coefficients R_p and R_s or the *ellipsometric angles* Ψ and Δ

$$\rho = \frac{R_p}{R_s} = \tan(\Psi)e^{j\Delta} \tag{9.7}$$

is measurable. Since ρ is the ratio of reflection coefficients, that is, the ratio of the intensities and the relative phase difference, it is not necessary to make absolute intensity and phase measurements.

The ellipsometric angles Ψ ($0° \leq \Psi \leq 90°$) and Δ ($0° \leq \Delta \leq 360°$) are the most commonly used variables in ellipsometry and are defined as

$$\Psi = \tan^{-1}\left(\frac{R_p}{R_s}\right) \qquad (9.8a)$$

$$\Delta = \text{differential phase change} = \Delta_p - \Delta_s \qquad (9.8b)$$

The determination of Ψ and Δ is the subject of Section 9.3.2. Ψ and Δ determine the differential changes in amplitude and phase, respectively, experienced upon reflection by the component vibrations of the electric field vectors parallel and perpendicular to the plane of incidence.

How are Ψ and Δ used to determine the sample's optical parameters? Let us use the simple example of light reflected at an air-solid-absorbing substrate interface shown in Fig. 9.9. The air is characterized by its index of refraction n_0, and the sample is characterized by its complex index of refraction $n_1 - jk_1$, where n_1 is the index of refraction and k_1 the extinction coefficient. Using Fresnel's equations, it can be shown that[20]

$$N_1 = n_1 - jk_1 = n_0 \tan(\phi)\sqrt{1 - \frac{4\rho}{(1+\rho)^2}\sin^2(\phi)} \qquad (9.9)$$

where ρ is given by Eq. (9.7). The complex index of refraction of the sample can be determined if n_0 is known and if the ellipsometric ratio ρ is measured at the incidence angle ϕ. This case is easily calculated.

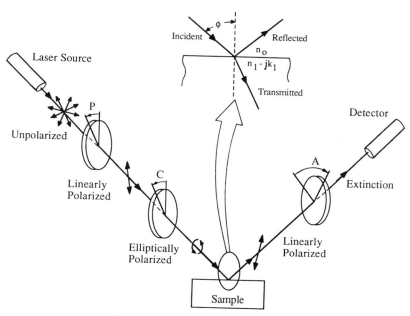

Fig. 9.9 Ellipsometer schematic.

Of considerable importance in ellipsometric measurements is a substrate covered by a thin film, frequently an insulator. For the air (n_0)-thin film (n_1)-substrate $(n_2 - jk_2)$ system, the equations become much more complicated because they are dependent on the refractive indices, the film thickness, the angle of incidence, and the wavelength. If n_2 and k_2 are known from an independent measurement and if the film is transparent (no absorption), then n_1 and film thickness may be calculated from the results of a single Ψ and Δ measurement. But the computation becomes very tedious. An entire book is devoted to ellipsometric tables and curves showing the dependence of Ψ and Δ on the oxide thickness and oxide refractive index for the air-SiO_2-Si system at selected mercury and He-Ne laser spectral lines.[21] The only satisfactory way of handling the equations is by use of numerical techniques, as is commonly done in computer-controlled ellipsometers today.

9.3.2 Null Ellipsometry

Most commercial ellipsometers are based on the null ellipsometric principle. The PCSA version (Polarizer-Compensator-Sample-Analyzer) is the most common null ellipsometer configuration.[20,22] The principle of operation is most easily explained with reference to the manual ellipsometer in Fig. 9.9. A collimated beam of unpolarized monochromatic light, typically from a laser, is linearly polarized by the polarizer. The Glan-Thompson prism, consisting of two sections of calcite cemented together, is a common polarizer. When unpolarized light is incident on such a polarizer, total internal reflection allows only linearly polarized light to exit. The compensator, also known as a *retarder*, changes the linearly polarized light to elliptically polarized light. The compensator contains a fast and a slow optical axis perpendicular to the direction of transmission. The component of incident polarized light whose electric field is parallel to the slow axis is retarded in phase relative to the component parallel to the fast axis as the light passes through the compensator. When the relative retardation is $\pi/2$, the compensator is called a *quarter-wave retarder* or *quarter-wave plate*. Linear quarter-wave retarders, made out of birefringent mica or quartz, are most commonly used in ellipsometers.

The angles P and C of the polarizer and the compensator can be adjusted to any state of polarization ranging from linear to circular. The aim of an ellipsometry measurement is a null at the detector. This is attained by choosing P and C to give light of elliptical polarization, which, when reflected from the sample, becomes linearly polarized to be extinguished by the analyzer. The linearly polarized light is passed through the analyzer, which is similar to the polarizer, and the angle A is adjusted for minimum photodetector output. The operator adjusts the polarizer and analyzer angles sequentially for minimum detector signal. Computer-controlled stepping motors have replaced manual adjustments on most modern instruments. The angular convention is that all angles are measured as positive counterclockwise from the plane of incidence when looking into the beam and that the polarizer angle is adjusted to zero when the plane of transmission is in the plane of incidence.

There are 32 combinations of P, C, and A that can result in a given pair of Ψ and Δ. Because any two angles of the polarizer, compensator, and analyzer that are 180° apart are optically identical, the number of combinations of P, C, and A settings giving any pair of Ψ and Δ can be reduced to 16 if all angles are restricted to less than 180°. The 16 pairs of linear equations relating Ψ and Δ to P and A at null are summarized as[23]

$$\Psi: \quad A, \quad 180° - A \tag{9.10a}$$

$$\Delta: \quad 2P - 90°, \quad 2P - 270°, \quad (2m-1)90° \pm 2P \tag{9.10b}$$

where $m = 1$, 2, 3, or 4.

The 16 equation pairs can be reduced to two pairs by restricting the compensator to one angle, for example, 45°, and the ranges of P and A to two zones, defined as[23]

$$\text{Zone 2:} \quad -45° \leq P_2 \leq 135°, \quad 0° \leq A_2 \leq 90°, \quad C = 45° \tag{9.11a}$$

$$\text{Zone 4:} \quad -135° \leq P_4 \leq 45°, \quad -90° \leq A_4 \leq 0°, \quad C = 45° \tag{9.11b}$$

With P and A determined in each of these two zones, their relationship to Ψ and Δ is

$$\Psi_2 = A_2, \quad \Psi_4 = -A_4 \tag{9.12a}$$

$$\Delta_2 = 270° - P_2, \quad \Delta_4 = 90° - 2P_4 \tag{9.12b}$$

Although equally valid measurements can be made in either zone, effects of imperfections in the compensator can be eliminated by making measurements in both zones and averaging the results to obtain the mean values

$$\bar{\Psi} = \frac{\Psi_2 + \Psi_4}{2} = \frac{A_2 - A_4}{2} \tag{9.13a}$$

$$\bar{\Delta} = \frac{\Delta_2 + \Delta_4}{2} = 180° - (P_2 + P_4) \tag{9.13b}$$

The angles are then used to calculate the optical parameters of the sample.

9.3.3 Rotating Analyzer Ellipsometry

The *rotating analyzer ellipsometer* falls within a class of ellipsometers known as *photometric ellipsometers*. In the rotating analyzer ellipsometer linearly polarized light is incident on the sample and becomes elliptically polarized upon reflection.[20,24-25] The reflected beam passes through the analyzer, rotating around the beam axis at a constant angular velocity, to be detected by an optical detector. If the light incident on the analyzer were linearly polarized, the detected light would be a sine-squared function with a maximum and *zero* minimum per half rotation of the analyzer. Unmodulated, uniform output results for circularly polarized light and sinusoidal output variations, similar to linearly polarized light, are observed for elliptically polarized light, but the maxima are smaller and the minima larger, making the amplitude variation smaller. The amplitude variation of the sinusoidal detector output is a function of the ellipticity of the reflected light. The output is generally Fourier analyzed to yield Ψ and Δ.

The major advantages of rotating analyzer ellipsometers lie in their higher speed and their increased accuracies. Effects of noise and random errors are reduced since hundreds or thousands of light intensity samples constitute a single measurement. The lack of a compensator improves the measurement since errors associated with commercial compensators do not affect the measurement. The demands on the optical system, however, are more stringent. Stray light must be carefully controlled, and the light source intensity should not change with time. The detector response must be linear to avoid generation of harmonics. Rotating analyzer ellipsometers are particularly suited to spectroscopic ellipsometric measurements because none of the wavelength-dispersive properties of the compensator can play a role if there is no compensator and because the data acquisition time can be short. Nevertheless, the majority of commercial instruments is of the null ellipsometer type.

9.3.4 Applications

Film Thickness The major application of ellipsometry is for measurements of thickness and index of refraction of thin, non-absorbing films on semiconductor substrates. There is, in principle, no limit to the thickness of the layer that can be determined. Films as thin as 10 Å have been measured, as have films many thousands of Ångströms thick. Although ellipsometry gives numeric values, the results for very thin films are questionable. The model assumes uniform optical properties and a sharp planar film/substrate boundary. Furthermore the ellipsometric equations are based on the macroscopic Maxwell's equations which generally do not apply to layers only a few atomic layers thick. Nevertheless, the measurements appear to give reasonable average thicknesses.

Thick layers have a different problem. The interpretation becomes more difficult due to optical path lengths. In Fig. 9.10 we show a thin transparent layer on a substrate. The two reflected rays interfere with one another, ranging from being completely in phase to being completely out of phase. A consequence of this interference is the cyclical nature of thickness measure-

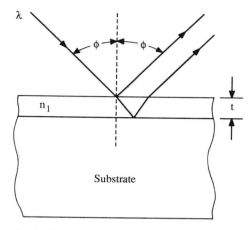

Fig. 9.10 Schematic showing multiple reflections.

ments, where Ψ and Δ are cyclic functions of the film thickness. They repeat for the full-cycle film thickness

$$t = \frac{\lambda}{2\sqrt{n_1^2 - \sin^2(\phi)}} \tag{9.14}$$

For example, at $\phi = 70°$ the full-cycle thickness of SiO_2 films with $n_1 = 1.465$ at $\lambda = 0.6328 \, \mu m$ is 2815 Å. If a 500-Å-thick SiO_2 film gives certain ellipsometric angles, the same angles will be measured for films of $(500 + 2815)$ Å, $(500 + 5630)$ Å, etc. Hence for films thicker than the full-cycle thickness, one must have independent knowledge of the film thickness to within one full-cycle thickness.

Substrates, Layer Growth   Although the major use of ellipsometry is for the analysis of non-absorbing, insulating films, the method has found some application in the characterization of semiconductors. In particular, it has been used to study semiconducting materials that have been modified in some way. For example, ion implantation damage has been correlated with ellipsometric measurements for Si, GaAs, and InP.[26–29] It is believed that the implant-induced damage, not the doping concentration, changes the refractive indices. Although quantitative results are difficult to obtain, the measurements do allow a rapid, non-destructive measurement of the crystal damage and the behavior of this damage with annealing treatments.

Ellipsometry has also found applications during crystal growth, where its non-contacting, real-time nature is particularly useful when used *in situ*. For example, it has been used to monitor the growth of superlattice structures grown by molecular beam epitaxy (MBE) and metalorganic chemical vapor deposition (MOCVD).[30] Wavelength variation gives *spectroscopic ellip-*

sometry an additional advantage much as varying the frequency is useful to determine the impedance of electronic circuits.[31] For example, using 3.25 and 4.25 eV light with very short penetration depth, it is possible to study the surface damage as a function of surface treatments.[32] Spectroscopic ellipsometry applied to heterostructures allowed seven material parameters to be determined.[33] Ellipsometry is a benign technique that is little affected by deposition methods if used to monitor the growth of a layer. It does not influence the deposition process if used as an *in situ* process monitor to monitor the growth of an MBE film or the growth of an insulating layer.

9.4 TRANSMISSION

9.4.1 Theory

Optical transmission or absorption measurements are routinely used by chemists to determine the constituents of chemical compounds. They are also used in the semiconductor industry, but only for certain specialized applications. For example, the optical absorption coefficients of semiconductors are determined by transmission measurements. However, once they have been determined for a given semiconductor, they are rarely remeasured. Deep-level impurities do not respond well to optical transmission measurements, although their concentrations and energy levels can be, and sometimes are, so determined. Shallow-level impurities respond much more effectively to optical measurements and are routinely measured by optical absorption measurements. We have discussed this in Section 2.7.2. Certain impurities possess characteristic absorption lines due to vibrational modes, for example, oxygen and carbon in silicon, and are routinely determined from such measurements. We discuss in this chapter the appropriate theory of optical transmission measurements and give some examples.

During transmission measurements light is incident on the sample, and the transmitted light is measured as a function of wavelength as illustrated in Fig. 9.11(a). The sample is characterized by reflection coefficients R_1, R_2, absorption coefficient α, complex refractive index $(n_1 - jk_1)$, and thickness d. Light of intensity I_i and wavelength λ is incident from the left. The transmitted light I_t can be measured absolutely, or the ratio of transmitted to incident light can be formed. As shown in Appendix 9.1, the transmittance T of a sample with identical front and back reflection coefficient and light incident normal to the sample surface is given by

$$T = \frac{(1 - R)^2 e^{-\alpha d}}{1 + R^2 e^{-2\alpha d} - 2R e^{-\alpha d} \cos(\phi)} \tag{9.15}$$

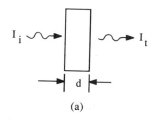

(a)

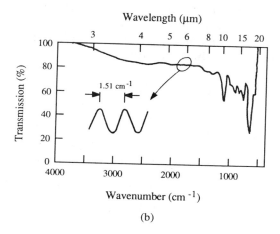

(b)

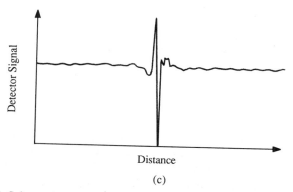

(c)

Fig. 9.11 (a) Schematic transmittance measurements; (b) normalized FTIR trans-mittance curve for a double-side polished Si wafer, $\Delta f = 4\,\mathrm{cm}^{-1}$ for transmittance curve, $\Delta f = 1\,\mathrm{cm}^{-1}$ for inset; (c) interferogram for the same wafer, $\Delta f = 4\,\mathrm{cm}^{-1}$. From period of $1.51\,\mathrm{cm}^{-1}$ in (b) the wafer thickness is $970\,\mu\mathrm{m}$ from Eq. (A9.8). Courtesy of N. S. Kang, Arizona State University.

where $\phi = 4\pi n_1 d/\lambda$, α is the absorption coefficient, and the reflectance R is given by

$$R = \frac{(n_0 - n_1)^2 + k_1^2}{(n_0 + n_1)^2 + k_1^2} \qquad (9.16)$$

The absorption coefficient is related to the extinction coefficient k_1 by $\alpha = 4\pi k_1/\lambda$. Absorption coefficients and refractive indices are shown in Figs. A9.2 and A9.3 in Appendix 9.2 for selected semiconductors. A normalized curve of I_t for polished Si is shown in Fig. 9.11(b).

Semiconductors are generally transparent for photon energies less than the band-gap energy, and the transmittance becomes

$$T = \frac{(1 - R)^2}{1 + R^2 - 2R\cos(\phi)} \qquad (9.17)$$

The "cos" term can be written as $\cos(f/f_1)$, where $f = 2\pi/\lambda$. $f_1 = 1/(2n_1 d)$ is a characteristic spatial frequency. If the resolution of the apparatus is sufficiently high, $\Delta f \leq 1/(2n_1 d)$, then an oscillatory transmittance curve is observed. For example, $\Delta f \leq 4.9\ \text{cm}^{-1}$ for a Si wafer with thickness $d = 300\ \mu\text{m}$ and refractive index $n_1 = 3.42$. If the resolution of the instrument is insufficient to resolve these fine-structure oscillations, then

$$T = \frac{(1 - R)^2}{1 - R^2} = \frac{1 - R}{1 + R} \qquad (9.18)$$

For the Si example this becomes $T = 0.54$ for $R = 0.3$. As shown in Appendix 9.1, the wafer thickness can be determined from the period of the oscillatory transmittance versus wave number curve with

$$d = \frac{1}{2n_1 \Delta(1/\lambda)} \qquad (9.19)$$

where $\Delta(1/\lambda)$ is the wave number interval between two maxima or two minima illustrated in Fig. 9.11(b). The transmittance curves are sometimes plotted as a function of *wavelength* and sometimes as a function of *wave number*, defined by wave number = 1/wavelength.

Certain impurities contained in a semiconductor sample exhibit absorption. Examples are interstitial oxygen and substitutional carbon in silicon. Their concentration is proportional to the absorption coefficient at those wavelengths. The transmittance with absorption but no "cos" oscillations, is given by

$$T = \frac{(1 - R)^2 e^{-\alpha d}}{1 - R^2 e^{-2\alpha d}} \qquad (9.20)$$

The absorption coefficient from Eq. (9.20) is[34]

$$\alpha = -\frac{1}{d} \ln\left[\frac{\sqrt{(1-R)^4 + 4T^2R^2} - (1-R)^2}{2TR^2} \right] \qquad (9.21)$$

R can be determined from the part of the transmittance curve where $\alpha \approx 0$. In some spectral regions of the transmittance curve, there may be some absorption due to lattice vibrations, and for heavily doped substrates, absorption due to free-carriers. The lattice absorption coefficient is about $0.85-1\,cm^{-1}$ for the wavelengths of oxygen in Si and about $6\,cm^{-1}$ for carbon in Si. This must be considered in the analysis.[35–36]

Complications in interpretation of transmission data occur when both surfaces are not polished. Most commercial semiconductor wafers are polished on only one side; the other side is lapped and etched. Due to surface roughness the transmittance becomes wavelength dependent, and T can vary significantly from wafer to wafer. If transmission is severely impaired, the signal to noise can be so poor that the measurement becomes meaningless. The transmission of waters with rough surfaces can be written as

$$T \sim e^{-C/\lambda^2} e^{-\alpha d} \qquad (9.22)$$

where C characterizes the scattering of the rough back surface.[35,37]

9.4.2 Instrumentation

Monochromator   There are two basic instrumentation approaches to measure the transmission spectrum. The older, more traditional one is based on the use of a *monochromator*; the more recent and now more popular one is based on an *interferometer*. The monochromator approach is illustrated in Fig. 9.12(a). The monochromator selects a narrow band of wavelengths $\Delta\lambda$ from a source of radiation. The spectral band is centered on a wavelength λ that can be varied. The monochromator can be thought of as a tunable filter with a band pass $\Delta\lambda$ and a resolution $\Delta\lambda/\lambda$. For a transmittance measurement, light from source S enters the monochromator through a narrow entrance slit. The light rays should be made parallel for optimum performance by passing the light through a collimating lens. When the light falls on the prism or grating, it is dispersed; that is, the prism or the grating breaks the light into its spectral components. A prism disperses and refracts white light by virtue of having a wavelength-dependent refractive index. Short wavelength light is refracted more than long wavelength light. A grating consists of many equidistant parallel lines inscribed on a polished substrate (glass or metal film on glass) with typically between 10,000 and 50,000 lines or grooves per inch. The dispersed light depends on the groove spacing and on the incident angle. Two advantages of gratings over prisms are (1) high resolution and dispersion and (2) a dispersion that is almost constant with wavelength. The chief disadvantages are that gratings are

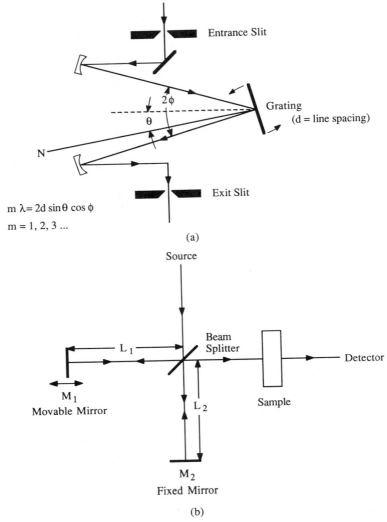

$m \lambda = 2d \sin \theta \cos \phi$

$m = 1, 2, 3 ...$

(a)

(b)

Fig. 9.12   (a) Monochromator schematic, (b) Fourier transform infrared spectrometer schematic.

slightly less rugged, and they generate slightly more scattered light, particularly at shorter wavelengths.

The dispersed light passes through a narrow exit slit, and only those wavelengths that pass through the slit are incident on the sample. It is this second slit that largely controls the spectral resolution; the narrower the slit, the narrower is the wavelength range that reaches the detector. The slits can be thought of as spectral band-pass filters. As the slit becomes narrower, however, the amount of light reaching the sample is likewise reduced. The wavelengths incident on the sample are varied by changing the angular position of the prism or the grating. In a monochromator only a narrow band of wavelengths is selected for the transmission measurement, avoiding simultaneous excitation of competing processes in the sample that can result from other wavelengths. For example, above band-gap light creates electron–hole pairs, which may interfere with measurements using below band-gap light. Monochromator transmission measurements avoid this problem by eliminating above band-gap light. A disadvantage of the monochromator is that only a small portion of the total spectrum is available at one time. Hence the signal is small and signal to noise may be a problem. Frequently this is overcome by using lock-in or signal-averaging techniques. A more serious disadvantage is the time-consuming nature of the measurement since all wavelengths of interest have to be scanned sequentially. It is for these reasons that interferometric techniques have become very popular.

For greater sensitivity and minimization of atmospheric attenuation, double-beam monochromators are frequently used. The beam is split into two similar paths, with the sample placed in one of these paths. Sometimes a reference sample is placed in the reference beam, and the sample transmission is compared to that of the reference.

A monochromator is inserted between the light source and the sample, ensuring that only selected wavelengths are incident on the sample at one time. It is also possible for all wavelengths to be incident on the sample at one time and to spectrally resolve the light *after* it has been transmitted through the sample. Then the instrument is known as a *spectrometer.* A spectrometer is more commonly used when the light is not transmitted through the sample but is emitted from the sample, whereas a monochromator is used to decompose white light into its spectral components for a subsequent spectral response measurement.

Fourier Transform Infrared Spectroscopy   The foundations of modern *Fourier transform infrared spectroscopy* (FTIR) were laid in the latter part of the nineteenth century by Michelson.[38] Soon after that Raleigh recognized the relationship of an interferogram to its spectrum by a Fourier transformation.[39] It was not until the advent of reasonably priced computers and the fast Fourier algorithm[40] that interferometry began to be applied to spectroscopic measurements in the 1970s.

The basic optical component of Fourier transform spectrometers is the Michelson interferometer shown in simplified form in Fig. 9.12(b).[41–43] Light from an infrared source—a heated element or a glowbar—is collimated and incident on a beam splitter. An ideal beam splitter creates two separate optical paths by reflecting 50% of the incident light and transmitting the remaining 50%. In the near and middle infrared region, germanium deposited on a KBr substrate is commonly used as a beam splitter. In one path the beam is reflected by a fixed-position mirror back to the beam splitter where it is partially transmitted to the source and partially reflected to the detector. In the other leg of the interferometer, the beam is reflected by the movable mirror that is translated back and forth and maintained parallel to itself. The beam from the movable mirror is also returned to the beam splitter where it too is partially reflected back to the source and partially transmitted to the detector. Although the light from the source is incoherent, when it is split into two components by the beam splitter, the components are *coherent* and can produce interference phenomena when the beams are combined. The detector is typically a deuterated triglycene sulfate pyroelectric detector, and the movable mirror rides on an air bearing for good stability. Other detectors such as cooled HgCdTe are also used.

The light intensity reaching the detector is the sum of the two beams. The two beams are in phase and reinforce each another when $L_1 = L_2$. When M_1 is moved, the optical path lengths are unequal, and an optical path difference δ is introduced. If M_1 is moved a distance x, the retardation is $\delta = 2x$ since the light has to travel an additional distance x to reach the mirror and the same additional distance to reach the beam splitter.

Consider the output signal from the detector when the source emits a single frequency or wavelength. For $L_1 = L_2$ the two beams reinforce each other because they are in phase, $\delta = 0$, and the detector output is a maximum. If M_1 is moved by $x = \lambda/4$, the retardation becomes $\delta = 2x = \lambda/2$. The two wave fronts reach the detector 180° out of phase, resulting in destructive interference or zero output. For an additional $\lambda/4$ movement by M_1, $\delta = \lambda$ and constructive interference results again. The detector output—the interferogram—consists of a series of maxima and minima that can be described by the equation

$$I(x) = B(f)[1 + \cos(2\pi xf)] \tag{9.23}$$

where $B(f)$ is the source intensity. $B(f)$ and $I(x)$ are shown in Fig. 9.13(a) for this simple case. When the source emits more than one frequency, Eq. (9.23) is replaced by the integration

$$I(x) = \int_0^\infty B(f)[1 + \cos(2\pi xf)]\, df \tag{9.24}$$

For example, consider the source spectral distribution, $B(f) = A$ for $0 \le f \le f_1$ in Fig. 9.13b. The interferogram is obtained by eliminating the unmodulated term from Eq. (9.24)

$$I(x) = \int_0^{f_1} A \cos(2\pi xf)\, df = Af_1\, \frac{\sin(2\pi xf_1)}{2\pi xf_1} \tag{9.25}$$

shown in Fig. 9.13(b). As f_1 is increased by including more frequencies, the interferogram becomes narrower.

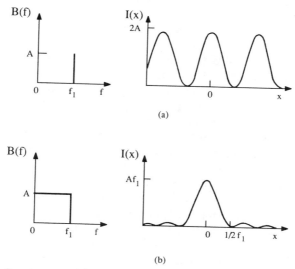

Fig. 9.13 (a) Spectrum and interferogram for a simple cosine wave, (b) spectrum and interferogram for a band-limited signal.

The interferogram always retains its maximum at $x = 0$ where $L_1 = L_2$ because *all* wavelengths interfere constructively for that and only that mirror position. For $x \neq 0$ waves interfere destructively, and the interferogram amplitude decreases from its maximum, as shown in the interferogram for a Si wafer in Fig. 9.11(c). The strong maximum at $x = 0$ is known as the *centerburst*. The higher resolution spectral information is contained in the wings of the interferogram, corresponding to larger mirror travel. There is a practical limit to the mirror displacement, represented by $x = \Delta$. The best spectral resolution is $\Delta f = 1/\Delta$. In practice other practical considerations reduce Δf below this value.

The measured quantity in FTIR is the interferogram. It contains not only the spectral information of the source, which we have considered so far, but also the transmittance characteristics of the sample. The interferogram, however, is of little direct interest. It is the spectral response that is of interest, which is calculated from the interferogram using the Fourier transformation

$$B(f) = \int_{-\infty}^{\infty} I(x) \cos(2\pi x f)\, dx \qquad (9.26)$$

Note that $B(f)$ in Eq. (9.26) contains the spectral content of the source, the sample, and the ambient in the path of the measurement. It is common practice to reduce atmospheric H_2O and CO_2 absorption lines by purging the apparatus with dry nitrogen. The effect of the source is eliminated by making one measurement without the sample—the background measurement—and one with the sample. Storing the two interferograms in a computer allows the ratio of the two to be calculated, thereby eliminating the background. It is this feature of FTIR—the use of a computer for signal processing—that has made the technique so powerful. For example, since the mirror travel is finite, irregularities are introduced into the interferogram. Some of these irregularities can be subsequently reduced by using *weighting* or *apodization* schemes in the computer analysis.[43]

FTIR has two major advantages over prism or grating monochromators. One is the multiplex gain or the *Fellget* advantage. In monochromator transmission measurements only a small fraction of the entire spectrum is observed at a given time, whereas in FTIR the entire spectrum is observed over the measurement period of a second or less. With N spectral elements, each $\Delta\lambda$ wide, the FTIR approach has a signal-to-noise advantage of $N^{1/2}$ over the monochromator approach when the detector is limited by noise other than photon noise.[42] A second major advantage is the optical throughput gain or *Jacquinot* advantage, referring to the amount of light one is able to pass through the instrument. Monochromators are limited by the entrance and exit slits, whereas FTIRs have relatively large circular entrance apertures. The optical throughput gain is typically about 100. The original major disadvantage of FTIR, the necessity of using a computer for the Fourier transformation, is no longer a disadvantage with the advent of computers of high calculation capacity and small size. In fact it has become an advantage since the computer can be used to do a number of signal "massaging" functions. For example, data stored in digital form can be added, subtracted, ratios can be formed, and other functions can be performed. Many interferograms of a given sample can be collected and averaged for higher signal-to-noise ratio.

9.4.3 Applications

Transmittance spectroscopy is used very comprehensively in the chemical industry. For semiconductors it is primarily used to detect certain impurities. The most prominent examples are oxygen and carbon in Si and EL2 concentrations in GaAs. Interstitial oxygen in silicon causes absorption at $\lambda = 9.05$ μm (1105 cm^{-1}) at 300 K and at 8.87 μm (1227.6 cm^{-1}) at 77 K due to the antisymmetric vibration of the SiO$_2$ complex.[44] Substitutional carbon has absorption peaks at $\lambda = 16.47$ μm (607.2 cm^{-1}) at 300 K and at 607.5 cm^{-1} at 77 K due to a local vibrational mode.[45] These absorption peaks are superimposed on phonon excitations of the silicon substrate and should be subtracted from the spectrum of a carbon- and oxygen-free reference sample. An example is shown in Fig. 9.14 where the transmission spectrum of a low oxygen and low carbon Si wafer is subtracted from the spectrum of a sample containing oxygen and carbon giving the spectrum of oxygen and carbon alone.

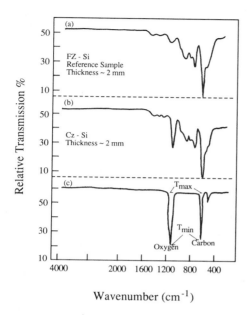

Fig. 9.14 Transmission spectra: (a) silicon wafer low in oxygen and carbon, (b) silicon wafer containing more oxygen and carbon, (c) difference between spectra in (a) and (b). Reprinted with permission of *Solid State Technology*.

The optical absorption coefficients are converted to concentrations by

$$N = C_1 \alpha \quad \text{cm}^{-3} \tag{9.27a}$$

$$N = C_2 \alpha \quad \text{ppma} \tag{9.27b}$$

for N in cm^{-3} and in ppma (parts per million atomic). C_1 and C_2 are given in Table 9.2. Also shown is the full width at half maximum (FWHM) line width. This latter value dictates the bandwidth of the measuring system. The

TABLE 9.2   Absorption Coefficient to Concentration Conversion Factors

Impurity	C_1 (cm^{-2})	C_2 (cm-ppma)	FWHM (cm^{-1})	Reference
Oxygen in Si (300 K)	4.81×10^{17}	9.62	34	"Old ASTM" [44]
Oxygen in Si (300 K)	2.45×10^{17}	4.9	34	"New ASTM" [44]
Oxygen in Si (77 K)	0.95×10^{17}	1.9	19	"New ASTM" [44]
Oxygen in Si (300 K)[a]	3.03×10^{17}	6.06	34	"JEIDA" [47]
Oxygen in Si (300 K)	2.45×10^{17}	4.9	34	"DIN" [48–49]
Oxygen in Si (300 K)[b]	3.14×10^{17}	6.28	34	IOC-88 [50]
Carbon in Si (300 K)	1×10^{17}	2	6	[45, 51]
Carbon in Si (77 K)	4.5×10^{16}	0.9	3	[45]
EL2 in GaAs (300 K)[c]	1.25×10^{16}	0.25		[52]

[a]JEIDA: Japan Electronic Industry Development Association.
[b]International Oxygen Coefficient 1988.
[c]At $\lambda = 1.1 \, \mu$m; EL2, being a deep-level impurity, has a broad absorption band.

oxygen conversion factors were obtained by calibrating the IR transmittance against oxygen concentrations determined by charged particle activation analysis, vacuum fusion gas analysis, and photon activation analysis. For oxygen in silicon measurements, it is necessary to specify which conversion factors are used, due to the diversity of these values. A good discussion of the state of oxygen-in-silicon measurements is given in Bullis et al.[46] Based on recent measurements it appears that a conversion factor $C_1 \approx 3 \times 10^{17}$ cm^{-2} is the preferred value.

The lower detection limit for oxygen in silicon by the IR technique is around 5×10^{15} cm^{-3}; for carbon in silicon, it is around 10^{16} cm^{-3} at room temperature and 5×10^{15} cm^{-3} at 77 K. Low carbon concentrations are particularly difficult to measure because there is a strong two-phonon lattice absorption band near the carbon band at 16 μm. Separation of these absorption bands requires either sample cooling to "freeze out" the lattice band or a comparison with a "carbon-lean" reference sample. A recent measurement method, based on low-temperature photoluminescence of samples subjected to a CF_4 reactive ion etch, suggests detection limits for C in Si as low as 10^{13} cm^{-3}.[53] Transmittance measurements are of course also used to determine the optical absorption coefficients of semiconductors[54] and have been used to determine the boron and phosphorus content of deposited glasses.[55] Microspot FTIR measurements use beams as small as 1 μm, but multiple scans may be necessary for acceptable signal-to-noise ratio.

9.5 REFLECTION

9.5.1 Theory

Reflectivity measurements are commonly made to determine layer thicknesses, both for insulating layers on semiconducting substrates and for epitaxial semiconductor films. The reflectance for the structure in Fig. 9.15(a), consisting of a non-absorbing layer of thickness t_1 on a non-absorbing substrate, is given by[56]

$$R = \frac{r_1^2 + r_2^2 + 2r_1 r_2 \cos(\phi_1)}{1 + r_1^2 r_2^2 + 2r_1 r_2 \cos(\phi_1)} \qquad (9.28)$$

where

$$r_1 = \frac{n_0 - n_1}{n_0 + n_1}, \quad r_2 = \frac{n_1 - n_2}{n_1 + n_2}$$

$$\phi_1 = \frac{4\pi n_1 t_1 \cos(\phi')}{\lambda}, \quad \phi' = \sin^{-1}\left[\frac{n_0 \sin(\phi)}{n_1}\right]$$

The reflectance exhibits maxima at the wavelengths

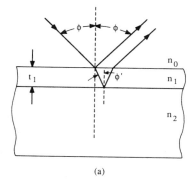

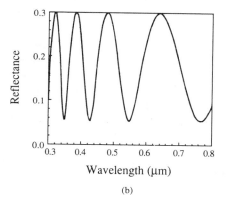

Fig. 9.15 (a) Reflection spectroscopy schematic, (b) calculated reflectance curve for SiO_2 on Si using $W_{ox} = 7500$ Å, $n_0 = 1$, $n_1 = 1.46$, and $n_2 = 3.42$.

$$\lambda(\text{max}) = \frac{2n_1 t_1 \cos(\phi')}{m} \tag{9.29}$$

where $m = 1, 2, 3, \ldots$. By a slight manipulation of Eq. (9.29), it can be shown that the layer thickness can be extracted by measuring the wavelengths at various maxima. The thickness is[57]

$$t_1 = \frac{i\lambda_0 \lambda_i}{2n_1(\lambda_i - \lambda_0)\cos(\phi')} \tag{9.30}$$

where i is the number of complete cycles from λ_0 to λ_i, the two wavelength peaks that bracket the i cycles. For two adjacent maxima $i = 1$, for a maximum and an adjacent minimum $i = \frac{1}{2}$, for two adjacent minima $i = 1$, etc. For example, in Fig. 9.15(b) for the first two peaks $i = 1$, $\lambda_0 = 0.32$ μm, and $\lambda_1 = 0.384$ μm, giving $t_1 = 0.75$ μm. A similar thickness is obtained by choosing any other two adjacent peaks or by using $i = 3$, $\lambda_0 = 0.32$ μm, and $\lambda_3 = 0.64$ μm. The quantity $2n_1 \cos(\phi')$ is a constant determined by the experimental arrangement and the film's index of refraction. This quantity is sometimes written in terms of the incidence angle ϕ as

$$2n_1 \cos(\phi') = 2\sqrt{n_1^2 - n_0^2 \sin^2(\phi)}$$

$n_0 = 1$ for air.

Instead of illuminating the sample with monochromatic light and changing the wavelength, it is possible to shine white light, containing many wavelengths, onto the sample and analyze the reflected light by passing it through a spectrometer. Small areas can be characterized by shining the light through a microscope. Once the various wavelengths have been dispersed by the spectrometer, they can be detected by a diode array, in which the various wavelengths fall on different diodes in the array, for automatic data acquisition.[58] Reflectance measurements are also used to determine the thickness of epitaxial semiconductor layers. But the technique only works if there is a substantial doping concentration change at the epitaxial-substrate interface because there must be a measurable index of refraction change at that interface.

Dielectric film thicknesses can also be measured using white light without a spectrometer. The white light is reflected both from a reference variable-thickness film and from the unknown sample onto a detector. The detector output is a maximum for $n_r t_r = n_x t_x$ where n_r, t_r are the refractive index and thickness of the reference and n_x, t_x are those of the unknown.[59] The variable-thickness reference can be a semicircular wedge of oxidized Si. For $n_r = n_x$ the maximum detector output corresponds to the unknown film thickness being equal to the reference film thickness.

An alternate method utilizes FTIR. As described in Section 9.4.2, a maximum in the interferogram is observed when both optical paths from the beam splitter to the mirrors are identical. For thickness measurements of a layer on a substrate, a secondary maximum is observed when the movable mirror has moved by a distance equal to the optical path through the layer. The thickness is determined from the location x of this secondary maximum relative to the centerburst on the interferogram by the relation[60]

$$t_1 = \frac{x}{2n_1 \cos(\phi)} \tag{9.31}$$

This relationship is not strictly correct because phase shifts in the reflected beam alter the shape and position of the side burst peaks. These phase shifts are not easy to include in the analysis because the detector sees a broad range of wavelengths and thus a broad range of phase shifts. In practice, an empirical relationship is established between side burst position and layer thickness.

9.5.2 Applications

Dielectrics   The reflectance method lends itself to dielectric film thickness measurements especially for thicker films, for SiO_2 films thicker than 2000 Å or so on Si. For thinner films ($t < 2000$ Å) it is difficult to find the first minimum unless extremely short wavelengths are used. For such films it is much easier to use ellipsometry. Instead of varying the wavelength, it is also possible to keep the wavelength constant and to vary the incident angle. The technique is then known as *variable-angle monochromator fringe observation* (VAMFO).[61-62]

When dielectric films on semiconductor substrates are viewed by eye or through a microscope, interference effects give the layer a characteristic color determined by the film thickness, its index of refraction, and the

TABLE 9.3 Colar Chart for Thermally Grown SiO$_2$ Films Observed Perpendicularly under Daylight Fluorescent Lighting

Film Thickness (Å)	Color	Film Thickness (Å)	Color
500	Tan	6300	Violet-red
700	Brown	6800	"Bluish" (Not blue but borderline
1000	Dark violet to red violet		between violet and blue-green. It
1200	Royal blue		appears more like a mixture
1500	Light blue to metallic blue		between violet-red and blue-green
1700	Metallic to very light yellow green		and looks greyish)
2000	Light gold or yellow	7200	Blue-green to green (quite broad)
	slightly metallic	7700	"Yellowish"
2200	Gold with slight yellow-orange	8000	Orange (rather broad for orange)
2500	Orange to mellon	8200	Salmon
2700	Red-violet	8500	Dull, light red-violet
3000	Blue to violet-blue	8600	Violet
3100	Blue	8700	Blue-violet
3200	Blue to blue-green	8900	Blue
3400	Light green	9200	Blue-green
3500	Green to yellow-green	9500	Dull yellow-green
3600	Yellow-green	9700	Yellow to "yellowish"
3700	Green-Yellow	9900	Orange
3900	Yellow	10,000	Carnation pink
4100	Light orange	10,200	Violet-red
4200	Carnation-pink	10,500	Red-violet
4400	Violet-red	10,600	Violet
4600	Red-violet	10,700	Blue-violet
4700	Violet	11,000	Green
4800	Blue-violet	11,100	Yellow-green
4900	Blue	11,200	Green
5000	Blue-green	11,800	Violet
5200	Green (broad)	11,900	Red-violet
5400	Yellow-green	12,100	Violet-red
5600	Green-yellow	12,400	Carnation pink to salmon
5700	Yellow to "yellowish" (not yellow	12,500	Orange
	but is in the position where	12,800	"Yellowish"
	yellow is to be expected. At	13,200	Sky blue to green-blue
	times it appears to be light	14,000	Orange
	creamy grey or metallic)	14,500	Violet
5800	Light orange or yellow to pink	14,600	Blue-violet
	borderline	15,000	Blue
6000	Carnation pink	15,400	Dull yellow-green

Source: Pliskin and Conrad.[61]

TABLE 9.4 Color Chart for Deposited Si$_3$N$_4$ Films Observed Perpendicularly under Daylight Fluorescent Lighting

Film Thickness (Å)	Color	Film Thickness (Å)	Color
100	Very light brown	950	Light blue
170	Medium brown	1050	Very light blue
250	Brown	1150	Light blue-brownish
340	Brown-pink	1250	Light brown-yellow
350	Pink-purple	1350	Very light yellow
430	Intense purple	1450	Light yellow
525	Intense dark blue	1550	Light to medium yellow
600	Dark blue	1650	Medium yellow
690	Medium blue	1750	Intense yellow

spectral distribution of the light source. For example, a SiO$_2$ film on Si has a different color when viewed under fluorescent light than under incandescent light. Using calibrated color charts, thicknesses can be judged accurate to 100 to 200 Å by the unaided eye. The eye is a very good discriminator against slight color changes. Such color charts are widely used to determine SiO$_2$ and Si$_3$N$_4$ films on Si substrates when viewed under fluorescent light. The method works best when calibrated samples, measured by ellipsometry, for example, are compared with the unknown sample.

A potential source of difficulty arises as the films becomes thicker than about 3000 Å for SiO$_2$ or about 2000 Å for Si$_3$N$_4$ because different orders have substantially the same color. A trained eye will be able to detect slight color changes for different orders. However, a more definite approach is to view the sample at an angle and compare it with the calibrated samples held at the same angle. The colors will not match unless they are both of the same order. A rough guide to the colors is given in Tables 9.3 and 9.4. Words, however, do not convey the colors precisely, and for more precise determination one should use calibrated samples.

The charts may also be used for films other than SiO$_2$ or Si$_3$N$_4$. In that case $t_x = t_0 n_0 / n_f$, where t_x is the unknown film thickness, t_0 the film thickness from color chart, n_0 the index of refraction of the original film (e.g., SiO$_2$), and n_x the index of refraction of the film to be measured.

Semiconductors Two types of semiconductor layers are of interest for thickness measurements: epitaxial layers and diffused or ion-implanted layers. Use of a spectrophotometer for reflectance measurements as given in Eqs. (9.29) and (9.30) poses several difficulties. There is only a small difference in the refractive index between the epitaxial layer and the substrate, leading to low-amplitude reflection from that interface. The refractive index difference increases at longer wavelengths, giving enhanced interference patterns at longer wavelengths. Typical wavelengths for epitaxial layer thickness measurements lie in the 5 to 50 μm range. The index of refraction difference also increases with substrate doping concentration increase. The ASTM recommendation calls for Si epitaxial layer resistivity $\rho_{epi} > 0.1$ Ω-cm and substrate resistivity $\rho_{substr} < 0.02$ Ω-cm.[63] An additional complication arises from the phase shift at the air–semiconductor interface being different from that at the epitaxial-substrate interface, leading to the modified thickness equation[57,63–64]

$$t_{epi} = \frac{(m - 1/2 + \theta_i/2\pi)\lambda_i}{2\sqrt{n_1^2 - \sin^2(\phi)}} \tag{9.32}$$

where m is the order of the maxima or minima in the spectrum, θ_i the phase shift at the epitaxial-substrate interface, and λ_i the wavelength of ith extrema in the spectrum. The $\frac{1}{2}$ comes from the phase-shift term. The phase shift at the epitaxial-substrate interface must be accurately known. Tabulated values for both n-Si and p-Si are given in Beadle et al.[60] For very thin layers or for layers on very thin buried structures, these phase-shift values are crucial.[65] The most common optical thickness characterization technique is the FTIR method of Eq. (9.31) with which Si epitaxial layer thicknesses from about 1 μm to about 100 μm can be determined.

9.5.3 Bevel-Stain and Other Methods

We discuss here, for completeness, several other thickness measuring methods that are based on optical interferometry and also non-optical methods. Junction depths are sometimes measured by the *bevel and stain* technique illustrated in Fig. 9.16(a). The sample is beveled and stained to delineate the diffused region. A partially reflecting reference plane over the beveled portion produces interference fringes when the sample is illuminated with monochromatic light.[66–67] The fringes represent lines of constant optical path difference, with adjacent fringes differing in optical path difference by one-half wavelength. The reference surface can be a small piece of cover glass carefully positioned on the sample, and sodium light, with a wavelength of 0.588 μm, is frequently used as the light source. With the reference plane in perfect coincidence with the original surface [Fig. 9.16(b)], the thickness is[57]

$$t = \frac{N\lambda}{2} \tag{9.33}$$

where N is the number of fringes from the edge of the bevel to the edge of the stained region. The depth determination is independent of the bevel angle since interference fringes occur every $\lambda/2$. Any metallization should be removed from the semiconductor, but dielectric layers like SiO_2 on the semiconductor are useful for acurate surface location.

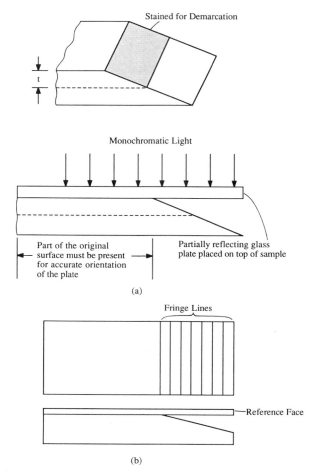

Fig. 9.16 (a) Bevel and stain geometry, (b) top view shows the interference fringes.

A variation of this method is the *groove and stain* technique illustrated in Fig. 9.17. The surface is grooved by a cylinder and stained [see Fig. 9.17(a)]. The junction depth is obtained from geometrical considerations as[68]

$$t = \sqrt{R^2 - \left(\frac{W_2}{2}\right)^2} - \sqrt{R^2 - \left(\frac{W_1}{2}\right)^2} \qquad (9.34)$$

where R is the radius of the grooving cylinder. Inaccuracies can occur when it is difficult to determine the edges of the groove. One way around that problem is to groove the sample by holding the grooving cylinder not parallel to the sample surface but at a small angle, as illustrated in Fig. 9.17(b). The thickness can be determined from Eq. (9.34) where now $W_2 = 0$ and $W_1 = W$, giving[69]

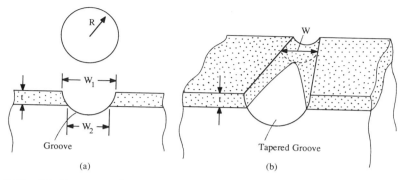

Fig. 9.17 (a) Groove and stain showing the grooved sample, (b) tapered groove obtained by grooving at an angle.

$$t = R - \sqrt{R^2 - \left(\frac{W}{2}\right)^2} \approx \frac{W^2}{8R} \qquad (9.35)$$

valid for $(W/2)^2 \ll R^2$. This method is accurate to within 100 to 200 Å for junction depths of 0.3 to 1.5 μm.[69]

Thickness measurements can also be made with a *profilometer* in which a mechanical stylus is drawn over the step. The stylus consists of a diamond attached directly to the core of a linear variable-differential transformer. The vertical motion of the stylus traversing the step produces a corresponding electrical output signal that is amplified, digitized, and displayed on a cathode ray tube. A step is easily produced for a dielectric layer by etching the layer using a suitable mask; it is more difficult to do for epitaxial or diffused layers because it is difficult to find an etch that stops at the epitaxial-substrate interface. Epitaxial layer thicknesses are sometimes determined by weighing the wafer before and after the epitaxial growth in the *weight differential* technique.[57] The wafer area and the semiconductor density must be known. The area is generally the wafer area, provided that growth proceeds only on one wafer side. *In situ* substrate etching prior to epitaxial deposition leads to errors by removing a portion of the wafer after it is weighed.

Stacking faults (SF), originating during epitaxial growth at the substrate/epitaxial interface, can be used to determine epitaxial film thicknesses. The method is based on the SF size related to the epitaxial layer thickness since the SF increases in size as the layer grows. On (111)-oriented substrates, defect-delineating etches, or Nomarski microscope images show triangular SF shapes with the epitaxial layer thickness given by $t = 0.816 \times$ length of the triangular SF side. For (100)-oriented substrates the relationship is $t = 0.707 \times$ length of the square SF side, the SF shape being a square.[70] Other orientations require other conversion factors.[71] This method works only if the SF's originate at the substrate/epitaxial layer interface. SF's

TABLE 9.5 Comparison of Junction Depth Measuring Methods

Method	Epitaxial Layers	Diffused or Implanted Layers	Repeatable Accuracy ±% Layer Thickness				Time/Test (min)	Operator Training Time
			0.5 μm	1–2 μm	2–5 μm	>5 μm		
Weight gain	Yes	No	a	a	a	a	10	1 day
Stacking fault	Yes	No	a	a	a	a	10	1 day
Angle lap/angle assumption	Yes	Yes	b	40	30	20	30–60	2 weeks
Angle lap/angle verification	Yes	Yes	40	3	1	1	30–60	2 weeks
Angle lap/interferometry	Yes	Yes	b	30	10	2	30–60	2 weeks
Groove/interferometry	Yes	Yes	b	30	10	2	5–10	1 week

[a] Generally not considered to be sufficiently reliable to be useful.
[b] Out of range.
Source: Alley and Turner.[72]

originating within the epitaxial layer during the growth give erroneous results.

A useful comparison of various aspects of a number of semiconductor thickness measurement techniques is given in Table 9.5. Other semiconductor layer thickness measurement techniques include *spreading resistance* and *secondary ion mass spectrometry*.

Angle lap/stain techniques require staining to delineate the thickness of the diffused or implanted layer. Various stains have been used. They generally stain the more heavily doped regions. The following stains work for silicon: *Silver stain* consists of 100 ml 49% HF, 2 ml HNO_3, and 1 ml $AgNO_3$ solution, composed of 2 g $AgNO_3$ in 100 ml H_2O.[64] The solution is applied under illumination, and as soon as bubbles begin to appear, the stain is quenched in water. *Copper stain* consisting of 8 g/liter $CuSO_4{:}5H_2O$ + 10 ml/liter 49% HF stains *n*-Si, provided that $N_D > 10^{14}$ cm^{-3}.[73] A recent study concluded that 2 g/liter $CuSO_4{:}5H_2O$ + 2.5 ml/liter 49% HF produced better results when stained for about 5 s.[74] Other stains consist of HF + HNO_3 or HF + H_3PO_4. Generally strong light is required during the staining operation. $KOH{:}K_3Fe(CN)_6{:}H_2O$ – 12 g:8 g:100 ml works for GaAs p^+n junctions. Staining is more an art than a science and frequently depends more on the operator than on the solution. Ambient light, temperature, volume of etchant, residual surface oxides, adsorbed impurities, and surface damage all play a role. Furthermore it is not clear whether the stain delineates the heavily doped region to the edge of the space-charge region or to the metallurgical junction. The space-charge region may also be distorted at the surface by the bevel. All of these effects introduce some uncertainties into regions delineated by staining techniques.

9.6   LINE WIDTH

Line widths and their measurement become increasingly important as device dimensions decrease. Line-width measurements are routinely used to evaluate key lithographic parameters and photomask quality, and a line-width measurement system should be able to measure the width of the line and be repeatable to less than the tolerance—typically 10%. For a line 1 μm wide, the measurement should be accurate to 0.1 μm. The measurement error should be 3 to 10 times smaller than the process error. Line widths are frequently called *critical dimensions* (CD), and their measurements are referred to as CD measurements. Line widths on photo masks are more easily measured than line widths on chips because line edges on a mask are well defined and there is good contrast from light transmitted through the mask. Measurements on processed wafers are more difficult due to the many factors that affect the properties of the reflected light. Most line-width measuring systems are capable of measuring line widths down to about

0.5 μm, but users place the limit closer to 1 μm. Optically based systems are limited by the resolution of the microscope optics as well as by the wavelength of light. Water-immersion optics allow somewhat lower critical dimensions to be measured. There are several terms related to line-width measurements. *Accuracy* is the deviation of a measured line width from the true line width; *short-term precision* is the distribution of errors due to the instrument in repeated measurements; *long-term stability* is the variation of the average measured line width over time.

9.6.1   Optical Methods

Most line-width measurement techniques are based on optical microscopy. They can be grouped into four general categories: video scan, slit scan, laser scan, and image shearing.[75] In the *video scanning* system the image of the line is viewed by a video camera through a microscope. The camera makes measurements on a video image of the feature taken through an optical microscope. This image consists of pixels (picture elements), with each pixel having a grey scale representative of the intensity of the light reflected from a point on the surface of the line. It is easy to manipulate digital data and to generate a waveform showing the intensity profile across the line. In the *scanning slit* method a 0.1-μm-wide scanning slit is mechanically stepped across the line. The reflected light is measured by a photomultiplier through a microscope. A line profile is generated in both methods by plotting reflected light intensity, represented by a voltage level, as a function of the lateral dimension. Most systems allow the user to select the voltage level from which the width is determined—the bottom, center, or top of the line. This of course assumes that the user knows at what level the most accurate measurement can be found, which is dependent on film thickness and type. Additional complications can arise from subsurface layer reflections.

In the *laser scan* method laser light is focused onto and scanned across the sample through a microscope and is sensed by detectors mounted beside the microscope objective. When the beam approaches the line edge, the light is scattered into the detector, measured, and digitized. In the *image shear* technique two identical images of a feature are superimposed, but the two outer sections of one image and the center section of the other image are suppressed. This creates one complete image. By shearing the center image through its full width, the line on one side of this image aligns with the line on the opposite side of the second image, resulting in a measurement of the structure.

The characteristics of the optical image formed in a microscope are determined by diffraction, aberrations in the optics, focus position, the spectral bandwidth of the illumination, and the coherence parameter defined as the ratio of the condenser to objective NA.[76] Even with a nearly vertical film edge, there is a gradual transition of the optical image from light to dark at the line edge due to diffraction. The width of the transition or blur

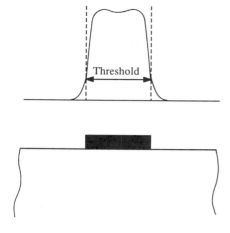

Fig. 9.18 The relationship between the line edges and the optical image showing the threshold corresponding to edge location.

region of the image is comparable to the diameter of the diffraction spot or Airy disc. To determine the line width optically to an accuracy better than this, one must use an optical threshold as illustrated in Fig. 9.18. The line itself is shown as ideal with vertical walls. This is rarely the case for real lines where the walls have varying degrees of slopes. In addition lines may not be entirely straight in the direction into the paper. Frequently lines have "corrugated" edges. For line-width measurement on wafers, one needs to correct for relative reflectance and phase at line edges because of partial coherence present in optical images.[77] All of these "practical" considerations make interpretation of line-width data difficult.

9.6.2 Electrical Methods

One electrical approach that has inherently higher resolution than optical systems is the scanning electron microscope (SEM).[78] It has not found wide acceptance due to its relatively low throughput. However, most SEM manufacturers offer software packages that allow intensity profiles to be generated to extract line widths. An entirely different approach is based on an electrical resistance measurement of a conductive line. This method is not suitable for insulator line-width measurements. The test pattern can be repeated over an entire test wafer or over only a portion of a product wafer, where the test structure may be located in the scribe lines or in "drop-in" test patterns. Maps and statistics can be easily generated with automated test equipment by providing an array of test structures across the entire wafer. Much like resistivity maps, line-width maps give an instant display of the process uniformity. Electrical line-width measurements achieve precisions of 0.005 μm, and line widths as narrow as 0.2 μm can be measured.

A test structure incorporating many of the requirements for line-width measurements is the split-cross-bridge resistor of Fig. 9.19.[79] The right

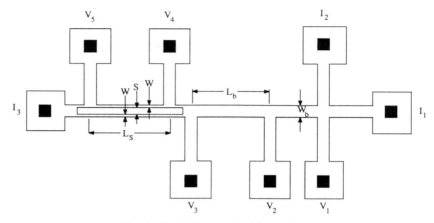

Fig. 9.19   Split-cross-bridge resistor.

portion is a cross resistor for van der Pauw sheet resistance measurements. The central portion is a bridge resistor, and the left portion is a split-bridge resistor. The cross resistor, discussed in detail in Section 1.2.2, gives the sheet resistance through the equation

$$\rho_s = \frac{\pi}{\ln(2)} \frac{V_{12}}{I_{12}} \tag{9.36}$$

where $V_{12} = V_1 - V_2$ and I_{12} is the current flowing into contact I_1 and out of contact I_2. The line width is determined from the bridge resistor by

$$W_b = \frac{\rho_s L_b I_{13}}{V_{23}} \tag{9.37}$$

where $V_{23} = V_2 - V_3$ and I_{13} is the current flowing into contact I_1 and out of contact I_3. The line width of the split-bridge resistor is

$$W_s = 2W = \frac{\rho_s L_s I_{13}}{V_{45}} \tag{9.38}$$

The line width $W_b = W_s + S$, where S is the line spacing, is given by

$$S = W_b - W_s = \rho_s \frac{(L_b V_{45} - L_s V_{23}) I_{13}}{V_{23} V_{45}} \tag{9.39}$$

The pitch $P_t = W + S$ can also be determined from this structure. A clever implementation of the line-width test structure allows oxide sidewall spacers to be determined.[80]

A test structure, not directly related to line width but nevertheless useful for determining mask superposition errors is the eight-contact structure in Fig. 9.20.[81] Two masking operations are used to delineate the pattern. The first mask level $M1$ creates the square body and sensor arms 2, 4, 6, and 8 as a window in an oxidized Si wafer, for example. The second mask level $M2$ is aligned to $M1$ in such a way to add sensor arms 1, 3, 5, and 7 at the midpoints of the square if $M1$ and $M2$ are correctly aligned. After both mask openings the actual device is made by ion implantation or diffusion into the oxide window. Subsequently contacts are made by a metallization process. The mask sequence might also be the following: $M1$ open oxide window and diffuse or implant; $M2$ deposit metal.

If mask level $M2$ has been incorrectly aligned with respect to $M1$, sensor arms 1 and 5 might be displaced a distance d_x from the midpoint of the side, and sensor arms 3 and 7 a distance d_y from its midpoint. The distances d_x and d_y are given by[81]

$$d_x = 0.316a \arcsin(\gamma_x) \quad \text{where} \quad \gamma_x = \frac{(V_{65}/I_{82}) - (V_{54}/I_{82})}{(V_{65}/I_{82}) + (V_{54}/I_{82})} \quad (9.40)$$

$$d_y = 0.316a \arcsin(\gamma_y) \quad \text{where} \quad \gamma_y = \frac{(V_{43}/I_{68}) - (V_{32}/I_{68})}{(V_{43}/I_{68}) + (V_{32}/I_{68})} \quad (9.41)$$

where a is the length of the side of square $M1$. Mask-to-mask registration errors of ± 0.1 μm have been detected with this method. The method lends itself not only to photolithographic registration errors due to mask misalignment but can also be used to measure process-induced wafer deformation.

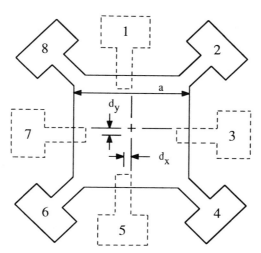

Fig. 9.20 Mask superposition error test structure.

9.7   PHOTOLUMINESCENCE

Photoluminescence (PL) provides a non-destructive technique for the determination of certain impurities in semiconductors.[82] It is particularly suited for the detection of shallow-level impurities but can also be applied to certain deep-level impurities, provided that radiative recombination events dominate nonradiative recombination.[83] We discuss PL only briefly by giving the main concepts and a few examples because PL is not routinely used for semiconductor characterization. *Identification* of impurities is easy with PL, but measurement of the *concentration* of impurities is more difficult. PL can provide simultaneous information on many types of impurities in a sample. However, only those impurities that produce radiative recombination processes can be detected. Fortunately many impurities fall within this category. A typical PL setup is illustrated in Fig. 9.21. The sample is placed in a cryostat and cooled to temperatures near liquid helium. Low-temperature measurements are necessary to obtain the fullest spectroscopic information by minimizing thermally-activated non-radiative recombination processes and thermal line broadening. The thermal distribution of carriers excited into a band contributes a width of approximately $kT/2$ to an emission line originating from that band. This makes it necessary to cool the sample to reduce the width. The thermal energy $kT/2$ is only 1.8 meV at $T = 4.2$ K. For many measurements this is sufficiently low, but occasionally it is necessary to reduce this broadening further by reducing the sample temperature below 4.2 K.

The sample is excited with an optical source, typically a laser with $hv > E_g$, generating electron–hole pairs (ehp's) that recombine by one of several mechanisms. Photons are emitted for *radiative* recombination. For *non-radiative* processes photons are not emitted. For good PL output one would like the majority of the recombination processes to be radiative. The photon energy depends on the recombination process illustrated in Fig. 9.22, where five of the most commonly observed PL transitions are shown.[84] Band-to-band recombination [Fig. 9.22(a)] dominates at room temperature but is rarely observed at low temperatures in materials with small effective masses due to the large electron orbital radii. Excitonic recombination is commonly observed, but what are excitons? When a photon generates an ehp, Coulombic attraction can lead to the formation of an excited state in which an electron and a hole remain bound to each other in a hydrogenlike state.[85] This excited state is referred to as a free *exciton* (FE). Its energy, shown in Fig. 9.22(b), is slightly less than the band-gap energy required to create a *separated* ehp. An exciton can move through the crystal, but being a *bound* ehp, both electron and hole move together and no photoconductivity results, for example. A free hole can combine with a neutral donor [Fig. 9.22(c)] to form a positively charged excitonic ion or *bound exciton* (BE).[86] The electron bound to the donor travels in a wide orbit about the donor. Similarly electrons combining with neutral acceptors also form bound excitons.

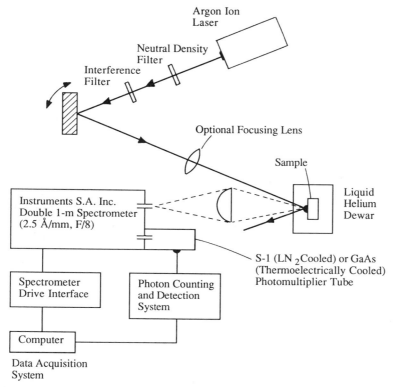

Fig. 9.21 Apparatus for photoluminescence measurements. Reprinted with permission after Stillman et al.[94] This paper was originally presented at the Fall 1988 Meeting of the Electrochemical Society, Inc. held in Chicago, Illinois.

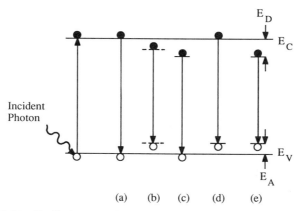

Fig. 9.22 Radiative transitions observed with photoluminescence.

If the material is sufficiently pure, free excitons form and recombine by emitting photons. The photon energy in direct band-gap semiconductors is[86]

$$hv = E_g - E_x \qquad (9.42a)$$

where E_x is the excitonic energy. In indirect band-gap semiconductors, momentum conservation requires the emission of a phonon, giving[86]

$$hv = E_g - E_x - E_p \qquad (9.42b)$$

where E_p is the phonon energy. Bound exciton recombination dominates over free exciton recombination for less pure material. A free electron can recombine with a hole on a neutral acceptor [Fig. 9.22(d)], and similarly a free hole can recombine with an electron on a neutral donor [Fig. 9.22(c)].

Lastly, an electron on a neutral donor can recombine with a hole on a neutral acceptor, the well-known donor-acceptor (D-A) recombination, illustrated in Fig. 9.22(e). The emission line has an energy modified by the Coulombic interaction between donors and acceptors[83]

$$hv = E_g - (E_A + E_D) + \frac{q^2}{K_s \varepsilon_0 r} \qquad (9.43)$$

where r is the distance between donor and acceptor. The photon energy in Eq. (9.43) can be larger than the band gap for small $E_A + E_D$. Such photons are generally re-absorbed in the sample. The full width at half maximum (FWHM) for bound exciton transitions are typically $\leq kT/2$ and resemble slightly broadened delta functions. This distinguishes them from donor-valence band transitions which are usually a few kT wide. Energies for these two transitions are frequently similar, and the line widths are used to determine the transition type. A PL spectrum showing BE-D, BE-A, and D-A recombination in InP is shown in Fig. 9.23.[87]

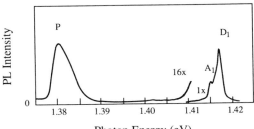

Photon Energy (eV)

Fig. 9.23 Low-temperature PL spectrum of InP. D_1 and A_1 are broadened donor and acceptor bound exciton lines, and P is a $D-A$ line due to carbon acceptors. Reprinted with permission after Dean et al.[87]

The optics in a PL apparatus are designed to ensure maximum light collection. The PL-emitted light from the sample is analyzed by a spectrometer and detected by a photodetector. PL radiation from shallow-level impurities in Si and GaAs can be detected with a photomultiplier tube. An S-1 photocathode is able to detect wavelengths from about 0.4 to 1.1 μm. Lower-energy light from deeper levels requires a PbS (1–3 μm) or doped germanium detector. Spectra may be recorded directly on an X-Y recorder, or they may be recorded digitally by digitizing the output from the photodetector. Sometimes the detector output signal is fed to a lock-in amplifier and synchronized to a chopper in the laser path for enhanced signal-to-noise ratio.

The volume analyzed in PL measurements is determined by the absorption depth of the exciting laser light and the diffusion length of the minority carriers. It is generally difficult to correlate the intensity of a given PL spectral line with the concentration of the impurity giving rise to that line. This is due to non-radiative bulk and surface recombination that can vary significantly from sample to sample and from location to location on a given sample. A novel approach to this problem is due to Tajima.[88–89] For Si samples of different resistivity, he found spectra with both intrinsic and extrinsic peaks as shown in Fig. 9.24. Higher resistivity samples showed higher intrinsic peaks. The ratio $X_{TO}(BE)/I_{TO}(FE)$ is proportional to the doping concentration, where $X_{TO}(BE)$ is the transverse optical phonon PL

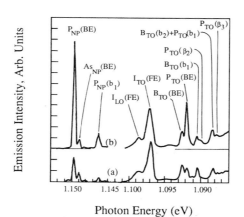

Photon Energy (eV)

Fig. 9.24 Photoluminescence spectra for Si at $T = 4.2$ K: (a) starting material, (b) after neutron transmutation doping. Base lines for measuring the peak heights are shown by the horizontal lines. Symbols: I = intrinsic, TO = transverse optical phonon, LO = longitudinal optical phonon, BE = bound exciton, FE = free exciton. The sample contains a residual amount of arsenic. Components labeled b_n and β_n are due to recombination of multiple bound excitons. Reprinted with permission after Tajima et al.[89] This paper was originally presented at the Spring 1981 Meeting of the Electrochemical Society, Inc. held in Minneapolis, MN.

intensity peak of the bound exciton for element X (boron or phosphorus) and I_{TO}(FE) is the transverse optical phonon intrinsic PL intensity peak of the free exciton.

Calibration curves of photoluminescence intensity ratio versus impurity concentration for Si are shown in Fig. 2.23(a). Quite good agreement is found between the resistivity measured electrically and the resistivity calculated from the carrier concentration measured by photoluminescence in Fig. 2.23(b). Very pure float-zone Si was used, and varying amounts of phosphorus were introduced, using neutron transmutation doping to generate calibration curves for the PL data.[90] It is estimated that for samples with areas of 0.3 cm^2 and 300 μm thickness, the detection limits for P, B, Al and As in Si are around 5×10^{10}, 5×10^{11}, 2×10^{11}, and 5×10^{11} cm^{-3}, respectively. Various impurities in Si have been catalogued.[91] The interpretation has also been applied to InP, where the donor concentration as well as the compensation ratio was determined.[92]

The ionization energies of donors in GaAs are typically around 6 meV, and the energy difference between the various donor impurities is too small to be observable by conventional PL. However, acceptors with their wider spread of ionization energies can be detected by using the free electron to neutral acceptor [Fig. 9.22(d)] and the electron on a neutral donor to hole on a neutral acceptor [Fig. 9.22(e)] transitions. Acceptors in GaAs determined with PL have also been catalogued.[93] Complications arise when the energy difference between the ground states of two or more acceptors is identical to the difference between their band-acceptor and donor-acceptor pair transitions. When this occurs, transitions can often be differentiated through variable excitation power measurements that cause a shift of the donor-acceptor pair transition to higher energies or through variable temperature measurements.[94] Donors in GaAs can be detected by *magneto-photoluminescence* measurements. The application of a magnetic field splits some of the spectral lines into several components due to magnetic field splitting of the bound exciton initial states.[95] Another technique allowing donors to be identified is the *photothermal ionization spectroscopy* method discussed in Section 2.7.2. Quantitative correlation of PL data to impurity concentrations can of course also be made by Hall measurements.

9.8 RAMAN SPECTROSCOPY

Raman spectroscopy is a vibrational spectroscopic technique that can detect both organic and inorganic species and measure the crystallinity of solids. It is free from charging effects that can influence electron and ion beam techniques and is most useful for the characterization of polymers, organic contaminants, corrosion, and metal oxides. It is not commonly used in semiconductor characterization. We mention it here briefly because it is likely to find increased use in the future. When light is scattered from the surface of a sample, the scattered light is found to contain mainly wavelengths that were incident on the sample (*Raleigh* scattering) but also different wavelengths at very low intensities (few parts per million) that represent an interaction of the incident light with the material. The interaction of the incident light with optical phonons is called *Raman scattering*, and the interaction with acoustic phonons results in *Brillouin* scattering. Optical phonons have higher energies than acoustic phonons giving larger photon energy shifts. Hence Raman scattering is easier to detect than Brillouin scattering. Since the intensity of Raman scattered light is very weak, Raman spectroscopy is only practical when an intense monochromatic light source like a laser is available.

Raman spectroscopy is based on the Raman effect, first reported by Raman in 1928.[96] If the incident photon gives part of its energy to the lattice in the form of a phonon (phonon emission) it emerges as a lower-energy photon. This down-converted frequency shift is known as *Stokes-shifted* scattering. *Anti-Stokes-shifted* scattering results when the photon absorbs a phonon and emerges with higher energy.[86] The anti-Stokes mode is usually much weaker than the Stokes mode, and it is Stokes-mode scattering that is usually monitored.

During Raman spectroscopy measurements a laser beam, referred to as the pump, is incident on the sample. The scattered light or signal is passed through a double monochromator, and the Raman-shifted wavelengths are detected by a photodetector. In the Raman microprobe, illustrated in Fig. 9.25, a laser illuminates the sample through a commercial microscope. Laser

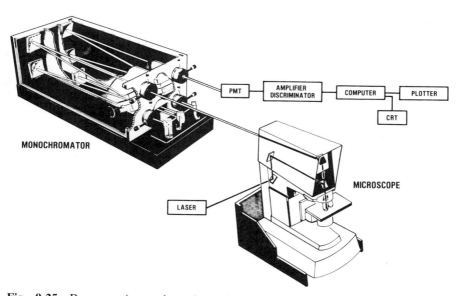

Fig. 9.25 Raman microprobe schematic. Reprinted with permission from Microelectronics Processing: Inorganic Materials Characterization Fig. 1, p. 232. Copyright 1986 American Chemical Society, after Adar.[103]

power is usually held below 5 mW to reduce sample heating and specimen decomposition.[97] Frequency-shifted photons scattered by the sample in an area as small as $1 \mu m \times 1 \mu m$ are directed into the monochromator and detected by the photodetector.[98] In order to separate the signal from the pump, it is necessary that the pump be a bright, monochromatic source. Detection is made difficult by the weak signal against an intense background of scattered pump radiation. The signal-to-noise ratio is enhanced if the Raman radiation can be observed at right angles to the pump beam. A major limitation in Raman spectroscopy is the interference caused by fluorescence, either of impurities or the sample itself. The fluorescent background problem is eliminated by combining Raman spectroscopy with FTIR dramatically demonstrated with the spectra in Fig. 9.26.[99]

By using lasers with different wavelengths and different absorption depths, it is possible to profile the sample to some depth. The technique is non-destructive and requires no contacts to the sample. Most semiconductors can be characterized by Raman spectroscopy. The wavelengths of the scattered light are analyzed and matched to known wavelengths for identification.

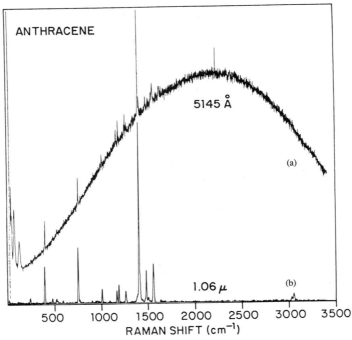

Fig. 9.26 (a) Conventional Raman spectrum of anthracene, (b) FTIR-Raman spectrum of anthracene. Reprinted with permission from *J. Am. Chem. Soc.* **108**, 7485 (1986). Copyright 1986 American Chemical Society.

Various properties of the sample can be characterized. Its composition can be determined. Raman spectroscopy is also sensitive to crystal structure. For example, different crystal orientations give slightly different Raman shifts. Scattering due to transverse optical (TO) phonons is forbidden in (100)-oriented GaAs. However, damage and structural imperfections induce scattering by the forbidden TO phonons allowing implant damage to be monitored, for example. The Stokes line shifts, broadens, and becomes asymmetric for microcrystalline Si with grain sizes below 100 Å.[100] The lines become very broad for amorphous semiconductors, allowing a distinction to be made between single-crystal, polycrystalline, and amorphous materials. The frequency is also shifted by stress and strain in thin film. For example, stresses in narrow Si gate MOSFET's were detected by Raman spectroscopy.[101] A summary of semiconductor applications including structural defects, ion damage, laser annealing, alloy fluctuations, interfaces, and heterojunctions is given by Pollack and Tsu.[102] The Raman microprobe is able to identify organic contaminants that appear as particles as small as 2 μm or as films as thin as 1 μm. The technique is most successful for organic materials because organic spectral data bases exist. For example, silicone films, teflon, cellulose, and other contaminants have been detected.[103] A good discussion of the effectiveness of Raman spectroscopy, especially when coupled with other characterization techniques, to problem solving in semiconductor processing is given by Ramsey[104] and Nakashima and Hangyo.[105] A thorough survey of the Raman spectroscopy literature between 1985 and 1987 is given by Gerrard and Bowley.[106]

APPENDIX 9.1   TRANSMISSION EQUATIONS

Consider the sample of Fig. A9.1, characterized by reflection coefficients R_1, R_2, absorption coefficient α, complex refractive index $(n_1 - jk_1)$, and thickness d. Light of intensity I_i and wavelength λ is incident from the left. $I_{r1} = R_1 I_i$ is reflected at point A and $(1 - R_1)I_i$ is transmitted into the sample where it is attenuated as it traverses the sample. At point B, just inside the sample at $x = d$, the intensity is $(1 - R_1)\exp(-\alpha d)I_i$. The fraction $R_2(1 - R_1)\exp(-\alpha d)I_i$ is reflected back into the sample at point B, and the fraction $I_{t1} = (1 - R_2)(1 - R_1)\exp(-\alpha d)I_i$ is transmitted through the sample. Some of the light reflected at B is reflected back into the sample at C and the component I_{r2} is back-reflected. As shown on the figure, light is reflected back and forth, and each time some of it is reflected to the left and some is transmitted to the right. When all the components are summed, it can be shown that the transmittance T is given by[56]

$$T = \frac{I_t}{I_i} = \frac{(1 - R_1)(1 - R_2)e^{-\alpha d}}{1 + R_1 R_2 e^{-2\alpha d} - 2\sqrt{R_1 R_2}e^{-\alpha d}\cos(\phi)} \tag{A9.1}$$

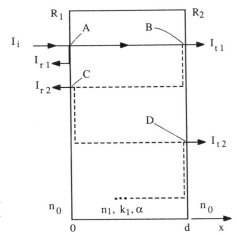

Fig. A9.1 Schematic showing the various reflected and transmitted components.

where $\phi = 4\pi n_1 d/\lambda$. For symmetrical samples, $R_1 = R_2 = R$, allowing Eq. (A9.1) to become

$$T = \frac{(1 - R)^2 e^{-\alpha d}}{1 + R^2 e^{-2\alpha d} - 2Re^{-\alpha d}\cos(\phi)} \tag{A9.2}$$

The "cos" term can be written as $\cos(f/f_1)$, where $f = 2\pi/\lambda$ and $f_1 = 1/(2n_1 d)$ is a spatial frequency. If the detector does not have sufficient spectral resolution, then the oscillations due to $\cos(\phi)$ average to zero. The effect of the reduced resolution can be calculated by averaging the transmitted intensity over a period of the cosine term as[107]

$$T = \frac{1}{2\pi} \int_{-\pi}^{\pi} \frac{(1 - R)^2 e^{-\alpha d}}{1 + R^2 e^{-2\alpha d} - 2Re^{-\alpha d}\cos(\phi)} \, d\phi \tag{A9.3}$$

Assuming α and n_1 to be constant over the wavelength interval, the transmittance becomes

$$T = \frac{(1 - R)^2 e^{-\alpha d}}{1 - R^2 e^{-2\alpha d}} \tag{A9.4}$$

where R is the reflectance given by

$$R = \frac{(n_0 - n_1)^2 + k_1^2}{(n_0 + n_1)^2 + k_1^2} \tag{A9.5}$$

The absorption coefficient α is related to the extinction coefficient k_1 by

$$\alpha = \frac{4\pi k_1}{\lambda} \tag{A9.6}$$

$\cos(\phi)$ has maxima when $m\lambda_0 = 2n_1 d$, where $m = 1, 2, 3, \ldots$, and can be used to determine the sample thickness through the relationship

$$d = \frac{m\lambda_0}{2n_1} = \frac{(m+1)\lambda_1}{2n_1} = \frac{(m+i)\lambda_i}{2n_1} \qquad \text{(A9.7)}$$

or $m = i\lambda_i / [\lambda_0 - \lambda_i]$, where i is the number of complete cycles from λ_0 to λ_i. For one cycle $i = 1$, and

$$d = \frac{1}{2n_1(1/\lambda_1 - 1/\lambda_0)} = \frac{1}{2n_1\Delta(1/\lambda)} \qquad \text{(A9.8)}$$

where $1/\lambda$ is the wave number and $\Delta(1/\lambda)$ is the wave number interval between two maxima or minima of the oscillatory transmittance curve, indicated on Fig. 9.11(b).

APPENDIX 9.2   ABSORPTION COEFFICIENTS AND REFRACTIVE INDICES FOR SELECTED SEMICONDUCTORS

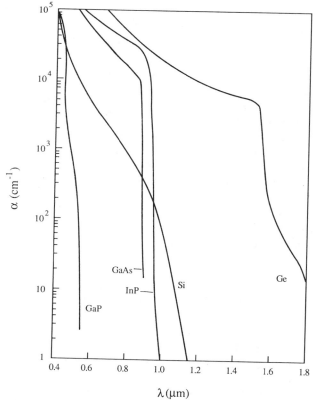

Fig. A9.2   Absorption coefficient as a function of wavelength for selected semiconductors. Adapted from data in Palik.[108]

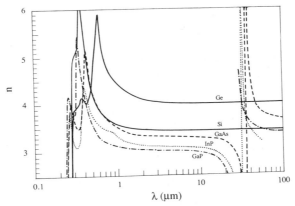

Fig. A9.3   Refractive index as a function of wavelength for selected semiconductors. Adapted from data in Palik.[108]

REFERENCES

[1] W. C. McCrone, R. G. Draftz, and J. G. Delly (eds.), *The Particle Atlas*, Ann Arbor Science Publ., Naples, FL, 1967.

[2] J. K. Beddow (ed.), *Particle Characterization in Technology*, Vol. 1 and 2, CRC Press, Boca Raton, FL, 1984.

[3] M. Spencer, *Fundamentals of Light Microscopy*, Cambridge University Press, Cambridge, 1982.

[4] T. G. Rochow and E. G. Rochow, *An Introduction to Microscopy by Means of Light, Electrons, X-Rays, or Ultrasound*, Plenum, New York, 1978.

[5] G. Nomarski, French Patents Nos. 1059124 and 1056361.

[6] D. C. Miller and G. A. Rozgonyi, "Defect Characterization by Etching, Optical Microscopy and X-Ray Topography," in *Handbook on Semiconductors*, Vol. 3 (S. P. Keller, ed.), North-Holland, Amsterdam, 1980, pp. 217–246.

[7] J. Hilton, *All You Ever Wanted to Know about Microscopy and Were Afraid to Ask*, Altitude Publ. Co., Littleton, CO, 1984.

[8] E. Sirtl and A. Adler, "Chromic Acid-Hydrofluoric Acid as Specific Reagents for the Development of Etching Pits in Silicon," *Z. Metallkd.* **52**, 529–534, Aug. 1961.

[9] W. C. Dash, "Copper Precipitation on Dislocations in Silicon," *J. Appl. Phys.* **27**, 1193–1195, Oct. 1956; "Evidence of Dislocation Jogs in Deformed Silicon," *J. Appl. Phys.* **29**, 705–709, April 1958.

[10] F. Secco d'Aragona, "Dislocation Etch for (100) Planes in Silicon," *J. Electrochem. Soc.* **119**, 948–951, July 1972.

[11] D. G. Schimmel, "Defect Etch for ⟨100⟩ Silicon Ingot Evaluation," *J. Electrochem. Soc.* **126**, 479–483, March 1979; D. G. Schimmel and M. J. Elkind, "An Examination of the Chemical Staining of Silicon," *J. Electrochem. Soc.* **125**, 152–155, Jan. 1978.

[12] M. W. Jenkins, "A New Preferential Etch for Defects in Silicon Crystals," *J. Electrochem. Soc.* **124**, 757–762, May 1977.

[13] K. H. Yang, "An Etch for Delineation of Defects in Silicon," *J. Electrochem. Soc.* **131**, 1140–1145, May 1984.

[14] H. Seiter, "Integrational Etching Methods," in *Semiconductor Silicon/1977* (H. R. Huff and E. Sirtl, eds.), Electrochem. Soc., Princeton, NJ, 1977, pp. 187–195.

[15] M. Ishii, R. Hirano, H. Kan, and A. Ito, "Etch Pit Observation of Very Thin (001)-GaAs Layer by Molten KOH," *Japan. J. Appl. Phys.* **15**, 645–650, April 1976; for a more detailed discussion of GaAs etching, see D. J. Stirland and B. W. Straughan, "A Review of Etching and Defect Characterisation of Gallium Arsenide Substrate Material," *Thin Solid Films* **31**, 139–170, Jan. 1976.

[16] D. T. C. Huo, J. D. Wynn, M. Y. Fan, and D. P. Witt, "InP Etch Pit Morphologies Revealed by Novel HCl-Based Etchants," *J. Electrochem. Soc.* **136**, 1804–1806, June 1989.

[17] ASTM Standard F47, "Standard Test Method for Crystallographic Perfection of Silicon by Preferential Etch Techniques," *1988 Annual Book of ASTM Standards*, Am. Soc. Test. Mat., Philadelphia, 1988.

[18] ASTM Standard F26, "Standard Test Method for Detection of Oxidation Induced Defects in Polished Silicon Wafers," *1988 Annual Book of ASTM Standards*, Am. Soc. Test. Mat., Philadelphia, 1988.

[19] ASTM Standard F154, "Standard Practices and Nomenclature for Identification of Structures and Contaminants Seen on Specular Silicon Surfaces," *1988 Annual Book of ASTM Standards*, Am. Soc. Test. Mat., Philadelphia, 1988.

[20] R. M. A. Azzam and N. M. Bashara, *Ellipsometry and Polarized Light*, North-Holland, Amsterdam, 1987.

[21] G. Gergely, ed., *Ellipsometric Tables of the Si-SiO$_2$ System for Mercury and He-Ne Laser Spectral Lines*, Akadémiai Kiadó, Budapest, 1971.

[22] R. H. Muller, "Principles of Ellipsometry," in *Adv. in Electrochem. and Electrochem. Eng.* (R. H. Muller, ed.), **9**, Wiley, New York, 1973, pp. 167–226.

[23] Rudolph Research Auto ER-III Instruction Manual, Rudolph Research, 1982.

[24] D. E. Aspnes and A. A. Studna, "High Precision Scanning Ellipsometer," *Appl. Opt.* **14**, 220–228, Jan. 1975.

[25] K. Riedling, *Ellipsometry for Industrial Applications*, Springer, Vienna, 1988.

[26] A. A. Immorlica, Jr., "Effect of Implant and Anneal Parameters on Boron-Implanted Si as Determined from Ellipsometric Etching Techniques," *Solid-State Electron.* **25**, 1141–1145, Nov. 1982.

[27] A. Moritani and C. Hamaguchi, "High-Speed Ellipsometry of Arsenic-Implanted Si during CW Laser Annealing," *Appl. Phys. Lett.* **46**, 746–748, April 1985.

[28] M. Erman and J. B. Theeten, "Multilayer Analysis of Ion Implanted GaAs Using Spectroscopic Ellipsometry," *Surf. and Interf. Analys.* **4**, 98–108, June 1982.

[29] R. Fremunt, Y. Hirayama, F. Arai, and T. Sugano, "Characterization of Laser Annealed InP with Ellipsometry and Hall Effect Measurements," *Appl. Phys. Lett.* **44**, 530–532, March 1984.

[30] F. Hottier, J. Hallais, and F. Simondet, "*In Situ* Monitoring by Ellipsometry of Metalorganic Epitaxy of GaAlAs–GaAs Superlattice," *J. Appl. Phys.* **51**, 1599–1602, March 1980.

[31] D. E. Aspnes, "The Characterization of Materials by Spectroscopic Ellipsometry," *Proc. SPIE* **452**, 60–70, 1983.

[32] D. E. Aspnes and A. A. Studna, "Optical Detection and Minimization of Surface Overlayers on Semiconductors Using Spectroscopic Ellipsometry," *Proc. SPIE* **276**, 227–232, 1981.

[33] J. L. Freeouf, "Application of Spectroscopic Ellipsometry to Complex Samples," *Appl. Phys. Lett.* **53**, 2426–2428, Dec. 1988.

[34] ASTM Standard F120, "Standard Practices for Determination of the Concentration of Impurities in Single Crystal Semiconductor Materials by Infrared Absorption Spectroscopy," *1988 Annual Book of ASTM Standards*, Am. Soc. Test. Mat., Philadelphia, 1988.

[35] P. Stallhofer and D. Huber, "Oxygen and Carbon Measurements on Silicon Slices by the IR Method," *Solid State Technol.* **26**, 233–237, Aug. 1983.

[36] H. J. Rath, P. Stallhofer, D. Huber, and B. F. Schmitt, "Determination of Oxygen in Silicon by Photon Activation Analysis for Calibration of the Infrared Absorption," *J. Electrochem. Soc.* **131**, 1920–1923, Aug. 1984.

[37] K. L. Chiang, C. J. Dell'Oca, and F. N. Schwettmann, "Optical Evaluation of Polycrystalline Silicon Surface Roughness," *J. Electrochem. Soc.* **126**, 2267–2269, Dec. 1979.

[38] A. A. Michelson, "Visibility of Interference Fringes in the Focus of a Telescope," *Phil Mag.* **31**, 256–259, 1891; "On the Application of Interference Methods to Spectroscopic Measurements," *ibid.* **31**, 338–346, 1891; *ibid.* **34**, 280–299, 1892.

[39] Lord Raleigh, "On the Interference Bands of Approximately Homogeneous Light; in a Letter to Prof. A. Michelson," *Phil Mag.* **34**, 407–411, 1892.

[40] J. W. Cooley and J. W. Tukey, "An Algorithm for the Machine Calculation of Complex Fourier Series," *Math. Comput.* **19**, 297–301, April 1965.

[41] R. J. Bell, *Introductory Fourier Transform Spectroscopy*, Academic Press, New York, 1972.

[42] G. Horlick, "Introduction to Fourier Transform Spectroscopy," *Appl. Spectrosc.* **22**, 617–626, Nov./Dec. 1968.

[43] W. D. Perkins, "Fourier Transform-Infrared Spectroscopy," *J. Chem. Educ.* **63**, A5–A10, Jan. 1986.

[44] ASTM Standard F121, "Standard Test Method for Interstitial Atomic Oxygen Content of Silicon by Infrared Absorption," *1988 Annual Book of ASTM Standards*, Am. Soc. Test. Mat., Philadelphia, 1988.

[45] ASTM Standard F123, "Standard Test Method for Substitutional Atomic Carbon Content of Silicon by Infrared Absorption," *1988 Annual Book of ASTM Standards*, Am. Soc. Test. Mat., Philadelphia, 1988.

[46] W. M. Bullis, M. Watanabe, A. Baghdadi, Y.-Z. Li, R. I. Scace, R. W. Series, and P. Stallhofer, "Calibration of Infrared Absorption Measurements of Interstitial Oxygen Concentration in Silicon," in *Semiconductor Silicon/1986* (H. R. Huff, T. Abe, and B. O. Kolbesen, eds.), Electrochem. Soc., Pennington, NJ, 1986, pp. 166–180.

[47] T. Iizuka, S. Takasu, M. Tajima, T. Arai, N. Inoue, and M. Watanabe, "Determination of Conversion Factor for Infrared Measurement of Oxygen in Silicon," *J. Electrochem. Soc.* **132**, 1707–1713, July 1985.

[48] K. Graff, E. Grallath, S. Ades, G. Goldach, and G. Tolg, "Determination of Parts Per Billion of Oxygen in Silicon by Measurement of the IR-Absorption of 77 K," *Solid-State Electron,* **16**, 887–893, Aug. 1973.

[49] Deutsche Normen DIN 50 438/1, "Bestimmung des Verunreinigungsgehaltes in Silicium Mittels IR Absorption: O_2 in Si," Beuth Verlag, Berlin, 1978.

[50] A. Baghdadi, W. M. Bullis, M. C. Croarkin, Y.-Z. Li, R. I. Scace, R. W. Series, P. Stallhofer, and M. Watanabe, "Interlaboratory Determination of the Calibration Factor for the Measurement of the Interstitial Oxygen Content of Silicon by Infrared Absorption," *J. Electrochem. Soc.* **136**, 2015–2024, July 1989.

[51] J. L. Regolini, J. P. Stoquert, C. Ganter, and P. Siffert, "Determination of the Conversion Factor for Infrared Measurements of Carbon in Silicon," *J. Electrochem. Soc.* **133**, 2165–2168, Oct. 1986.

[52] G. M. Martin, "Optical Assessment of the Main Electron Trap in Bulk Semi-insulating GaAs," *Appl. Phys. Lett.* **39**, 747–748, Nov. 1981.

[53] J. Weber and M. Singh, "New Method to Determine the Carbon Concentration in Silicon," *Appl. Phys. Lett.* **49**, 1617–1619, Dec. 1986.

[54] G. G. MacFarlane, T. P. McClean, J. E. Quarrington, and V. Roberts, "Fine Structure in the Absorption-Edge Spectrum of Si," *Phys. Rev.* **111**, 1245–1254, Sept. 1958.

[55] W. Kern and G. L. Schnable, "Chemically Vapor-Deposited Borophosphosilicate Glasses for Silicon Device Applications," *RCA Rev.* **43**, 423–457, Sept. 1982.

[56] H. Anders, *Thin Films in Optics*, The Focal Press, London, 1967, ch. 1.

[57] W. R. Runyan, *Semiconductor Measurements and Instrumentation*, McGraw-Hill, New York, 1975.

[58] P. Burggraaf, "How Thick Are Your Thin Films?" *Semicond. Int.* **11**, 96–103, Sept. 1988.

[59] J. R. Sandercock, "Film Thickness Monitor Based on White Light Interference," *J. Phys. E: Sci. Instrum.* **16**, 866–870, Sept. 1983.

[60] W. E. Beadle, J. C. C. Tsai, and R. D. Plummer, *Quick Reference Manual for Silicon Integrated Circuit Technology*, Wiley-Interscience, New York, 1985, pp. 4–23.

[61] W. A. Pliskin and E. E. Conrad, "Nondestructive Determination of Thickness and Refractive Index of Transparent Films," *IBM J. Res. Develop.* **8**, 43–51, Jan. 1964.

[62] W. A. Pliskin and R. P. Resch, "Refractive Index of SiO_2 Films Grown on Silicon," *J. Appl. Phys.* **36**, 2011–2013, June 1965.

[63] ASTM Standard F95-88a, "Standard Test Method for Thickness of Lightly Doped Silicon Epitaxial Layers on Heavily Doped Silicon Substrates by an Infrared Dispersive Spectrophotometer," *1988 Annual Book of ASTM Standards*, Am. Soc. Test. Mat., Philadelphia, 1988.

[64] P. A. Schumann, Jr., "The Infrared Interference Method of Measuring Epitaxial Layer Thickness," *J. Electrochem. Soc.* **116**, 409–413, March 1969.

[65] B. Senitsky and S. P. Weeks, "Infrared Reflectance Spectra of Thin Epitaxial Silicon Layers," *J. Appl. Phys.* **52**, 5308–5313, Aug. 1981.

[66] W. L. Bond and F. M. Smits, "The Use of an Interference Microscope for Measurement of Extremely Thin Surface Layers," *Bell System Tech. J.* **35**, 1209–1221, Sept. 1956.

[67] ASTM Standard F110, "Standard Test Method for Thickness of Epitaxial or Diffused Layers in Silicon by the Angle Lapping and Staining Technique," *1988 Annual Book of ASTM Standards*, Am. Soc. Test. Mat., Philadelphia, 1988.

[68] B. McDonald and A. Goetzberger, "Measurement of the Depth of Diffused Layers on Si by the Grooving Method," *J. Electrochem. Soc.* **109**, 141–144, Feb. 1962.

[69] S. Prussin, "Junction Depth Measurement for VLSI Structures," *J. Electrochem. Soc.* **130**, 184–187, Jan. 1983.

[70] W. C. Dash, "A Method for Measuring the Thickness of Epitaxial Silicon Films," *J. Appl. Phys.* **33**, 2395–2396, July 1962.

[71] S. Mendelson, "Stacking Fault Nucleation in Epitaxial Silicon on Variously Oriented Silicon Substrates," *J. Appl. Phys.* **35**, 1570–1581, May 1964.

[72] C. L. Alley and R. S. Turner, "Junction Depth Measuring Methods: The Pros and Cons," *Semicond. Int.* **3**, 25–31, May 1980.

[73] C. P. Wu, E. C. Douglas, C. W. Mueller, and R. Williams, "Techniques for Lapping and Staining Ion-Implanted Layers," *J. Electrochem. Soc.* **126**, 1982–1988, Nov. 1979.

[74] S. T. Ahn and W. A. Tiller, "A Staining Technique for the Study of Two-Dimensional Dopant Diffusion in Silicon," *J. Electrochem. Soc.* **135**, 2370–2373, Sept. 1988.

[75] P. H. Singer, "Linewidth Measurement Aids Process Control," *Semicond. Int.* **8**, 66–73, Feb. 1985.

[76] D. Nyyssonen, "Linewidth Measurement Spotlight," *Semicond. Int.* **3**, 39–56, March 1980.

[77] D. Nyyssonen, "Calibration of Optical Systems for Linewidth Measurements on Wafers," *Opt. Eng.* **21**, 882–887, Sept./Oct. 1982.

[78] W. H. Arnold, B. Singh, and K. Phan, "Linewidth Metrology Requirements for Submicron Lithography," *Solid State Technol.* **32**, 139–145, April 1989.

[79] M. G. Buehler and C. W. Hershey, "The Split-Cross-Bridge Resistor for Measuring the Sheet Resistance, Linewidth, and Line Spacing of Conducting Layers," *IEEE Trans. Electron Dev.* **ED-33**, 1572–1579, Oct. 1986.

[80] T. Y. Huang, "Using the Cross-Bridge Structure to Monitor the Effective Oxide Sidewall-Spacer Width in LDD Transistors," *IEEE Electron Dev. Lett.* **EDL-6**, 208–210, May 1985.

[81] D. S. Perloff, "A van der Pauw Resistor Structure for Determining Mask Superposition Errors on Semiconductor Slices," *Solid-State Electron.* **21**, 1013–1018, Aug. 1978.

[82] H. B. Bebb and E. W. Williams, "Photoluminescence I: Theory," in: *Semiconductors and Semimetals* (R. K. Willardson and A. C. Beer, eds.), Academic Press, New York, **8**, pp. 181–320, 1972; E. W. Williams and H. B. Bebb, "Photoluminescence II: Gallium Arsenide," *ibid.*, pp. 321–392.

[83] P. J. Dean, "Photoluminescence as a Diagnostic of Semiconductors," *Prog. Crystal Growth Charact.* **5**, 89–174, 1982.

[84] K. K. Smith, "Photoluminescence of Semiconductor Materials," *Thin Solid Films* **84**, 171–182, Oct. 1981.

[85] J. P. Wolfe and A. Mysyrowicz, "Excitonic Matter," *Sci. Am.* **250**, 98–107, March 1984.

[86] J. I. Pankove, *Optical Processes in Semiconductors*, Dover, New York, 1975.

[87] P. J. Dean, D. J. Robbins, and S. G. Bishop, "Dye Laser Selective Spectroscopy in Bulk-Grown Indium Phosphide," *J. Phys. C: Solid State Phys.* **12**, 5567–5575, Dec. 1979.

[88] M. Tajima, "Determination of Boron and Phosphorus Concentration in Silicon by Photoluminescence Analysis," *Appl. Phys. Lett.* **32**, 719–721, June 1978.

[89] M. Tajima, T. Masui, T. Abe, and T. Iizuka, "Photoluminescence Analysis of Silicon Crystals," in *Semiconductor Silicon/1981* (H. R. Huff, R. J. Kriegler, and Y. Takeishi, eds.), Electrochem. Soc., Pennington, NJ, 1981, pp. 72–89.

[90] M. Tajima, "Recent Advances in Photoluminescence Analysis of Si: Application to an Epitaxial Layer and Nitrogen in Si," *Japan. J. Appl. Phys.* **21**, Supplement 21-1, 113–119, 1982.

[91] P. J. Dean, J. R. Haynes, and W. F. Flood, "New Radiative Recombination Processes Involving Neutral Donors and Acceptors in Silicon and Germanium," *Phys. Rev.* **161**, 711–729, Sept. 1967.

[92] G. Pickering, P. R. Tapster, P. J. Dean, and D. J. Ashen, "Determination of Impurity Concentration in *n*-Type InP by a Photoluminescence Technique," in *GaAs and Related Compounds* (G. E. Stillman, ed.) Conf. Ser. No. 65, Inst. Phys., Bristol, 1983, pp. 469–476.

[93] D. J. Ashen, P. J. Dean, D. T. J. Hurle, J. B. Mullin, and A. M. White, "The Incorporation and Characterization of Acceptors in Epitaxial GaAs," *J. Phys. Chem. Solids* **36**, 1041–1053, Oct. 1975.

[94] G. E. Stillman, B. Lee, M. H. Kim, and S. S. Bose, "Quantitative Analysis of Residual Impurities in High Purity Compound Semiconductors," in *Diagnostic Techniques for Semiconductor Materials and Devices* (T. J. Shaffner and D. K. Schroder, eds.), Electrochem. Soc., Pennington, NJ, 1988, pp. 56–70.

[95] S. S. Bose, B. Lee, M. H. Kim, and G. E. Stillman, "Identification of Residual Donors in High-Purity GaAs by Photoluminescence," *Appl. Phys. Lett.* **51**, 937–939, Sept. 1987.

[96] C. V. Raman and K. S. Krishna, "A New Type of Secondary Radiation," *Nature* **121**, 501–502, March 1928.

[97]   J. Raptis, E. Liarokapis, and E. Anastassakis, "Effect of Temperature Gradients on the First-Order Raman Spectrum of Si," *Appl. Phys. Lett.* **44**, 125–127, Jan. 1983.

[98]   P. M. Fauchet, "The Raman Microprobe: A Quantitative Analytical Tool to Characterize Laser-Processed Semiconductors," *IEEE Circ. Dev. Mag.* **2**, 37–42, Jan. 1986.

[99]   B. D. Chase, "Fourier Transform Raman Spectroscopy," *J. Am. Chem. Soc.* **108**, 7485–7488, Nov. 1986.

[100] H. Richter, Z. P. Wang, and L. Ley, "The One Phonon Raman Spectrum in Microcrystalline Silicon," *Solid State Commun.* **39**, 625–629, Aug. 1981.

[101] S. R. J. Brueck, B. Y. Tsaur, J. C. C. Fan, D. V. Murphy, T. F. Deutsch, and D. J. Silversmith, "Raman Measurements of Stress in Silicon-on-Sapphire Device Structures," *Appl. Phys. Lett.* **40**, 895–898, May 1982.

[102] F. H. Pollack and R. Tsu, "Raman Characterization of Semiconductors Revisited," *Proc. SPIE* **452**, 26–43, 1983.

[103] F. Adar, "Application of the Raman Microprobe to Analytical Problems in Microelectronics," in *Microelectronic Processing: Inorganic Materials Characterization* (L. A. Casper, ed.), American Chemical Soc., ACS Symp. Series **295**, 1986, pp. 230–239.

[104] J. N. Ramsey, "Microelectronics Processing Problem Solving: The Synergism of Complementary Techniques," in *Microelectronic Processing: Inorganic Materials Characterization* (L. A. Casper, ed.), American Chemical Soc., ACS Symp. Series **295**, 1986, pp. 398–425.

[105] S. Nakashima and M. Hangyo, "Characterization of Semiconductor Materials by Raman Spectroscopy," *IEEE J. Quant. Electron.* **25**, 965–975, May 1989.

[106] D. L. Gerrard and H. J. Bowley, "Raman Spectroscopy," *Anal. Chem.* **60**, 368R–377R, June 1988.

[107] A. Baghdadi, "Multiple-Reflection Corrections in Fourier Transform Spectroscopy," in *Defects in Silicon* (W. M. Bullis and L. C. Kimerling, eds.), Electrochem. Soc., Pennington, NJ, 1983, pp. 293–302.

[108] E. D. Palik (ed.), *Handbook of Optical Constants of Solids*, Academic Press, Orlando, FL, 1985.

CHAPTER 10

CHEMICAL AND PHYSICAL CHARACTERIZATION

10.1 INTRODUCTION

Those chemical and physical characterization techniques most commonly used in the semiconductor industry are discussed in this chapter. In particular, we stress electron beam, ion beam, and X-ray methods and briefly describe a few others. These characterization techniques are generally more specialized and require more complicated and more expensive equipment than those of the previous chapters. Some methods are used a great deal by a few specialists or are offered as services. Others are used less frequently, but when they are used, they give important information, often not obtainable by other methods. Due to the specialized nature of each of the techniques in this chapter, only a brief description is given of the principles of each method, the instrumentation, and the most important areas of application. The specialist using any of the methods is already familiar with the details, and the non-specialist is usually not interested in the details but may be interested in an overview, in the detection limits, the required sample size, etc.

Many papers, review papers, chapters in books, and books have been written describing these characterization techniques in great detail. For this chapter I have drawn heavily on references 1–11 and found the *Metals Handbook*[6] to be one of the most comprehensive treatments of analytical characterization techniques. The contributing authors of that volume discuss not only the methods of this chapter but many others as well. Very extensive bibliographies, covering all of the methods in this chapter and containing more than 8000 references, have been published by Larrabee and others[12–17] and by Turner and coworkers.[18–21] The most salient features of the characterization techniques are summarized in Appendix 10.1

The characterization of semiconductor materials and devices frequently requires a measurement of an impurity spatially in the x and y as well as in the z-dimension. Typical x-y resolution capabilities of the three techniques are shown in Fig. 10.1. Only electron beam methods are suitable for diameters less than 1 μm. Electron beams can be focused to diameters as small as 2 Å. Ion beams cover the 1 to 100 μm range, and X-rays typically have diameters of 1000 μm and above. There is a dichotomy in the characterization of materials at small dimensions: high-sensitivity and small volume sampling are mutually exclusive.[1] Typically, as the beam diameter is decreased, the sensitivity becomes poorer. High sensitivity requires large excitation beam diameters.

All analytical techniques are based on similar principles. Primary electrons, ions, or photons on the sample cause backscattering of the incident particles/waves or the emission of secondary particles/waves. The mass, energy, or wavelength of the emitted entities is characteristic of the target element or compound from which it originated. The distribution of the unknown can be mapped in the x-y plane and frequently also in depth. Each of the techniques has particular strengths and weaknesses, and frequently more than one method must be utilized for unambiguous identification. Differences between the various techniques include sensitivity, elemental or molecular information, spatial resolution in x-, y-, and z-directions, destructiveness, matrix effects, speed, imaging capability, and cost.

The term *spectroscopy* is used for characterization techniques that are primarily qualitative in their ability to determine concentrations even though

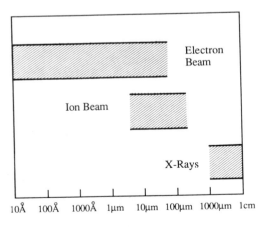

Analytical Diameter

Fig. 10.1 Diameter capabilities of electron beam, ion beam, and X-ray characterization techniques. Reprinted with permission after Larrabee.[1]

they may be very quantitative for identifying impurities; *spectrometry* is used for quantitative methods.

10.2 ELECTRON BEAM TECHNIQUES

Electron beam techniques are summarized in the diagram in Fig. 10.2. Incident electrons are absorbed, emitted, reflected, or transmitted and can in turn cause light or X-ray emission. An electron beam of energy E_i incident on the sample surface causes emission of electrons from the surface over a wide range of energies, as illustrated in Fig. 10.3, where the number of electrons emitted by a sample is plotted against the electron energy E. Three groups of electrons can be distinguished. The electron yield $N(E)$ shows a broad maximum for low energy or *secondary electrons* (indicated by I). The interaction of an electron beam with a solid can lead to the ejection of loosely bound electrons from the conduction band. These are the secondary electrons with energies below about 50 eV with a maximum $N(E)$ at 2 to 3 eV. Backscattered electrons that have undergone inelastic collisions (shown as II) have a low yield with sharp features on a broad background. This is the energy region where *Auger electrons* originate. $N(E)$ is high for *backscattered electrons* that have undergone large-angle elastic collisions, leaving the sample with essentially the same energy as the incident electrons (region III).

Electrons can be focused, deflected, and accelerated by appropriate potentials; they can be efficiently detected and counted, their energy and angular distribution can be measured, and they do not contaminate the vacuum system. Because they are charged, they can cause sample charging, which may distort the measurement.[22]

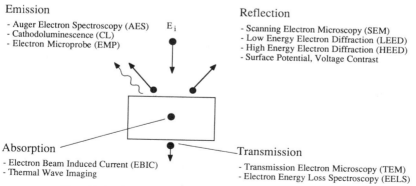

Emission

- Auger Electron Spectroscopy (AES) E_i
- Cathodoluminescence (CL)
- Electron Microprobe (EMP)

Reflection

- Scanning Electron Microscopy (SEM)
- Low Energy Electron Diffraction (LEED)
- High Energy Electron Diffraction (HEED)
- Surface Potential, Voltage Contrast

Absorption

- Electron Beam Induced Current (EBIC)
- Thermal Wave Imaging

Transmission

- Transmission Electron Microscopy (TEM)
- Electron Energy Loss Spectroscopy (EELS)

Fig. 10.2 Electron beam characterization techniques.

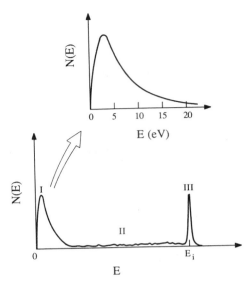

Fig. 10.3 Energy distribution of electrons reflected from a solid surface.

10.2.1 Scanning Electron Microscopy

Principle An *electron microscope* utilizes an electron beam (e-beam) to produce a magnified image of the sample. There are three principal types of electron microscopes: *scanning, transmission,* and *emission.* In the scanning and transmission electron microscope an electron beam incident on the sample produces an image, whereas in the field-emission microscope the specimen itself is the source of electrons. A good discussion of the history of electron microscopy is given by Cosslett.[23] *Scanning electron microscopy* (SEM) is similar to light microscopy, with the exception that electrons are used instead of photons. This has two main advantages: much larger magnifications are possible since electron wavelengths are much smaller than photon wavelengths and the depth of field is much larger.

De Broglie proposed in 1923 that particles can also behave as waves.[24] The electron wavelength λ_e depends on the electron velocity v or the accelerating voltage V as

$$\lambda_e = \frac{h}{mv} = \frac{h}{\sqrt{2qmV}} = \frac{12.2}{\sqrt{V}} \quad (\text{Å}) \tag{10.1}$$

The wavelength is $0.12\,\text{Å}$ for $V = 10,000\,\text{V}$—a wavelength significantly below the 4000- to 7000-Å wavelength range of visible light. Hence the resolution of an SEM can be much higher than that of an optical microscope.

The image in an SEM is produced by scanning the sample with a focused electron beam and detecting the secondary and/or backscattered electrons. We will not concern ourselves with the details of forming a focused electron beam because this is discussed in appropriate books and papers.[25] Electrons and photons are emitted at each beam location and subsequently detected. Secondary electrons form the conventional SEM image, backscattered electrons can also form an image, X-rays are used in the electron microprobe, emitted light is known as cathodoluminescence, and absorbed electrons are measured as electron beam induced current. All of these signals can be detected and amplified to control the brightness of a cathode ray display tube (CRT) scanned in synchronism with the sample beam scan in the SEM. A one-to-one correspondence is thus established between each point on the display and each point on the sample. Magnification M results from the mapping process according to the ratio of the dimension scanned on the CRT to the dimension of the scanned sample.

$$M = \frac{\text{length of CRT display}}{\text{length of sample scan}} \qquad (10.2)$$

For a 10-cm-wide CRT displaying a sample scanned over a 100 μm length, the magnification is 1000×. Magnifications of 100,000× or slightly higher are possible in SEMs. It is obvious that high magnifications are easily achieved with SEMs, but low magnifications are more difficult. For a magnification of 10× the scanned length on the sample is one centimeter, only 10× smaller than the CRT scan. An SEM typically has one large viewing CRT and a high-resolution CRT with typically 2500 lines resolution for photography.

The contrast in an SEM depends on a number of factors, For a flat, uniform sample the image shows no contrast. If, however, the sample consists of materials with different atomic numbers, a contrast is observed if the signal is obtained from backscattered electrons, because the backscattering coefficient increases with the atomic number Z. The secondary electron emission coefficient, however, is not a strong function of Z, and atomic number variations give no appreciable contrast. Contrast is also influenced by surface conditions and by local electric fields. But the main contrast-enhancing feature is the sample topography. Secondary electrons are emitted from the top 100 Å or so of the sample surface. When the sample surface is tilted from normal beam incidence, the electron beam path lying within this 100 Å is increased by the factor $1/\cos\theta$, where θ is the angle from normal incidence ($\theta = 0°$ for normal incidence). The interaction of the incident beam with the sample increases with path length and the secondary electron emission coefficient increases. The contrast C depends on the angle as[25]

$$C = \tan\theta \, d\theta \qquad (10.3)$$

For $\theta = 45°$ a change in angle of $d\theta = 1°$ produces a contrast of 1.75%, and at

$60°$ the contrast increases to 3% for $d\theta = 1°$. The sample stage is an important component in SEMs. It must allow precise movement in tilt and rotation to be able to view the sample at the appropriate angle.

The angle effect is responsible for the striking three-dimensional nature of SEM images. But the striking pictures come about also due to the signal collection. Secondary electrons are attracted and collected by the detector even if they leave the sample in a direction away from the detector. This does not happen in optical microscopes, where light reflected away from the detector (the eye) is not observed. An SEM forms its picture in an entirely different manner than an optical microscope, where light reflected from a sample passes through a lens and is formed into an image. In an SEM no true image exists. The secondary electrons that make up the conventional SEM image are collected, and their density is amplified and displayed on a CRT. Image formation is produced by the mapping which transforms information from specimen space to CRT space.

Instrumentation   A schematic representation of an SEM is shown in Fig. 10.4. Electrons emitted from an electron gun pass through a series of lenses to be focused and scanned across the sample. The electron beam should be bright with small energy spread. The most common electron gun is a tungsten "hairpin" filament, emitting electrons thermionically with an energy spread of around 2 eV. Tungsten sources have been replaced to some extent by lanthanum hexaboride (LaB_6) sources with higher brightness, lower energy spread (~ 1 eV) and longer life, and field-emission guns with an energy spread of about 0.2 to 0.3 eV. Field-emission guns are about $100\times$ brighter than LaB_6 sources and $1000\times$ brighter than tungsten sources.

The incident or primary electron beam causes secondary electrons to be emitted from the sample and these are ultimately accelerated to 10 to 12 kV. They are most commonly detected with an Everhart-Thornley (ET) detector.[27] The basic component of this detector is a scintillation material that emits light when struck by energetic secondary electrons accelerated from the sample to the detector. The light from the scintillator is channeled through a light pipe to a photomultiplier, where the light incident on a photocathode produces electrons that are multiplied creating the very high gains necessary to drive the CRT. High potentials of 10 to 12 kV are necessary for efficient light emission by the scintillator. For the electron beam not to be influenced by the high ET detector potential, the scintillator is surrounded by a Faraday cage at a few hundred volt potential.

The beam diameter in SEMs is in the range of 50 to 100 Å. Yet the resolution of e-beam measurements is not always that good. Why is that? It has to do with the shape of the electron-hole cloud generated in the semiconductor. When electrons impinge on a solid, they lose energy by elastic scattering (change of direction with negligible energy loss) and inelastic scattering (energy loss with negligible change in direction). Elastic scattering is caused mainly by interactions of electrons with nuclei and is

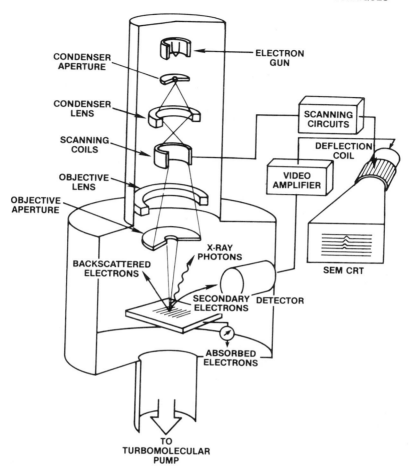

Fig. 10.4 Schematic of a scanning electron microscope. Reprinted with permission from Microelectronics Processing: Inorganic Materials Characterization Fig. 1, p. 51. Copyright 1986 American Chemical Society, after Young and Kalin.[26]

more probable in high atomic number materials and at low beam energies. Inelastic scattering is caused mainly by scattering from valence and core electrons. The result of these scattering events is a broadening of the original nearly collimated, well-focused electron beam within the sample. The generation volume is a function of the e-beam energy and the atomic number Z of the sample. Secondary electrons, backscattered electrons, characteristic and continuum X-rays, Auger electrons, photons of various energies, and electron-hole pairs are produced. For low-Z samples most electrons penetrate deeply into the sample and are absorbed. For high Z-samples there is considerable scattering near the surface and a large fraction of the incident electrons is backscattered. The shape of the electron distribution within the sample depends on the atomic number. For low-Z material ($Z \le 15$) the distribution is "teardrop"-shaped, as shown in Fig. 10.5. As Z increases ($15 < Z < 40$), the shape becomes more spherical, and for $Z \ge 40$, it becomes hemispherical. Teardrop shapes have actually been observed by Everhart et al. by exposing polymethylmethacrylate to an electron beam and etching the exposed portion of the material.[25] Electron trajectories, calculated with Monte Carlo techniques, also agree with these shapes.[28]

The depth of electron penetration is the *electron range* R_e, defined as the average total distance from the sample surface that an electron travels in the sample along a trajectory. A number of empirical expressions have been derived for R_e. One such expression is[29]

$$R_e = \frac{4.28 \times 10^{-6} E^{1.75}}{\rho} \quad (\text{cm}) \qquad (10.4)$$

where ρ is the sample density (g/cm^3) and E is the electron energy (keV). The electron ranges for Si ($\rho = 2.33$ g/cm^3), Ge (5.32 g/cm^3), GaAs (5.35 g/ cm^3), and InP (4.7 g/cm^3) are

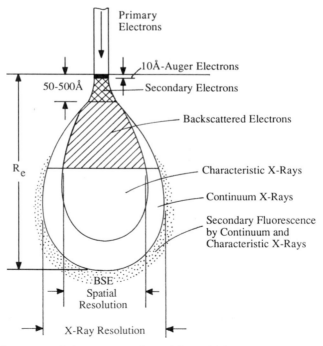

Fig. 10.5 Summary of the range and spatial resolution of backscattered electrons, secondary electrons, X-rays, and Auger electrons for electrons incident on a solid. Reprinted with permission after Goldstein et al.[25]

$$R_e(\text{Si}) = 1.84 \times 10^{-6} E^{1.75}, \quad R_e(\text{Ge}) = 8.05 \times 10^{-7} E^{1.75}$$

$$R_e(\text{GaAs}) = 8.0 \times 10^{-7} E^{1.75}, \quad R_e(\text{InP}) = 9.1 \times 10^{-7} E^{1.75}$$

Equation (10.4) is sufficiently accurate for $20 < E < 200$ keV, but underestimates the range slightly compared to a more accurate expression provided by Everhart and Hoff.[30]

Applications The most common use of SEMs for semiconductor applications, when used as a microscope, is to view the surface of the device, frequently during failure analysis and for cross-sectional analysis to determine device dimensions—for example, MOSFET channel length or junction depth. In the past the SEM was mainly a research tool, but more recently it has moved closer to the wafer-processing production line for on-line inspection and line-width measurement. When inspecting integrated circuits, it is important to reduce or eliminate surface charging, which happens when electrons landing on insulators cannot discharge to ground potential. Surface charging is eliminated by coating the surface with a thin conductive layer or by reducing the beam energy until the number of primary electrons is roughly equal to the number of secondary and backscattered electrons. The energy for this balance is around 1 keV, which is also sufficiently low to minimize electron beam damage to devices. The reduced signal-to-noise ratio of low-energy beams is optimized by using high beam brightness and digital frame storage for signal enhancement.

10.2.2 Auger Electron Spectroscopy

Principle *Auger electron spectroscopy* (AES) is based on the Auger (pronounced something like "O-J") effect, discovered by Auger in 1925.[31] It has become a powerful surface characterization method for the study of chemical and compositional properties of materials. All elements except hydrogen and helium can be detected. Data interpretation is simplified by the large data base of available literature for identification of elemental species.[32–34] Spectra from individual elements do not interfere with one another. Recent advances in electron spectrometers allow chemical binding state information to be obtained from Auger transition energy shifts as well. Although the basic Auger technique samples a depth of typically 5 to 50 Å, it is possible to obtain depth information by sputter etching the sample. Important from an application point of view is the availability of commercial equipment incorporating computer-controlled data acquisition systems.

The process of Auger electron emission is illustrated in Fig. 10.6 for a semiconductor. The energy band diagram of the semiconductor shows the vacuum level, the conduction and valence bands, and the lower lying core levels not usually shown on semiconductor energy band diagrams. In particular, we assume a material with a K level at energy E_K and two L

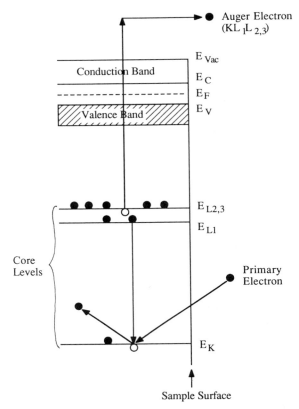

Fig. 10.6   Electronic processes in Auger electron spectroscopy.

levels (E_{L1} and $E_{L2,3}$). For Si $E_K = 1839$ eV, $E_{L1} = 149$ eV, $E_{L2} = 100$ eV, and $E_{L3} = 99$ eV below the Fermi level.[2,8] A primary electron from an electron gun ejects an electron from the K shell. The K-shell vacancy is filled by an outer shell electron (L_1 in this case). The energy $E = E_{L1} - E_K$ is transferred to a third electron—the Auger electron—originating in this case at the $L_{2,3}$ level. The atom remains in a doubly ionized state and the entire process is labeled "$KL_1L_{2,3}$" or sometimes simply as "KLL." Since the Auger process is a three-electron process, it is obvious why hydrogen and helium cannot be detected; both have less than three electrons. The dominant Auger energy transitions are as follows: for $3 < Z < 14$: KLL transitions, for $14 < Z < 40$: LMM transitions, and for $40 < Z < 82$: MNN transitions.[4]

The energy of the Auger electron, characteristic of the emitting atom of atomic number Z, for the $KL_1L_{2,3}$ transition is given by[2–3]

$$E_{KL1L2,3} = E_K(Z) - E_{L1}(Z) - E_{L2,3}(Z + 1) - q\phi \qquad (10.5a)$$

where $q\phi$ is the sample work function. In general, when an electron is excited from level A, the vacancy is filled by an electron from level B, and an electron from level C is ejected, the kinetic energy of the Auger electron is[3-4]

$$E_{ABC} = E_A(Z) - E_B(Z) - E_C(Z + \Delta) - q\phi \qquad (10.5b)$$

where Δ is included to account for the energy of the final doubly-ionized state being larger than the sum of the energies for individual ionization of the same levels. Δ lies between 0.5 and 0.75.[32] Auger electron energies are characteristic of the sample and are independent of the incident electron energy.

Instrumentation Auger electron spectroscopy instrumentation consists of an electron gun, electron beam control, an electron energy analyzer, and data analysis electronics. The incident electron beam of 1- to 2-mm diameter has typically 1- to 5-keV energy; higher-energy beams produce electrons deeper within the sample that have little chance of escaping. The beam is focused to a diameter as small as 1000 Å in scanning Auger systems. The emitted Auger electrons are detected with a retarding potential analyzer, a cylindrical mirror analyzer, or a hemispherical analyzer. The most common analyzer is the cylindrical mirror analyzer (CMA) of Fig. 10.7.[35-36] A coaxial configuration with the analyzer wrapped around the electron gun reduces shadowing and allows room for positioning the ion sputter gun. Auger electrons enter the inlet aperture between the two concentric cylinders. They are focused by a negative potential applied to the outer cylinder so that only those electrons with energies within a certain small range pass through the analyzer to arrive at the electron multiplier detector. The electrons are focused by a negative potential which creates a cylindrical electric field between the coaxial electrodes. The CMA allows electrons with $E \sim V_a$ and energy spread ΔE to pass through the exit slit. Ramping the analyzing potential V_a provides the electron energy spectrum. The energy resolution is defined by[4]

$$R = \frac{\Delta E}{E} \qquad (10.6)$$

where ΔE is the pass energy of the analyzer and E the range of electron energies. $R \approx 0.005$ for the CMA.

Auger electrons have energies of typically 30 to 3000 eV. The analysis area is chiefly determined by the diameter of the primary electron beam. Although the interaction volume of an electron beam in a solid is large compared to the electron beam diameter, as illustrated in Fig. 10.5, Auger electrons escape from a very shallow surface layer. Metals tend to have the shortest escape depth, followed by semiconductors and insulators. Modern

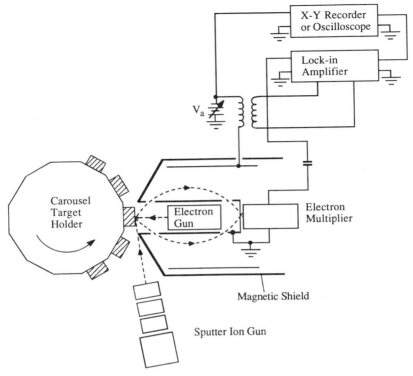

Fig. 10.7 Layout of an AES system with a cylindrical mirror analyzer detector. Reprinted with permission after Davis et al.[32]

scanning AES systems have lateral diameters as small as 1000 Å for samples with flat surfaces. The resolution of non-scanning systems is poorer. It is determined by the electron beam size that is typically 0.5 to 2 mm. AES can be operated in a point analysis mode that detects many elements in a small sample area, or a map of a selected element can be generated by scanning the beam across the sample with the detector tuned to one element. A high, oil-free vacuum (10^{-9} torr or lower) is required to protect the sample from carbon contamination. The presence of surface contamination interferes with the Auger signal.

Auger as well as non-Auger electrons are emitted when a solid is struck by an electron beam. The number of electrons emitted $N(E)$ is plotted in Fig. 10.8 as a function of energy for a silver sample. The Auger electron peaks appear as small perturbations on a high background consisting of beam electrons that have lost varying amounts of energy before being backscattered, as well as Auger electrons that have lost energy propagating through the sample. Those Auger electrons formed deeper in the sample will lose energy and be unrecognizable in the background. In order to

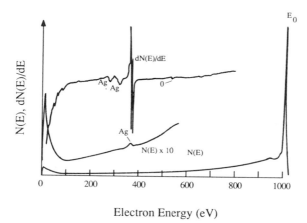

Fig. 10.8   Auger electron spectra $N(E)$ and $dN(E)/dE$ of silver. Incident beam energy is 1 keV. Reprinted with permission after Goldstein et al.[25]

enhance the Auger peaks, it is common practice to differentiate the Auger signal and present it as $dN(E)/dE$ versus E, also shown in Fig. 10.8. The introduction of signal differentiation led to the rapid growth of AES.[37]

The Auger energy position is indicated by the peak in the $N(E)$ spectrum or by the maximum negative excursion of the $dN(E)/dE$ peak in the differentiated spectrum. The Auger signal is differentiated electronically by passing the signal through a lock-in amplifier or by digitizing the detector output signal and then differentiating it. In the lock-in amplifier approach a small ac voltage is superimposed on the slowly changing energy-selecting voltage V_a applied to the outer cylinder of the CMA. The modulated AES signal and the reference signal are both fed to a lock-in amplifier. The first derivative of the synchronously detected in-phase signal is the amplitude of the first harmonic. The second derivative is the amplitude of the second harmonic and is sometimes used to obtain additional information of the sample signature. The $N(E)$ spectrum can also be differentiated by a computer, usually when the AES signal is obtained from a double-pass CMA. The signal-to-background ratio is enhanced by differentiation, but the signal-to-noise ratio is degraded. Recently the trend has been toward the undifferentiated $N(E)$ versus E curve with background suppression, using computer techniques to lift the signal above the background.

The AES detection limit is about 0.1%, but it varies greatly from element to element. Davies et al. give the relative Auger sensitivities for the elements for 3 keV, 5 keV, and 10 keV primary electron energies.[32] Detection limits are also influenced by the beam current and by the analysis time. Quantitative AES analysis is difficult.[4] Reported accuracies are currently about 20–50%, with 5% precision for simple semiconductor samples.[38] The most common correction scheme relies on published Auger intensities or sensitivity factors for the elements. The analyst corrects measured peak intensities by weighting the spectrum with each element's sensitivity factor. Peak-to-peak values of differentiated spectra are commonly used for intensities, a practice that has come under recent criticism.[39] A more accurate method is one of measuring the area under the peaks in the integrated, not the differentiated, spectrum.

It is important to ascertain that the incident electron beam does not alter the sample. Insulators sometimes show artifacts due to sample charging.[40] Depth profiles, generated by alternately sputtering with an inert ion beam and acquiring the Auger signal, are displayed as Auger electron intensity versus sputtering time. Depth can be correlated with sputtering time by measuring the crater depth after the analysis. When the surface is sputtered during AES depth profiling, sputter-induced artifacts may appear.[41] These include crater edge effects, redeposition of sputtered material, surface roughness, preferential sputtering, varying sputter rates, atomic mixing, charging effects, and specimen damage.[42]

Applications AES has found applications in measuring semiconductor composition, oxide film composition, phosphorus-doped glasses, silicides, metallization, bonding pad contamination, lead frame failures, particle analysis, and the effects of surface cleaning.[4,39,43–46] Elemental scans give a rapid means of identifying surface elements. *Scanning Auger microscopy (SAM)* allows the sample to be mapped for one selected element at a time. AES is not suitable for trace element analysis because its sensitivity lies in the 0.1 to 1% range. Traditionally only *elements* were detected with AES, but recent improvements in AES systems allow *chemical information* to be obtained. When elements combine to form compounds, there is a characteristic energy shift and shape change in the Auger spectrum as illustrated in Fig. 10.9. Chemical information had traditionally been the domain of X-ray photoelectron spectroscopy.

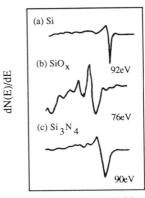

Fig. 10.9 Chemical information in Auger spectra indicated by the shape and energy of the LMM Si peaks from (a) elemental Si, (b) SiO_x, (c) Si_3N_4. Reprinted with permission after Kazmerski.[4]

Electron Energy (eV)

Thompson summarizes AES chemical, metallurgical, and semiconductor applications.[47] Publications for the period 1925 to 1975 numbering over 2000 references are given by Hawkins.[48]

10.2.3 Electron Microprobe

Principle The *electron microprobe* (EMP), also known as *electron probe microanalysis* (EPM or EPMA) was first described by Castaing in his doctoral thesis in 1948.[49-50] The principle of the method consists of electron bombardment of the sample and X-ray emission from the sample. An EMP is usually a part of a scanning electron microscope equipped with appropriate X-ray detectors.[51] Of all the signals generated by the interaction of the primary electron beam with the sample in the SEM, X-rays are most commonly used for material characterization. The X-rays have energies characteristic of the element from which they originate leading to elemental identification. The X-ray intensity can be compared with intensities from known samples, and the ratio of the sample intensity to the intensity of the standard can be considered a measure of the amount of the element in the sample. The correlation, however, is not entirely straightforward. There are factors that tend to complicate the interpretation. Most important of these factors is the influence of other elements in the sample that absorb some of the X-rays generated by the primary electron beam and release other X-rays of their own characteristic energy, known as *fluorescence*. If the energy of the characteristic radiation from element A exceeds the absorption energy for element B in a sample containing A and B, a characteristic fluorescence of B by A will occur. Additionally not all X-rays leaving the sample are captured by the detector. Best accuracy in quantitative concentration determination is obtained if the standards are identical or at least very similar in composition to the unknown. Pure elemental standards can also be used but may lead to inaccuracies. Fortunately quantitative analysis is not always necessary; simple, qualitative, and semi-quantitative analyses are frequently sufficient.

EMP is not a true surface technique because X-rays are emitted from within the sample volume as shown in Fig. 10.5. The method is illustrated with the aid of the band diagram in Fig. 10.10. A primary electron beam of typically 5 to 20 keV strikes the sample. The electron beam energy should be approximately three times the X-ray energy. X-rays are generated by electron bombardment of a target by two distinctly different processes: (1) deceleration of electrons in the Coulombic field of the atom core leads to formation of a continuous spectrum of X-ray energies from zero to the incident electron energy. This is the X-ray continuum or *Bremsstrahlung* (German for braking radiation), sometimes called white radiation by analogy with white light of the visible spectrum. (2) The interaction of the primary electrons with inner-shell electrons. Incident electrons eject electrons from one of the inner atomic shells and electrons from higher-lying

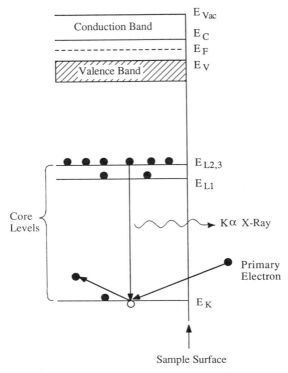

Fig. 10.10   Electronic processes in the electron microprobe.

shells drop into the vacancies created by the ejected electrons. These are the characteristic X-rays. If the X-ray emission is the result of an L→K transition, the X-rays are known as Kα X-rays. Kβ X-rays are the result of M→K transitions, Lα X-rays are due to M→L transitions, etc. There is but one K level, but the other levels are subdivided. The L shell is split into a triple fine structure, and the M shell has five levels. This leads to further subdivision of the nomenclature. For example, the $L_2 \to K$ transition is known as $K\alpha_2$, and the $L_3 \to K$ transition results in $K\alpha_1$ X-rays.[50] Not all possible transitions occur with equal probability, and some are so improbable to have earned the name "forbidden" transitions. For example, the $L_1 \to K$ transition does not occur.

X-ray detectors frequently lack the resolving power to separate X-ray lines close to one another (doublets). The unresolved doublets are measured in such cases as if they were a single line. This is indicated by dropping the subscript; the notation Kα refers to the unresolved doublet $K\alpha_1 + K\alpha_2$. Sometimes the term $K\alpha_{1,2}$ is used. The X-ray photon energy for an L→K transition in the EMP is

$$E_{\text{EMP}} = E_K(Z) - E_{L2,3}(Z) \qquad (10.7)$$

The energy between the K and L levels is much higher than that between the L and M levels, which in turn is higher than that between the M and N levels. For example, for silicon $E(K\alpha_1) = 1.74$ keV; for copper $E(K\alpha_1) = 8.048$ keV and $E(L\alpha_1) = 0.93$ keV, whereas for gold $E(K\alpha_1) = 68.794$ keV, $E(L\alpha_1) = 9.713$ keV, and $E(M\alpha_1) = 2.123$ keV.[25] The most common EMP X-ray lines are the $K_{\alpha1,2}$ the $K_{\beta1}$, the $L_{\alpha1,2}$ and the $M_{\alpha1,2}$.[50] A graphical representation of all X-ray lines observed in high quality X-ray spectra in the 0.7 to 10 keV energy range is given by Fiori and Newbury.[52] A detailed discussion of both qualitative and quantitative EMP spectra interpretation is given in Goldstein et al.[25] The relationship between X-ray energy E and wavelength λ is

$$\lambda = \frac{hc}{E} = \frac{12.398}{E} \quad (\text{Å}) \tag{10.8}$$

with E in keV. Both wavelength and energy are used in X-ray analysis.

The only possible outcomes of an ionization event involving the K shell are the emission of a K-line X-ray photon or of an Auger electron. The fraction of the total number of ionizations leading to the emission of X-rays is the *fluorescence yield*. The sum of the probability of X-ray emission and that of Auger electron emission is unity. For low Z material Auger emission is dominant, but X-ray emission dominates for high Z material. The two probabilities are about equal for $Z \approx 33$, as shown in Fig. 10.11.

Although the electron beam diameter in the EMP may be less than one micron, the emitted X-rays originate from a larger area; EMP does not have the high resolution associated with AES, nor is it a true surface-sensitive technique. EMP is a good example where the size of the exciting beam bears little relation to the resolution of the measurement. Reduction of the beam diameter leads to slightly higher resolution at the expense of signal reduction. As in AES the electron beam can be stationary, and an elemental scan gives the sample impurities. The use of beam rastering to scan a limited sample area leads to imaging and elemental mapping. The EMP sensitivity is better than that of AES because a larger volume is probed. The sensitivity is 10^3 to 10^4 ppm (parts per million) for $Z = 4$ to 10, 10^3 ppm for $Z = 11$ to 22,

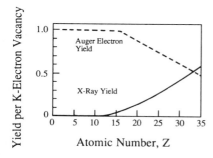

Fig. 10.11 Auger electron and X-ray yields per K vacancy as a function of atomic number.

and 100 ppm for $Z = 23$ to 100 but varies with instrumental and sample parameters.[4]

Instrumentation EMP utilizes the electron beam, focusing lenses, and deflection coils of an SEM. Only the X-ray detector is added, and many SEMs have EMP capability. Two different types of detectors are used: *energy-dispersive* spectrometers (EDS) and *wavelength-dispersive* spectrometers (WDS), illustrated in Fig. 10.12. The two spectrometers complement each other, making it desirable to have both on an SEM. The EDS is commonly used for rapid sample analysis and the WDS for high-resolution measurements.

The X-ray detector in the EDS is a reverse-biased semiconductor (usually Si) pin or Schottky diode. X-rays are absorbed in a solid according to the equation[50]

$$I(x) = I_0 \exp\left[-\left(\frac{\mu}{\rho}\right)(\rho x) \right] \tag{10.9}$$

where (μ/ρ) is the mass absorption coefficient, ρ the detector material

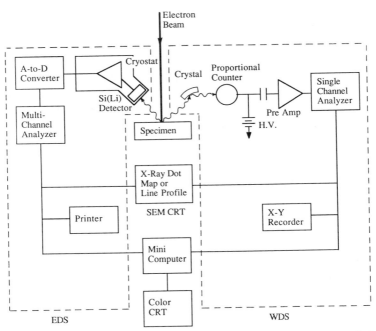

Fig. 10.12 EDS and WDS X-ray detector systems. Reprinted with permission from *Microelectronics Processing: Inorganic Materials Characterization* Fig. 8, p. 60. Copyright 1986 American Chemical Society, after Young and Kalin.[26]

density, $I(x)$ the X-ray intensity in the detector, and I_0 the incident X-ray intensity. The mass absorption coefficient is characteristic of a given element at specified X-ray energies. Its value varies with the photon wavelength and with the atomic number of the target element, generally decreasing smoothly with energy. It exhibits discontinuities in the energy region immediately above the "absorption edge" energy, corresponding to the energy necessary to eject an electron from a shell. For Si $\mu/\rho = 6.533$ cm^2/g for Mo Kα ($E = 17.44$ keV) and 65.32 cm^2/g for Cu Kα ($E = 8.05$ keV) X-rays.[8] The absorption equation for Cu Kα X-rays incident on a Si detector with ρ(Si) = 2.33 g/cm^3 is

$$\frac{I(x)}{I_0} = \exp(-152.2x) \qquad (10.10)$$

with x in cm. The thickness for 50% absorption is 46 μm, and for 90% absorption it is 151 μm. The X-ray penetration depth in Si is quite large, and it is important that the space-charge region (scr) of the reverse-biased diode be sufficiently wide to absorb the X-rays. With the scr width $W \sim 1/N_D^{1/2}$, this requires either very pure Si or lithium drifting to produce an effectively intrinsic region.[53–54]

The X-rays from the sample pass through a thin beryllium window (not shown on Fig. 10.12) onto a lithium-drifted Si detector which should be cooled at all times. Liquid nitrogen cooling prevents lithium diffusion and also reduces the diode leakage current. Detector bias voltage should never be applied to a non-cooled Li-drifted detector because the electric field causes the Li ions to drift even at room temperature. Each absorbed X-ray creates many electron-hole pairs that are swept out of the diode by the high electric field in the space-charge region. The charge pulse is converted to a voltage pulse by a cooled charge-sensitive preamplifier, and the signal is further amplified and shaped and then passed to a multichannel analyzer (MCA). The MCA measures and sorts the pulses from the preamplifier and assigns them to the appropriate channel (memory location) in the display, with the channel location or number calibrated to correspond to X-ray photon energy. The amplitude versus channel number or energy spectrum is displayed on a CRT or on an X-Y recorder. Most analyzers are equipped with computers for additional signal analysis. The pulse from each absorbed X-ray should not interfere with the pulse from the next absorbed X-ray. If say two 5 keV pulses coincide, the detector output will be that of one 10 keV pulse. The likelihood of such an occurrence is rare, however. Pulse pileup can occur if the spacing between pulses is so small that they overlap and cause erroneous amplitude measurements.

An energetic particle or photon of energy E absorbed in a semiconductor generates N_{eh} electron-hole pairs (ehp), given by[55]

$$N_{eh} = \left(\frac{E}{E_{eh}}\right)\left(1 - \frac{\alpha E_{bs}}{E}\right) \qquad (10.11)$$

where E_{eh} is the average energy necessary to create one ehp, E_{bs} the mean energy of the backscattered electrons, and α the backscattering coefficient ($\alpha \approx 0.1$ for Si in the 2 to 60 keV energy range). $E_{eh} \approx 3.2 E_g$ and for Si it is about 3.6 to 3.8 eV.[56] A 5-keV X-ray photon generates about 1250 ehp's or a charge of 2×10^{-16} C in Si.

The energy of incident X-rays is determined in such semiconductor detectors by the number of ehp's those X-rays produce. Elements from Na to U can be detected with EDS. Lower Z elements are difficult to detect due to the Be window that isolates the cooled detector from the vacuum system. Windowless systems allow lower Z elements to be detected (see Table 10.1). It is possible for X-rays from the sample absorbed in the Si detector to generate Si Kα X-rays that are subsequently absorbed in the detector. These X-rays, which do not originate from the sample, appear in the spectrum as the so-called silicon internal fluorescence peak. A good discussion of the factors affecting EDS is found in Goldstein et al.[25]

WDS is based on an entirely different principle. X-rays from the sample are directed onto an analyzing crystal. Only those X-rays that strike the crystal at the proper angle are diffracted through a polypropylene window into the detector, usually a gas proportional counter. The proportional counter consists of a gas-filled tube with a thin tungsten wire in the center of the tube held at a 1- to 3-kV potential. The gas (usually 90% argon–10% methane) flows through the tube because it is difficult to seal the thin entrance window. An absorbed X-ray creates a shower of electrons and positive ions. The electrons are attracted to the wire and produce a charge pulse, much as ehp's are generated and collected in semiconductor detector.

X-ray diffraction is determined by Bragg's law

$$n\lambda = 2d \sin \theta_B \tag{10.12}$$

where $n = 1, 2, 3, \ldots$, λ is the X-ray wavelength, d the interplanar spacing of the analyzing crystal, and θ_B the Bragg angle. The detector signal is amplified, converted to a standard pulse size by a single-channel analyzer,

TABLE 10.1   Comparison between X-Ray Spectrometers[25–26]

Operating Characteristics	WDS Crystal Diffraction	EDS Si Energy Dispersive
Quantum efficiency	Variable, <30%	~100% for 2–16 keV
Elements detected	$Z \geq 5$ (B)	$Z \geq 11$ (Na) for Be window
		$Z \geq 6$ (C) windowless
Resolution	Crystal dependent	Energy dependent
	~5 eV	150 eV at 5.9 keV
Data collection time	Minutes to hours	Minutes
Sensitivity	0.01–0.1%	0.1–1%

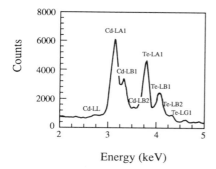

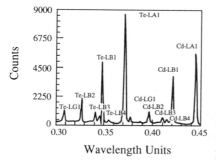

Fig. 10.13 (a) EDS and (b) WDS spectra for a CdTe polycrystalline thin film. Reprinted with permission after Kazmerski.[4]

and then counted or displayed. The analyzing crystals are curved to focus the X-rays onto the detector. More than one crystal is necessary to span an appreciable wavelength range. Common crystal materials with varying lattice spacing are: α-quartz, KAP, LiF, PbSt, PET, and RAP.[25]

WDS detectors have larger collection areas and are located at longer distances from the sample, giving them lower collection efficiencies than EDS detectors. WDS has higher energy resolution since only a small range of wavelengths is detected at one time, allowing greater peak-to-background ratios and higher count rates for individual elements. This gives approximately 1 to 2 orders of magnitude better sensitivity but makes the method slow. Table 10.1 summarizes the major features of the two techniques. EDS and WDS spectra are shown in Fig. 10.13, clearly illustrating the higher resolution of WDS.

Applications   Electron microprobe analysis with EDS spectrometers is used for quick surveys and for spatial maps of individual elements. It is frequently one of the first techniques tried to solve a problem or diagnose a failure. WDS spectrometers are less frequently used because their use is more tedious. Impurities are identified from either EDS or WDS by matching the experimental spectra to known X-ray energies. This can be done manually or by software residing in a dedicated computer. The comparison can be done automatically by appropriate software routines, or it is possible to display the experimental spectrum and also to display known

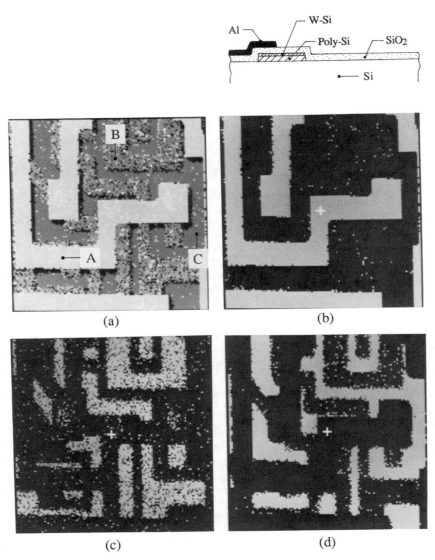

Fig. 10.14 EDS map of a Si circuit: (a) composite EDS map of Al (A), W (B), and Si (C): (b) aluminum map from the Al Line; (c) tungsten map from the W-Si line; (d) silicon map from the substrate. A schematic cross section is shown in the upper right. Courtesy of J. B. Mohr, Arizona State University.

spectra for best match with the experimental data. EMP is not a trace analysis method due to its poor sensitivity; it is particularly insensitive to light elements in a heavy matrix (see Fig. 10.11 and Table 10.1). It has reasonably good spatial resolution of 1 to 10 μm determined by the electron interaction volume in the sample and is well suited for quantitative measures of metals on semiconductors, alloy compositions, etc. Detection of carbon, oxygen, and nitrogen is difficult due to the low X-ray yield and the fact that these are common contaminants in vacuum systems. A picture of an elemental map is shown in Fig. 10.14

10.2.4 Other Electron Beam Techniques

Transmission Electron Microscopy *Transmission electron microscopy* (TEM) was originally used for highly magnified sample images. Later, analytical capabilities such as electron energy loss detectors and light and X-ray detectors were added to the instrument, and the technique is now also known as *analytical transmission electron microscopy* (AEM).[57–59] The "M" in TEM and AEM stands for either "microscopy" or "microscope." Transmission electron microscopes are, in principle, similar to optical microscopes; both contain a series of lenses to magnify the sample. The main strength of TEM lies in its extremely high resolution approaching 1.8 to 2 Å. The reason for this high resolution can be found in the resolution equation (Eq. (9.2a)], $s = 0.61\lambda/\text{NA}$. In optical microscopy the numerical aperture $\text{NA} \approx 1$ and $\lambda \approx 5000$ Å, giving $s = 3000$ Å. In electron microscopy NA is approximately 0.01 due to larger electron lens imperfections, but the wavelength is much shorter.[57] According to Eq. (10.1), $\lambda_e \approx 0.04$ Å for $V = 100,000$ V, giving a resolution of $s = 2.5$ Å, which is significantly better than optical microscopy. The actual resolution expression is more complicated, and this simple calculation should only be taken as a coarse estimate.

A schematic of a transmission electron microscope is shown in Fig. 10.15. Electrons from an electron gun are accelerated to high voltages—typically 100 to 400 kV—and focused on the sample by the condenser lenses. The static beam has a diameter of a few microns. The sample must be sufficiently thin (a few hundred to a few thousand angstroms) for electrons to be transmitted through it. This circumvents the resolution problem of Fig. 10.5 because the beam does not have a chance to "balloon" in the sample when the sample is so thin. The transmitted and forward scattered electrons form a diffraction pattern in the back focal plane and a magnified image in the image plane. With additional lenses either the image or the diffraction pattern is projected onto a fluorescent screen for viewing or for photographic recording.

Image contrast does not depend very much on absorption as it does in optical transmission microscopy, but rather on scattering and diffraction of electrons in the sample. Images formed using only the transmitted electrons are bright-field images and images formed using a specific diffracted beam are dark-field images. Few electrons are absorbed in the sample. Absorbed

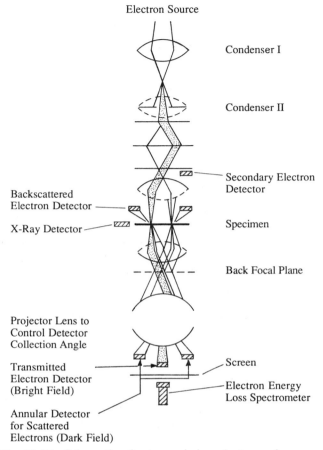

Electron Source

Condenser I

Condenser II

Secondary Electron
Detector

Backscattered
Electron Detector

X-Ray Detector

Specimen

Back Focal Plane

Projector Lens to
Control Detector
Collection Angle

Transmitted
Electron Detector
(Bright Field)

Screen

Electron Energy
Loss Spectrometer

Annular Detector
for Scattered
Electrons (Dark Field)

Fig. 10.15 Schematic of a transmission electron microscope.

electrons lead to sample heating which can change the sample during the measurement.

Consider an amorphous sample consisting of atoms A with inclusions of atoms B, where $Z_B > Z_A$ (Z is the atomic number). Electrons experience very little scattering from atoms A but are more strongly scattered by atoms B. The more strongly scattered electrons are not transmitted by the image forming lenses and do not reach the fluorescent screen, but the weakly scattered electrons do. Hence the heavier elements do not appear on the screen. In other words, the image brightness is determined by the intensity of those electrons transmitted through the sample that pass through the image forming lenses. For crystalline specimen the wave nature of electrons must be considered, and Bragg diffraction of electrons by the sample crystal planes occurs. Electrons "make it" to the screen if they are not deflected by

Bragg diffraction. Contrast comes about by diffraction instead of by scattering.

A stationary, parallel, coherent electron beam passes through the sample in TEM forming a magnified image in the image plane which is then simply projected onto a fluorescent screen. In *scanning transmission electron microscopy* (STEM) a fine beam (diameter < 100 Å) is scanned across the sample in a raster fashion. The objective lens recombines the transmitted electrons from all points scanned by the probe beam to a fixed region in the back focal plane where they are detected by an electron detector. The detector output modulates the brightness of a CRT, much as secondary electrons do in an SEM. The primary electrons in a STEM also produce secondary electrons, backscattered electrons, X-rays, and light (cathodoluminescence) above the sample much as in SEMs. Below the sample, inelastically scattered transmitted electrons can be analyzed for electron energy loss, making the instrument truly an analytical electron microscope. X-ray analysis has become an important aspect of transmission electron microscopy at magnifications much higher than possible for EMP in an SEM. However, the volume for X-ray generation is much smaller giving much weaker X-ray intensity than in SEMs. The integration time for each picture element in STEM is limited since the data are collected serially. This makes for poorer images than obtained in TEM, where all picture elements are integrated simultaneously on a photographic plate.

Electron energy loss spectroscopy (EELS) is the analysis of the distribution of electron energies for electrons transmitted through the sample.[58–62] EELS complements EDS as it is more sensitive to low Z elements ($Z \leq 10$), whereas EDS can only detect elements with $Z > 10$. Theoretically hydrogen should be detectable, but boron is a more practical limit. EELS is concerned with the measurement of electron energy loss due to inelastic collisions and is more sensitive than EDS for several reasons. (1) It is a primary event that does not rely on a secondary event of X-ray emission when an excited atom returns to its ground state making it a more efficient process especially for low Z elements. (2) Only a fraction of the emitted X-rays are detected, while most of the transmitted electrons are detected. EELS is primarily used to provide microanalytical and structural information approaching the very high resolution of the electron beam. The EELS spectrum generally consists of three distinct groupings of spectral peaks: the zero-loss peak containing no useful analytical information, the low-energy-loss peaks due primarily to plasmons, and the high-energy-loss peaks due to inner-shell ionization. EELS maps can be generated by displaying the intensity of a particular energy of the spectrum. Limitations due to specimen thickness and difficulties in quantification have prevented EELS from attaining its full potential. It is primarily used for qualitative determination of precipitates containing elements too light for EDS such as carbides, nitrides, and oxides.

In addition to structural information, diffraction information is also available in AEM. This is very important for crystalline samples, where *selected area diffraction* may be used to identify crystalline phases, amorphous regions, crystal orientations, defects such as stacking faults or dislocations. Diffraction patterns can be obtained from areas as small as 1 μm × 1 μm.

TEM pictures have taken on an important role in semiconductor integ-

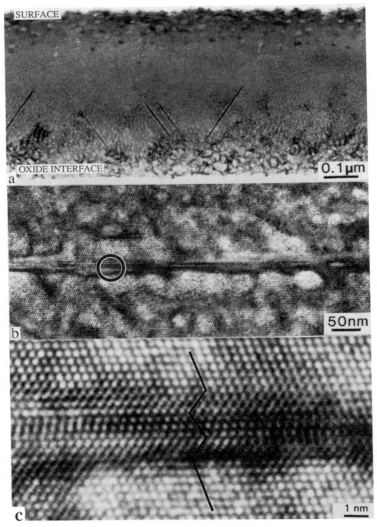

Fig. 10.16 TEM micrographs: (a) upper silicon layer of an oxygen-implanted silicon-on-insulator implanted at 200 keV, 1.5×10^{18} cm^{-2}, 600°C; (b) HREM image of one of the line defects in (a); (c) HREM image showing multiple stacking faults of the defect circled in (b). Courtesy of S. Visitserngtrakul and S. J. Krause, Arizona State University.

rated circuit development. In particular, cross-sectional images through semiconductor devices have brought out many aspects of process-induced features not obtainable with other techniques. An excellent book with many examples of cross-sectional pictures of semiconductor devices by Marcus and Sheng gives very striking evidence of this.[63] AEM success is very much dependent on sample preparation. Semiconductor devices to be analyzed contain semiconductors, insulators, and metals, and it is difficult to use chemical etches to thin all layers simultaneously. Additionally the devices are small, and it is very difficult to locate particular features of interest. Sheng gives a good discussion of the difficulties for successful sample preparation.[63–64]

High-resolution TEM (HREM) has come of age during the past decade or so.[65–67] It gives structural information of the atomic size level, is known as *lattice imaging*, and has become very important for interface analysis. For example, oxide-semiconductor, metal-semiconductor, and semiconductor-semiconductor interfaces have benefited a great deal from HREM images. In lattice imaging a number of different diffracted beams are combined to give an interference image. Many HREM examples are found in Batson[68] and Graham.[69] In Fig. 10.16 we show intermediate- and high-resolution images of a Si sample with a buried oxide layer formed by ion implantation. Only a very small volume is investigated in any one AEM measurement. It has been estimated that the total volume of matter investigated in the entire world until 1979 was less than 1 mm^3.[57]

Electron Beam Induced Current   *Electron beam induced current* (EBIC) has already been discussed with reference to minority carrier diffusion length measurements in Chapter 8. Here we extend the EBIC discussion to other applications.[70] The term EBIC was coined by Everhart.[71] The technique is also known as *charge collection scanning electron microscopy*.[29] In contrast to most of the techniques in this chapter, EBIC does not identify impurities but measures electrically active impurities. The method relies on collection of minority carriers generated by a scanned electron beam in a junction device. The electron beam generates N_{eh} electron-hole pairs, where the dependence of N_{eh} on the beam energy is given in Eq. (10.11). The generation rate of ehp's is, according to Eq. (8.73)

$$G = \frac{I_b N_{eh}}{q\text{Vol}}$$

(10.13)

where Vol is the volume of the ehp's. Two cases are of interest: Vol $\approx (4/3)\pi R_e^3$ when the minority carrier diffusion length L is less than the electron range R_e, and Vol $\approx (4/3)\pi L^3$ when $L \gg R_e$. The minority electron concentration in a p-substrate is

$$n = G\tau_n$$

(10.14)

Equation (10.14) expresses the essence of EBIC measurements. Minority carriers are generated by a scanned electron beam. They are collected by a junction (*pn* junction, Schottky barrier, MOSFET, MOS capacitor, electrolyte–semiconductor junction) and measured as a current—the *electron beam induced current*. The carrier concentration is dependent on the minority carrier lifetime, which in turn depends on the defect distribution of the sample. The interaction of an electron beam with the semiconductor sample can take place in a variety of geometries as shown in Chapter 8. Changes in the photocurrent collected by the junction can be effected by moving the beam in the *x*- and/or *y*-direction. Changes in the *z*-direction are produced by changing the beam energy. The e-beam creates ehp's at a distance *d* from the edge of the scr. Some of the minority carriers diffuse to the junction to be collected. We showed in Chapter 8 how the diffusion length is de-

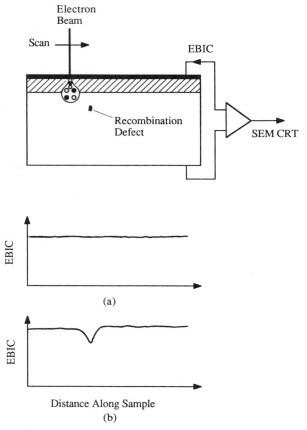

Fig. 10.17   EBIC measurement schematic: (a) EBIC scan for uniform material, (b) EBIC scan for nonuniform material.

termined by measuring the current as the electron beam is moved away from the collecting junction.

To determine the defect or recombination center distribution, one generally forms a large area collecting junction and scans the e-beam along the junction, as shown in Fig. 10.17. The current is constant for uniform material. In this figure we assume a recombination defect at some depth. For low beam energies, where the ehp's are generated near the top surface, most of the minority carriers are collected by the space-charge region and the current does not vary with distance [Fig. 10.17(a)]. The beam penetration at higher energies is sufficient that some of the ehp's recombine at the defect causing the collected current to decrease in the vicinity of the defect [Fig. 10.17(b)]. This example illustrates lateral as well as depth uniformity measurements by sweeping the beam and by varying the beam energy.

The current can be displayed as a line scan along one line on the sample, or it can be displayed as a brightness map on the SEM CRT. It can also be displayed as a pseudo three dimensional plot in which the z-axis represents the current and other two axes represent the x- and y-directions of the sample. An EBIC brightness map and a line scan across the sample are shown in Fig. 10.18.

Typical applications for EBIC include the measurement of minority carrier diffusion length and lifetime, recombination sites (dislocations, precipitates, grain boundaries), doping concentration inhomogeneities, and junction location. Due to the noncontacting nature of the electron beam, it is possible to scan small regions of the sample. For example, in a study of recombination behavior of twin planes in dendritic web silicon, the Schottky contact was formed on the 100 μm-thick wafer cross section, and the beam was scanned across it to reveal recombination activity at the twin planes.[72]

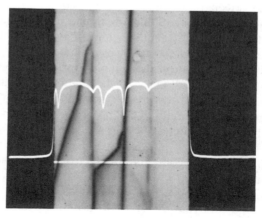

Fig. 10.18 EBIC map of polycrystalline Si showing high recombination grain boundaries. A line scan is taken along the horizontal marker line and the EBIC along that line is displayed. Courtesy of J. B. Mohr, Arizona State University.

Cathodoluminescence   *Cathodoluminescence* (CL) is a contactless technique based on the emission of light from a sample excited by an electron beam.[73–76] CL is related to both EMP (e-beam excitation, X-ray emission) and photoluminescence (light excitation, light emission). Photoluminescence is generally a non-imaging technique in which light is collected from the entire illuminated sample. CL's strength lies in its imaging capability. The electron beam is scanned across the sample, and the emitted light is detected and displayed on a CRT. The chief difference between EMP and CL, with both employing e-beams for excitation, is that EMP X-rays originate from electronic transitions between inner-core energy levels whereas CL photons originate from transitions between conduction and valence bands.

CL brightness maps can be related to sample recombination behavior through the external photon quantum efficiency η (emitted photons per incident electron) given by[74]

$$\eta = \frac{(1 - R)(1 - \cos \theta_c)}{(1 + \tau_{rad}/\tau_{nrad})} \, e^{-\alpha d} \tag{10.15}$$

where $1 - R$ accounts for reflection losses at the semiconductor-vacuum interface, $1 - \cos \theta_c$ accounts for internal reflection losses, $\exp(-\alpha d)$ accounts for internal absorption losses where d is the photon path length, and τ_{rad}, τ_{nrad} are the radiative and nonradiative minority carrier lifetimes.

All of the factors in Eq. (10.15) can be spatially dependent and can contribute to CL image contrast, making quantitative interpretation difficult.[77] For example, there may be local reflectance variations, and surface morphology can produce shadowing or enhanced light emission through changes in the $1 - \cos \theta_c$ term. Samples consisting of semiconductors with varying band gaps can have varying absorption coefficients. The major CL contrast, however, is usually due to lifetime variations. Mechanisms causing enhanced or reduced light emission include dopant concentrations, temperature, recombination centers (metallic impurities, dislocations, stacking faults, precipitates), and the presence of electric fields.

In the simplest implementation the sample is kept at room temperature, and the CL light is collected. This panchromatic technique is useful for quick data collection. Distinct advantages are gained by cooling the sample and resolving the light spectrally. Cooling to liquid helium temperature reduces thermal line broadening and raises the signal/noise ratio. Resolving the light into its spectral components can lead to impurity identification. CL resolution is determined by a combination electron beam diameter, electron range R_e, and minority carrier diffusion length L. For $L \ll R_e$ the resolution is essentially R_e, and for $L \gg R_e$ it is essentially L.

CL is mainly used for III-V materials with its high radiative recombination. It is more difficult to use for Si due to its low luminescence efficiency. Of course CL emission can be enhanced with higher beam current, but that leads to sample heating. CL has found its main application in defect studies, such as dislocations and precipitates.[76-77] Time-resolved CL is useful for lifetime measurements, with both bulk and surface recombination contributing to the effective lifetime.[76] The technique can be combined with other methods in the SEM (EBIC, scanning electron microscopy, EMP) for a more complete analysis. CL implementation is also possible in transmission electron microscopes. But space limitations in the instrument and small sample size make light collection more difficult.

Low-Energy, High-Energy Electron Diffraction *Low-energy electron diffraction* (LEED), first demonstrated in 1927 by Davisson and Germer,[78] is one of the oldest surface characterization techniques for investigating the crystallography of sample surfaces.[7,79-82] It provides structural, not elemental information, and is illustrated in Fig. 10.19(a). A low-energy (20 to 300 eV), narrow-energy spread electron beam incident on the sample penetrates only the first few atomic layers. Electrons are diffracted by the periodic atomic arrangement of the sample. The elastically scattered, diffracted electrons emerge from the surface in directions, satisfying interference conditions from the crystal periodicity and strike a fluorescent screen

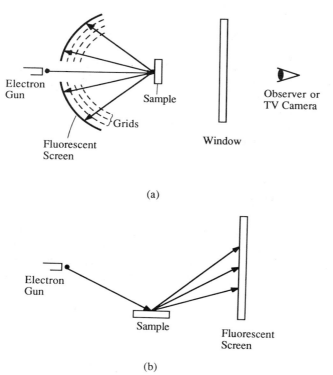

(a)

(b)

Fig. 10.19 (a) LEED diffractometer, (b) RHEED diffractometer.

forming a distinct array of diffraction spots due to the orientation of the crystal lattice of the sample. The diffraction pattern is viewed through a window behind the screen. The pattern can also be photographed or viewed with a television camera. A series of grids filter the scattered electrons from the sample.

LEED provides information on the atomic arrangement and is sensitive to crystallographic defects. The diffraction conditions can be most easily studied using a reciprocal lattice and an Ewald sphere.[81] Sample preparation is important. To study the properties of the surface, it is important that there be few contaminants on the surface because contaminated surfaces generally do not give diffraction patterns. Consequently LEED measurements are generally made in an ultra-high vacuum (UHV) of less than 10^{-10} torr. A monolayer of contamination takes about one second to form at a pressure of 10^{-6} torr but takes about one day at 10^{-10} torr. Even a fraction of a monolayer is sufficient to prevent accurate surface crystallography measurements. Samples should be cleaved in vacuum, if at all possible, to expose the appropriate surfaces that have not been subjected to ambient contamination.

Electron diffraction by high-energy electrons is known as *reflection high-energy electron diffraction* (RHEED).[80–81] As shown in Fig. 10.19(b) 1 to 100-keV electrons are incident on the sample, but because such energetic electrons penetrate deeply, they are made to strike the sample at a shallow, glancing angle of typically less than 5°. Forward-scattered electrons are utilized as there is little backscattering. Molecular beam epitaxial growth (MBE) has done much to foster the use of RHEED by allowing continuous monitoring of the growth of epitaxial films.[83] The experimental arrangement of Fig. 10.19(b) leaves the front of the sample clear for growth beams. Additionally, since the electron beam strikes the sample at a glancing angle and picks out surface irregularities more effectively than LEED, it is a more critical characterization method.

10.3   ION BEAM TECHNIQUES

Ion beam characterization techniques are illustrated in Fig. 10.20. Incident ions are absorbed, emitted, scattered, or reflected and can in turn cause light, electron, or X-ray emission. Aside from characterization, ion beams are also used for processing as in ion implantation. We discuss two main ion beam material characterization methods: *secondary ion mass spectrometry* (SIMS) and *Rutherford backscattering spectrometry* (RBS).

10.3.1   Secondary Ion Mass Spectrometry

Principle   *Secondary ion mass spectrometry* (SIMS), also known as *ion microprobe* and *ion microscope*, is one of the most powerful and versatile

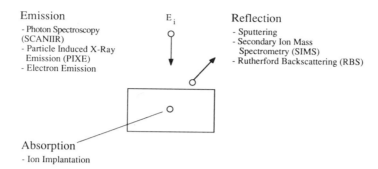

Fig. 10.20   Ion beam characterization techniques.

analytical techniques for semiconductor characterization.[1–10,84–85] It was developed independently by Castaing and Slodzian at the University of Paris in France and by Herzog and collaborators at the GCA Corp. in the United States in the early 1960s but did not become practical until Benninghoven showed that it was possible to maintain the surface integrity for periods well in excess of the analysis time.[86–89] Benninghoven did much to further the evolution and advances of SIMS. The technique is element specific and is capable of detecting all elements as well as isotopes and molecular species. Of all the beam techniques it is the most sensitive, with detection limits for some elements in the 10^{14} to 10^{15} cm^{-3} range. Lateral resolution is typically 100 μm but can be as small as 1 μm, and depth resolution is 50 to 100 Å.

The basis of SIMS, shown in Fig. 10.21, is the destructive removal of material from the sample by sputtering and the analysis of that material by a mass spectrometer. A primary ion beam impinges on the sample, and atoms from the sample are sputtered or ejected from the sample. Most of the ejected atoms are neutral and cannot be detected by conventional SIMS. But a small fraction is ejected as positive or negative ions. This fraction was estimated as about 1% of the total in 1910,[90] an estimate that is still considered reasonable today.[91] The mass/charge ratio of the ions is analyzed, detected as a mass spectrum, as a count, or displayed on a fluorescent screen or on a CRT.

Sputtering is a process in which incident ions lose their energy mainly by momentum transfer as they come to rest within the solid. In the process they displace atoms within the sample. The incident projectiles need not be ions; neutral beam bombardment causes sputtering also, but ions are used in SIMS. Sputtering takes place when atoms near the surface receive sufficient energy from the incident ions to be ejected from the sample. The escape depth of the sputtered atoms is generally a few monolayers for primary energies of 10 to 20 keV typically used in SIMS. The primary ion loses its energy in the process and comes to rest tens to hundreds of angstroms below the sample surface. Ion bombardment leads not only to sputtering but also

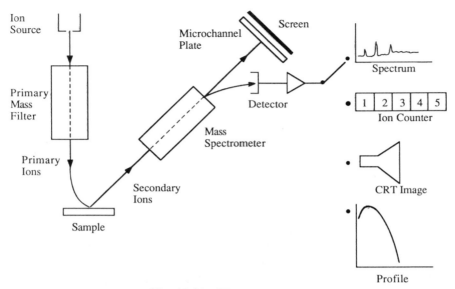

Fig. 10.21 SIMS schematic.

to ion implantation and lattice damage. The sputtering yield is the average number of atoms sputtered per incident primary ion; it depends on the sample or target material, its crystallographic orientation, and the nature, energy, and incidence angle of the primary ions. Selective or preferential sputtering can occur in multicomponent or polycrystalline targets when the components have different sputtering yields. The component with lowest yield becomes enriched at the surface, whereas that with the highest yield becomes depleted. However, once an equilibrium situation is reached, the sputtered material leaving the surface has the same composition as the bulk material and preferential sputtering is not a problem in SIMS analysis.[92]

The yield for SIMS measurements with 1- to 20-keV energy Cs^+, O_2^+, O^-, and Ar^+ ions ranges from 1 to 20. What is important, however, is the not the total yield, but the yield of ionized ejected atoms or the *secondary ion yield* because only ions can be detected. The secondary ion yield is significantly lower than the total yield but can be influenced by the type of primary ion. Electronegative oxygen (O_2^+) is a secondary ion yield-enhancing species for electropositive elements (e.g., B and Al in Si) which produce predominantly positive secondary ions. The situation is reversed for electronegative elements (e.g., P, As, and Sb in Si), having greater yields when sputtered with electropositive ions like cesium (Cs^+). The secondary ion yield for the elements varies over five to six orders of magnitude.[93]

SIMS has not only a wide variation in secondary ion yield between different elements, it also shows strong variations in the secondary ion yield from the same element in different samples or matrices. The latter is the well-known *matrix effect*. For example, the secondary ion yield for oxidized surfaces is higher than for bare surfaces by as much as 1000.[93] A striking example is a SIMS profile of B or P implanted into oxidized Si obtained by sputtering through an oxidized Si wafer. The yield of Si in SiO_2 is about 100 times higher than the yield of Si from the Si substrate. A plot of yield versus sputtering time shows a sharp drop when the sample is sputtered through the SiO_2-Si interface.

SIMS can give three types of results. For a low sputtering rate (~1 Å per hour), a complete mass spectrum can be recorded for surface analysis of the outer 5 Å or so. This mode of operation is known as *static* SIMS. The intensity of one peak for one particular mass can be recorded as a function of time as the sample is sputtered at a higher sputter rate (~10 μm/h), yielding a depth profile. This is known as *dynamic* SIMS. It is also possible to display the intensity of one peak as a two-dimensional image. The various output signals are illustrated in Fig. 10.21.

Quantitative depth profiling is unquestionably the major strength of SIMS. It yields one selected mass plotted as secondary ion yield versus sputtering time. Such a plot must be converted to concentration versus depth. The conversion of signal intensity to concentration can, in principle, be calculated knowing the primary ion beam current, the sputter yield, the ionization efficiency, the atomic fraction of the ion to be analyzed, and an instrumental factor. Some of these factors are generally poorly known, and a successful technique for routine quantitative SIMS analysis has not yet emerged.[84] The usual approach is one of using standards with composition and matrices identical or similar to the unknown. Ion-implanted standards are very convenient and also very accurate. The implant dose of an ion-implanted standard can be controlled to an accuracy of 5% or better. When such a standard is measured, one calibrates the SIMS system by

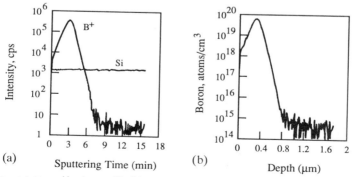

Fig. 10.22   (a) Raw $^{11}B^+$ and $^{30}Si^{2+}$ secondary ion signals versus sputtering time, (b) boron profile for a boron implant into a silicon substrate. Reprinted after Pantano,[84] with permission from *Metals Handbook*, Vol. 10, Materials Characterization, American Society for Metals, Metals Park, OH 1986.

integrating the secondary ion yield signal over the entire profile. Calibrated standards are therefore very important for accurate SIMS measurements. The time-to-depth conversion is usually made by measuring the sputter crater depth after the analysis is completed. An example of the conversion of yield or intensity versus time to concentration versus depth profile is given in Fig. 10.22, where both the raw SIMS plot and the concentration profile are shown.

Instrumentation   There are two instrumentation approaches to SIMS: the *ion microprobe* and the *ion microscope*. A good discussion of SIMS instrumentation is given by Bernius and Morrison.[94] The ion microprobe is an ion analog of the electron microprobe. The primary ion beam is focused to a fine spot and rastered over the sample surface. The secondary ions are mass analyzed, and the mass spectrometer output signal is displayed on a CRT in synchronism with the primary beam to produce a map of secondary ion intensity across the surface. The spatial resolution is determined by the spot size of the primary ion beam, and resolutions of a few microns are possible. An energy-filtered, quadrupole mass spectrometer is commonly used with the ion microprobe.[95] It consists of four parallel rods, with dc and rf signals allowing ions of one given charge/mass ratio entering the quadrupole to follow a stable oscillatory trajectory between the rods to reach the detector. All other ions strike the spectrometer walls and are unable to reach the detector. The ion mass is typically scanned by holding the frequency and the dc/rf ratio constant while varying the dc and rf amplitudes.[96] An energy filter placed in front of the quadrupole analyzer to reduce the energy spread to a few electron volts ensures good mass resolution.

The ion microscope is a direct imaging system, analogous to an optical microscope or a TEM. The primary ion flood beam illuminates the sample and secondary ions are simultaneously collected over the entire imaged area. The spatial distribution of the secondary image is preserved through the system using a magnetic sector mass spectrometer, amplified by a microchannel plate and displayed on a fluorescent screen. A small aperture may be inserted to select an area for analysis with a resolution of 1 μm or better. In the magnetic sector mass spectrometer the ions are energy filtered, accelerated by a potential, and injected into a uniform magnetic field where they follow circular trajectories and only those ions with a certain value of charge/mass are able to reach the detector.[93] Appropriate masses are selected by changing the magnetic field. The mass-analyzed signal can also be detected with a Faraday cup and displayed as an amplitude for both types of instruments.

Proper mass resolution is essential for unambiguous SIMS analysis. For example, a SIMS mass/charge (m/e) spectrum for high-purity Si obtained with an O_2^+ primary ion beam is shown in Fig. 10.23(a). The spectrum shows

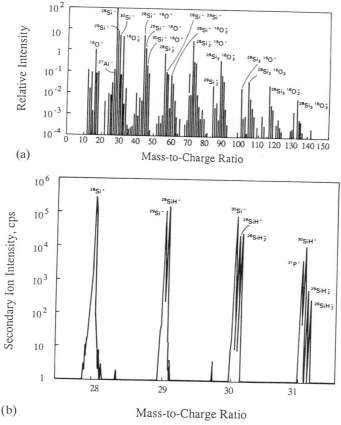

Fig. 10.23 SIMS spectrum for (a) high-purity Si, (b) phosphorus-doped Si. Both spectra were obtained with an O_2^+ primary ion beam. Reprinted after Pantano,[84] with permission from *Metals Handbook*, Vol. 10, Materials Characterization, American Society for Metals, Metals Park, OH 1986.

the $^{28}Si^+$, $^{29}Si^+$, and $^{30}Si^+$ isotopes, polyatomic Si_2^+ and Si_3^+, as well as many molecular species involving oxygen. The latter are not from the sample itself but are due to the oxygen primary beam causing oxygen implantation and subsequent sputtering. This plethora of signals requires a high-resolution spectrometer. For example, if iron is a suspected contaminant in silicon, it is difficult to detect because Si_2^+ (m/e = 55.9539) interferes with Fe^+ (m/e = 55.9349). An interference problem exists also for phosphorus in silicon, as shown in Fig. 10.23(b) where the interference at m/e ≈ 31 between Si–H complexes and $^{31}P^+$ is clearly obvious. The hydrogen may be in the sample or in the residual vacuum. The kind of resolution required for phosphorus detection requires a double-focusing magnetic sector spectrometer; a quadrupole mass spectrometer has insufficient resolution.

Another instrumentation effect that complicates SIMS analysis is the *edge effect*. To obtain good depth resolution, it is important that only the signal from the flat, bottom portion of the sputtered crater be analyzed. In ion-implanted samples, atoms are ejected from the crater bottom as well as from the sidewalls as sputtering proceeds. But the sidewalls, especially near the top surface, contain a much higher doping concentration than the crater bottom. Using electronic gating of the secondary ion yield signal or a lens system, it is possible to detect only those ions from the central part of the crater.[42]

A major source of the limited sensitivity of SIMS is the fact that most of the sputtered material is neutral and cannot be detected. In *secondary neutral mass spectrometry* (SNMS) the neutral atoms are ionized by a laser or by an electron gas and then detected.[97-98] Significant sensitivity enhancements over conventional SIMS are achieved. The primary ion beam in SIMS is replaced by a pulsed laser in the *laser microprobe mass spectrometer* (LAMMA) or *laser ionization mass analysis* (LIMA).[99-101] The pulsed laser volatizes and ionizes a small volume of the sample, and the ions are analyzed in a time-of-flight mass spectrometer. LAMMA has high sensitivity, high speed of operation, is applicable to inorganic as well as organic samples, and has microbeam capability with a spatial resolution of ~1 μm.

Applications   SIMS has found its greatest utility in semiconductor characterization. In particular, it has found widespread application in dopant profiling of implanted and diffused layers.[9,42,91,102-103] for a more detailed discussion and comparison with spreading resistance measurements, see Section 2.8. SIMS measurements are well suited for semiconductor applications, because matrix effects are of minor consequence and ion yields can be assumed to be linearly proportional to concentrations up to 1%. Furthermore the substrate sputters very uniformly, at least for Si. SIMS has also found applications in metal, biology, and geology studies.[91]

Factors that need to be considered in data analysis are crater edge effects, ion knock-on, atomic mixing, diffusion, preferential sputtering, and surface roughness. Some of these are instrumental and can be alleviated to some extent, but others are intrinsic to the sputtering process. Knock-on causes atomic mixing and permanent displacement of atoms, changing their original distribution and contributing to slightly deeper junctions when junction depths measured by SIMS are compared to junction depths measured by spreading resistance.[104] A high vacuum is very important for SIMS. The arrival rate of gaseous species from the vacuum chamber should be less than that of the primary ion beam; otherwise, it is vacuum contamination that is measured, not the sample. This is particularly important for low mass species like hydrogen. A very thorough discussion of these effects can be found in the papers by Zinner listing 35 factors affecting SIMS depth profiling.[42,105]

10.3.2   Rutherford Backscattering Spectrometry

Principle   *Rutherford backscattering spectrometry* (RBS) is based on back-scattering of ions or projectiles incident on a sample.[8,106–109] The technique is also known as *high-energy ion (back)-scattering spectrometry* (HEIS). It is quantitative without recourse to calibrated standards. Experiments by Rutherford and his students in the early 1900s proved the existence of nuclei and scattering from these nuclei.[110] The field of ion interactions in solids was very intensively researched and developed following the discovery of fission and nuclear weapons development. But it was not until the late 1950s that nuclear backscattering was put to practical use to detect a variety of elements.[111] Further developments in the 1960s led to identification of minerals[112] and determination of properties of thin films as well as thick samples. More recently the concepts have been adopted and implemented for fairly routine materials characterization.

RBS is based on bombarding a sample with energetic ions—typically He ions of 1 to 3 MeV energy—and measuring the energy of the backscattered ions. It allows determination of the *masses* of the elements in a sample, their *depth distribution* over distances from 100 Å to a few microns from the surface, and the *crystalline structure* in a non-destructive manner. The depth resolution is on the order of 100 Å. The use of ion backscattering as a quantitative materials analysis tool depends on an accurate knowledge of the nuclear and the atomic scattering processes. Fortunately these are generally very well known.

The method is illustrated in Fig. 10.24. Ions of mass M_1, atomic number Z_1, and energy E_0 are incident on a solid sample or target composed of atoms of mass M_2 and atomic number Z_2. Most of the incident ions come to rest within the solid losing their energy through interactions with valence electrons. A small fraction—around 10^{-6} of the number of incident ions—undergoes elastic collisions and is backscattered from the sample at various angles. The incident ions lose energy, traversing the sample until they experience a scattering event and then lose energy again as they travel back

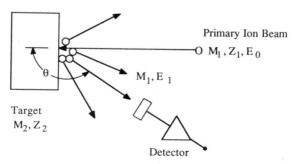

Fig. 10.24   Rutherford backscattering schematic.

to the surface leaving the sample with reduced energy. For those incident ions scattered by surface atoms, conservation of energy and momentum leads to a relationship of the energy *after* scattering E_1 to the incident energy E_0 through the *kinematic factor* K[113-114]

$$K = \frac{E_1}{E_0} = \frac{[\sqrt{1 - (R \sin \theta)^2} + R \cos \theta]^2}{(1 + R)^2} \approx 1 - \frac{2R}{(1 + R)^2} (1 - \cos \theta)$$

(10.16)

where $R = M_1/M_2$ and θ is the scattering angle. The approximation in Eq. (10.16) holds for $R \ll 1$ and θ close to 180°. Equation (10.16) is the key RBS equation. It takes a particularly simple form for $\theta = 90°$ and $\theta = 180°$:

$$K(\theta = 90°) = \frac{1 - R}{1 + R} \approx (1 - 2R)$$

$$K(\theta = 180°) = \frac{(1 - R)^2}{(1 + R)^2} \approx (1 - 4R)$$

where the approximations hold for $R \ll 1$.

The kinematic factor is a measure of the primary ion energy loss. The scattering angle should be as large as possible. It is obviously impossible for the ions to be scattered by 180° (ions scattered back into the source), but angles of 100° to 170° are commonly used. The unknown mass M_2 is calculated from the measured energy E_1 through the kinematic factor.

We illustrate the use of RBS with the two examples in Fig. 10.25. In Fig. 10.25(a) we show a silicon substrate with a very thin film (approximately one monolayer coverage) of nitrogen, silver, and gold. The atomic weight and calculated R, K, and E_1 in Table 10.2 are for $\theta = 170°$ and incident helium ions ($M_1 = 4$) with $E_0 = 2.5$ *MeV*. Helium ions with energies of 0.78, 1.41,

TABLE 10.2   Calculated R, K, and E_1 (for 2.5 MeV He Ions)

Target Atom (M_2)	Atomic Weight	R	K	E_1 (MeV)
N	14	0.256	0.311	0.78
O	16	0.25	0.363	0.91
Si	28.1	0.142	0.566	1.41
Cu	63.6	0.063	0.779	1.95
Ag	107.9	0.037	0.863	2.16
Au	197	0.020	0.923	2.31

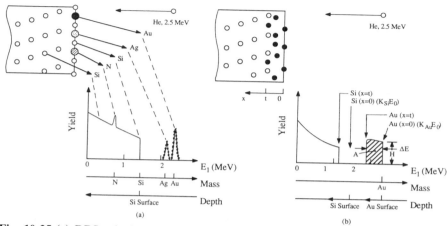

Fig. 10.25 (a) RBS calculated spectrum for N, Ag, and Au on Si. The experimental data are adapted from Feldman and Mayer.[4] Schematic spectrum for a Au film on Si; '*A*' is the area under the curve.

2.16, and 2.31 MeV are scattered from the N, Si, Ag, and Au atoms at the sample surface. Since N, Ag, and Au is only at the surface in this example, RBS signals from these elements have a narrow spectral distribution confirmed by experimental data. The yield is not to scale on this figure.

Figure 10.25(a) brings out two important properties of RBS plots: (1) the RBS yield increases with element atomic number, and (2) the RBS signal of elements lighter than the substrate or matrix rides on the matrix background while elements heavier than the matrix are displayed by themselves. This makes the nitrogen signal more difficult to detect because it rides on the Si signal. The Si background count represents the "noise" and the signal-to-noise ratio is degraded compared to heavy elements on a light matrix.

RBS plots become slightly more complicated for layers of finite thicknesses. In Fig. 10.25(b) we consider a gold film of thickness t on a silicon substrate. He ions are backscattered from surface gold atoms with $E_{1,Au} = 2.31$ MeV as in Fig. 10.25(a). However, those ions backscattered from deeper within the Au film emerge with lower energies due to additional losses within the film. These losses come from coulombic interactions between gold ions and electrons with He ions. Consider a scattering event from those Au atoms at the Si–Au interface at $x = t$. The He ion loses energy ΔE_{in} traveling through the Au film before the scattering event at the back gold surface. Upon scattering, it loses additional energy $(E_0 - \Delta E_{in})(1 - K_{Au})$. To reach the detector, it must traverse the film a second time, losing energy ΔE_{out}. The total energy loss is the sum of these three losses. The energy of He ions scattered from the sample at depth t is

$$E_1(t) = (E_0 - \Delta E_{in})K_{Au} - \Delta E_{out} \tag{10.17}$$

The energy losses are slightly energy dependent and are listed in tables of stopping powers.[115] The energy difference of the ions backscattered from the surface and from the interface ΔE can be related to the film thickness by the relation

$$\Delta E = \Delta E_{in} K_{Au} + \Delta E_{out} = [S_0]t \qquad (10.18)$$

where $[S_0]$ is the backscattering energy loss factor; it is in units of eV/Å and is tabulated for pure-element samples. $[S_0] = 133.6 \, \text{eV/Å}$ for gold films.

The *backscattering yield* A also designated as the total number of detected ions or counts is given by

$$A = \sigma \Omega Q N_s \qquad (10.19)$$

where σ is the average scattering cross section in cm^2/sr, Ω the detector solid angle in steradians (detector area/(detector-sample distance)^2), Q the total number of ions incident on the sample, and N_s the sample atom concentration/cm^2. The total count A is also the area under the experimental yield–energy curve or the total number of detected He ions backscattered from the element of interest or the sum of the counts in each channel, shown on Fig. 10.25(b) as A. $N_s = Nt$ for a thin film where N is the atom concentration/cm^3. Q is determined by the time integration of the current of charged particles incident on the target, but it is difficult to determine accurately due to secondary sample electron emission. The scattering cross section is[116–117]

$$\sigma = \left(\frac{Z_1 Z_2 q^2}{4E}\right)^2 \frac{4}{\sin^4\theta} \frac{[\sqrt{1 - (R\sin\theta)^2} + \cos\theta]^2}{\sqrt{1 - (R\sin\theta)^2}} \qquad (10.20)$$

E is the energy of the projectile immediately before scattering. Values for σ are tabulated for all elements for He probe ions. Typical values of the scattering cross section are 1 to $10 \times 10^{-24} \, \text{cm}^2/\text{sr}$. The yield increases with increasing atomic number leading to higher RBS sensitivity for high Z elements than for low Z elements. However, due to the kinematics of scattering, high-mass elements are more difficult to distinguish from one another than low-mass elements.

The areal density N_s is determined from the yield according to Eq. (10.19) which can be cast in a different form because it may be difficult or tedious to determine Q accurately. Furthermore the detector solid angle Ω may change if the detector develops "dead" spots after prolonged exposure to energetic projectiles.

The energy width of a single channel in the multichannel analyzer δE_1 corresponds to a depth uncertainty δx

$$\delta E_1 = [S_0]\delta x = N[\varepsilon]\delta x \qquad (10.21)$$

where $[\varepsilon] = (1/N)\, dE/dx$ is the backscattering stopping cross section.[118–119] δE_1 is determined by the detector and the electronic system and is typically 2 to 5 keV. The total count A can be written as

$$A = H \frac{\Delta E}{\delta E_1} \tag{10.22}$$

where H is the height (count/channel) of the spectrum and ΔE is the width of the spectrum shown in Fig. 10.25(b). Solving for ΩQ from Eqs. (10.19), (10.21), and (10.22) and substituting into Eq. (10.19) gives

$$N_s = \frac{A\, \delta E_1}{H[\varepsilon]} \tag{10.23}$$

An unknown impurity on a known substrate, for example, impurity X on a Si substrate, is determined from the relationship[106]

$$(N_s)_X = \frac{A_X}{H_{Si}} \frac{\sigma_{Si}}{\sigma_X} \frac{\delta E_1}{[\varepsilon]_{Si}} \tag{10.24}$$

To find the unknown concentration, it is only necessary to determine the RBS spectrum area, the height of the Si spectrum, and look up the two cross sections and the Si stopping cross section. Typical values for the backscattering stopping cross section lie in the 10 to 100 eV/(10^{15} atoms/cm^2) range with $[\varepsilon]_{Si} = 49.3$ eV/(10^{15} atoms/cm^2) and $[\varepsilon]_{Au} = 115.5$ eV/(10^{15} atoms/cm^2) for 2 MeV He ions.[8]

The RBS spectrum of the thick Si substrate in Fig. 10.25 has a characteristic slope with the yield increasing at lower energies. This comes about from scattering within the target. The yield is inversely proportional to the ion energy at depth t_1. The yield at energy E_1, the energy of those ions backscattered from atoms at depth t_1, is proportional to $(E_0 + E_1)^{-2}$. Hence the yield increases as E_1 decreases at greater depths into the target.

Instrumentation   An RBS system consists of an evacuated chamber that contains the He ion generator, the accelerator, the sample, and the detector. Negative He ions are generated in the ion accelerator at close to ground potential. In a tandem accelerator these ions are accelerated to 1 MeV, traversing a gas-filled tube or "stripper canal," where either two or three electrons are stripped from the He$^-$ to form He$^+$, respectively.[114] These ions with energies of around 1 MeV are accelerated a second time to ground potential, at which point the He$^+$ ions have 2 MeV and the He^{2+} ions have 3 MeV energy. A magnet separates the two high-energy species.

In the sample chamber the He ions are incident on the sample and the backscattered ions are detected by a Si surface barrier detector, that operates much like the X-ray EDS detector described in Section 10.2.3. The energetic ions generate many electron–hole pairs in the detector, resulting in output voltage pulses from the detector. The pulse height, proportional to the incident energy, is detected by a pulse height or multichannel analyzer that stores pulses of a given magnitude in a given voltage bin or channel. The spectrum is displayed as yield or counts versus channel number with channel number proportional to the incident energy. The energy resolution of Si detectors, set by statistical fluctuations, is around 10 to 20 keV for typical RBS energies. The sample is mounted on a goniometer for precise sample/beam alignment or channeling measurements. Typical RBS runs take 15 to 30 minutes.

Applications   Typical semiconductor applications include measurements of thickness, thickness uniformity, stoichiometry, nature, amount, and distribution of impurities in thin films, such as silicides and Si- and Cu-doped Al. The technique is also very useful to determine the crystallinity of a sample to determine if a sample is single crystal or amorphous. Backscattering is strongly affected by the alignment of atoms in a single-crystal sample with the incident He ion beam. If the atoms are well aligned with the beam, those He ions falling between atoms in the channels, penetrate deeply into the sample and have a low probability of being backscattered. Those He ions that encounter sample atoms "head on" are, of course, scattered. The yield from a well-aligned single-crystal sample can be two orders of magnitude less than that from a randomly aligned sample. This effect is referred to as *channeling* and has been extensively used to study ion implantation damage in semiconductors, with the yield decreasing as the single-crystal nature of an implanted sample is restored by annealing.[120]

RBS is particularly suited for heavy elements on light substrates. Contacts to semiconductors generally fall into this category. Consequently RBS has been used extensively in the study of such contacts. For example, Fig. 10.26 shows RBS spectra for platinum and platinum silicide on silicon. Initially a Pt film is deposited on a Si substrate. The RBS spectrum clearly shows the Pt film. The Si signal is consistent, with E_1 taking into account the loss into and out of the Pt film. As the film is heated, PtSi forms. Note the formation from the Pt-Si interface indicated by the Pt yield decrease for that part of the film near the Si substrate. At the same time the Si signal moves to higher energies, indicative of Si moving into the Pt film. When stoichiometry is attained the Pt signal is uniform, but reduced and the Si signal has risen. It would have been very difficult to obtain these data with other techniques nondestructively.

RBS can provide both atomic composition and depth scales to accuracies of 5% or better. The detection limit lies in the 5×10^{20} cm^{-3} range but depends on the element and on energy. The sensitivity to light elements in the presence of heavier ones is poor. However, RBS is particularly good at detecting small amounts of surface impurities in the channeling mode. For example, a layer of selenium of 1.6×10^{13} atoms/cm^2 (~0.01 monolayer) on a Si substrate that led to poor epitaxial layers was detected and traced to a

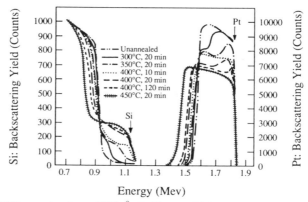

Fig. 10.26 RBS spectra for a 2000 Å Pt film on Si before and after heat treatment. Platinum silicide is formed first at the interface and then throughout the film, $E_0 = 2$ MeV. Reprinted with permission from Science **177**, 841 (1972), Nicolet et al.[121] Copyright 1972 by the AAS.

defective rubber O-ring in a sink, allowing the wafers rinsed there to be contaminated.[114]

Typical depth resolutions are 100 to 200 Å for film thicknesses of ≤ 2000 Å. The penetration depth of 2 MeV He ions is about 10 μm in silicon and 3 μm in gold. Beam diameters are commonly around 1 to 2 mm,[122] but microbeam backscattering with beam diameters as small as one micron has been reported.[123] Lateral nonuniformities over the area of the analyzing beam cannot be resolved. A particular difficulty is the ambiguity of RBS spectra, because the horizontal axis is simultaneously a depth and a mass scale. A light mass at the surface of a sample generates a signal that may be indistinguishable from that of a heavier mass located within the sample. Through the use of tabulated constants, experimental techniques such as beam tilting, detector angle changes, and incident energy variations as well as good analytical reasoning, sample analysis is usually successful, but additional information may have to be provided to resolve ambiguities. Computer programs are extensively used in spectrum analysis.[124] As with some other physical and chemical characterization techniques, the more is known about the sample before the analysis, the less ambiguous are the results. A comparison of RBS with SIMS is given by Magee.[125]

10.4 X-RAY AND GAMMA-RAY TECHNIQUES

X-ray interactions with a solid are illustrated in Fig. 10.27. Incident X-rays are absorbed, emitted, reflected, or transmitted and can in turn cause electron emission. We will discuss *X-ray fluorescence* and *X-ray photoelectron spectroscopy*, useful for chemical characterization, and briefly mention

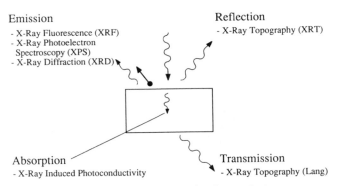

Fig. 10.27   X-ray characterization techniques.

X-ray topography, useful for structural characterization. Gamma rays, detected in *neutral activation analysis*, are included in this chapter for completeness.

10.4.1   X-Ray Fluorescence

Principle   In *X-ray fluorescence* (XRF), also known as *X-ray fluorescence spectroscopy* (XRFS), *X-ray fluorescence analysis* (XRFA), and *X-ray secondary emission spectroscopy*, primary X-rays incident on the sample are absorbed by ejecting electrons from the atomic K-shell as illustrated in Fig. 10.28.[2,126–131] Electrons from higher lying levels, for example, the L-shell, drop into the K-shell vacancies, and the energy liberated in the process is given off as characteristic secondary X-rays with energy

$$E_{XRF} = E_K(Z) - E_{L2,3}(Z) \tag{10.25}$$

The X-ray energy *identifies* the impurity, and the intensity gives its *concentration*. XRF allows nondestructive elemental analysis of solids and liquids and quantitative thin film analysis is readily obtained. It is not a high resolution method, as X-rays are difficult to focus, but the instrumentation is relatively inexpensive. Typical analysis areas are $1 \, cm^2$, although in recent instruments it is possible to analyze areas as small as 10^{-6} to $10^{-4} \, cm^2$.[132] The method is suitable for conductors as well as for insulators since neutral X-rays cause no charging of insulators.

Conventional XRF is not a surface-sensitive technique. As discussed in Section 10.2.3, X-rays penetrate some distance into the sample determined by the X-ray absorption coefficient. In Si the penetration depth is typically microns or tens of microns. For example, to detect X-rays emerging from the sample, it is reasonable to find the depth for 50% absorption since X-rays have to penetrate the sample and generate characteristic X-rays, which in turn have to be emitted to be detected. The 50% penetration depth

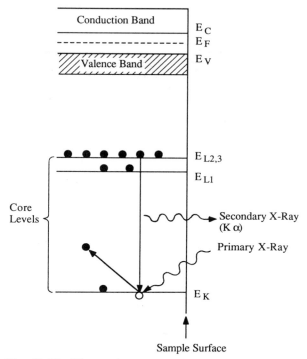

Fig. 10.28 Electronic processes in X-ray fluorescence.

of Cu Kα primary X-rays is 46 μm in Si according to Eq. (10.10). *Grazing incidence XRF* is surface sensitive. X-rays strike the sample at a very shallow angle and penetrate only a small distance into the sample.[133]

Instrumentation A beam of primary X-rays illuminates the sample, and secondary X-rays are detected by an energy dispersive (EDS) or a wavelength dispersive spectrometer (WDS). Both detectors are described in Section 10.2.3. EDS allows sodium and heavier elements to be detected. X-rays for the lighter elements are absorbed in the beryllium window. Conventional XRF sensitivity is around 0.01% or 5×10^{18} cm^{-3}, and the analysis area is on the order of 1 cm^2. Grazing incidence XRF is sensitive to surface contamination as low as 10^{11} cm^{-2}.[133]

Applications XRF is ideally suited for rapid initial sample survey to define subsequent, more detailed analyses. It is nondestructive and can be used in air for conductors, semiconductors, as well as insulators. It gives the average sample composition over the X-ray absorption depth rapidly but has no profiling capability. The technique has also found use for film thickness measurements. By establishing standards of a given film in which the thickness is measured independently, thicknesses of unknown films are easily determined by measuring the intensity of the secondary X-rays. Film thicknesses as thin as 100 Å can be determined. Standards are important for quantitative measurements, since XRF is subject to a *matrix effect*, which is the absorption of secondary X-rays by the sample itself. The standards should be well matched to the sample matrix. XRF requires no standards in the thin film approximation.[134]

XRF has also found application in determining the constituents of mixed conductors. For example, it is common practice in Si technology to add a small fraction of copper to aluminum to increase its electromigration resistance. The Cu fraction can be easily detected by XRF. Similarly glasses used to passivate Si chips are frequently doped with phosphorus to increase the ability to "flow" at moderate process temperatures. The phosphorus content of such glasses can be determined by XRF.

10.4.2   X-Ray Photoelectron Spectroscopy

Principle   *X-ray photoelectron spectroscopy* (XPS), also known as *electron spectroscopy for chemical analysis* (ESCA), is the high-energy version of the photoelectric effect proposed by Einstein in the early 1900s. It is primarily used for identifying chemical species at the sample surface, allowing all elements except hydrogen and helium to be detected. Hydrogen and helium can, in principle, also be detected but that requires a very good spectrometer. When photons of low energy (≤ 50 eV) are incident on a solid, they can eject electrons from the valence band; the effect is known as *ultraviolet photoelectron spectroscopy*, (UPS). When the photons are X-rays, the method is known as XPS.[135–140] Electrons can be emitted from any orbital, and photoemission occurs as long as the X-ray energy exceeds the binding energy. Although the principle of XPS was known for a long time, implementation had to await the introduction of a high-resolution spectrometer for the detection of the low-energy XPS electrons in the 1950s by Siegbahn and coworkers in Sweden.[141–142] He coined the term "electron spectroscopy for chemical analysis," but since other methods also give chemical information, it is more commonly known as XPS today. The early history and development of XPS has been well chronicled by Jenkin.[143–145]

The method is illustrated with the energy band diagram in Fig. 10.29 and the schematic in Fig. 10.30. Primary X-rays of 1 to 2 keV energy eject photoelectrons from the sample. The measured energy of the ejected electron at the spectrometer E_{sp} is related to the binding energy E_b, referenced to the Fermi energy E_F, by

$$E_b = h\nu - E_{sp} - q\phi_{sp} \tag{10.26}$$

where $h\nu$ is the energy of the primary X-rays and ϕ_{sp} the work function of the spectrometer (3 to 4 eV). The spectrometer and the sample are con-

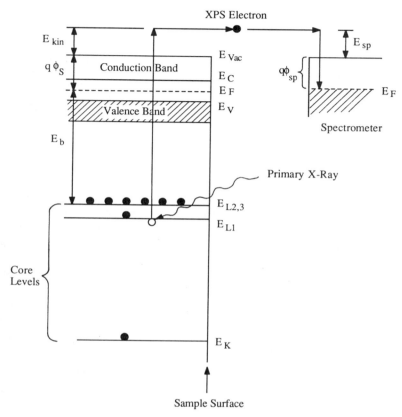

Fig. 10.29 Electron processes in X-ray photoelectron spectroscopy.

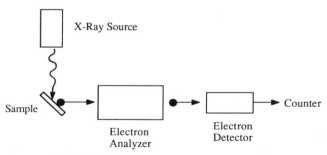

Fig. 10.30 XPS measurement schematic.

nected forcing their Fermi levels to line up. The Fermi energy of metals is well defined. Care must be taken in analyzing XPS data from semiconductors and insulators because E_F can vary from sample to sample.

The electron binding energy is influenced by its chemical surroundings, making E_b suitable for determining chemical states. Handbooks and graphs of binding energies for elements and compounds are available.[135,138,140,146] This leads to a major strength of XPS; it allows *chemical*, not only *elemental*, identification. Additionally X-rays tend to be less destructive, making the method more suitable for organics and oxides than AES. It is sometimes claimed that XPS causes no charging. Although it is true that X-rays possess no charge, electron emission from the sample may cause positive sample charging, especially for insulators. This can be compensated with an electron flood gun. X-ray induced Auger electron emission also occurs during XPS. Although such Auger lines can interfere with XPS lines, they can also be used to advantage. For example, varying the incident X-ray energy changes the energy of XPS electrons but not the energy of Auger electrons.

XPS is a surface-sensitive method because the emitted photoelectrons originate from the upper 5–50 Å of the sample, just as Auger electrons do, despite the deeper penetration of the primary X-rays compared to a primary electron beam.[147] The depth is governed by the electron escape depth or the related electron mean free path. Those electrons excited deeper within the sample are unable to exit the surface. Depth profiling is possible by ion beam sputtering or by sample tilting.[148] However, sputtering can alter oxidation states of the compound being measured. Sample tilting is the basis of *angle-resolved XPS* in which the sampling depth is $\lambda \sin \theta$, where θ is the angle between the sample surface and the trajectory of the emitted photoelectrons.[149]

The major use of XPS is for identification of compounds using energy shifts due to changes in the chemical structure of the sample atoms. For example, an oxide exhibits a different spectrum than a pure element. Care must be exercised in correctly interpreting the data. Unexpected peaks may appear for a variety of reasons.[4]

Instrumentation   The three basic components of XPS, shown in Fig. 10.30, are the X-ray source, the spectrometer, and a high vacuum even though such beam-induced chemistry as carbonization is minimized. X-ray line widths are proportional to atomic number of the target in the X-ray tube. The X-ray line width in XPS should be as narrow as possible; hence light elements like Al ($E_{K\alpha} = 1.4866 \, \text{keV}$) or Mg ($E_{K\alpha} = 1.2536 \, \text{keV}$) are common X-ray sources. Some XPS systems come equipped with multiple anode X-ray sources. X-ray generation from low Z materials also has reduced background radiation. The primary X-rays may be filtered by crystal dispersion to remove X-ray satellites and continuum radiation, but filtering reduces the X-ray intensity substantially. The XPS electrons are detected by one of several types of detectors. The cylindrical mirror analyzer (CMA) discussed in Section 10.2.2 is one of them. A double-pass CMA is generally used to discriminate against small energy shifts.

Chemical compounds or elements are identified by the location of energy peaks on the undifferentiated XPS spectrum. Concentration determination is more difficult. Peak heights and peak areas can be used with appropriate correction factors to obtain concentrations, but the method is primarily used for identification. X-ray techniques are generally large-area ($\sim 1\,cm^2$) methods. The analyzed sample area in XPS has been reduced over the years. Today 150 μm spot size is about the smallest that can be analyzed.[150] This has come about by either focusing the X-rays with a monochromator crystal or using a large-area X-ray beam but only allowing electrons from a small sample area to enter the electron analyzer. XPS sensitivity is around 0.1% or $5 \times 10^{19}\,cm^{-3}$.

Applications   XPS is used primarily for chemical surface information. It is particularly useful for analysis of organics, polymers, and oxides. For example, it has been used to follow the oxidation of elements. In Fig. 10.31 we show XPS spectra of lead in its pure form, and the spectral changes when the Pb oxidizes, first in the form PbO and later in the form of PbO_2. XPS has been extensively used in the semiconductor industry for a variety of problem solving. It has played a major role in understanding the chemistry and reaction mechanisms in the development of plasma etching. XPS has been applied to die attachment problems, adhesion of resins to metal surfaces, and interdiffusion of nickel through gold.[151] It is known that the vacuum environment can have a substantial influence on silicides in Si technology through oxide formation. XPS was very effective in investigating the effect of various vacuum environments on Ti/Si and Pt/Si formation.[152]

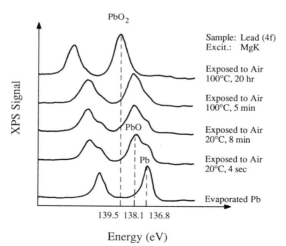

Fig. 10.31   XPS binding energy shifts of lead as an oxide forms. Reprinted with permission after Buckley.[80]

10.4.3   X-Ray Topography

X-Ray topography (XRT) is a nondestructive technique for determining structural crystal defects.[153–157] It requires little sample preparation and gives structural information over entire semiconductor wafers, but it does not identify impurities as most of the other techniques in this chapter do. The XRT image is not magnified because no lenses are used. It is therefore not a high-resolution technique. But it does give microscopic information through photographic enlargement of the topograph.

Consider a perfect crystal arranged to diffract monochromatic X-rays of wavelength λ from lattice planes spaced d. The X-rays are incident on the sample at an angle α, as shown in Fig. 10.32(a). The primary beam is

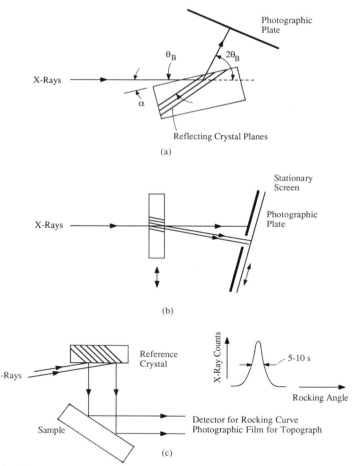

Fig. 10.32   (a) Berg-Barrett reflection topography, (b) Lang transmission topography, (c) double-crystal topography with a rocking curve.

absorbed by or transmitted through the sample; only the diffracted beam is recorded on the film. The diffracted beam emerges at twice the Bragg angle θ_B defined by

$$\theta_B = \sin^{-1}\left(\frac{\lambda}{2d}\right) \qquad (10.27)$$

The diffracted X-rays are detected on a high-resolution, fine-grained photographic plate or film held as close as possible to the sample without intercepting the incident beam. The plate should be held perpendicular to the secondary X-rays for highest resolution. If the lattice spacing or lattice plane orientation vary locally due to structural defects, Eq. (10.27) no longer applies simultaneously to the perfect and the distorted regions. Consequently there is a difference in X-ray intensity from the two regions. For example, the diffracted beam from dislocations is more intense than from an area without defects caused by the mitigation of extinction and by Bragg defocusing. Dislocations produce a more heavily exposed image on the film. The image is formed as a result of diffraction from an anomaly such as strain in the crystal but does not image the defect directly.

The reflection method illustrated in Fig. 10.32(a), known as the *Berg-Barrett* method, is based on the original work of Berg, modified by Barrett and further refined by Newkirk.[158–160] It is the simplest X-ray topography method. There are neither lenses nor moving parts except for the goniometer for sample alignment. Reflection XRT probes a thin sample region near the surface since the shallow incident angle α confines X-ray penetration to the near-surface region. This method is used to determine dislocations, for example, and is useful for dislocation densities up to about $10^6\,\mathrm{cm}^{-2}$. The resolution is about one micron, and areas as large as 75 mm diameter wafers can be examined with the Berg-Barrett method.

Transmission XRT, illustrated in Fig. 10.32(b), introduced by Lang, is by far the most popular XRT technique.[161–162] Monochromatic X-rays pass through a narrow slit and strike the sample aligned to an appropriate Bragg angle. The tall and narrow primary beam is transmitted through the sample and strikes a lead screen. The diffracted beam falls on the photographic plate through a slit in the screen. X-rays are absorbed in a solid according to Eq. (10.9). However, absorption is considerably reduced when the X-rays are aligned for diffraction along certain crystal planes.[155] A topograph is generated by scanning the sample and the film in synchronism, holding the screen stationary. Scanning combined with oscillation is effective when extreme sample warpage prevents imaging a large area. While the crystal is scanned, both crystal and film are also oscillating simultaneously around the normal to the plane containing the incident and reflected beam.[163] Entire, large-area wafers can be imaged. Large-diameter wafers become warped during processing, making it necessary to adjust the specimen continuously

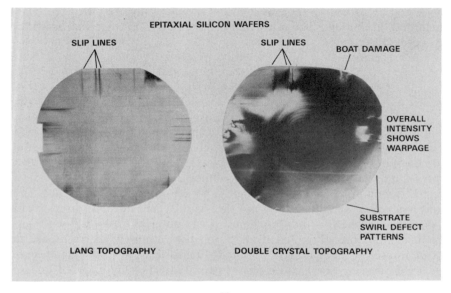

(a)

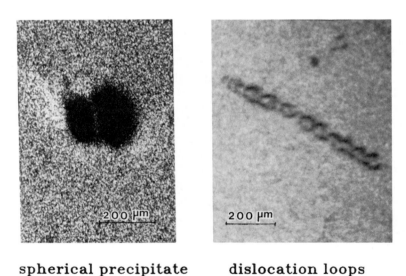

spherical precipitate   **dislocation loops**

(b)

Fig. 10.33 (a) X-ray topography of a 7 µm, (100)-oriented epitaxial silicon wafer using the Lang and double crystal topography methods; (b) crystal defects by the Lang transmission method. (a) reprinted with permission after Shaffner.[167] Courtesy of T. J. Shaffner, Texas Instruments.

during topography measurements to ensure that it stays on the chosen Bragg angle.

To "photograph" defects, one usually chooses a weakly diffracting plane. A uniform sample gives a featureless image. Structural defects cause stronger X-ray diffraction, thus providing film contrast or topographic features. The Lang technique has also been adapted to reflection topography. Scanning provides for considerably more flexibility than is possible with the Berg-Barrett technique. For semiconductors the Lang method is used primarily to study defects introduced during crystal growth or during wafer processing.[164] Transmission topographs provide information on defects through the entire sample; reflection topographs provide information of 10 to 30 μm depth from the surface. X-ray topographs of silicon wafers are shown in Fig. 10.33. It is even possible to take stereo topographs and display them in a manner to show a three-dimensional depth effect.[165]

In *section topography* the sample and film are stationary, and a narrow "section" of the sample—the cross section—is imaged.[154,156,166] The stationary sample is illuminated by a narrow X-ray beam, and the sample cross section is imaged on the film. The method is like that in Fig. 10.32(b), except both sample and photographic plate are stationary. Section topography has proved to be very valuable for defect depth information. For example, it is common in integrated circuit fabrication to precipitate oxygen in silicon wafers. Section topography is a convenient method to obtain a nondestructive cross-sectional picture through the wafer clearly showing the precipitated regions.[157]

Double-crystal topography provides higher accuracy because the beam is more highly collimated than is possible with single crystal topography.[154,157] The technique, shown in Fig. 10.32(c), consists of two successive Bragg reflections from reference and sample crystals. Reflection from the first, carefully selected "perfect" crystal produces a monochromatic and highly parallel beam to probe the sample. The double-crystal technique is used not only for topography but also for *rocking curve* determination. To record a rocking curve, the sample is slowly rotated or "rocked" about an axis normal to the diffraction plane, and the scattered intensity is recorded as a function of the angle. Such a rocking curve is shown in Fig. 10.32(c). The rocking curve width is a measure of crystal perfection. The narrower the curve, the more perfect is the material. For epitaxial layers it provides data on lattice mismatch, layer thickness, layer and substrate perfection, and wafer curvature.[157]

10.4.4 Neutron Activation Analysis

Neutron activation analysis (NAA) is a trace analysis method in which nuclear reactions lead to the production of radioactive isotopes from stable isotopes of the elements in the sample followed by measurement of the radiation emitted by the desired radio isotopes.[131,168–170] We mention it briefly because it is a technique with high sensitivity to certain elements important to semiconductors. The technique has not found wide use in the semiconductor community. It is generally offered as a service since few semiconductor laboratories have the required nuclear reactor.

The sample is placed into a nuclear reactor. Those elements that absorb neutrons find themselves in a highly excited state that relaxes by beta- and gamma-ray emission. The sample may also become radioactive. Gamma-ray emission is analogous to X-ray emission from orbital electron transitions. Beta rays have a continuous spectrum and are not an attractive tool for elemental determination. Gamma rays have well-defined, tabulated energies that are usually measured with a germanium detector.[53–54,171] The γ-ray energy identifies the element, and their intensity determines the concentration. The detection system is usually calibrated against standards for quantitative measurements. Typical detection limits for elements in silicon are shown in Fig. 10.34.

NAA is not a surface-sensitive technique since uncharged incident neutrons penetrate deeply into the sample. Similarly emitted γ-rays are also very penetrating. A disadvantage of NAA is the attendant radioactivity of the sample. The key to successful NAA use for Si is the short 2.6 h half life of Si and the longer half life of many contaminating elements. It is common

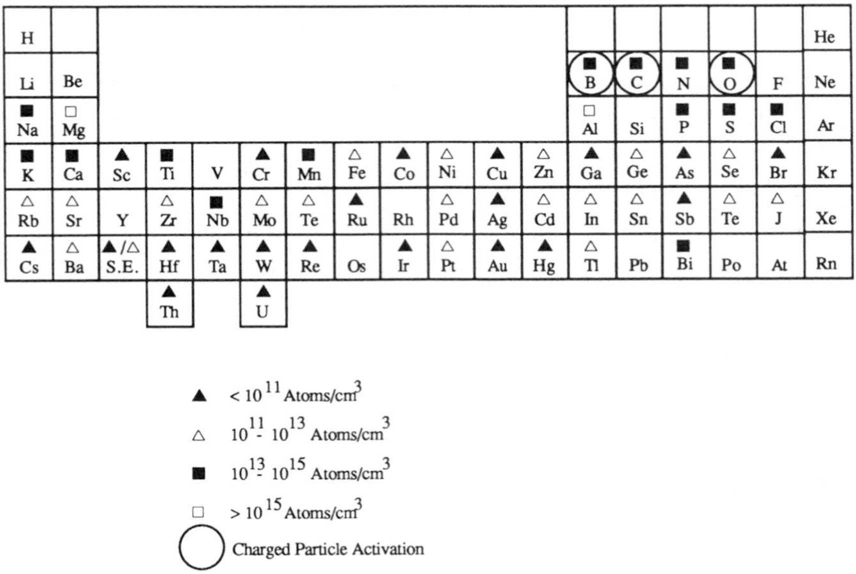

▲ $< 10^{11}$ Atoms/cm^3

△ $10^{11} - 10^{13}$ Atoms/cm^3

■ $10^{13} - 10^{15}$ Atoms/cm^3

□ $> 10^{15}$ Atoms/cm^3

◯ Charged Particle Activation

Fig. 10.34 Practical detection limits for elements in silicon detected by neutron activation analysis. Radionuclides have half lives >2 h; sample volume 1 cm^3; neutron flux 10^{14} thermal and 3×10^{13} fast neutrons/cm$^2 \cdot$s; irradiation time 1 to 5 days. Reprinted with permission after Haas and Hofmann.[172]

to irradiate a Si sample and then measure it 24 h later, when the Si activity has decayed to insignificant levels. NAA sensitivity can be extremely high. For example, gold in silicon can be determined at concentrations as low as 10^9 to 10^{10} cm^{-3} [169–170,172] But other elements are much less sensitive, as shown in Fig. 10.34. The method is most sensitive if the sample is destroyed after irradiation by etching. The more convenient *instrumental NAA*, in which the sample is measured as irradiated, is a powerful survey method but is not as sensitive.

NAA measures the purity of silicon during and after crystal growth and determines impurities introduced during processing.[9,172–173] Usually the total impurity content of the sample is measured. Profiling is possible by etching or lapping thin layers of the sample and measuring the activity in the removed material. NAA is not sensitive to boron, carbon, or nitrogen. Phosphorus does not emit gamma rays; instead β-ray decay must be measured. The method is not suitable for heavily doped wafers. For example, Sb and As form radioactive species.[131] For quantitative measurements careful calibration must be performed.[174] A good summary of NAA application to semiconductor problems can be found in the work of Haas and Hofmann.[172] They have used the method for the past 25 years for impurity monitoring during crystal growth, device processing, and detection of impurities in supplies like aluminum and even in plastic pipes used for water. They examined a number of materials for uranium and thorium, two impurities that cause data upsets in memory chips. They also use *autoradiography* in which an impurity is imaged to show its spatial distribution.

A method related to NAA is *neutron depth profiling* (NDP).[175] It is based on thermal neutron absorption and subsequent alpha particle emission and detection. A well-collimated thermal neutron beam is directed at the sample, and the emitted α-particles are detected. Their energy depends on the depth of generation since they lose energy while traveling through the sample, much as He ions do in RBS. By analyzing the energy of the detected α-particles it is possible to construct a depth profile of the element. NDP lends itself to only a few elements, Li, Be, B, and Na being the dominant ones. It has been used to determine the implanted boron profile in silicon for comparison with SIMS and spreading resistance measurements.[176] It is only rarely used since there are few NDP facilities. In the U.S. facilities exist at Brookhaven National Laboratories, the University of Michigan, the National Bureau of Standards, and North Carolina State University.

APPENDIX 10.1   SELECTED FEATURES OF SOME ANALYTICAL METHODS

Technique	Detectable Elements	Lateral Resolution	Depth Resolution	Detection Limit (atoms/cm^3)[a]	Analyzed Volume (cm^3)[b]	Type of Information[c]	Technique Destructive?	Depth Profiling?	Analysis Time	Matrix Effect
AES	≥ Li	100 μm	20 Å	10^{19}–10^{20}	10^{-11}	elem	Yes[d]	Yes	30 m	sputter mixing
SAM	≥ Li	1000 Å	20 Å	10^{21}	10^{-15}	elem	Yes[d]	Yes	30 m	sputter mixing
EMP	≥ Na	1–10 μm	1 μm	10^{19}–10^{20}	10^{-12}	elem	No	No	30 m	correctable
SIMS	All	1–100 μm	10–100 Å	10^{15}–10^{18}	10^{-11}	elem	Yes	Yes	1 h	severe
RBS	≥ Li	0.1 cm	200 Å	10^{19}–10^{20}	10^{-8}	elem	No	No	30 m	free
XRF	≥ C	0.1–1 cm	50 Å–10 μm	10^{17}–10^{18}	10^{-3}	elem	No	No	30 m	correctable
XPS	≥ Li	0.1 cm	20 Å	10^{19}–10^{20}	10^{-9}	elem/chem	Yes[d]	Yes	30 m	chemical shift
XRT	—	1–10 μm	100–500 μm	—	Whole wafer	cryst. struct.	No	Yes	45 m	strain interaction
PL	shallow level	0.1 cm	1–10 μm	10^{11}–10^{15}	10^{-5}	elem/band gap	No	No	1 h	bound excitons
FTIR	funct. grps.	1–1000 μm	1–10 mm	10^{12}–10^{16}	10^{-5}	molecule	No	No	15 m	molecular interaction
Raman	funct. grps.	1–1000 μm	1–10 μm	10^{19}	10^{-11}	molecule	No	No	1 h	molecular stress
EELS	≥ B	100 μm	20 Å	10^{19}–10^{20}	10^{-11}	elem/chem	No	No	30 m	free
LEED	—	0.1–100 μm	50 Å	—	10^{-12}	cryst. struct.	No	No	30 m	—
RHEED	—	0.1 μm–1 mm	50 Å	—	10^{-12}	cryst. struct.	No	No	30 m	—
NAA	selective	1 cm	1 μm	10^{10}–10^{18}	10^{-4}	elem	No	No	2 d	free

Note: AES: Auger electron spectroscopy, SAM: scanning Auger microprobe, EMP: electron microprobe, SIMS: secondary ion mass spectrometry, RBS: Rutherford backscattering spectrometry, XRF: X-ray fluorescence, XPS: X-ray photoelectron spectroscopy, XRT: X-ray topography, PL: photoluminescence, FTIR: Fourier transform infrared spectroscopy, Raman: Raman microprobe, EELS: electron energy loss spectroscopy, LEED: low-energy electron diffraction, RHEED: reflection high-energy electron diffraction, NAA: neutron activation analysis.

[a] Depends on element to be detected.
[b] Area × depth resolution.
[c] elem: elemental composition, chem: chemical composition.
[d] If profiled, otherwise "No."

APPENDIX 10.2   SENSITIVITY, DEPTH, AND SPOT SIZE OF ANALYTICAL METHODS

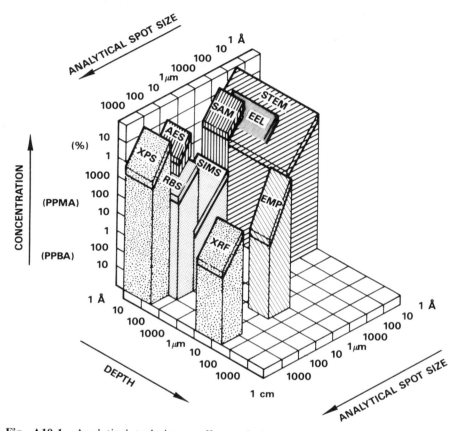

Fig. A10.1 Analytical techniques offer a choice between sensitivity, depth, and analyzing area. AES: Auger electron spectroscopy, EEL: electron energy loss spectroscopy, EMP: electron microprobe, RBS: Rutherford backscattering spectrometry, SAM: scanning Auger microscopy, SIMS: secondary ion mass spectrometry, STEM: scanning TEM, XRF: X-ray fluorescence, XPS: X-ray photoelectron spectroscopy. Reprinted with permission after Shaffner.[177]

REFERENCES

[1]   G. B. Larrabee, "Material Characterization for VLSI," in *VLSI Electronics: Microstructure Science* (N. G. Einspruch, ed.), Vol. 2, Academic Press, New York, 1981, pp. 37–65.

[2]   R. E. Honig, "Surface and Thin Film Analysis of Semiconductor Materials," *Thin Solid Films* **31**, 89–122, Jan. 1976.

[3]   H. W. Werner and R. P. H. Garten, "A Comparative Study of Methods for Thin-Film and Surface Analysis," *Rep. Progr. Phys.* **47**, 221–344, March 1984.

[4]   L. L. Kazmerski, "Advanced Materials and Device Analytical Techniques," in *Advances in Solar Energy* (K. W. Böer, ed.), Vol. 3, American Solar Energy Soc. Boulder, CO, 1986, pp. 1–123.

[5]   L. A. Casper (ed.), *Microelectronic Processing: Inorganic Material Characterization*, American Chemical Soc., Symp. Series 295, Washington, DC, 1986.

[6]   R. E. Whan (coord.), *Metals Handbook, Ninth Ed.*, Vol. 10, *Materials Characterization*, Am. Soc. Metals, Metals Park, OH, 1986.

[7]   D. P. Woodruff and T. A. Delchar, *Modern Techniques of Surface Science*, Cambridge University Press, Cambridge 1986.

[8]   L. C. Feldman and J. W. Mayer, *Fundamentals of Surface and Thin Film Analysis*, North-Holland, Amsterdam, 1986.

[9]   M. Grasserbauer, G. Stingeder, H. Pötzl, and E. Guerrero, "Analytical Science for the Development of Microelectronic Devices," *Fresenius Z. Anal. Chem.* **323**, 421–449, 1986.

[10]   J. M. Walls (ed.), *Methods of Surface Analysis*, Cambridge University Press, Cambridge, 1989.

[11]   K. N. Tu and R. Rosenberg (eds.), *Analytical Techniques for Thin Film Analysis*, Academic Press, San Diego, CA, 1988.

[12]   P. F. Kane and G. B. Larrabee, "Surface Characterization," *Anal. Chem.* **49**, 221R–230R, April 1977; "Surface Characterization," *Anal. Chem.* **51**, 308R–317R, April 1979.

[13]   G. B. Larrabee and T. J. Shaffner, "Surface Characterization," *Anal. Chem.* **53**, 163R–174R, April 1981.

[14]   R. A. Bowling and G. B. Larrabee, "Surface Characterization," *Anal. Chem.* **55**, 133R–156R, April 1983.

[15]   R. A. Bowling, T. J. Shaffner, and G. B. Larrabee, "Surface Characterization," *Anal. Chem.* **57**, 130R–175R, April 1985.

[16]   G. E. McGuire, "Surface Characterization," *Anal. Chem.* **59**, 294R–308R, June 1987.

[17]   J. E. Fulghum, G. E. McGuire, I. H. Musselman, R. J. Nemanich, J. M. White, D. R. Chopra, and A. R. Chourasia, "Surface Characterization," *Anal. Chem.* **61**, 243R–269R, June 1989.

[18]   N. H. Turner and R. J. Colton, "Surface Analysis: X-Ray Photoelectron Spectroscopy, Auger Electron Spectroscopy, and Secondary Ion Mass Spectrometry," *Anal. Chem.* **54**, 293R–322R, April 1982.

[19]   N. H. Turner, B. I. Dunlap, and R. J. Colton, "Surface Analysis: X-Ray Photoelectron Spectroscopy, Auger Electron Spectroscopy, and Secondary Ion Mass Spectrometry," *Anal. Chem.* **56**, 373R–416R, April 1984.

[20]   N. H. Turner, "Surface Analysis: X-Ray Photoelectron Spectroscopy and Auger Electron Spectroscopy," *Anal. Chem.* **58**, 153R–165R, April 1986.

[21]   N. H. Turner, "Surface Analysis: X-Ray Photoelectron Spectroscopy and Auger Electron Spectroscopy," *Anal. Chem.* **60**, 377R–387R, June 1988.

[22] T. J. Shaffner and J. W. S. Hearle, "Recent Advances in Understanding Specimen Charging," *Scanning Electron Microscopy*, 61–70, 1976.

[23] V. E. Cosslett, "Fifty Years of Instrumental Development of the Electron Microscope," in *Advances in Optical and Electron Microscopy* (R. Barer and V. E. Cosslett, eds.), Vol. 10, Academic Press, San Diego, CA, 1988, pp. 215–267.

[24] L. de Broglie, "Waves and Quanta (in French)," *Compt. Rend.* **177**, 507–510, Sept. 1923.

[25] J. I. Goldstein, D. E. Newbury, P. Echlin, D. C. Joy, C. Fiori, and E. Lifshin, *Scanning Electron Microscopy and X-Ray Microanalysis*, Plenum, New York, 1984.

[26] R. A. Young and R. V. Kalin, "Scanning Electron Microscopic Techniques for Characterization of Semiconductor Materials," in *Microelectronic Processing: Inorganic Material Characterization* (L. A. Casper, ed.) American Chemical Soc., Symp. Series 295, Washington, DC, 1986, pp. 49–74.

[27] T. E. Everhart and R. F. M. Thornley, "Wide-Band Detector for Micro-Microampere Low-Energy Electron Currents," *J. Sci. Instrum.* **37**, 246–248, July 1960.

[28] K. Murata, T. Matsukawa, and R. Shimizu, "Monte Carlo Calculations on Electron Scattering in a Solid Target," *Japan. J. Appl. Phys.* **10**, 678–686, June 1971.

[29] H. J. Leamy, "Charge Collection Scanning Electron Microscopy," *J. Appl. Phys.* **53**, R51–R80, June 1982.

[30] T. E. Everhart and P. H. Hoff, "Determination of Kilovolt Electron Energy Dissipation vs. Penetration Distance in Solid Materials," *J. Appl. Phys.* **42**, 5837–5846, Dec. 1971.

[31] P. Auger, "On the Compound Photoelectric Effect (in French)," *J. Phys. Radium* **6**, 205–208, June 1925.

[32] L. E. Davis, N. C. MacDonald, P. W. Palmberg, G. E. Riach, and R. E. Weber, *Handbook of Auger Electron Spectroscopy*, Physical Electronics Industries Inc., Eden Prairie, MN, 1976.

[33] G. E. McGuire, *Auger Electron Spectroscopy Reference Manual*, Plenum Press, New York, 1979.

[34] *Handbook of Auger Electron Spectroscopy*, JEOL Ltd., Tokyo, 1980.

[35] H. Hapner, J. A. Simpson, and C. E. Kuyatt, "Comparison of the Spherical Deflector and the Cylindrical Mirror Analyzers," *Rev. Sci. Instrum.* **39**, 33–35, Jan. 1968.

[36] P. W. Palmberg, G. K. Bohn, and J. C. Tracy, "High Sensitivity Auger Electron Spectrometer," *Appl. Phys. Lett.* **15**, 254–255, Oct. 1969.

[37] L. A. Harris, "Analysis of Materials by Electron-Excited Auger Electrons," *J. Appl. Phys.* **39**, 1419–1427, Feb. 1968.

[38] E. Minni, "Assessment of Different Models for Quantitative Auger Analysis in Applied Surface Studies," *Appl. Surf. Sci.* **15**, 270–280, April 1983.

[39] T. J. Shaffner, "Surface Characterization for VLSI," in *VLSI Electronics: Microstructure Science* (N. G. Einspruch and G. B. Larrabee, eds.) Vol. 6, Academic Press, Orlando, FL, 1983, pp. 497–527.

[40] T. J. Shaffner, "Shadowing Technique for Reducing Charge Buildup in Scanning Auger Microanalysis," *Rev. Sci. Instrum.* **49**, 1748–1749, Dec. 1978.

[41] T. Adachi and C. R. Helms, "The Effect of Ion Induced Roughness on the Depth Resolution of Auger Spectra Profiling of MNOS Devices," *J. Vac. Sci. Technol* **19**, 119–122, May/June 1981.

[42] E. Zinner, "Sputter Depth Profiling of Microelectronic Structures," *J. Electrochem. Soc.* **130**, 199C–222C, May 1983.

[43] P. H. Holloway and G. E. McGuire, "Characterization of Electronic Devices and Materials by Surface-Sensitive Analytical Techniques," *Appl. Surf. Sci.* **4**, 410–444, April 1980.

[44] G. E. McGuire and P. H. Holloway, "Applications of Auger Spectroscopy in Materials Analysis," in *Electron Spectroscopy: Theory, Techniques and Applications* (C. R. Brundle and A. D. Baker, eds.), Vol. 4, Academic Press, San Diego, CA, 1981, pp. 2–74.

[45] J. Keenan, "TiSi$_2$ Chemical Characterization by Auger Electron and Rutherford Backscattering Spectroscopy," *TI Tech. J.* **5**, 43–49, Sept./Oct. 1988.

[46] L. A. Files and J. Newsom, "Scanning Auger Microscopy: Applications to Semiconductor Analysis," *TI Tech. J.* **5**, 89–95, Sept./Oct. 1988.

[47] M. Thompson, "Applications of Auger-Electron Spectroscopy," *Talanta* **24**, 399–415, July 1977.

[48] D. T. Hawkins, *Auger Electron Spectroscopy*, Plenum, New York, 1977.

[49] R. Castaing, Thesis, Univ. of Paris, France, 1948; "Electron Probe Microanalysis," in *Advances in Electronics and Electron Physics* (L. Marton, ed.), Academic Press, New York, **13**, 317–386, 1960.

[50] K. F. J. Heinrich, *Electron Beam X-Ray Microanalysis*, Van Nostrand Rheinhold, New York, 1981.

[51] K. F. J. Heinrich and D. E. Newbury, "Electron Probe Microanalysis," in *Metals Handbook, Ninth Ed.* (R. E. Whan, coord.), Am. Soc. Metals, Metals Park, OH, **10**, 516–535, 1986.

[52] C. E. Fiori and D. E. Newbury, "Artifacts Observed in Energy-Dispersive X-Ray Spectrometry in the Scanning Electron Microscope," *Scanning Electron Microscopy*, 401–422, 1978.

[53] F. S. Goulding and Y. Stone, "Semiconductor Radiation Detectors," *Science* **170**, 280–289, Oct. 1970.

[54] A. H. F. Muggleton, "Semiconductor Devices for Gamma Ray, X-Ray and Nuclear Radiation Detectors," *J. Phys. E: Scient. Instrum.* **5**, 390–404, May 1972.

[55] J. F. Bresse, "Quantitative Investigations in Semiconductor Devices by Electron Beam Induced Current Mode: A Review," in *Scanning Electron Microscopy* **1**, 717–725, 1978.

[56] C. A. Klein, "Band Gap Dependence and Related Features of Radiation Ionization Energies in Semiconductors," *J. Appl. Phys.* **39**, 2029–2038, March 1968.

[57] M. von Heimendahl, *Electron Microscopy of Materials*, Academic Press, New York, 1980.

[58] D. B. Williams, *Practical Analytical Electron Microscopy in Materials Science*, Philips Electron. Instrum., Deerfield Beach, FL, 1984.

[59] D. C. Joy, A. D. Romig, Jr., and J. I. Goldstein (eds.), *Principles of Analytical Electron Microscopy*, Plenum, New York, 1986.

[60] A. D. Romig, "Analytical Transmission Electron Microscopy," in *Metals Handbook, Ninth Ed.* (R. E. Whan, coord.), Am. Soc. Metals, Metals Park, OH, **10**, 429–489, 1986.

[61] R F. Egerton, *Electron Energy-Loss Spectroscopy in the Electron Microscope*, Plenum, New York, 1986.

[62] C. Colliex, "Electron Energy Loss Spectroscopy in the Electron Microscope," in *Advances in Optical and Electron Microscopy* (R. Barer and V. E. Cosslett, eds.), Academic Press, San Diego, CA, **9**, 65–177, 1986.

[63] R. B. Marcus and T. T. Sheng, *Electron Microscopy of Silicon VLSI Circuits and Structures*, Wiley, New York, 1983.

[64] T. T. Sheng, "Cross-Sectional Transmission Electron Microscopy of Electronic and Photonic Devices," in *Analytical Techniques for Thin Film Analysis* (K. N. Tu and R. Rosenberg, eds.), Academic Press, San Diego, CA, 1988, pp. 251–296.

[65] J. M. Cowley, *Diffraction Physics*, North-Holland, Amsterdam, 1975.

[66] J. C. H. Spence, *Experimental High-Resolution Electron Microscopy*, 2d ed, Oxford University Press, Oxford, 1988.

[67] D. Cherns, "High-Resolution Transmission Electron Microscopy of Surface and Interfaces," in *Analytical Techniques for Thin Film Analysis* (K. N. Tu and R. Rosenberg, eds.), Academic Press, San Diego, CA, 1988, pp. 297–335.

[68] P. E. Batson, "Scanning Transmission Electron Microscopy," in *Analytical Techniques for Thin Film Analysis* (K. N. Tu and R. Rosenberg, eds.), Academic Press, San Diego, CA, 1988, pp. 337–387.

[69] R. J. Graham, "Characterization of Semiconductor Materials and Structures by Transmission Electron Microscopy," in *Diagnostic Techniques for Semiconductor Materials and Devices* (T. J. Shaffner and D. K. Schroder, eds.), Electrochem. Soc., Pennington, NJ, 1988, pp. 150–167.

[70] J. I. Hanoka and R. O. Bell, "Electron-Beam-Induced Currents in Semiconductors," in *Annual Review of Materials Science* (R. A. Huggins, R. H. Bube, and D. A. Vermilya, eds.), Annual Reviews, Palo Alto, CA, **11**, 353–380, 1981.

[71] T. E. Everhart, O. C. Wells, and R. K. Matta, "A Novel Method of Semiconductor Device Measurements," *Proc. IEEE* **52**, 1642–1647, Dec. 1964.

[72] K. Joardar, C. O. Jung, S. Wang, D. K. Schroder, S. J. Krause, G. H. Schwuttke, and D. L. Meier, "Electrical and Structural Properties of Twin Planes in Dendritic Web Silicon," *IEEE Trans. Electron Dev.* **ED-35**, 911–918, July 1988.

[73] S. M. Davidson, "Semiconductor Material Assessment by Scanning Electron Microscopy," *J. Microsc.* **110**, 177–204, Aug. 1977.

[74] G. Pfefferkorn, W. Bröcker, and M. Hastenrath, "The Cathodoluminescence

Method in the Scanning Electron Microscope," *Scanning Electron Microscopy*, SEM, AMF O-Hare, IL, 1980, pp. 251–258.

[75] D. B. Holt and S. Datta, "The Cathodoluminescent Mode as an Analytical Technique: Its Development and Prospects," *Scanning Electron Microscopy*, SEM, AMF O-Hare, IL, 1980, pp. 259–278.

[76] B. G. Yacobi and D. B. Holt, "Cathodoluminescence Scanning Electron Microscopy of Semiconductors," *J. Appl. Phys.* **59**, R1–R24, Feb. 1986.

[77] R. J. Roedel, S. Myhajlenko, J. L. Edwards, and K. Rowley, "Cathodoluminescence Characterization of Semiconductor Materials," in *Diagnostic Techniques for Semiconductor Materials and Devices* (T. J. Shaffner and D. K. Schroder, eds.), Electrochem. Soc., Pennington, NJ, 1988, pp. 185–196.

[78] C. Davisson and L. H. Germer, "Diffraction of Electrons by a Crystal of Nickel," *Phys. Rev.* **30**, 705–740, Dec. 1927.

[79] J. B. Pendry, *Low Energy Electron Diffraction*, Academic Press, New York, 1974.

[80] D. H. Buckley, *Surface Effects in Adhesion, Friction, Wear and Lubrication*, Elsevier, Amsterdam, 1981, pp. 73–78.

[81] M. G. Lagally, "Low-Energy Electron Diffraction," in *Metals Handbook*, *Ninth Ed.* (R. E. Whan, coord.), Am. Soc. Metals, Metals Park, OH, **10**, 536–545, 1986.

[82] K. Heinz, "Structural Analysis of Surfaces by LEED," *Progr. Surf. Sci.* **27**, 239–326, 1988.

[83] B. F. Lewis, F. J. Grunthaner, A. Madhukar, T. C. Lee, and R. Fernandez, "Reflection High Energy Electron Diffraction Intensity Behavior During Homoepitaxial Molecular Beam Epitaxy Growth of GaAs and Implications for Growth Kinetics," *J. Vac. Sci. Technol.* **B3**, 1317–1322, Sept./Oct. 1985.

[84] C. G. Pantano, "Secondary Ion Mass Spectroscopy," in *Metals Handbook*, *Ninth Ed.* (R. E. Whan, coord.), Am. Soc. Metals, Metals Park, OH, **10**, 610–627, 1986.

[85] A. Benninghoven, F. G. Rüdenauer, and H. W. Werner, *Secondary Ion Mass Spectrometry: Basic Concepts, Instrumental Aspects, Applications and Trends*, Wiley, New York, 1987.

[86] R. Castaing, B. Jouffrey, and G. Slodzian, "On the Possibility of Local Analysis of a Specimen Using Its Secondary Ion Emission (in French)," *Compt. Rend.* **251**, 1010–1012, Aug. 1960.

[87] R. Castaing and G. Slodzian, "First Attempts at Microanalysis by Secondary Ion Emission (in French)," *Compt. Rend.* **255**, 1893–1895, Oct. 1962.

[88] R. K. Herzog and H. Liebl, "Sputtering Ion Source for Solids," *J. Appl. Phys.* **34**, 2893–2896, Sept. 1963.

[89] A. Benninghoven, "The Analysis of Monomolecular Solid State Surface Layers with the Aid of Secondary Ion Emission (in German)," *Z. Phys.* **230**, 403–417, 1970.

[90] J. J. Thomson, "Rays of Positive Electricity," *Phil. Mag.* **20**, 752–767, Oct. 1910.

[91] P. Williams, "Secondary Ion Mass Spectrometry," in *Applied Atomic Collision Spectroscopy*, Academic Press, Orlando, FL, 1983, 327–377.

[92] D. E. Sykes, "Dynamic Secondary Ion Mass Spectrometry," in *Methods of Surface Analysis* (J. M. Walls, ed.), Cambridge University Press, Cambridge, 1989, pp. 216–262.

[93] A. Benninghoven, "Surface Analysis by Means of Ion Beams," *Crit. Rev. Solid State Sci.* **6**, 291–316, 1976.

[94] M. T. Bernius and G. H. Morrison, "Mass Analyzed Secondary Ion Microscopy," *Rev. Sci. Instrum.* **58**, 1789–1804, Oct. 1987.

[95] K. Wittmaack, "Design and Performance of Quadrupole-Based SIMS Instruments: A Critical Review," *Vacuum* **32**, 65–69, 1982.

[96] S. R. Goldfarb, "Mass Spectrometry in IC Fabrication," *Semicond. Int.* **9**, 55–60, Oct. 1986.

[97] M. J. Pellin, C. E. Young, W. F. Calaway, J. W. Burnett, and D. M. Gruen, "Secondary Neutral Mass Spectrometry: The Application of Laser Post Ionization to Trace Surface Analysis in Semiconductor Materials," in *Diagnostic Techniques for Semiconductor Materials and Devices* (T. J. Shaffner and D. K. Schroder, eds.), Electrochem. Soc., Pennington, NJ, 1988, pp. 73–85.

[98] N. Kelly, U. Kaiser, and H. Peters, "Quantitative Depth Profiling Analysis of Semiconductors and Superconductors by Secondary Neutral Mass Spectrometry (SNMS)," in *Diagnostic Techniques for Semiconductor Materials and Devices* (T. J. Shaffner and D. K. Schroder, eds.), Electrochem. Soc., Pennington, NJ, 1988, pp. 86–96.

[99] E. Denoyer, R. van Grieken, F. Adams, and D. F. S. Natusch, "Laser Microprobe Mass Spectrometry 1: Basic Principles and Performance Characteristics," *Anal. Chem.* **54**, 26A–41A, Jan. 1982.

[100] M. C. Arst, "Identifying Impurities in Silicon by LIMA Analysis," in *Emerging Semiconductor Technology* (D. C. Gupta and P. H. Langer, eds.), **STP 960**, Am. Soc. Test. Mat., Philadelphia, 1987, pp. 324–335.

[101] H. J. Heinen, S. Meier, H. Vogt, and R. Wechsung, "LAMMA 1000, A New Laser Microprobe Mass Analyzer for Bulk Samples," *Int. J. Mass Spectr. and Ion Phys.* **47**, 19–22, 1983.

[102] M. Ryan-Hotchkiss, "Applications of Secondary Ion Mass Spectroscopy to Characterization of Microelectronic Materials," in *Microelectronic Processing: Inorganic Material Characterization* (L. A. Casper, ed.), American Chemical Soc., Symp. Series 295, Washington, DC, 1986, pp. 96–117.

[103] P. Leta, G. H. Morrison, G. L. Harris, and C. A. Lee, "SIMS Determination of Ion-Implanted Depth Distributions," *Int. J. Mass Spectr. Ion Phys.* **34**, 147–157, June 1980.

[104] S. Clayton, L. Springer, B. Offord, T. Sedgwick, R. Reedy, A. Michel, and G. Scilla, "Formation and Analysis of Shallow Arsenic Profiles," *Electron. Lett.* **24**, 831–833, July 1988.

[105] E. Zinner, "Depth Profiling by Secondary Ion Mass Spectrometry," *Scanning* **3**, 57–78, 1980.

[106] W. K. Chu, J. W. Mayer, M.-A. Nicolet, T. M. Buck, G. Amsel, and P. Eisen, "Principles and Applications of Ion Beam Techniques for the Analysis of Solids and Thin Films," *Thin Solid Films* **17**, 1–41, July 1973.

[107] W. K. Chu, J. W. Mayer, and M.-A. Nicolet, *Backscattering Spectroscopy*, Academic Press, New York, 1978.

[108] W. K. Chu, "Rutherford Backscattering Spectrometry," in *Metals Handbook*, *Ninth Ed.* (R. E. Whan, coord.), Am. Soc. Metals, Metals Park, OH, **10**, 628–636, 1986.

[109] T. G. Finstad and W. K. Chu, "Rutherford Backscattering Spectrometry on Thin Solid Films," in *Analytical Techniques for Thin Film Analysis* (K. N. Tu and R. Rosenberg, eds.), Academic Press, San Diego, CA, 1988, pp. 391–447.

[110] E. Rutherford and H. Geiger, "Transformation and Nomenclature of the Radio-Active Emanations," *Phil. Mag.* **22**, 621–629, Oct. 1911.

[111] S. Rubin, T. O. Passell, and L. E. Bailey, "Chemical Analysis of Surfaces by Nuclear Methods," *Anal. Chem.* **29**, 736–743, May 1957.

[112] J. H. Patterson, A. L. Turkevich, and E. J. Franzgrote, "Chemical Analysis of Surfaces Using Alpha Particles," *J. Geophys. Res.* **70**, 1311–1327, March 1965.

[113] L. C. Feldman and J. M. Poate, "Rutherford Backscattering and Channeling Analysis of Interfaces and Epitaxial Structures," in *Annual Review of Materials Science* (R. A. Huggins, R. H. Bube, and D. A. Vermilya, eds.), Annual Reviews, Palo Alto, CA, **12**, 149–176, 1982.

[114] C. W. Magee and L. R. Hewitt, "Rutherford Backscattering Spectrometry: A Quantitative Technique for Chemical and Structural Analysis of Surfaces and Thin Films," *RCA Rev.* **47**, 162–185, June 1986.

[115] J. F. Ziegler, *Helium Stopping Powers and Ranges in all Elemental Matter*, Pergamon, New York, 1977.

[116] J. F. Ziegler and R. F. Lever, "Calculations of Elastic Scattering of ^{4}He Projectiles," *Thin Solid Films* **19**, 291–296, Dec. 1973.

[117] J. W. Mayer, M.-A. Nicolet, and W. K. Chu, "Backscattering Analysis with ^{4}He Ions," in *Nondestructive Evaluation of Semiconductor Materials and Devices* (J. N. Zemel, ed.), Plenum, New York, NY, 1979, pp. 333–366.

[118] J. F. Ziegler and W. K. Chu, "Stopping Cross Sections and Backscattering Factors for ^{4}He Ions in Matter: $Z = 1$–92, $E(^4He) = 400$–4000 keV," in *Atomic Data and Nuclear Data Tables* **13**, 463–489, May 1974.

[119] J. F. Ziegler, R. F. Lever, and J. K. Hirvonen, in *Ion Beam Surface Analysis* (O. Mayer, G. Linker, and F. Käppeler, eds.), Vol. 1, Plenum, New York, 1976, p. 163.

[120] L. C. Feldman, J. W. Mayer, and S. T. Picraux, *Materials Analysis by Ion Channeling*, Academic Press, San Diego, CA, New York, 1982.

[121] M.-A. Nicolet, J. W. Mayer, and I. V. Mitchell, "Microanalysis of Materials by Backscattering Spectrometry," *Science* **177**, 841–849, Sept. 1972.

[122] J. A. Keenan, "The Characterization of Semiconductor Materials by Backscattering Spectroscopy," *Nucl. Instrum. and Meth.* **B10/11**, 583–587, May 1985.

[123] W. G. Morris, H. Bakhru, and A. W. Haberl, "Materials Characterization with a He⁺ Microbeam," *Nucl. Instrum. and Meth.* **B10/11**, 697–699, May 1985.

[124] J. A. Keenan, "Backscattering Spectroscopy for Semiconductor Materials," in *Diagnostic Techniques for Semiconductor Materials and Devices* (T. J. Shaffner and D. K. Schroder, eds.), Electrochem. Soc., Pennington, NJ, 1988, pp. 15–26.

[125] C. W. Magee, "Secondary Ion Mass Spectrometry and Its Relation to High-Energy Ion Beam Analysis Techniques," *Nucl. Instrum. and Meth.* **191**, 297–307, Dec. 1981.

[126] E. P. Berlin, "X-Ray Secondary Emission (Fluorescence) Spectrometry," in *Principles and Practice of X-Ray Spectrometric Analysis*, Plenum, New York, 1970, Ch. 3.

[127] R. O. Muller, *Spectrochemical Analysis by X-Ray Fluorescence*, Plenum, New York, 1972.

[128] J. V. Gilfrich, "X-Ray Fluorescence Analysis," in *Characterization of Solid Surfaces* (P. F. Kane and G. B. Larrabee, eds.), Plenum, New York, 1974, Ch. 12.

[129] D. S. Urch, "X-Ray Emission Spectroscopy," in *Electron Spectroscopy: Theory, Techniques and Applications* (C. R. Brundle and A. D. Baker, eds.), Vol. 3, Academic Press, New York, 1978, pp. 1–39.

[130] W. E. Drummond and W. D. Stewart, "Automated Energy-Dispersive X-Ray Fluorescence Analysis," *Am. Lab.* **12**, 71–80, Nov. 1980.

[131] J. A. Keenan and G. B. Larrabee, "Characterization of Silicon Materials for VLSI," in *VLSI Electronics: Microstructure Science* (N. G. Einspruch and G. B. Larrabee, eds.), **6**, Academic Press, Orlando, FL, 1983, pp. 1–72.

[132] M. C. Nichols, D. R. Boehme, R. W. Ryon, D. Wherry, B. Cross, and D. Aden, "Parameters Affecting X-Ray Microfluorescence (XRMF) Analysis," in *Advances in X-Ray Analysis* (C. S. Barrett et al., eds.) **30**, Plenum, New York, 1987, pp. 45–51.

[133] R. S. Hockett, S. M. Baumann, and E. Schemmel, "An Evaluation of Ultra-surface (3 nm), Trace (10^{11} cm^{-2}) Impurity Analysis of Silicon Using a New X-Ray Technique," in *Diagnostic Techniques for Semiconductor Materials and Devices* (T. J. Shaffner and D. K. Schroder, eds.), Electrochem. Soc., Pennington, NJ, 1988, pp. 113–130.

[134] R. D. Giauque, F. S. Goulding, J. M. Jaklevic, and R. H. Pehl, "Trace Element Detection with Semiconductor Detector X-Ray Spectrometers," *Anal. Chem.* **45**, 671–681, April 1973.

[135] T. A. Carlson, *Photoelectron and Auger Spectroscopy*, Plenum, New York, 1975.

[136] C. S. Fadley, "Basic Concepts in X-Ray Photoelectron Spectroscopy," in *Electron Spectroscopy: Theory, Techniques and Applications* (C. R. Brundle and A. D. Baker, eds.), Vol. 2, Academic Press, New York, 1978, pp. 2–156.

[137] P. K. Gosh, *Introduction to Photoelectron Spectroscopy*, Wiley-Interscience, New York, 1983.

[138] D. Briggs and M. P. Seah (eds.), *Practical Surface Analysis by Auger and X-Ray Photoelectron Spectroscopy*, Wiley, New York, 1983.

[139] J. B. Lumsden, "X-Ray Photoelectron Spectroscopy," in *Metals Handbook, Ninth Ed.* (R. E. Whan, coord.), Am. Soc. Metals, Metals Park, OH, **10**, 568–580, 1986.

[140] N. Mårtensson, "ESCA," in *Analytical Techniques for Thin Film Analysis* (K. N. Tu and R. Rosenberg, eds.), Academic Press, San Diego, CA, 1988, pp. 65–109.

[141] C. Nordling, S. Hagström, and K. Siegbahn, "Application of Electron Spectroscopy to Chemical Analysis," *Z. Phys.* **178**, 433–438, 1964.

[142] S. Hagström, C. Nordling, and K. Siegbahn, "Electron Spectroscopic Determination of the Chemical Valence State," *Z. Phys.* **178**, 439–444, 1964.

[143] J. G. Jenkin, R. C. G. Leckey, and J. Liesegang, "The Development of X-Ray Photoelectron Spectroscopy: 1900–1960," *J. Electron Spectr. Rel. Phen.* **12**, 1–35, Sept. 1977.

[144] J. G. Jenkin, J. D. Riley, J. Liesegang, and R. C. G. Leckey, "The Development of X-Ray Photoelectron Spectroscopy (1900–1960): A Postscript," *J. Electron Spectr. Rel. Phen.* **14**, 477–485, Dec. 1978.

[145] J. G. Jenkin, "The Development of Angle-Resolved Photoelectron Spectroscopy: 1900–1960," *J. Electron Spectr. Rel. Phen.* **23**, 187-273, June 1981.

[146] C. D. Wagner, W. M. Riggs, L. E. Davies, J. F. Moulder, and G. E. Muilenberg, *Handbook of X-Ray Photoelectron Spectroscopy*, Perkin Elmer, Eden Prairie, MN, 1979.

[147] S. Tanuma, C. J. Powell, and D. R. Penn, "Proposed Formula For Electron Inelastic Mean Free Paths Based on Calculations for 31 Materials," *Surf. Sci.* **192**, L849–L857, Dec. 1987.

[148] K. L. Smith and J. S. Hammond, "Destructive and Nondestructive Depth Profiling Using ESCA," *Appl. Surf. Sci.* **23/23**, 288–299, 1985.

[149] C. S. Fadley, "Angle-Resolved X-Ray Photoelectron Spectroscopy," *Progr. Surf. Sci.* **16**, 275–388, 1984.

[150] W. F. Stickle and K. D. Bomben, "X-Ray Photoelectron Spectroscopy Applied to Microelectronic Materials," in *Microelectronic Processing: Inorganic Material Characterization* (L. A. Casper, ed.), American Chemical Soc., Symp. Series 295, Washington, DC, 1986, pp. 144–162.

[151] A. Torrisi, S. Pignataro, and G. Nocerino, "Applications of ESCA to Fabrication Problems in the Semiconductor Industry," *Applic. Surf. Sci.* **13**, 389–401, Sept./Oct. 1982.

[152] G. Berman, C. C. Shen, and B. Gnade, "XPS Characterization of Ti/Si and Pt/Si Systems as Affected by the Base Deposition Vacuum," *TI Tech. J.* **5**, 50–58, Sept./Oct. 1988.

[153] A. R. Lang, "Recent Applications of X-Ray Topography," in *Modern Diffraction and Imaging Techniques in Materials Science* (S. Amelinckx, G. Gevers, and J. Van Landuyt, eds.), North-Holland, Amsterdam, 1978, pp. 407–479.

[154] B. K. Tanner, *X-Ray Diffraction Topography*, Pergamon, Oxford, 1976.

[155] D. C. Miller and G. A. Rozgonyi, "Defect Characterization by Etching, Optical Microscopy and X-Ray Topography," in *Handbook on Semiconduc-*

tors (S. P. Keller, ed.), Vol. 3, North-Holland, Amsterdam, 1980, pp. 217–246.

[156] R. N. Pangborn, "X-Ray Topography," in *Metals Handbook, Ninth Ed.* (R. E. Whan, coord.), Am. Soc. Metals, Metals Park, OH, **10**, 365–379, 1986.

[157] B. K. Tanner, "X-Ray Topography and Precision Diffractometry of Semiconductor Materials," in *Diagnostic Techniques for Semiconductor Materials and Devices* (T. J. Shaffner and D. K. Schroder, eds.), Electrochem. Soc., Pennington, NJ, 1988, pp. 133–149.

[158] W. F. Berg, "An X-Ray Method for the Study of Lattice Disturbances of Crystals (in German)," *Naturwissenschaften* **19**, 391–396, 1931.

[159] C. S. Barrett, "A New Microscopy and Its Potentialities," *Trans. AIME* **161**, 15–64, 1945.

[160] J. B. Newkirk, "Subgrain Structure in an Iron Silicon Crystal as Seen by X-Ray Extinction Contrast," *J. Appl. Phys.* **29**, 995–998, June 1958.

[161] A. R. Lang, "Direct Observation of Individual Dislocations by X-Ray Diffraction," *J. Appl. Phys.* **29**, 597–598, March 1958.

[162] A. R. Lang, "Studies of Individual Dislocations in Crystals by X-Ray Diffraction Microradiography," *J. Appl. Phys.* **30**, 1748–1755, Nov. 1959.

[163] G. H. Schwuttke, "New X-Ray Diffraction Microscopy Technique for the Study of Imperfections in Semiconductor Crystals," *J. Appl. Phys.* **36**, 2712–2721, Sept. 1961.

[164] B. K. Tanner and D. K. Bowen, *Characterization of Crystal Growth Defects by X-Ray Methods*, Plenum, New York, 1980.

[165] A. E. Jenkinson, "Projection Topographs of Dislocations," *Philips Tech. Rev.* **23**, 82–88, Jan. 1962.

[166] Y. Epelboin, "Simulation of X-Ray Topographs," *Mat. Sci. Eng.* **73**, 1–43, Aug. 1985.

[167] T. J. Shaffner, "A Review of Modern Characterization Methods for Semiconductor Materials," *Scanning Electron Microscopy*, 11–23, 1986.

[168] P. Kruger, *Principles of Activation Analysis*, Wiley-Interscience, New York, 1971.

[169] R. M. Lindstrom, "Activation Analysis of Electronic Materials," in *Microelectronic Processing: Inorganic Material Characterization* (L. A. Casper, ed.), American Chemical Soc., Symp. Series 295, Washington, DC, 1986, pp. 294–307.

[170] R. M. Lindstrom, "Neutron Activation Analysis in Electronic Technology," in *Diagnostic Techniques for Semiconductor Materials and Devices* (T. J. Shaffner and D. K. Schroder, eds.), Electrochem. Soc., Pennington, NJ, 1988, pp. 3–14.

[171] G. Erdtmann, *Neutron Activation Tables*, Verlag Chemie, Weinheim, 1976.

[172] E. W. Haas and R. Hofmann, "The Application of Radioanalytical Methods in Semiconductor Technology," *Solid-State Electron.* **30**, 329–337, March 1987.

[173] P. F. Schmidt and C. W. Pearce, "A Neutron Activation Analysis Study of the Sources of Transition Group Metal Contamination in the Silicon Device Manufacturing Process," *J. Electrochem. Soc.* **128**, 630–637, March 1981.

[174]   M. Grasserbauer, "Critical Evaluation of Calibration Procedures for Distribution Analysis of Dopant Elements in Silicon and Gallium Arsenide," *Pure Appl. Chem.* **60**, 437–444, March 1988.

[175]   R. G. Downing, J. T. Maki, and R. F. Fleming, "Application of Neutron Depth Profiling to Microelectronic Materials Processing," in *Microelectronic Processing: Inorganic Material Characterization* (L. A. Casper, ed.), American Chemical Soc., Symp. Series 295, Washington, DC, 1986, pp. 163–180.

[176]   J. R. Ehrstein, R. G. Downing, B. R. Stallard, D. S. Simons, and R. F. Fleming, "Comparison of Depth Profiling [10]B in Silicon Using Spreading Resistance Profiling, Secondary Ion Mass Spectrometry, and Neutron Depth Profiling," in *Semiconductor Processing*, *ASTM* **STP 850** (D. C. Gupta, ed.) Am. Soc. Test. Mat., Philadelphia, PA, 1984, pp. 409–425.

[177]   T. J. Shaffner, "Exploring the Frontiers of Surface Analysis," *Semicond. Int.* **12**, 98–104, Feb. 1989.

APPENDIX A

LIST OF SYMBOLS

A area (cm^2)

A^{**} modified Richardson's constant $(A/cm^2 \cdot K^2)$

A_c contact area (cm^2)

A_G gate area (cm^2)

A_J junction area (cm^2)

B magnetic field strength (G or T)

B radiative recombination coefficient (cm^3/s)

b mobility ratio μ_n/μ_p

C capacitance $(F \text{ or } F/cm^2)$

c velocity of light $(2.998 \times 10^{10} \text{ cm/s})$

C_B bulk capacitance (F/cm^2)

C_f final capacitance $(F \text{ or } F/cm^2)$

C_{HF} high-frequency capacitance (F)

C_{HF} high-frequency capacitance (F/cm^2)

C_{it} interface trap capacitance (F/cm^2)

C_{LF} low-frequency capacitance (F)

C_{LF} low-frequency capacitance (F/cm^2)

C_N inversion (electron) capacitance (F/cm^2)

C_n Auger recombination coefficient for n-type (cm^6/s)

c_n electron capture coefficient (cm^3/s)

C_{ox} oxide capacitance (F)

C_{ox} oxide capacitance (F/cm^2)

C_P accumulation (hole) capacitance (F/cm^2)

C_p Auger recombination coefficient for p-type (cm^6/s)

c_p hole capture coefficient (cm^3/s)

C_s semiconductor capacitance (F)

$C_{S,DD}$ deep-depletion semiconductor capacitance (F/cm^2)

$C_{S,HF}$	high-frequency semiconductor capacitance (F/cm^2)
$C_{S,LF}$	low-frequency semiconductor capacitance (F/cm^2)
C_{sf}	final semiconductor capacitance (F)
D	diffusion constant (cm^2/s)
d	contact spacing (cm)
d	crystal plane spacing (cm)
d	wafer diameter (cm)
D_{it}	interface trapped charge density ($cm^{-2} \cdot eV^{-1}$)
D_n	electron diffusion constant (cm^2/s)
E	energy (eV)
$\mathscr{E}$	electric field (V/cm)
E_A	acceptor energy level (eV)
E_C	conduction band edge (eV)
E_D	donor energy level (eV)
$\mathscr{E}_{eff}$	effective electric field (V/cm)
E_{eh}	mean energy to generate one electron-hole pair (eV)
E_F	Fermi energy (eV)
E_g	band gap (eV)
E_{it}	interface trapped charge energy (eV)
e_n	electron emission coefficient (s^{-1})
E_p	phonon energy (eV)
e_p	hole emission coefficient (s^{-1})
E_T	trap energy (eV)
E_V	valence band edge (eV)
F	dimensionless electric field defined in Eq. (6.8)
F	Faraday constant (9.64×10^4 C)
F	van der Pauw F-function, defined in Eq. (5.21)
f	frequency (Hz)
G	bulk generation rate ($cm^{-3} \cdot s^{-1}$)
G	conductance (S)
g	conductance (S)
g_D	drain conductance (S)
g_m	transconductance (S)
G_p	parallel conductance (S)
G_s	surface generation rate ($cm^{-2} \cdot s^{-1}$)
h	Planck's constant (6.626×10^{-34} J·s)
I	current (A)
I_B	base current (A)
I_b	electron beam current (A)
I_C	collector current (A)
I_D	drain current (A)
I_d	displacement current (A)
I_{dk}	dark current (A)
I_{DS}	drain current (A)
I_E	emitter current (A)
I_e	emission current (A)

I_{ph}	photocurrent (A)
I_{sc}	short-circuit current (A)
J	current density (J)
J_{sc}	short-circuit current density (A/cm^2)
K	kinematic factor, defined in Eq. (10.15)
k	Boltzmann's constant (8.617×10^{-5} eV/K)
k	extinction coefficient
K_{ox}	oxide dielectric constant
K_s	semiconductor dielectric constant
L	channel length (cm)
L	contact or sample length (cm)
L	minority carrier diffusion length (cm)
L_D	Debye length, defined in Eq. (2.11)
L_{Di}	intrinsic Debye length, defined in Eq. (6.10)
L_n	electron diffusion length (cm)
L_p	hole diffusion length (cm)
L_T	$= \sqrt{\rho_c/\rho_s}$, transfer length (cm)
$L_{T'}$	$= \sqrt{\rho_c/(\rho_{sc} + \rho_m)}$, transfer length (cm)
L_{Tc}	$= \sqrt{\rho_c/\rho_{sc}}$, transfer length (cm)
M	elemental mass (kg)
M	molecular weight (g)
m	electron mass (9.11×10^{-31} kg)
m^*	effective mass (kg)
m_n	electron effective mass (kg)
N	electron density (cm^{-2})
n	diode ideality factor
n	electron concentration (cm^{-3})
n	index of refraction
N_A	acceptor doping concentration (cm^{-3})
NA	numerical aperture
N_C	effective density of states in the conduction band (cm^{-3})
N_D	donor doping concentration (cm^{-3})
N_f	density of fixed oxide charges (cm^{-2})
n_i	intrinsic carrier concentration (cm^{-3})
N_{it}	density of interface trapped charges (cm^{-2})
N_m	density of mobile oxide charges (cm^{-2})
n_0	equilibrium electron concentration (cm^{-3})
N_{ot}	density of oxide trapped oxide charges (cm^{-2})
n_{p0}	equilibrium minority electron concentration (cm^{-3})
N_T	deep-level impurity concentration (cm^{-3})
n_T	deep-level impurity concentration occupied by electrons (cm^{-3})
N_V	effective density of states in the valence band (cm^{-3})
p	hole concentration (cm^{-3})
p_0	equilibrium hole concentration (cm^{-3})
p_T	deep-level impurity concentration occupied by holes (cm^{-3})
q	magnitude of electron charge (1.6×10^{-19} C)

Q_B	bulk charge density (C/cm^2)
Q_f	fixed oxide charge density (C/cm^2)
Q_{it}	interface state charge density (C/cm^2)
Q_m	mobile oxide charge density (C/cm^2)
Q_N	electron charge density (C/cm^2)
Q_N	inversion charge density (C/cm^2)
Q_{ot}	oxide trapped charge density (C/cm^2)
Q_P	hole charge density (C/cm^2)
Q_S	semiconductor charge density (C/cm^2)
Q_s	semiconductor charge (C)
R	reflectivity, reflectance
r	contact radius (cm)
r	Hall scattering factor
r	wafer radius (cm)
r_B	base resistance (ohms)
r_{Bi}	internal base resistance (ohms)
r_{Bx}	external base resistance (ohms)
R_c	contact resistance (ohms)
r_C	collector resistance (ohms)
r_{dk}	dark resistance (ohms)
R_e	electron range (cm)
R_e	end resistance (ohms)
r_E	emitter resistance (ohms)
R_{geom}	geometry-dependent resistance (ohms)
R_H	Hall coefficient (cm^3/C)
R_{Hs}	sheet Hall coefficient (cm^2/C)
R_k	measured contact resistance (ohms)
R_m	metal or polysilicon resistance (ohms)
r_{ph}	photo resistance (ohms)
R_s	semiconductor resistance (ohms)
r_s	series resistance (ohms)
r_{sh}	shunt resistance (ohms)
R_{sp}	spreading resistance (ohms)
R_T	total resistance (ohms)
s	contact spacing (cm)
s, s_r	surface recombination velocity (cm/s)
s_g	surface generation velocity (cm/s)
T	sample thickness (cm)
T	temperature (K)
T	transmissivity, transmittance
t	time (s)
t	wafer thickness (cm)
t_d	drift time (s)
t_s	storage time (s)
U	bulk recombination rate $(cm^{-3} \cdot s^{-1})$
U_F	$= q\phi_F/kT$

U_S	surface recombination rate $(\text{cm}^{-2} \cdot \text{s}^{-1})$
U_s	$= q\phi_s/kT$
V	voltage (V)
v	velocity (cm/s)
V_0	defined in Eq. (8.97)
V_B	substrate voltage (V)
V_b	Dember potential (V)
V_{bi}	built-in potential (V)
V_{BS}	$V_B - V_S$
V_{CE}	collector-emitter voltage (V)
V_D	drain voltage (V)
V_{DS}	$V_D - V_S$
V_{EB}	emitter-base voltage (V)
V_{FB}	flatband voltage (V)
V_G	gate voltage (V)
V_{GS}	$V_G - V_S$
V_H	Hall voltage (V)
V_j	junction voltage (V)
v_n	electron velocity (cm/s)
V_{oc}	open-circuit voltage (V)
V_{ox}	oxide voltage (V)
V_S	source voltage (V)
V_T	threshold voltage (V)
v_{th}	thermal velocity (cm/s)
W	diffusion window width (cm)
W	line width (cm)
W	space-charge region width (cm)
w	width (cm)
W_f	$= (4K_s\varepsilon_0\phi_F/qN_A)^{1/2}$ final space-charge region width (cm)
W_{ox}	oxide thickness (cm)
Y	ratio of photocurrent to absorbed photon flux
Z	atomic number
Z	contact or sample width (cm)
z	dissolution valency
α	absorption coefficient (cm^{-1})
α	common-base current gain
α_F	forward common-base current gain
α_R	reverse common-base current gain
β	common-emitter current gain
χ	semiconductor electron affinity (eV)
δ	$= W - Z$
Δn	excess electron concentration (cm^{-3})
Δp	excess hole concentration (cm^{-3})
ε_0	permittivity of free space $(8.854 \times 10^{-14} \text{ F/cm})$
Φ	photon flux density $(\text{photons/s} \cdot \text{cm}^2)$
ϕ	work function (V)

ϕ_B	Schottky diode barrier height (V)
ϕ_F	Fermi potential (V)
$\phi_{F'}$	defined in Eq. (5.56)
ϕ_M	metal work function (eV)
ϕ_{MS}	metal-semiconductor work function (V)
ϕ_S	semiconductor work function (V)
ϕ_s	surface potential (V)
γ	body factor, defined in Eq. (5.56)
λ	wavelength (cm)
λ_e	electron wavelength (cm)
λ_p	plasma resonance wavelength (cm)
μ	mobility (cm^2/V·s)
μ/ρ	mass absorption coefficient (cm^2/g)
μ_{eff}	effective mobility (cm^2/V·s)
μ_{FE}	field-effect mobility (cm^2/V·s)
μ_{GMR}	geometric magnetoresistance mobility (cm^2/V·s)
μ_H	Hall mobility (cm^2/V·s)
μ_n	electron mobility (cm^2/V·s)
μ_p	hole mobility (cm^2/V·s)
μ_{sat}	saturation mobility (cm^2/V·s)
ν	frequency of light (Hz)
ρ	density (g/cm^3)
ρ	resistivity (ohm·cm)
ρ_c	specific contact resistance (ohm·cm^2)
ρ_i	specific interface resistance (ohm·cm^2)
ρ_m	metal or polysilicon sheet resistance (ohms/square)
ρ_s	sheet resistance (ohms/square)
ρ_{sc}	sheet resistance under a contact (ohms/square)
σ	conductivity (ohm^{-1}·cm^{-1} or S/cm)
σ_n	electron capture cross section (cm^2)
σ_p	hole capture cross section (cm^2)
σ_s	sheet conductance (1/ohms·square)
τ	time constant (s)
τ_B	bulk lifetime (s)
τ_c	capture time constant (s)
τ_e	$= 1/e_n$, electron emission time constant (s)
τ_g	generation lifetime (s)
τ_n	electron lifetime (s)
τ_{nrad}	nonradiative lifetime (s)
τ_p	hole lifetime (s)
τ_r	recombination lifetime (s)
τ_{rad}	radiative lifetime (s)
ω	radial frequency (s^{-1})
ξ	magnetoresistance scattering factor
Ψ	ellipsometric angle; see Eq. (9.7)
Δ	ellipsometric angle; see Eq. (9.7)

APPENDIX B

ABBREVIATIONS AND ACRONYMS

AEM	analytical transmission electron microscopy
AES	Auger electron spectroscopy
BJT	bipolar junction transistor
$C\text{-}V$	capacitance-voltage
CBKR	cross-bridge Kelvin resistor
CC-DLTS	constant-capacitance DLTS
CCD	charge-coupled device
CD	critical dimension
CER	contact end resistor
CL	cathodoluminescence
CMA	cylindrical mirror analyzer
CRT	cathode ray oscilloscope
cw	continuous wave
DLTS	deep-level transient spectroscopy
DUT	device under test
EBIC	electron beam induced current
EDS	energy dispersive spectroscopy
ehp	electron-hole pair
EMP	electron microprobe
EPM	electron probe microanalysis
ESCA	electron spectroscopy for chemical analysis
ESR	electron spin resonance
FE	field emission
FET	field-effect transistor
FTIR	Fourier transform infrared spectroscopy
G-R	generation-recombination

HEIS	high-energy ion backscattering spectrometry
HREM	high resolution transmission electron microscopy
I-V	current-voltage
JFET	junction field-effect transistor
LBIC	light beam induced current
LDD	lightly doped drain/source
LEED	low energy electron diffraction
MBE	molecular beam epitaxy
MCA	multichannel analyzer
MESFET	metal-semiconductor field-effect transistor
MOCVD	metalorganic vapor deposition
MODFET	modulation-doped field effect transistor
MOS	metal oxide semiconductor
MOS-C	MOS capacitor
MOSFET	MOS field effect transistor
NA	numerical aperture
NAA	neutron activation analysis
NDP	neutron depth profiling
OCVD	open circuit voltage decay
opd	optical path difference
PC	photoconductive
PCD	photoconductive decay
PICTS	photo-induced current transient spectroscopy
PITS	photo-induced current transient spectroscopy
PL	photoluminescence
ppm	parts per million
PTIS	photothermal ionization spectroscopy
Q-V	charge-voltage
qn	quasi-neutral
qnr	quasi-neutral region
RBS	Rutherford backscattering spectrometry
RHEED	reflection high-energy electron diffraction
RR	reverse recovery
S-DLTS	scanning DLTS
SAM	scanning Auger microscopy
SCCD	short circuit current decay
scr	space-charge region
SEM	scanning electron microscope or microscopy
SF	stacking fault
SI	semi-insulating
SIMS	secondary ion mass spectrometry
SPV	surface photovoltage
SRH	Shockley-Read-Hall
SRP	spreading resistance probe or profiling
STEM	scanning transmission electron microscopy

TE	thermionic emission
TEM	transmission electron microscope or microscopy
TFE	thermionic-field emission
TLM	transmission line model *or* transfer length method
TSC	thermally stimulated current
TSCAP	thermally stimulated capacitance
TVS	triangular voltage sweep
UHV	ultra-high vacuum
UV	ultraviolet
WDS	wavelength dispersive spectroscopy
XPS	X-ray photoelectron spectroscopy
XRF	X-ray fluorescence
XRFA	X-ray fluorescence analysis
XRFS	X-ray fluorescence spectroscopy
XRT	X-ray topography

INDEX

三